Pantelis Sopasakis

Control Systems: An introduc

Applied Mathemati$\mathcal{X}$ Press

Citation information:
Sopasakis, P. (2023). *Control Systems: An introduction.* 1.0.1. ed. Belfast, UK: Applied Mathematix Press.

Paperback ISBN: 978-1-7391386-6-0
First paperback edition February 2023.

Author:
Pantelis Sopasakis
School of Electronics, Electrical Eng. & Computer Science (EEECS)
Centre for Intelligent Autonomous Manufacturing Systems (i-AMS)
Queen's University Belfast

Classifications. LLC: TJ216. DDC: 629.8. BIC: TJFM, GPFC.
Mathematics Subject Classification (2010). 93B52, 93C80, 34H05, 93C05, 44A10.

This book contains 315 exercises, 182 examples, more than 55 definitions and 84 results (theorems, lemmata, propositions and corollaries). The whole book was typeset in LaTeX; all plots were made using Tikz.

Copyright © 2023 Pantelis Sopasakis. All rights reserved. No part of this book may be reproduced or used in any manner without the prior written permission of the copyright owner, except for the use of brief quotations in a book review.

The graffiti on the cover page of this book was made by Wild Drawing (WD), https://wdstreetart.com/; used with permission. The photograph of the graffiti on the cover page was taken by Marilena Grispou; used with permission. Cover design by Dimitris Lykos, https://www.nineteendesign.gr/.

The links (URLs) provided in this book are for information purposes only. The author does not endorse the content of third-party links, does not guarantee that their content will remain decent, relevant and appropriate, and does not bear any responsibility for the accuracy of any Internet websites referenced in this book.

MATLAB® is a registered trademark of The MathWorks Inc, 3 Apple Hill Drive, Natick MA. The Control Systems Toolbox™, the Symbolic Math Toolbox™, the Systems Identification Toolbox™, the Robotics System Toolbox™, and the Reinforcement Learning Toolbox™ are trademarks of The MathWorks Inc. The MathWorks does not warrant the accuracy of the text or exercises in this book. Segway® is a registered trademark of Segway Inc.

Book website (for updates, errata, and contact information):
https://am-press.github.io/ControlSystemsBook/

Applied Mathemati𝒳 Press
https://am-press.github.io
Belfast, UK

Preface

In March 2019 I started to prepare lecture notes for an undergraduate control engineering module at Queen's University Belfast, which a few months later started to take the form of a textbook. My aim was to offer a rigorous treatment of classical control theory with properly and clearly stated definitions, theorems and their proofs, accompanied by plenty of examples and exercises to make the material accessible to readers who are not familiar with the topic. Let me, therefore, welcome you all to this journey in control theory, which I hope you will enjoy.

Why study control theory?

Control is ubiquitous: from home appliances (washing machines, thermostats, etc) to aerospace (aircraft attitude control) and from miniature robots to huge radio telescopes and kilometre-long particle colliders. Control theory encompasses a wide gamut of areas such as

1. **Modelling** systems using, for example, differential equations: for example, using Newton's law of motion to model mechanical systems, or basic principles of mass transfer and reaction kinetics to model physicochemical processes, or the Lotka-Volterra differential equations to describe the evolution of populations of species.

2. **Simulating**: dynamical models can be used to perform simulations and predict the future behaviour of the system.

3. **Filtering**: whenever a sensor is used, the obtained measurements will always be corrupted with noise, which needs to be filtered out.

4. **Estimating**: Often certain variables of interest cannot be measured directly. Instead, we may only have access to indirect information

that can be combined to provide estimates of those "hidden" variables. For example, the pitch, roll and yaw angles of an aircraft cannot be measured directly; instead measurements of linear acceleration angular velocity and GPS coordinates are used to estimate its orientation.

5. **Analysing**: Often, we need to reveal certain qualitative characteristics of a dynamical system. Given a model that describes the population of a certain species in an area, we may want to tell whether that species is threatened with extinction. Given a model for a chemical reaction, we may need to assess whether it may lead to an explosion, or given the model of an autonomous car, we need to assess whether it can crash.

6. **Designing** control systems that can alter the behaviour of a system to achieve desirable results,

7. **Guaranteeing** that the system will work as expected even under adverse conditions which may involve exposure to noise, component failures or even cyber-attacks.

The obvious reason why control theory is part of all (electrical, mechanical, chemical, etc) engineering curricula is because control is part of the majority of engineering applications [C1]. But there are more reasons why studying control theory is useful even to those graduates whose job description will not involve the design of control systems:

1. When working with any engineering system, it is important to have a good understanding of all the components it involves. Even if we do not have to design or tune a controller, it is important to know how it works and what it does.

2. Control theory and its methods find applications in domains as diverse as biology, medicine and economics. In fact, feedback loops — the staple of control theory — are naturally occurring structures. For example, the human body is replete with feedback loops for controlling the levels of blood sugar, heart rate, cell reproduction and so on.

3. Control theory addresses practical problems by introducing a number of abstractions and this offers different perspectives. A control-theoretic way of thinking can prove useful in other disciplines such as machine learning, statistics, computer science, numerical analysis and more.

4. Control theory exposes you to an array of mathematical concepts and tools that find applications in other domains. For example, the Laplace transform can be used to solve initial value problems, which are used to simulate dynamical systems in all engineering fields.

Moreover, the Laplace transform finds applications in probability theory, statistics and other fields of applied mathematics. Likewise, the applications of the fundamental theorem of calculus, Taylor's approximation theorem and the argument principle are practically countless.

5. Control theory equips the student with several valuable skills that involve problem-solving, abstract thinking, programming and software development and integrating different systems at a hardware and software level.

6. Lastly, many next-generation applications call for advanced control approaches. Examples include, but are not limited to autonomous ground, aerial, marine, submarine and subterrestrial vehicles, robotics, aerospace and satellite applications, and microgrids and energy management systems. This book is the first stepping stone before you can study more advanced control theory.

What you will find in this book

This book is an introduction to *classical* control theory for single-input single-output continuous-time systems.

The material is organised in twelve chapters, plus an appendix with some necessary mathematical background the reader can refer to if necessary. The first chapter offers a high-level math-free introduction to the topic of dynamical systems and control theory. A good part of the book is spent in Chapter 2 on modelling where we use basic principles of physics and engineering to derive differential equations that describe common systems. Some of these models, such as Newton's second law of motion, may be already known to the reader, but we put them in the context of systems theory. In Chapter 3 we use Taylor's theorem to approximate complicated system dynamics that we often encounter in practice by simpler, linear, differential equations.

In Chapter 4 we introduce the *Laplace transform*: a mathematical construct of central importance in systems and control theory. The Laplace transform takes us from the time domain, where systems are described in terms of (difficult) differential equations, to the complex frequency domain, where linear dynamical systems turn into (easy) algebraic equations. This facilitates our analysis and lays the foundations of classical control theory. In Chapter 5 we learn how to go from the complex frequency domain back to the time domain. This allows us to solve and simulate dynamical systems.

The Laplace transform leads naturally to the definition of the transfer function of a dynamical system which we introduce in Chapter 6 and discuss how we can combine multiple such transfer functions to model multiple interacting dynamical systems. Afterwards, it is time for a well-deserved break with two easy chapters on first and second-order systems where we apply the theory we have learned this far to two common types of systems, namely, systems of the first and second order (see Chapters 7 and 8). First and second-order systems deserve special attention for two reasons: firstly, a great many physical systems can be modelled adequately well by such simple dynamics and, secondly, because we can often reduce the study of higher-order linear systems to such first and second-order components.

When designing control systems, it is important to guarantee that they lead to a "stable" behaviour. Alongside Chapter 4, Chapter 9 is the most important in the book as it introduces the (or better, "a") concept of stability, namely, stability in the bounded-input bounded-output sense. We give Routh's stability criterion that allows us to tell whether a linear dynamical system is stable. Afterwards, we use Routh's criterion as well as certain properties of the Laplace transform to design stable and well-behaving PID controllers in Chapter 10. PID controllers have dominated industrial practice over the last century as they can be easily designed and implemented. In fact, according to a 2017 survey by Tariq Samad (Honeywell), the PID controller ranks first in industrial impact[1].

In Chapter 11 we study how linear stable systems respond when excitated with a sinusoidal input of a certain frequency. In particular, we study how the input frequency affects and magnitude and lag of the oscillations we observe at the output of the system. These correspondences can be described in certain logarithmic plots known as Bode plots. If you happen to have an audio amplifier, it is likely you can find its Bode plot in its manual. Apart from the apparent value of such a frequency-based analysis for signal processing applications, it turns out we can use Bode plots to tell whether a system is stable.

Lastly, both Routh's and Bode's stability criteria come with certain limitations that we can overcome using Nyquist's method that we introduce in Chapter 12. Moreover, Bode's criterion follows from the much more general Nyquist criterion.

Most theorems are followed by their proof which we consider to be necessary for their deep understanding. Routh's stability criterion (see Theorem 9.12) is an exception: we have omitted the proof as it is quite lengthy and involved. Simpler versions of the original proof have proposed, and the interested reader can refer to [C23] and [C24].

Given the time constraints of an eleven-week semester (22 lectures), certain important topics had to be sat out (οὐκ ἐν τῷ πολλῷ τὸ εὖ[2]). These include the Bromwich inversion formula, loop shaping, some controller tuning methods (such as IMC, pole placement and integral criteria) and many more.

Most exercises, unless trivial, come with a hint (denoted with ✌) or an answer (look for ※) so that the reader can verify their solutions. The symbol □ marks the end of a proof. Sections marked with an asterisk (*)

---

[1] See T. Samad, A Survey on Industry Impact and Challenges Thereof, IEEE Control Systems 37, 2017.

[2] Ancient Greek, roughly translated as "less is more."

are more advanced.

Software for Control Engineers

The control engineer's toolbox should contain three things: paper, pencils/pens and a computer with appropriate software installed.

The last few decades, MATLAB®[3] by The Mathworks Inc. has dominated the field of control engineering. MATLAB® is a proprietary numeric computing platform and programming language that facilitates computations that involve matrices, allows the fast prototyping of algorithms and plotting of data. Moreover, MATLAB® comes with a rich collection of toolboxes for control applications such as the Control Systems Toolbox™, the Symbolic Math Toolbox™, the Systems Identification Toolbox™, the Robotics System Toolbox™, the Reinforcement Learning Toolbox™ and other useful toolboxes.

The popularity of Python for machine learning applications could not go unnoticed in other fields, including that of control and automation. Python is a free and open-source software with excellent documentation and a large community of users. It ranked first in IEEE Spectrum's list of top programming languages in 2019[4] and second "most loved, dreaded and wanted" programming language in StackOverflow's 2019 survey. Python's control systems library as well as `numpy`, `scipy`, `sympy` and `matplotlib` are some of the most relevant libraries in control engineering. Python can be used for prototyping, but it can also run on single-board computers, such as Raspberry Pi and Beaglebone, which can be used in control applications. Of course other than MATLAB and Python, a lot more programming languages and frameworks, such as Julia, C/C++, Rust, ROS (robot operating system), and more can be used for the design and implementation of control systems.

At the end of most chapters there is a section on how to use Python [S1, S2] and MATLAB® [S3] to perform certain tasks. Although the reader is expected to have some basic knowledge of MATLAB or Python, the code snippets provided in this book are self-explanatory, and the interested reader can copy, paste and run them.

The code examples in this book have been tested in Python 3.8 and MATLAB® 2020b.

---

[3]see `https://uk.mathworks.com/products/matlab.html` and [S3]. MATLAB® is a registered trademark of The MathWorks, Inc.

[4]see `https://spectrum.ieee.org/computing/software/the-top-programming-languages-2019`

Prerequisites

A basic level in mathematics is required; in particular, calculus (limits, differentiation and integration), polynomials (roots of quadratic equation, factorisation), complex numbers, and some basic trigonometry. Appendix A gives an overview of some key results that are used in this book. The reader may find links to some useful resources in Appendix B.

Acknowledgements

I would like to thank my colleagues Dr Nikolaos Athanasopoulos (Queen's University Belfast), Dr Shane Trimble (EquipmentShare), and Dr Christian A. Hans (TU Berlin) for going through the material and providing suggestions and critical feedback. I would also like to thank Dr Emmanuel Prempain (University of Leicester) for his constructive comments on Chapters 4 and 11. I am also grateful to my students (AY 2019-2020) Ruairi Moran, Owen Fitzgerald, Matthew Cooke, Mark Meggary, and Andrew Barber, (AY 2020-2021) Allwyn Bino, John Boden, Jiaming Zhang, Compannage Vihan Dilmith Fonseka, James Ferguson, Lorcan Quail, Jonathan Lockhart, Hyehyeon Han, and Jonathan Tutty, (AY 2021-22) George Littlewood, and (AY 2022-23) Orla Quail, for finding typos and for their constructive recommendations that improved the quality of earlier drafts of this book. Without their help, this book would certainly contain many more typos than it already probably does. I would like to thank Wild Drawing (WD), Marilena Grispou and Dimitris Lykos for the artwork and the cover design of this book.

This book would not have been written was it not for a number of people in my life whom I want to thank — first and foremost, my sweet partner in crime, Marietta Dalamanga, for her love and support. I am also grateful to my close friends from Greece who have remained by my side even ten years after I left, and to all the new friends I have made along the way.

Pantelis Sopasakis,
Belfast, May 23, 2023

Contents

1 Introduction ... 17
- **1.1 A few words about systems and control** — 17
  - 1.1.1 Open-loop control systems — 17
  - 1.1.2 Feedback control — 18
  - 1.1.3 A few examples — 20
  - 1.1.4 Systems and control theory — 21
- **1.2 Structure of this book** — 26
- **1.3 Exercises** — 27

2 Modelling ... 29
- **2.1 Modelling in the time domain** — 29
  - 2.1.1 State space representation — 29
  - 2.1.2 Systems with input derivatives* — 34
- **2.2 Mechanical systems** — 36
  - 2.2.1 Laws of motion — 36
  - 2.2.2 Friction and drag — 38
  - 2.2.3 Hooke's law — 39
  - 2.2.4 Planar rotation — 45
- **2.3 Circuits** — 59
  - 2.3.1 Resistance, capacitance, inductance — 59
  - 2.3.2 Kirchhoff's laws — 60

	2.3.3	Dynamics of RLC circuits	60
2.4		**Mass transfer**	**64**
	2.4.1	Independent cylindrical tanks	65
	2.4.2	Unsteady flow*	68
	2.4.3	Independent tanks with non-constant cross-section area	70
	2.4.4	Interconnected tanks	71
2.5		**Simulating dynamical systems**	**72**
	2.5.1	Python	72
	2.5.2	MATLAB	74
2.6		**One-minute round-up**	**76**
2.7		**Exercises**	**78**

3 Linearisation 87

3.1	**Taylor's theorem**	**87**
3.2	**Equilibrium points and deviation variables**	**93**
3.3	**Linearisation in several variables**	**99**
3.4	**Higher-order systems**	**105**
3.5	**Integro-differential systems**	**112**
3.6	**Quality of linearisation**	**114**
3.7	**Symbolic differentiation**	**116**
	3.7.1 Python	116
	3.7.2 MATLAB	119
3.8	**One-minute round-up**	**123**
3.9	**Exercises**	**125**

4 Laplace transform 131

4.1		**Definition**	**131**
4.2		**Laplace transforms of elementary functions**	**135**
	4.2.1	Constant function	135
	4.2.2	Heaviside function	136
	4.2.3	Exponential function	137
	4.2.4	Powers	138
	4.2.5	Trigonometric functions	139
4.3		**Properties of the Laplace transform**	**141**
	4.3.1	Linearity	141
	4.3.2	Functions defined piecewise	141

4.3.3	Functions with multiple branches	146
4.3.4	Time scaling	147
4.3.5	Translation in the s-domain	147
4.3.6	First-order derivatives	148
4.3.7	Higher-order derivatives	149
4.3.8	Time integrals	150
4.4	**Finer properties of the Laplace transform**	**152**
4.4.1	Periodic functions	152
4.4.2	Derivative and integral of Laplace transform	157
4.4.3	Asymptotic analysis	161
4.4.4	Dirac's delta	168
4.5	**Symbolic computations**	**172**
4.5.1	Symbolic computations in Python	172
4.5.2	Symbolic computations in MATLAB	172
4.6	**One-minute round-up**	**173**
4.7	**Exercises**	**174**
5	**Inverse Laplace transform**	**185**
5.1	**General properties**	**185**
5.2	**Rational functions**	**188**
5.2.1	Distinct simple real poles	188
5.2.2	Multiple real poles	191
5.2.3	Simple complex conjugate poles	192
5.2.4	Simple complex conjugate poles – revisited**	197
5.2.5	Multiple complex conjugate poles	197
5.2.6	One-size-fits-all approach	201
5.2.7	Numerator and denominator have equal degrees	202
5.2.8	Rational times exponential	203
5.2.9	An alternative approach	204
5.3	**Convolution theorem**	**208**
5.4	**Solving initial value problems**	**216**
5.5	**Symbolic computations**	**221**
5.5.1	Symbolic computations in Python	221
5.5.2	Symbolic computations in MATLAB	222
5.6	**One-minute round-up**	**223**
5.7	**Exercises**	**225**

6 Transfer functions and block diagrams 237

- 6.1 Transfer functions — 237
- 6.2 System response — 242
- 6.3 Solving IVPs using transfer functions* — 250
- 6.4 Block diagrams — 251
  - 6.4.1 An algebra of blocks — 251
  - 6.4.2 Reduction of block diagrams — 256
- 6.5 More transfer functions — 260
  - 6.5.1 Transfer functions to time domain — 260
  - 6.5.2 Interlude: Solving IVPs with the method of linear differential operators* — 261
  - 6.5.3 Transfer functions to state representation — 269
  - 6.5.4 Transfer functions of MIMO systems — 270
- 6.6 Impedance — 272
- 6.7 Software for control systems — 276
  - 6.7.1 MATLAB's control systems toolbox — 276
  - 6.7.2 Python's control systems module — 281
- 6.8 One-minute round-up — 283
- 6.9 Exercises — 284

7 First-order systems 293

- 7.1 Introduction — 293
- 7.2 Impulse response — 296
- 7.3 Step response — 297
- 7.4 Frequency response — 299
  - 7.4.1 Frequency response and its asymptotic behaviour — 299
  - 7.4.2 Amplitude ratio and its asymptotes — 302
- 7.5 One-minute round-up — 307
- 7.6 Exercises — 309

8 Second-order systems 313

- 8.1 Introduction — 313
- 8.2 Impulse response — 315
- 8.3 Step response — 316
  - 8.3.1 Overdamped systems ($\zeta > 1$) — 316
  - 8.3.2 Critically damped systems ($\zeta = 1$) — 318

8.3.3	Underdamped systems ($\zeta < 1$)		319
8.4	**Characteristics of underdamped systems**		**320**
8.4.1	Peak time		320
8.4.2	Overshoot		321
8.4.3	Rise time		323
8.4.4	Decay ratio		324
8.4.5	Upper and lower envelopes		324
8.4.6	Settling time		325
8.5	**Frequency response**		**329**
8.6	**One-minute round-up**		**332**
8.7	**Exercises**		**333**

9 BIBO stability 341

9.1	**The concept of BIBO stability**	**341**
9.2	**BIBO stability criterion based on the poles of G**	**348**
9.3	**Routh's method**	**354**
9.3.1	Routh's stability criterion	354
9.3.2	A zero in the first column	358
9.3.3	A zero row	360
9.3.4	Poles with sufficiently negative real part	364
9.3.5	Poles in desirable region	366
9.4	**Stability of systems with time delays**	**369**
9.5	**One-minute round-up**	**372**
9.6	**Exercises**	**373**

10 PID controllers 383

10.1	**Intuition**	**383**
10.1.1	Conceptual construction of the PID controller	383
10.1.2	Additional material	387
10.1.3	Effect of PID parameters	388
10.1.4	Outlook	390
10.2	**Regulation and disturbance rejection**	**390**
10.2.1	Structure of feedback control systems	390
10.2.2	Regulation	392
10.2.3	Disturbance rejection	402

10.3	Tuning	**409**
10.3.1	Feeling lucky?	409
10.3.2	Ziegler-Nichols first method	410
10.3.3	Ziegler-Nichols ultimate sensitivity method	411
10.4	**Implementation of PID controllers**	**416**
10.5	**One-minute round-up**	**418**
10.6	**Exercises**	**420**

11 Frequency response and design of stable closed loops 427

11.1	Frequency response of general linear systems	**427**
11.1.1	Frequency response theorem	427
11.1.2	Special case: systems with a pole at zero	435
11.1.3	Special case: simple pairs of imaginary poles	440
11.2	**Bode plots**	**444**
11.2.1	Elementary Bode plots	444
11.2.2	Construction of Bode plots	454
11.3	**Bode stability criterion**	**461**
11.3.1	Motivation for Bode's stability criterion	461
11.3.2	Bode stability criterion	462
11.3.3	Stability margins	473
11.3.4	Necessary and sufficient stability conditions	480
11.4	**Effect of PID parameters on Bode plot**	**481**
11.4.1	Effect of proportional gain	481
11.4.2	Effect of derivative time	482
11.4.3	Effect of integral time	484
11.5	**Offset detection in Bode plots**	**485**
11.6	**Bode plots using software**	**488**
11.6.1	Python	488
11.6.2	MATLAB	489
11.7	**One-minute round-up**	**490**
11.8	**Exercises**	**491**

12 Nyquist plot and stability criterion 505

12.1	Plotting $G(j\omega)$ on the complex plane	**505**
12.1.1	Integrators	506

12.1.2	First-order systems	507
12.1.3	Second-order systems	508
12.1.4	Delay systems	510
12.1.5	PID controllers	510
12.1.6	Higher-order systems	511
12.2	**Nyquist stability criterion**	**514**
12.2.1	Cauchy's principle of argument	514
12.2.2	Nyquist stability criterion	519
12.2.3	Nyquist plots	522
12.3	**Stability margins**	**537**
12.3.1	Gain, phase, and delay margins	537
12.3.2	Sensitivity	540
12.4	**One-minute round-up**	**543**
12.5	**Exercises**	**544**

A Mathematical background 553

A.1	**Complex numbers**	**553**
A.1.1	Definitions	553
A.1.2	Basic operations with complex numbers	553
A.1.3	Modulus, argument and exponential form	554
A.1.4	Operations via the exponential form	555
A.1.5	Complex functions	556
A.2	**Real logarithm**	**558**
A.3	**Polynomials**	**559**
A.3.1	Basic concepts	559
A.3.2	Horner's scheme	560
A.4	**Trigonometric identities**	**563**
A.4.1	Sines, cosines and tangents	563
A.4.2	Inverse trigonometric functions	564
A.5	**Limits**	**566**
A.5.1	Limits of sequences and functions	566
A.5.2	Limits of rational functions at infinity	567
A.6	**Differentiation and integration**	**567**
A.6.1	Derivative and its properties	568
A.6.2	Definite integrals	570
A.6.3	Improper integrals	572

A.6.4	Integrals of sequences (and series)	574
A.6.5	Antiderivative & the fundamental theorem of calculus	575
A.6.6	L'Hôpital's rule	577
A.7	**Asymptotic notation**	**577**
A.7.1	Big-O notation	578
A.7.2	Little-o notation	580
A.8	**Hyperbolic functions**	**580**
A.9	**Infimum and supremum**	**581**
A.10	**One-minute round-up**	**582**
B	**Bibliography**	**583**
B.1	Control	583
B.2	Physics	586
B.3	Mathematics	586
B.4	Software	588
	Index	589

1 Introduction

This first chapter introduces the reader to the fundamental problems of control theory and control engineering. Control systems fall into two large categories: open and closed loop systems; our attention focuses almost exclusively on the latter. We define the components of a closed-loop (aka feedback) control system and give a few examples from everyday life.

1.1 A few words about systems and control

Heraclitus of Ephesus was a pre-Socratic Greek philosopher known for his theory of eternal change and *flow*. He advocated that one cannot walk into the same river twice: "new waters flow on those who step into the same rivers." Indeed, everything changes. The question is *how*. Once posed, this question quickly decomposes into a several more: What changes? What makes a system change? How can we describe this change? How can we effect desirable changes? This is what this book is about.

Control theory sits at the confluence of mathematics, engineering and computer science and studies how systems evolve and change in time and how the flow of dynamical systems can be shaped and directed towards a desired mode of operation.

1.1.1 Open-loop control systems

An average city dweller who commutes to work will spend a good deal of time every day waiting for green. Why is that? Traffic lights give priorities to vehicles and pedestrians according to some assumed traffic flow. In particular, conventional traffic control systems assume (and trust) a traffic model and take actions to avoid congestion and facilitate the movement of vehicles and pedestrians irrespective of the actual state of the traffic network. Conventional traffic control systems will, therefore, work in exactly the same way regardless of weather conditions, traffic or any other parameters as they do not receive information from the network at the time of operation. That said, it is unlikely you have never waited at a traffic light for no good reason.

Traffic lights are **open-loop** control systems: they will operate exactly the same way regardless of the actual traffic on the streets. Another such example is the washing machine: it will run exactly the same washing cycle regardless of how dirty the clothes are (it will even run exactly the same sequence of actions even if not loaded with clothes at all). The same goes for toasters, hand dryers, dishwashers, clothes dryers and many other simple appliances where the achievement of the control objective is not of critical importance.

Essentially, open-loop control systems are "blind" with respect to the quantities they are supposed to control and will take no corrective action and as a result, they are not suitable for tasks with high reliability and precision requirements. System-model mismatches and external disturbances may lead to results other than what the designer planned. On the other hand, **feedback control systems** keep an eye on the controlled system and adapt their actions accordingly.

1.1.2 Feedback control

Unlike open-loop control systems, feedback control systems (also known as closed-loop control systems), instead of blindly implementing a predefined sequence of actions, use measurements and adjust their decisions and actions continuously. This is what people do in most situations. Let us give a relevant example.

Imagine you are having a shower in a bath tub. There is one variable you care to control: the temperature of the water, which you sense with your skin. Your skin acts as a **sensor** and the temperature — let us denote it by T — is called the **output**. At the same time, you have a desired temperature at which you like to bathe; this is called the **set point**, which we shall denote by T^{sp}. The difference between the temperature you feel, T, and the desired temperature, T^{sp}, defines the **error**, that is

$$e = T^{\text{sp}} - T. \tag{1.1}$$

The error is then understood by your brain: if $e = 0$, the temperature is just right! If $e > 0$, the water is colder than you want it to be, so you will turn the tap accordingly. Likewise, if $e < 0$, the water is warmer, and you will turn the tap the other way. Therefore, your brain serves as a **controller** which translates the error to **control actions**. A control action is often referred to as a **manipulated input** (or simple an *input*). Our control actions affect the output variable, which is the temperature of

Chapter 1. Introduction

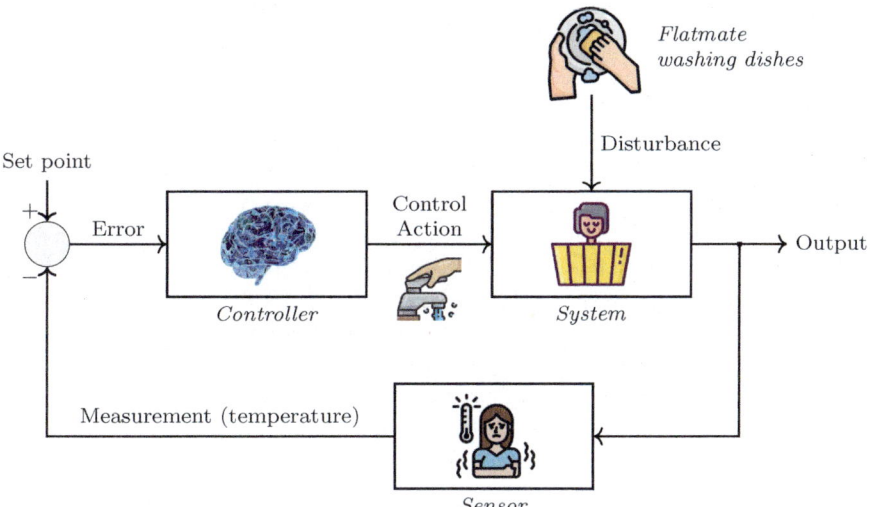

Figure 1.1: Taking a shower: a complete feedback control system.

the water. This defines a **feedback control loop** (see Figure 1.1[1]).

> **Flatmate** (n): a person who shares an apartment with another person (Cambridge Dictionary); a person who washes the dishes while you are trying to have a nice warm shower (Our definition).

However, as dictated by Murphy's Law and Finagle's Corollary[2], once you have adjusted perfectly the water temperature, your flatmate will find it a brilliant idea to wash the dishes. Such an unknown, unforeseen and uncontrolled external quantity that can affect the output is called a **disturbance**. Both manipulated inputs and disturbances affect the system output; we distinguish between them because we can prescribe the values of the former, but we cannot manipulate the latter.

Overall, in control theory and control engineering we try to address the following two fundamental problems:

- (Regulation problem). The output of a controlled system needs to be steered towards a desired set point; the controlled system should track the (externally provided) set point; in other words, the controlled system should follow our commands.

- (Disturbance rejection problem). The control system should be able to attenuate disturbances; the system should be able to operate in a realistic context where it will be exposed to unmodelled disturbances.

---

[1] Sources: Hot tub icon by @Konkapp on flaticon.com; dish washing icon, tap by @Umeicon and sensor icon by @photo3idea_studio on www.flaticon.com (Flaticon licence: free for personal and commercial purpose with attribution); Brain icon by @geralt on pixabay.com

[2] See the Wikipedia article at https://en.wikipedia.org/wiki/Finagle's_law

1.1.3 A few examples

Let us give a few examples of feedback control systems that we may encounter in our everyday life.

(Temperature control) In the third century CE, Roman emperor (from 218 to 222) *Elagabalus* employed a hugely inefficient and unfair technology to cool down his palace: he would command slaves to move tons of snow from the Apennine Mountains into his backyard. A few thousand years and several laughable attempts later, people now use what is known as an air conditioner. The principle, though, remains roughly the same: the user specifies a desired temperature (the *set point*), the air conditioner *measures* the ambient temperature (the system *output*), defines the *control error* and a *controller* decides the rate of heat flow from the room to the environment (the *control action*). If at the same time we open a window, the heat that will flow into the room will act as a *disturbance*, which the controller will need to attenuate.

(Cruise control) Many cars are nowadays equipped with what is known as a cruise control system: it is a speed control system that maintains a constant speed on the highway. The driver defines a desired speed v^{sp}, a speed *set point*, the cruise control system measures the speed, v, of the car (the *output*), computes the control error $e = v^{\text{sp}} - v$ and provides this value to a speed *controller*. When $e = 0$, the car moves at exactly the desired speed, when $e > 0$ the car moves with $v < v^{\text{rm}}$, so the controller knows it needs to speed up, so it will push down the accelerator accordingly (*control action*). The inclination of the road acts as a *disturbance* and may make the car speed up or slow down. The controller's job is to counteract the effect of this disturbance to maintain a constant speed.

(Artificial pancreas) Diabetes is a group of metabolic disorders related to the regulation of the levels of blood sugar, leading to high levels of blood sugar. Roughly, diabetes may occur either because of inadequate production of insulin from the pancreas (type I) or resistance to insulin (type II). An artificial pancreas is a device responsible for pumping insulin as a response to increased levels of blood glucose[3]. A closed-loop artificial pancreas uses a sensor to continuously monitor the concentration of blood glucose (output), c, and compares it to a certain *set point*, c^{sp}, thus defining the error $e = c^{\text{sp}} - c$. If the level of glucose is higher than the set point ($e < 0$), the controller may decide to pump an appropriate amount of insulin (*control action*) into the blood stream. The food or drink that the person takes acts as a disturbance, as it may alter the blood glucose levels, which the artificial pancreas needs to control.

(Rocket launch) This example comes from aerospace engineering. When a rocket launches, the primary objective is that it takes off upright. What we need to control is the "tilt", a, of the rocket with respect to its upright position. This defines the *output* of the system and the upright position is the *set point*. We can influence the tilt of the spacecraft by manipulating

---

[3] Read more at https://www.diabetes.co.uk/artificial-pancreas.html.

the thrust vector which defines our *control action*. This is illustrated in Figure 1.2[4].

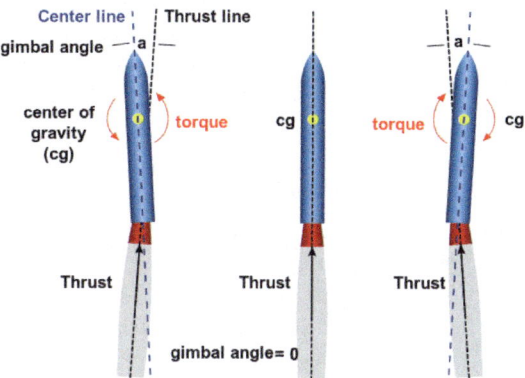

Figure 1.2: *Thrust vectoring control system. By manipulating the gimbal angle, a, we can affect the orientation of the rocket.*

Feedback control systems are by no means a human invention; nature abounds with control loops. Look no further than the human body: a series of feedback loops continuously monitor and adjust our body temperature, heart rate, blood sugar levels, stomach pH, blood pH, fluid balance, energy balance, gene expression and more to provide the necessary physicochemical conditions for life (known as homeostasis). For example, in healthy individuals, when blood sugar increases above its regular level, the hormone *insulin* is secreted by the beta cells found in the pancreas, which causes the liver to covert the excess of insulin to glycogen and signals body cells to absorb glucose and remove it from the blood stream. Conversely, if the levels of blood sugar drop significantly, the pancreatic alpha cells release the hormone *glucagon* which mediates the conversion of glycogen back into glucose. This is a natural control loop.

1.1.4 Systems and control theory

Systems and Control Theory is an inherently multidisciplinary field; it is impossible to study it without "encroaching" on other domains such as mathematics, physics, engineering (electrical, mechanical, chemical, civil), computer science, and machine learning. In Figure 1.3 we have tried to chart some relationships among these disciplines. The figure is all but exhaustive, but it is enough to demonstrate the strong interplay between Systems and Control and other disciplines. At its conception, in the 19th century, control theory had a purely industrial focus, but developed into a theory of *decision-making* with applications in almost all fields of science and engineering.

---

[4]Image taken from Wikipedia (public domain image uploaded by User Brian0918). This image was created by NASA. See https://www.grc.nasa.gov/www/k-12/rocket/gimbaled.html. Accessed on 27 December 2022.

Systems and Control Theory is concerned with two main tasks: the **analysis** of the qualitative properties of systems, their transient and long-term behaviour and the **design** of (typically closed-loop) systems with desired properties. Mathematics provides the theory and methods to analyse dynamical systems, while engineering sets the system requirements so that systems behave in a desired (safe, productive, reliable, predictable) way. Physics provides the principles that are used to model physical systems. Newton's laws of motions and Hooke's law of elastic springs are featured in most control engineering books. Newtonian mechanics provides the foundations for deriving dynamical models for a wide range of applications including robotics, aircraft dynamics, autonomous vehicles and many more. Similarly, chemistry, biology and other natural sciences provide **mathematical models** that describe how physical systems work.

The two main questions that systems and control theory addresses are those of **stability** and **robustness**. Although there exist quite a few formal definitions of stability (Bounded-input bounded-output stability, Lyapunov stability, input-to-state stability, mean-square stability, and more), roughly speaking, a system is said to be stable if it resists to change, therefore, small changes of its input or small disturbances will not cause it to go haywire. In this book we introduce the notion of **bounded-input bounded-output** (BIBO) stability. A system is said to be BIBO-stable if its output remains within certain bounds whenever its input is bounded. In other words, the output of a BIBO-stable system will not diverge so long as its input remains bounded. Robustness refers to the property of systems that are able to continue to function in spite of the presence of disturbances. A robust control system is able to operate in uncertain contexts or if its parameters change within reasonable bounds.

Systems and control theory underpins numerous applications and may serve one or more of the following purposes: (i) it reduces the effort that should be otherwise put by humans, (ii) leads to a safer and more reliable operation compared to the manual operation of processes, (iii) increases productivity, (iv) reduces the production cost by managing optimally the available resources, (v) reduces the maintenance cost by avoiding operating conditions that fatigue the machinery and equipment, (v) provides constant operating conditions and as a result constant product characteristics, (vi) leads to a higher precision operation.

In this book we focus on systems of a single input and a single output (SISO). Dynamical system models are typically available in the form of differential equations which are difficult to deal with directly. **Classical control theory** overcomes this difficulty by using the Laplace transform: an integral transform that transforms differential equations to simple polynomial equations of one complex variable, which are easier to work with. The application of the Laplace transform on the differential equations that describe the system dynamics yields an algebraic equation that describes the input-output dynamics; the input-output correspondence is then described by a complex function of one complex variable referred to as the **transfer**

function of the system. However, classical control theory focuses almost exclusively on *linear* SISO systems. **Modern control theory** casts the system equations into a standard form known as the **state space** representation and does not resort to the Laplace transform; the modern approach can deal with systems with multiple inputs and outputs (MIMO systems) and nonlinearities. Despite its limitations, classical control theory is very popular in engineering practice.

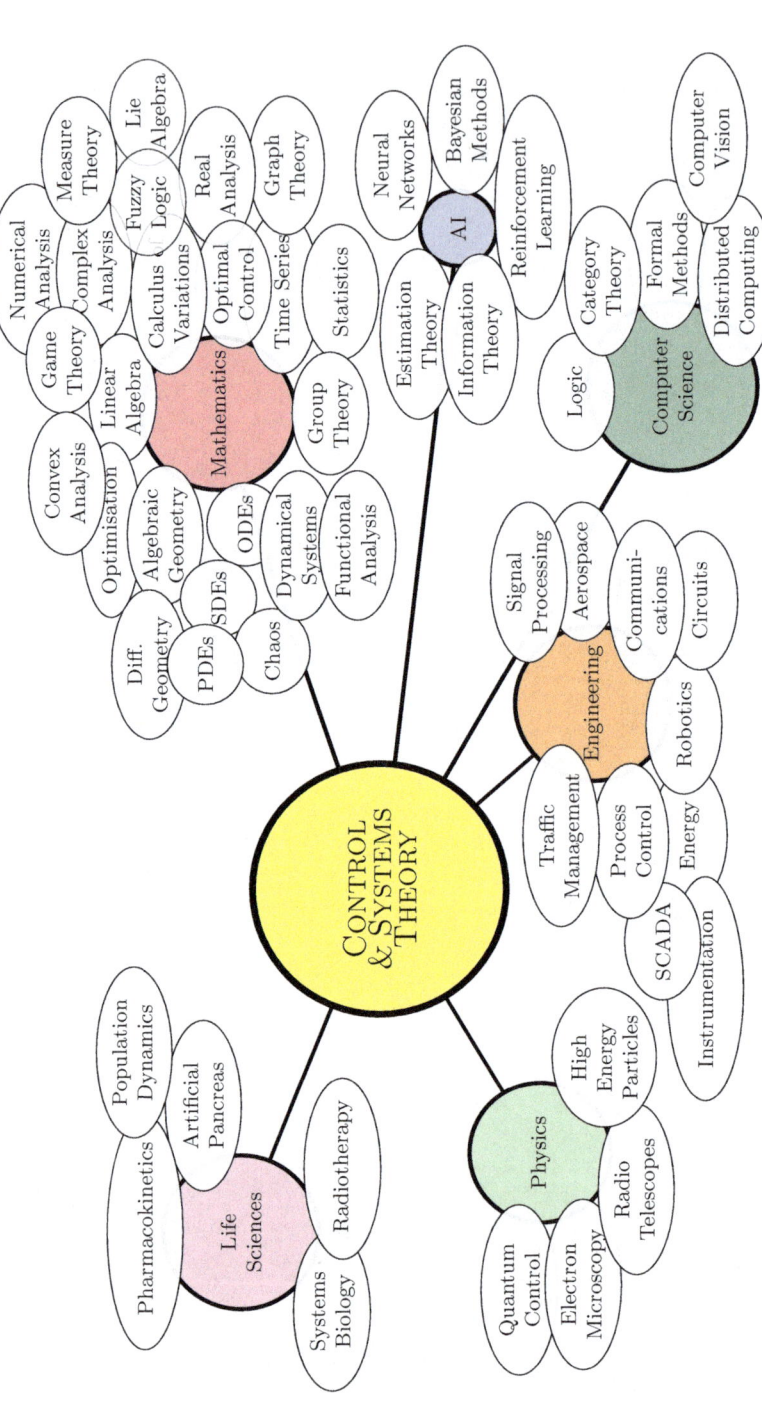

Figure 1.3: Control theory and links to related disciplines and topics.

In order to fully understand the material, we will often need to use software. Python and MATLAB are commonly used in practice to analyse dynamical systems and design and evaluate controllers. Although MATLAB had been the control engineer's holy grail for quite some time because of its support for control-specific functionality, researchers and practitioners are gradually transitioning to Python as a free, open-source and powerful alternative for controller design, simulations and prototyping. In this book we give examples in both Python and MATLAB as they are both essential for engineers.

1.2 Structure of this book

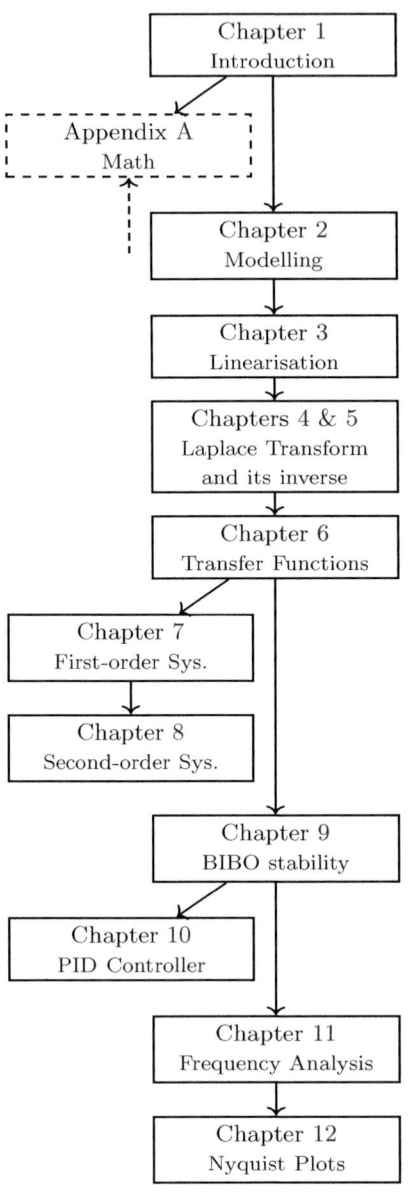

This book is structured as follows:

1. Chapter 1 — this chapter — is an introduction to the basic concepts of control theory.

2. In Chapter 2 we see how we can use first principles from physics and engineering to derive mathematical models to describe dynamical systems. These will have the form of ordinary differential equations and integro-differential equations. We also discuss how one can use such system equations to simulate dynamical systems in Python and MATLAB.

3. In Chapter 3 we derive linear approximations of dynamical systems, and we introduce the notion of deviation variables. These approximations will be used to design stabilising controllers in later chapters.

4. In Chapter 4 we introduce the Laplace transform: the basic methodological tool that we will be using throughout the course. All subsequent discussions will draw from the concept of the Laplace transform. In a nutshell, the Laplace transform allows us to transform differential equations into simple algebraic ones.

5. In Chapter 5 we introduce the inverse Laplace transform that will allow us to solve any linear initial value problem explicitly.

6. In Chapter 6 we introduce the concept of a transfer function; this construct represents elegantly the input-output dynamic relationship of systems. We also introduce the concept of *impedance*; a concept that facilitates the dynamical modelling of circuits.

7. In Chapters 7 and 8 we study two special cases of dynamical systems: first and second-order systems. Many physical systems fall in one of these two categories. Using the theoretical foundations that we studied in the previous chapters, we analyse the dynamic characteristics of these systems.

8. In Chapter 9 we introduce the concept of BIBO stability which is a notion of central importance in control theory.

9. Chapter 10 introduces the PID controller; to date, the most popular and widely used controller in engineering practice. We also discuss how to select the parameters of a controller to achieve good dynamic characteristics (this procedure is called **tuning**)

10. In Chapter 11 we study what happens when a dynamical system is excited with a sinusoidal signal. This is the so-called frequency response of dynamical systems. We present the Bode plot which is important in the design of frequency filters (high-pass, low-pass and band-pass) and finds applications in the analysis and design of electric circuits for sound processing, telecommunications and more. Moreover, Bode plots offer an advanced controller design methodology: they allow us to design "adequately" stable control systems and to quantify how robust a control system is.

11. Chapter 12 presents the Nyquist plot and the associated stability criterion which has its foundations in Cauchy's argument principle — an important result in complex analysis. The Nyquist criterion is the definite way to tell whether a linear closed-loop SISO system is stable in the BIBO sense and lifts some limitations of the Bode criterion.

12. Appendix A is a brief summary of essential results from mathematics that is necessary for this course. You might already know most of it from your first year Mathematics course. You can skim through it in the beginning and consult it later, as necessary.

At the end of each chapter there is a summary of the main takeaway points followed by some exercises you can try to solve to practice.

1.3 Exercises

Exercise 1.1 (Everyday control systems) Describe two feedback control systems we encounter in everyday life. What are the associated set points, outputs, control actions and disturbances?

Exercise 1.2 (Smartphones) What sensors are available on a typical smartphone?

✳ ....

Answer. Modern smartphones are equipped with IMUs (inertial sensors that measure linear accelerations and angular velocities), a magnetometer (compass), barometer, thermometer, humidity sensor — often a lot more.

Exercise 1.3 (Control systems at home) Describe some feedback control systems that can be found in an average house.

Exercise 1.4 (Jet pack) Watch the video at https://youtu.be/EAJM5L9hhBs.

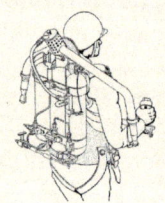

How do you believe jetpacks work?[a] What are the involved variables (inputs, outputs, disturbances, set points)? What are the objectives of the control system of a jetpack?

[a]The jetsuit sketch is a public domain image taken from Wikipedia and refers to U.S. Patent 3,243,144. John K. Hulbert, Wendell F. Moore, 1966.

2
Modelling

*In this chapter we introduce the concept of a dynamical model in the time domain. The aim of this chapter is to demonstrate that most physical phenomena which evolve in time, such as the motion of objects, the voltages and currents in an electric circuit, the flows and pressures in a network of tanks and so on, can be described by differential or integro-differential equations which can be derived from first principles. The objective here is to **model** the dynamics of various physical systems in terms of differential or integro-differential equations and point out to the involved **inputs**, **outputs** and **states**.*

2.1 Modelling in the time domain

2.1.1 State space representation

A **dynamical system** is a mathematical formalisation that describes the time evolution of a vector of variables of interest, $x(t)$, called *state* variables. This may be driven by an input signal[1], $u(t)$, and/or disturbance signals. Dynamical systems that can be manipulated using such a signal, $u(t)$, are called *control systems*. Hereafter, we denote the *derivative* of a signal $x(t)$ by $\dot{x}(t) = \mathrm{d}x(t)/\mathrm{d}t$. The signal $x(t)$ can be n-dimensional, that is, it can be $x(t) = (x_1(t), x_2(t), \ldots, x_n(t))$. In that case, its derivative, $\dot{x}$, is meant in the sense $\dot{x}(t) = (\dot{x}_1(t), \dot{x}_2(t), \ldots, \dot{x}_n(t))$. We denote the second derivative of $x(t)$ by $\ddot{x}(t) = \mathrm{d}^2 x(t)/\mathrm{d}t^2$. Third and higher-order derivatives will be denoted by $x^{(3)}, x^{(4)}$ and so on.

A dynamical system with **state** $x \in \mathbb{R}^n$ and input $u \in \mathbb{R}^m$ can be described by an **ordinary differential equation** (ODE), which is a func-

---
[1] The term *signal* means a function of time, $t \geq 0$

tional equation of the form

$$F(x, \dot{x}, \ddot{x}, \ldots, x^{(p)}, u, \dot{u}, \ldots, u^{(q)}) = 0, \qquad (2.1)$$

where typically $p > q$, which involves the state, the input, and a finite number of their time derivatives. For example, the equation

$$(\dot{x})^2 - 2x + \sqrt{u} = 0, \qquad (2.2)$$

with $x, u \in \mathbb{R}$, is an ODE.

Given an input signal $u(t)$, a signal $\phi(t)$ which satisfies (2.1), that is,

$$F(\phi(t), \dot{\phi}(t), \ldots, \phi^{(p)}(t), u(t), \dot{u}(t), \ldots, u^{(q)}(t)) = 0,$$

is called a **solution** of the ODE. For example, the function $\phi(t) = e^{-t}$ is a solution of the ODE $\dot{x} + x = 0$ as the reader can easily verify.

We often encounter problems where we need to determine a solution, $\phi(t)$ satisfying an **initial condition** of the form

$$\phi(0) = x_0. \qquad (2.3)$$

Such problems are termed **initial value problems**.

Remark: *Solving differential equations or initial value problems can be very difficult. However, the Laplace transform, which we will introduce later in this book (see Chapters 4 and 5), facilitates the procedure of solving certain differential equations and initial value problems systematically. We can use various software to solve them and build simulators for dynamical systems, which are valuable design tools. There is, however, a lot more to dynamical systems than just solving them. In fact, control theory does not focus on solving differential equations. But for details you will have to wait until Chapter 9.* •

The highest order state derivative determines the **order** of the system. More often than not, systems of order 2 or higher can be written as systems of order 1 which have the following form

$$\dot{x} = f(x, u), \qquad (2.4)$$

by introducing additional variables. Systems in the form of Equation (2.4) are referred to as **state space** systems. But before we see how, one might wonder why bother do that in the first place. Firstly, by casting the system dynamics in the form of Equation (2.4) we can clearly identify the *state*, x, and the input, u. Secondly, typically software tools and libraries expect us to specify the system dynamics in this standard form. Lastly, most of the theory has been developed around this standard representation.

Writing a high-order system into a state space representation is easy. Here is a first example:

Example 2.1 (Second-order ODE to first-order ODE) A dynamical system is described by the ODE

$$\ddot{x} + 3\dot{x}^2 - x = u, \qquad (2.5)$$

with $x, u \in \mathbb{R}$. Let us define the variables $x_1 = x$ and $x_2 = \dot{x}_1 = \dot{x}$. Then, $\ddot{x} = \dot{x}_2$ and Equation (2.5) yields

$$\dot{x}_1 = x_2, \qquad (2.6a)$$
$$\dot{x}_2 = -3x_2^2 + x_1 + u. \qquad (2.6b)$$

This is a dynamical system with a two-dimensional state $z = (x_1, x_2) \in \mathbb{R}^2$ and input $u \in \mathbb{R}$. In other words, we can write the system in the simple form

$$\dot{z} = f(z, u), \qquad (2.6c)$$

where $f : \mathbb{R}^2 \times \mathbb{R} \to \mathbb{R}^2$ is the function

$$f(z, u) = \begin{bmatrix} x_2 \\ -3x_2^2 + x_1 + u \end{bmatrix}. \qquad (2.6d)$$

Equation (2.6c) is the state space representation of the given ODE in Equation (2.5).

Let us give another example where we have a fourth-order system

Example 2.2 (State space representation of fourth-order ODE) Consider the following differential equation

$$x^{(4)} + x^2 x^{(3)} - \sin \dot{x} = u. \qquad (2.7)$$

Since this is a fourth-order system, we need to introduce *four new variables*, namely, $x_1 = x$, $x_2 = \dot{x}_1$, $x_3 = \dot{x}_2$ and $x_4 = \dot{x}_3$ (note the pattern: $x_1 = x$ and each next variable we introduce is the derivative of the previous one). Now note that $\dot{x} = x_2$, $\ddot{x} = x_3$, $x^{(3)} = x_4$ and $x^{(4)} = \dot{x}_4$. That said, we may write Equation (2.7) as follows

$$\dot{x}_4 + x_1^2 x_4 - \sin x_2 = u$$
$$\Leftrightarrow \dot{x}_4 = u - x_1^2 x_4 + \sin x_2. \qquad (2.8)$$

Let us define the state variable $z = (x_1, x_2, x_3, x_4)$. Then,

$$\dot{z} = \begin{bmatrix} \dot{x}_1 \\ \dot{x}_2 \\ \dot{x}_3 \\ \dot{x}_4 \end{bmatrix} = \begin{bmatrix} x_2 \\ x_3 \\ x_4 \\ u - x_1^2 x_4 + \sin x_2 \end{bmatrix} =: f(z, u). \qquad (2.9)$$

This is the state space representation of Equation (2.7).

Exercise 2.1 (High-order ODEs to first-order ones) Following the same procedure as in Example 2.1 introduce appropriate variables to write the following high-order systems in the form $\dot{z} = f(z, u)$

1. $\ddot{x} = u + 2020$

2. $x^{(3)} = u$

3. $x^{(3)} + \ddot{x}^2 + x \tan \dot{x} = \cos u + xu$

4. $x^{(4)} + \ddot{x} = -u$

5. $\ddot{x} + 3x\dot{x}^2 = u$

6. $5x^{(3)} + 10\ddot{x} - 3\dot{x} + 8x - u = 0$

7. $x^{(4)} - 2x^{(3)}\ddot{x}^2 x + \ddot{x}^2 + x = u^2$

8. $\sum_{i=0}^{n} a_i x^{(i)} = u$, $a_n \neq 0$

✂ ....

Hint: If an ODE involves derivatives up to order n, that is, terms $x, \dot{x}, \ldots, x^{(n)}$, then introduce the variables $x_1 = x$, $x_2 = \dot{x}, \ldots, x_n = x^{n-1}$. Then, note that $x^{(n)} = \dot{x}_n$. By convention $x = x_{(0)}$.

Certain systems, such as electric circuits with capacitors, are often described by **integro-differential equations** (IDEs). These are functional equations involving the state, the input, and a finite number of their time derivatives and time integrals

$$F\left(x, \dot{x}, \ddot{x}, \ldots, x^{(p)}, \int_0^t x(\tau)d\tau, \right.$$
$$\left. \int_0^t \int_0^\xi x(\tau)d\tau d\xi, \ldots, u, \dot{u}, \ldots, u^{(q)}\right) = 0. \quad (2.10)$$

Often, IDEs can be reduced to ODEs by introducing additional variables in a similar fashion as in Example 2.1.

Remark: *To avoid certain pathological cases, when dealing with IDEs we shall assume that (i) $x(t)$ is defined and finite over all intervals $[0, t]$ (ii) all involved integrals exist for all t and are finite. In particular, we assume that $\int_0^t |x(\tau)|d\tau < \infty$ for all $t \geq 0$. It then follows that the function $I(t) = \int_0^t x(\tau)d\tau$ is **continuous** in t (see Appendix A). This assumption is to avoid cases such as $x(t) = 1/t$ where $\int_0^t |1/\tau|d\tau$ does not exist and cases where x escapes to infinity at some finite time $t \geq 0$.* •

Similar to ODEs, IDEs can often be written in the common form of a state space representation using the fundamental theorem of calculus (see Theorem A.22 in the appendix). Let us give an example.

Example 2.3 (IDE as ODE) A series RLC circuit (see Section 2.3 for details regarding the derivation of such models) is described by the

Chapter 2. Modelling

following IDE

$$RI + L\dot{I} + \frac{1}{C}\int_0^t I(\tau)d\tau = V, \qquad (2.11)$$

where V is the input voltage signal, I is the current that runs through the circuit, and R, L and C are constant parameters. We may introduce the variable

$$V_c(t) = \frac{1}{C}\int_0^t I(\tau)d\tau. \qquad (2.12)$$

By applying a time derivative on both sides of Equation (2.12), and using the fact that $V_c(0) = \frac{1}{C}\int_0^0 I(\tau)d\tau = 0$, we arrive at the following initial value problem (see the remark after this example for a detailed discussion)

$$\dot{V}_c(t) = \frac{I(t)}{C}, \qquad (2.13a)$$
$$V_c(0) = 0. \qquad (2.13b)$$

Note that Condition (2.13b) is an important one. If we do not include it, Equation (2.13) gives

$$V_c(t) = \frac{1}{C}\int_0^t I(\tau)d\tau + c, \qquad (2.14)$$

for any constant $c \in \mathbb{R}$. Equation (2.11) becomes

$$RI + L\dot{I} + V_c = V. \qquad (2.15)$$

Indeed, the original IDE can be equivalently written as the following ODE

$$\dot{I} = L^{-1}V - L^{-1}RI - L^{-1}V_c, \qquad (2.16a)$$
$$\dot{V}_c = C^{-1}I, \qquad (2.16b)$$

which is a dynamical system with state variable $x = (I, V_c)$, input variable V, and the initial condition $V_c(0) = 0$.

Remark: In (2.13) we used the fundamental theorem of calculus (see Theorem A.22 in the appendix). The theorem tells us what happens if we differentiate the integral $\int_0^t I(\tau)d\tau$ of the continuous function $I(t)$. It states that $\int_0^t I(\tau)d\tau$ is continuous and differentiable and its derivative is equal to $I(t)$. Therefore, by differentiating both sides of Equation (2.12) we obtain (2.13a).

However, the fact that two functions, f and g, have equal derivatives does not mean that the functions are equal (indeed, the functions $f(t) = 1$

and $f(t) = 2$ have equal derivatives). If $f' = g'$ we can conclude that $f(x) = g(x) + c$ for some constant c. As a result, although (2.12) implies (2.13a), (2.13a) is not equivalent to (2.12). The remedy is to note that $V_c(0) = 0$. •

Exercise 2.2 (IDE to first-order ODE) Consider the following integro-differential equation

$$x + 4\dot{x} + \ddot{x} + \int_0^t x(\tau)\mathrm{d}\tau = u. \qquad (2.17)$$

Introduce appropriate variables to write it in the form of a first-order ODE.

✂ ⋯⋯⋯

Hint. Work as in Exercise 2.1 for the derivatives and as in Example 2.3 for the integral.

2.1.2 Systems with input derivatives*

So far we have focused on differential and integro-differential equations where no derivatives of the input are present. Equations with derivatives on the input variables are often encountered in mechanical systems, circuits and other systems of practical interest and call for a special treatment. Let us give an example where the first derivative of the input is present in the system dynamics

Example 2.4 (State space representation of ODE with $\dot{u}$) Consider the following dynamical system

$$15\ddot{x} + 4\dot{x} + 3x + 8\dot{u} = -6u. \qquad (2.18)$$

The idea is to combine the term that involves $\dot{u}$ with the second-order term under the same differentiation operation as follows

$$\frac{\mathrm{d}}{\mathrm{d}t}(15\dot{x} + 8u) + 4\dot{x} + 3x = -6u. \qquad (2.19)$$

We now define $x_1 = x$ and

$$x_2 = 15\dot{x} + 8u. \qquad (2.20)$$

This way, Equation (2.19) can be written as

$$\dot{x}_2 + 4\dot{x}_1 + 3x_1 = -6u. \qquad (2.21)$$

From Equation (2.20) we can write $\dot{x} = \dot{x}_1 = \frac{1}{15}(x_2 - 8u)$, so Equation (2.21) becomes

$$\dot{x}_2 + \tfrac{4}{15}(x_2 - 8u) + 3x_1 = -6u. \tag{2.22}$$

Rearranging and solving for $\dot{x}_2$ yields

$$\dot{x}_2 = -\tfrac{58}{15}u - 3x_1 - \tfrac{4}{15}x_2. \tag{2.23}$$

The state space representation of the given system is

$$\begin{bmatrix} \dot{x}_1 \\ \dot{x}_2 \end{bmatrix} = \begin{bmatrix} \tfrac{1}{15}x_2 - \tfrac{8}{15}u \\ -\tfrac{58}{15}u - 3x_1 - \tfrac{4}{15}x_2 \end{bmatrix} = \begin{bmatrix} 0 & \tfrac{1}{15} \\ -3 & -\tfrac{4}{15} \end{bmatrix} \begin{bmatrix} x_1 \\ x_2 \end{bmatrix} + \begin{bmatrix} -\tfrac{8}{15} \\ -\tfrac{58}{15} \end{bmatrix} u.$$

This is the state space representation of the system in terms of the state variable $z = (x_1, x_2)$ and the input variable u.

Following the procedure we introduced in Example 2.4 the reader can solve the following exercise which involves a linear and three nonlinear systems. It will become obvious that the above methodology can be applied to nonlinear systems too (sometimes).

Exercise 2.3 (Dynamical systems involving $\dot{u}$) Write the following dynamical system in their state space representation

1. $\ddot{x} + \dot{u} = x + u$
2. $5\ddot{x} = 3\dot{u} + 5\sin(x) + u$
3. $\ddot{x} + x\dot{u} = u + x$
4. $x^{(3)} + \dot{u} = x + u$

※ ....

Answer. 1. Introduce the variables $x_1 = x$ and $x_2 = \dot{x}_1 + u$ in: the state space representation of the system is $\dot{x}_1 = x_2 - u$, $\dot{x}_2 = x_1 + u$. 2. Define $x = x_1$ and $x_2 = \dot{x}_1 + u$; then, it is $\dot{x}_1 = \frac{3}{5}x_2 + \frac{2}{5}u$, $\dot{x}_2 = \sin x_1 + u$. 3. This one is a bit more tricky: first add and subtract $x\dot{u}$ on the left-hand side: $\ddot{x} + x\dot{u} + \dot{x}u - \dot{x}u = u + x$ which can be written as $(\dot{x} + xu)\frac{d}{dt} = -(\dot{x} + xu) - \dot{x}u + u + x$. 4. Define $x_3 = x + u$; can you take it from there?

When we have dynamical systems that involve higher order derivatives on their input variables, things are somewhat more complicated, but we will be able to address this question in Section 6.5.3.

2.2 Mechanical systems

The position, x, and velocity, $v = \dot{x}$, of a point particle can be defined relative to a given frame of reference. Our objective is to derive equations that describe the trajectories of a point particle upon the exercise of a force. The understanding of the effect of a force on the motion of a point particle is highly relevant for anything that can move, including applications of robotics, aircraft, ships, spacecraft, cars and all sorts of vehicles, motors, pulleys, cranes and so on. The main apparatus we shall use to describe the motion of point particles is Newton's laws of motion. These are used to describe the *state* of a point particle, which typically is its position and velocity, (x, v), upon the action of *input* signals and *disturbances*, which typically are externally applied forces.

2.2.1 Laws of motion

Newton's three laws of motion describe the translational motion of point particles. Newtonian kinematics holds true under the assumption that the motion is studied with respect to an inertial frame of reference, that is, a frame of reference where Newton's first law holds:

> **Newton's First "Law".** In an *inertial* frame of reference, an object either remains at rest or continues to move at a constant velocity, unless acted upon by a force.

The first law can be seen as a *definition* of an inertial frame of reference leading to the second law of motion. Perhaps Newton's intention was to refute Aristotle's understanding of motion where an *effect* (velocity) required a *cause* (force).

Newton's second law of motion is the main apparatus to describe the motion of an object upon the action of an external force:

> **Newton's Second Law.** Suppose that an object has a constant mass m. Then, in an inertial frame of reference, the total force F acting on that object is equal to its mass m multiplied by its acceleration:
> $$F = ma, \qquad (2.24)$$

Note that both F and a in Equation (2.24) are vectors. Newton's second law can be written as
$$F(t) = m\ddot{x} = m\dot{v}, \qquad (2.25)$$
where $v = \dot{x}$ is the object's velocity relative to the said inertial frame of reference.

Remark: *Newton's second law holds true in an inertial frame of reference (defined by the first law). In a non-inertial frame, i.e., one that accelerates with respect to an inertial frame of reference, the second law does not hold.*

For example, if you are in an accelerating vehicle, Newton's second law, in the above form, does not hold with respect to a frame of reference affixed to the car. In a non-inertial frame we need to account for the "fictitious force" that is associated with the acceleration of the frame with respect to an inertial frame. •

Newton's Third Law describes the forces that develop when two objects interact. It states that

> **Newton's Third Law.** When one body exerts a force on a second body, the second body simultaneously exerts a force equal in magnitude and opposite in direction on the first body.

It is often referred to as the *action-reaction law* (where one force is called the *action* and the other the *reaction*).

Example 2.5 (Book on the table) A book of mass $m = 1$ kg is placed on a table and does not move (relative to the table). The acceleration of gravity in Belfast is $g = 9.8147$ m/s², therefore, the weight of the book is $mg = 9.8147$ N. Since the book does not move, by Newton's first law of motion, we conclude that the total force applied to it is zero, therefore, there must be another force applied to it, which is of opposite direction to its weight and of equal magnitude. This is an upward force exercised by the table to the book as provisioned by Newton's third law of motion.

Example 2.6 (Free fall) An object of mass m is falling in vacuum.

Figure 2.1: *Anvil in free fall. The only force acting on the object is its weight.*

The only force exercised on the particle is its weight as shown in the figure above. Let us denote the altitude of the object by y. Then, its acceleration is $\ddot{y}$. By Newton's second law of motion,

$$\not{m}\ddot{y} = -\not{m}g \Rightarrow \ddot{y} = -g.$$

This ODE describes the system dynamics.

2.2.2 Friction and drag

Dry Friction is a force which acts to stop the relative motion of objects sliding against each other. Dry friction is subdivided into **static friction** and **kinetic friction**.

Imagine there is an anvil at rest on a level surface as in Figure 2.2. If we apply a **small** horizontal force, the box will not move. According to Newton's first law of motion, there must be a force that prevents it from moving and this is equal in magnitude to the force we apply and has the opposite direction (see Figure 2.2).

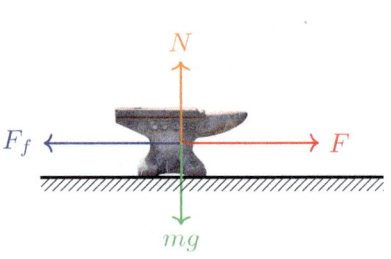

Figure 2.2: *Forces acting on a static object. Horizontally: (i) the (small) force F that we apply, and (ii) the static friction, F_f. Vertically: (iii) the weight, mg, (iv) the supporting force from the ground, N. The photo of the anvil was created by @Momentmal on* `pixabay.com` *and is a royalty-free image distributed under the terms of the pixabay licence (Free for commercial use, No attribution required).*

If we keep increasing the magnitude of the force we apply (F), we will eventually move the object. There is a *maximum* static friction that the two surfaces can put up. The maximum static friction that can be developed is given by

$$F_{f,\max} = \mu_s N, \tag{2.26}$$

where μ_s is called the **static friction coefficient**. Any force larger than $F_{f,\max}$ will cause sliding to occur. Once the surfaces start to slide, static friction is no longer applicable. Then, the friction is

$$F_f = \mu_k N, \tag{2.27}$$

where μ_k is called the **kinetic friction coefficient**. This type of friction is called **kinetic friction**.

When an object moves in a fluid (liquid or gas), then there is a force exercised from the medium to the body with direction opposite to that of its motion. This force is known as **drag** and its magnitude depends on the properties of the medium and the size, shape and speed of the object. If the velocity, v, of the object relative to the medium is large, then the drag is given by the **drag equation**

$$F_D = -\beta v^2, \tag{2.28}$$

where β is a constant[2].

---

[2]This is given by $\beta = \frac{1}{2}\rho C_D A$, where ρ is the density of the medium, A is the cross-sectional area of the object and C_D is the drag coefficient, which depends on the size and shape of the object and the speed of the object relative to the medium.

If the velocity of the object relative to the medium is very low, we may use the following linear approximation

$$F_D = -bv, \qquad (2.29)$$

where b is a constant[3].

Figure 2.3: *Drag force on a moving submarine. The drag force, F_D, depends on the speed v of the submarine relative to the water. The submarine image was downloaded from* `iconspng.com` *and is a royalty-free photo.*

2.2.3 Hooke's law

Hooke's law describes the restoring force exercised by a spring when displaced (extended or compressed) by some distance x. It states that

> **Hooke's law.** The restoring force, F_{spring}, exerted by a spring is proportional to the displacement (or *deformation*), x, and opposite in direction,
>
> $$F_{\text{spring}} = -kx. \qquad (2.30)$$

A spring that is governed by Hooke's law is called **elastic**. Constant k in Hooke's law is called the **stiffness** of the spring. Hooke's law is a good approximation of the behaviour of an elastic spring insofar as x is considerably smaller than the maximum deformation of the spring.

> **Example 2.7 (Horizontal spring-mass system)** In this example we will derive a model for a spring-mass system.
>
>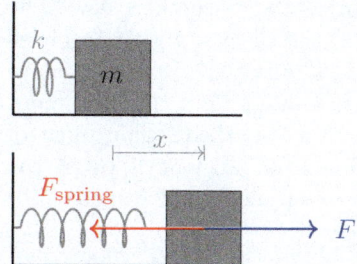
>
> In the figure on the left, an object of mass m is attached to a spring in a horizontal position. We assume that there is no friction between the object and the ground. The spring satisfies Hooke's law and its stiffness is k. An external force F is exerted to the mass leading to a displacement x.
>
> According to Hooke's law — see Equation (2.30) — the restoring force exercised by the spring on the object is $F_{\text{spring}} = -kx$, where x is the displacement with respect to the rest position (where $F_{\text{spring}} =$

---

[3]This approximation is valid for objects moving in a medium at rather low speeds without turbulent flow. Coefficient b is proportional to the viscosity of the medium and depends on the shape and size of the object.

0) and by Newton's second law of motion:

$$F - kx = m\ddot{x}. \tag{2.31}$$

This is the ODE that describes this system.

> **Exercise 2.4 (Action and reaction)** In Exercise 2.7, what are the forces acting on the floor and the wall?
>
> ※ ....
>
> *Answer.* The weight acts on the floor and, by Newton's third law, this induces an upward reaction force, N, acting on the object. The force F_{spring} induces a reaction force $-F_{\text{spring}} = kx$ acting on the wall. In turn, the wall exerts a force $-kx$ acting on the left end of the spring.

In practice, springs are not ideal and, therefore, dissipate energy. In particular, we often assume that there exists an internal damping force which is proportional to the linear speed of the spring and has the opposite direction, that is

$$F_{\text{damping}} = -b\dot{x}, \tag{2.32}$$

where b is a constant called the **viscous damping coefficient**. Next, we give an example of a damped spring-mass system.

Example 2.8 (Vertical spring-mass-damper system) In this example we will derive a model for a spring-mass system in presence of a damping force.

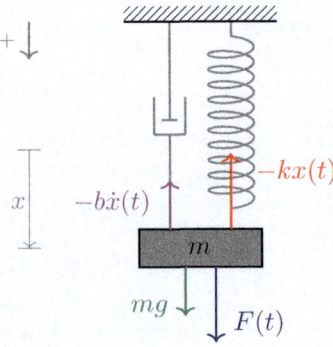

In the figure on the left, an object of mass m is attached to a vertical spring of stiffness k. We assume that the spring satisfies Hooke's law. The weight of the object, mg, as well as an external force, F, are exerted to the mass leading to a displacement x. Note that x is the displacement of the spring from its length when the total force applied to it is zero (not the length of the spring).

According to Hooke's law — see Equation (2.30) — the restoring force exercised by the spring on the object is $F_{\text{spring}} = -kx$. A damping force, $F_{\text{damping}} = -b\dot{x}$ is also exerted on the object. By Newton's second law of motion:

$$F + mg - b\dot{x} - kx = m\ddot{x}. \tag{2.33}$$

This is the kinematic equation that describes this system.

It is important to underline that x is not the length of the spring! It is the displacement of the spring from it initial length, which is its length when the total force applied to it is equal to zero. This refers to the length of the spring when $F = 0$ and there is no object attached to it. We will revisit this example in the context of deviation variables in Chapter 3.

In Example 2.8 we used an inertial reference frame whose origin was placed at the initial length of the spring, that is, position $x = 0$ corresponds to a zero displacement of the spring. When $F = 0$, the object equilibrates at a displacement x^{eq} where its velocity and acceleration are equal to 0, therefore, from Equation (2.33) we have that

$$\cancel{F} + mg - \cancel{b\dot{x}} - kx^{\text{eq}} = \cancel{m\ddot{x}} \Rightarrow mg - kx^{\text{eq}} = 0. \tag{2.34}$$

We subtract (2.34) and (2.33) by parts to obtain

$$F + \cancel{mg} - b\dot{x} - kx - \cancel{mg} + kx^{\text{eq}} = m\ddot{x}$$
$$\Rightarrow F - b\dot{x} - k(x - x^{\text{eq}}) = m\ddot{x}. \tag{2.35}$$

Let us define the displacement from the equilibrium as $y = x - x^{\text{eq}}$. We observe that $\dot{y} = \dot{x}$ and $\ddot{y} = \ddot{x}$. We may write the dynamics as

$$F - b\dot{y} - ky = m\ddot{y}, \tag{2.36}$$

so we have derived a model which is independent of the acceleration of gravity.

Exercise 2.5 (Spring-mass-damper system) Consider the following system consisting of a body of mass m, two springs and two dampers.

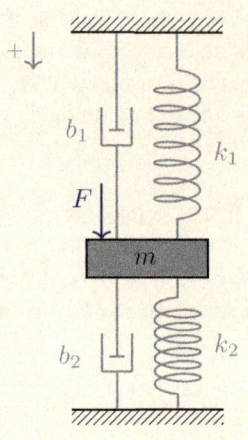

In the figure on the left, an object of mass m is attached to two vertical springs with stiffness coefficients k_1 and k_2. We assume that both springs satisfy Hooke's law. The object is attached to two dampers as shown in the figure with damping coefficients b_1 and b_2 respectively. An external force, F, is applied to the object.

1. Using Newton's laws of motion and Hooke's law, write the equation of motion of the system

2. What is the vertical position at which the object can equilibrate?

3. Can you follow a procedure similar to the one above to eliminate all constant terms from the ODE you derived in step 1?

✂ .... ✂ ....

Answer. 1. It is $m\ddot{y} = -k_1(y - y_{0,1}) - k_2(\dot{y} - y_{0,2}) - b_1\dot{y} - b_2\dot{y} + mg + F$. *Hint.* Use the frame of reference provided. Suppose that the upper spring is at its natural length when the object is at position $y_{0,1}$ and the lower spring is at its natural length at $y_{0,2}$; then, the restoring force from the upper spring is $-k_1(y - y_{0,1})$, while the restoring force from the lower spring is $-k_2(y - y_{0,2})$. The forces from the two dampers are $-b_1\dot{y}$ and $-b_2\dot{y}$. 2. At the equilibrium point the velocity is zero, i.e., $\dot{y} = 0$. Therefore, $0 = -k_1(y_{eq} - y_{0,1}) - k_2(y_{eq} - y_{0,2}) + mg + F_{eq}$. 3. Define $\tilde{y} = y - y_{eq}$, $\tilde{F} = F - F_{eq}$ and subtract by parts the system dynamics you obtained in step 1 and the equation from step 2 to obtain $m\ddot{\tilde{y}} = -(k_1 + k_2)\tilde{y} - (b_1 + b_2)\dot{\tilde{y}} + \tilde{F}$.

Example 2.9 (Car towing a caravan) In this example, the spring is not attached to a fixed point, but instead connects two moving objects as shown below.

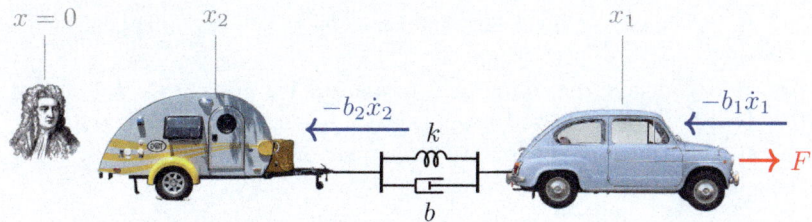

Figure 2.4: A car towing a caravan. The car can exercise a forward force F. The two objects are connected by a winch which can be modelled by a spring and a damper. Sources: The car and caravan are royalty-free images distributed under the terms of the pixabay licence (Free for commercial use, no attribution required), created by @Emslichter and @rauschenberger, respectively.

Problem statement. The positions of the car and the trailer with respect to a fixed point on the ground are denoted by x_1 and x_2 respectively. We denote the mass of the car by m_1 and the mass of the caravan by m_2. Lastly, the spring has a *natural length* where it does not apply any forces; this corresponds to a distance between the two vehicles which we denote by L.

We will further assume that there is a drag force acting on each vehicle. The drag force on the car is given by $-b_1\dot{x}_1$ and the drag force of the caravan is $-b_2\dot{x}_2$.

Revision: how the spring works. Let us introduce the auxiliary variable $y = x_1 - x_2$ which denotes the (signed) distance between the car and the caravan. When $y = L$ the spring applies no forces. When the two objects come closer to one another ($y < L$), the spring applies a force that pushes them away. When the distance between the car and the caravan is larger than L ($y > L$), then the spring pulls them together. We know that the spring force is equal to its constant, k, times its displacement. The displacement from its natural length is equal to $y - L$, so the restoring force acting on the car

$$F_{\text{spring}} = -k(y - L), \qquad (2.37)$$

and, by Newton's third law, the force $-F_{\text{spring}}$ acts on the caravan.

Revision: how the damper works: The damping force acting on the car is proportional to its speed *relative to the caravan* which is $\dot{y} = \dot{x}_1 - \dot{x}_2$, and has an opposite direction to that of $\dot{y}$, that is

$$F_{\text{damper}} = -b\dot{y} = -b(\dot{x}_1 - \dot{x}_2). \quad (2.38)$$

Again, by Newton's third law, the force F_{damper} acts on the caravan.

In order to derive the dynamical model of this system we will study the dynamics of each body separately. To that end, we sketch each object and the forces that are exercised on it[a]. This is called the **free body diagram**.

The forces acting on the car are:

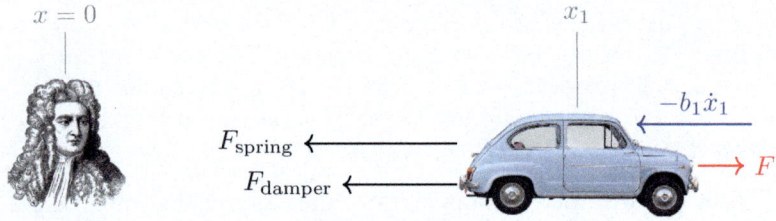

and the forces on the caravan are:

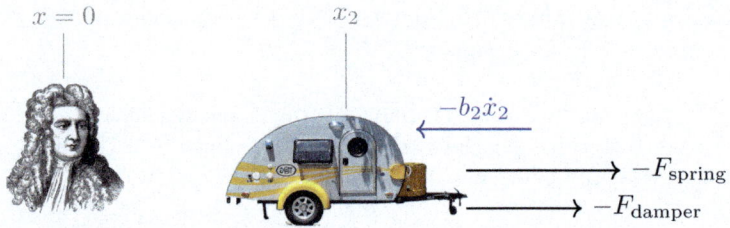

By invoking Newton's second law on each object, we have

$$\text{(car)} \quad m_1 \ddot{x}_1 = F - b_1 \dot{x}_1 - k(y - L) - b\dot{y}$$
$$\Rightarrow m_1 \ddot{x}_1 = F - b_1 \dot{x}_1 - k(x_1 - x_2 - L) - b(\dot{x}_1 - \dot{x}_2) \quad (2.39a)$$

and similarly,

$$\text{(caravan)} \quad m_2 \ddot{x}_2 = -b_2 \dot{x}_2 + F_{\text{spring}} + F_{\text{damper}}$$
$$\Rightarrow m_2 \ddot{x}_2 = -b_2 \dot{x}_2 + k(x_1 - x_2 - L) + b(\dot{x}_1 - \dot{x}_2). \quad (2.39b)$$

Equations (2.39) describe the (interconnected) dynamics of the two objects.

---

[a] We omit the weight of the objects and the supporting force from the ground as they do not contribute to the system dynamics

2.2.4 Planar rotation

In Section 2.2.1 we studied the laws that govern the *linear* motion of bodies and used Newton's laws of motion for our analysis. In this section we study rotational motions such as that of the Earth around its axis, the motion of a door opening as it rotates around its hinges, and even the motion of our legs as we walk (indeed, our legs and feet rotate around our several joints as we walk).

In Section 2.2.1, all linear motions were studied relevant to a certain reference system. In order to define a rotation, we need two systems of axes known as *frames*: the one is fixed at some point in space and the other has its origin affixed at the point of rotation and is allowed to rotate as shown in Figure 2.5. In what follows we will study only rotations on a two-dimensional plane.

Pairs of such frames are given in Figure 2.5; the frame denoted by (x, y) is the fixed frame and (x', y') is the rotating frame. The angle between the two frames is denoted by θ; this is the rotational counterpart of position (see Figure 2.6). As shown in Figure 2.5, the angle of a counterclockwise rotation is by convention *positive*. The rate of change of this angle, $\dot{\theta}$, is called the **angular velocity** of the rotating body and $\ddot{\theta}$ is called its **angular acceleration**.

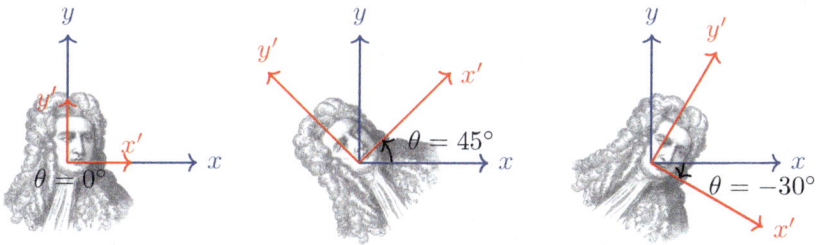

Figure 2.5: *The orientation of an object is defined with respect to two frames of reference. The global frame of reference, (x, y), is assumed to be fixed on earth, while the body-fixed frame rotates with the body.*

The above example, is similar to the spin of the Earth around its axis. In Figure 2.6 we see another example where an object is rotated around a point O, similar to the revolution of the Earth around the sun. The rotation is again described by two frames of reference.

In the context of the exposition of Newton's laws of motion, the attentive reader might have noticed that a *force* is interpreted as the cause for the (linear) acceleration of an object. The cause of an angular acceleration is a *torque*: a **torque** relevant to a given point of rotation O generated by a force F is defined as the product of the force with the distance between F and O. This is illustrated in Figure 2.7.

Newton's first and second laws of motion have the following rotational counterparts (we consider a global frame and a body-fixed frame as shown in Figure 2.7):

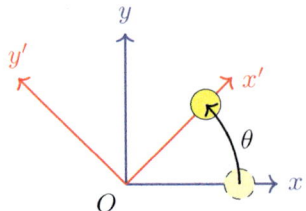

Figure 2.6: *The angle of rotation, θ, is the angle between the two frames of reference.*

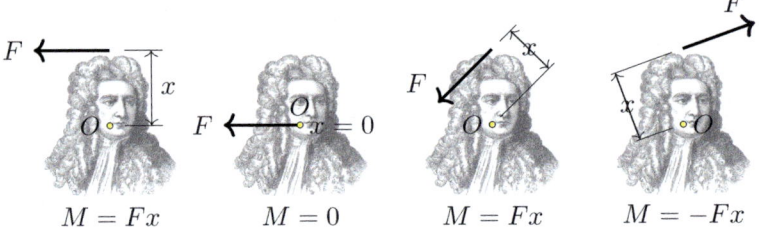

Figure 2.7: *Torque M with respect to the point of rotation O. By convention, torques that tend to rotate the object counterclockwise are positive, whereas those that tend to rotate the object clockwise are negative.*

Counterpart of Newton's first law of rotational motion. The angle θ between the (inertial) global frame and the body-fixed frame remains constant or changes with constant rate, unless a nonzero torque, relevant to the that point of rotation, is exercised.

Counterpart of Newton's second law of rotational motion. Suppose that the global frame of reference is an inertial frame of reference and an object has a constant moment of inertia. The rotational version of Newton's second law is

$$M = I\ddot{\theta}, \qquad (2.40)$$

where M is the net torque, relevant to a point of rotation O, $\ddot{\theta}$ is the angular acceleration of the object and I is its moment of inertia relevant to point O.

The moment of inertia of an object, relevant to a point O, gives the rotational inertia of this object relevant to O. Similar to how mass determines the force needed to achieve a certain (linear) acceleration, the moment of inertia of a body gives the torque needed to achieve a certain angular acceleration. The moment of inertia of certain simple planar objects can be easily computed and depends on their mass and geometry. Let us give just a couple of examples that we may often encounter in practice:

the moment of inertia of (i) a point-mass relevant to a point at a distance R from it and (ii) an isotropic rod of length L (the distribution of mass throughout the rod is uniform, and its width is negligible compared to its length) relevant to its centre of mass and (iii) an isotropic disc of radius R with respect to its centre of mass.

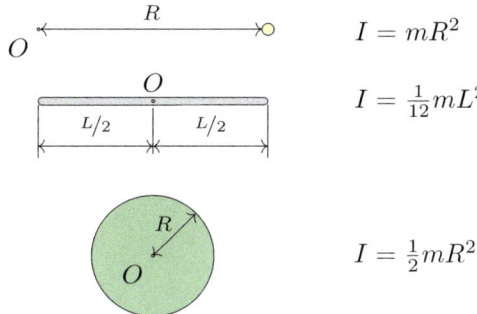

Figure 2.8: *Moments of inertia of various objects. The moment of inertia of standard geometric shapes can be found in [Φ1]. The moment of inertia of irregular shapes can be determined either numerically or experimentally.*

The analogy between linear and rotational motion is summarised in the following table

Linear Motion	Rotational Motion
Position, x	Angle, θ
Velocity, $\dot{x}$	Angular Velocity, $\dot{\theta}$
Acceleration, $\ddot{x}$	Angular Acceleration, $\ddot{\theta}$
Mass, m	Moment of Inertia, I
Force, F	Torque, M
Newton's Second Law, $F = m\ddot{x}$	$M = I\ddot{\theta}$

Example 2.10 (Simple pendulum) Consider the following system of a point mass m which is suspended from the ceiling with a massless string of length L. Note the frame of reference and the convention of positive rotations that we are using in the example. We start by drawing the forces that are applied on the object:

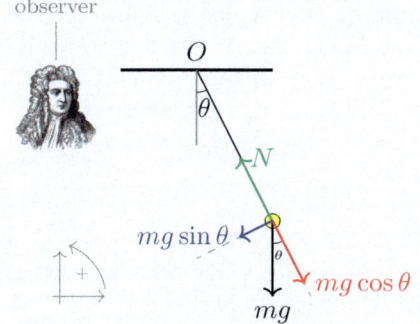

Its weight, mg, which acts at its centre of gravity and has a downward direction. We can decompose the weight into two forces, one of which is parallel to the string and has magnitude $mg \cos \theta$, and a force that is perpendicular to the string and has magnitude $mg \sin \theta$ as shown in the figure on the left

The object does not move along the axis defined by the string, therefore, by Newton's first law of motion, the total force along this direction must be zero. Indeed, by Newton's third law of action and reaction, there is a supporting force N exerted by the string to the object so that $N = mg\cos\theta$.

The force $mg\sin\theta$ that is perpendicular to the string causes the following torque around point O

$$M = -mgL\sin\theta. \tag{2.41}$$

The negative sign is because this torque tends to rotate the object in the clockwise direction which is the negative direction of rotation. By the rotational counterpart of Newton's second law of motion (with respect to point O), we have

$$M = I\ddot{\theta}. \tag{2.42}$$

The moment of inertia of the object of mass m around point O is $I = mL^2$, (see Figure 2.8) therefore, Equation (2.42) becomes

$$-\cancel{m}gL\sin\theta = \cancel{m}L^{\cancel{2}}\ddot{\theta} \Leftrightarrow -g\sin\theta = L\ddot{\theta}. \tag{2.43}$$

Equation (2.43) describes the motion of the object. It is worth noting that the motion of the object is independent of its mass. This means that if we suspend two objects of different mass from strings of the same length, they will perform identical motions.

Exercise 2.6 (Pendulum with isotropic rod) Follow similar steps as in Example 2.10 to solve this exercise.

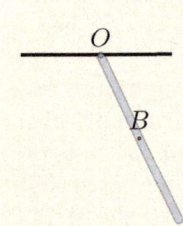

An *isotropic* rod, that is, a rod with its mass equally distributed along its length, is hanging from the ceiling as shown in the figure on the left. The centre of mass of the rod is at its geometric centre. Its moment of inertia with respect to point O is $I = \frac{1}{3}mL^2$, where L is the rod's length. Derive the equations of motion of the rod.

✂ . . . .

Hint. (i) introduce an inertial frame of reference, (ii) apply Newton's second law of motion and its rotational counterpart with respect to point O.

Chapter 2. Modelling

Exercise 2.7 (Simple pendulum with non-isotropic rod) Consider the non-isotropic rod of total length $L = 2\,\text{m}$ and mass $m = 1\,\text{kg}$ shown below.

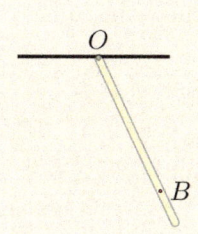

Its moment of inertia with respect to point O is $I = 3.3\,\text{kg}\,\text{m}^2$ and its centre of mass is located at a distance of $1.6\,\text{m}$ from point O. Derive its equation of motion.

✂ ....

Hint: Sketch all forces acting on the rod and determine the torque with respect to point O. Note that the weight is exerted on the centre of mass of the rod.

The following example is commonly used in control engineering; it is a system of an inverted pendulum, and we model it using Newton's laws of translational and rotational motion.

Example 2.11 (Inverted pendulum on a vehicle) An inverted pendulum is a pendulum that has its centre of mass above its pivot point. The inverted pendulum, and its several variants, has numerous applications in engineering including space rockets and humanoid robots. Here is a list of relevant videos:

1. https://youtu.be/AuAZ5zOPOyQ: An inverted pendulum on a cart like the one in this example

2. https://youtu.be/JhQ4s2_pG1o: Segways are inverted pendula

3. https://youtu.be/15DIidigArA: An inverted pendulum on a quadcopter

4. https://youtu.be/o2sHf-udJI8: Vertical take off and landing is based on very similar principles

5. https://youtu.be/YvbAqwOsk6M: Walking robots are instances of the inverted pendulum

In this example we will study the dynamics of an inverted pendulum on a moving cart, which can move along the x axis as shown in Figure 2.9.

Frame of reference. The x axis points to the right and the y axis upwards. The upright position corresponds to $\theta = 0$ and angles increase in the clockwise direction.

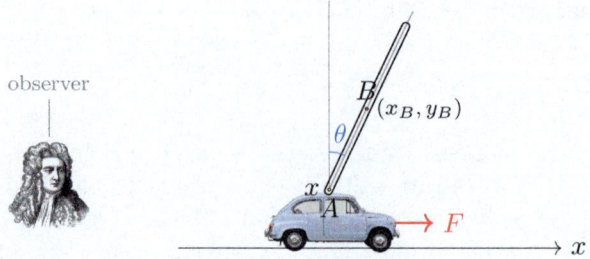

Figure 2.9: *Inverted pendulum*

Problem statement. The moving part of the pendulum is a rod of mass m and total length L. The rod is isotropic (this means that its mass is uniformly distributed) and its centre of gravity (point B) coincides with its geometric centre (at distance $L/2$ from its ends). The fulcrum of the pendulum (point A) is attached on top of the car and the friction at that joint is negligible. The angle between the rod and the vertical axis, y, is denoted by θ. We can move the car along the x axis by applying a horizontal force F. The mass of the car is denoted by M. We need to derive a model that describes the dynamics of θ and explains how this is affected by F.

The coordinates of point A are $x_A(t) = x(t)$ and (by convention) $y_B(t) = 0$, and the coordinates of point B are

$$x_B(t) = x(t) + \tfrac{L}{2}\sin\theta(t), \qquad (2.44\text{a})$$
$$y_B(t) = \tfrac{L}{2}\cos\theta(t). \qquad (2.44\text{b})$$

Approach. We will study the car and the rod *separately*: we will first draw each individual object and the forces that act on them. The car's motion is purely translational, so we will use Newton's second law of motion. The rod does both a translational (back and forth) and a rotational motion, therefore, we need to invoke both Newton's second law of motion and its rotational counterpart.

The sketch of the individual objects and the forces that act on them in known as the **free body diagram**, and it is shown below.

Forces on the rod. There are three forces acting on the rod:

Chapter 2. Modelling

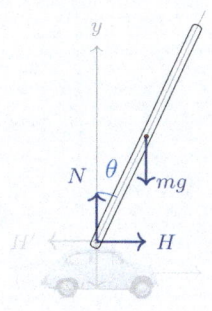

1. Its weight, mg, which acts at its centre of gravity and has a downward direction,

2. A supporting force N which props the rod up which is exerted from the car to the rod at the fulcrum point and has an upward direction,

3. A force H which is exercised from the car to the rod at its fulcrum point.

Forces on the car. The following forces act on the car:

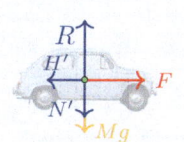

1. An externally applied horizontal force, F

2. The reaction to H, which we denote by H'; by Newton's third law of motion, $H' = -H$. This force is exerted by the rod to the car

3. Its weight, Mg, which is downward force

4. The reaction to N, which we denote by N'; by Newton's third law of motion, $N' = -N$.

5. The supporting force R which is applied from the ground to the car.

Rod dynamics. The rod moves in a translational as well as rotational manner. The **translational** dynamics of its centre of gravity, at point (x_B, y_B), can be described by Newton's second law of motion according to which

$$m\ddot{x}_B = H, \tag{2.45}$$
$$m\ddot{y}_B = N - mg, \tag{2.46}$$

and substituting Equations (2.44), we obtain

$$m\tfrac{d^2}{dt^2}(x + \tfrac{L}{2}\sin\theta) = H, \tag{2.47a}$$
$$m\tfrac{d^2}{dt^2}\left(\tfrac{L}{2}\cos\theta\right) = N - mg. \tag{2.47b}$$

Since θ is a function of t, the left-hand side of Equation (2.47a) becomes

$$\tfrac{d^2}{dt^2}(x + \tfrac{L}{2}\sin\theta) = \ddot{x} + \tfrac{L}{2}\tfrac{d}{dt}(\dot\theta\cos\theta) = \ddot{x} + \tfrac{L}{2}(\ddot\theta\cos\theta - \dot\theta^2\sin\theta).$$

Likewise, the left-hand side of Equation (2.47b) becomes

$$\tfrac{d^2}{dt^2}\left(\tfrac{L}{2}\cos\theta\right) = \tfrac{L}{2}\tfrac{d}{dt}(-\dot\theta\sin\theta) = -\tfrac{L}{2}(\ddot\theta\sin\theta + \dot\theta^2\cos\theta).$$

Equations (2.47a) and (2.47b) become

$$m\ddot{x} + m\tfrac{L}{2}(\ddot{\theta}\cos\theta - \dot{\theta}^2\sin\theta) = H, \qquad (2.48a)$$
$$-m\tfrac{L}{2}(\ddot{\theta}\sin\theta + \dot{\theta}^2\cos\theta) = N - mg. \qquad (2.48b)$$

The **rotational** dynamics of the rod can be described by the rotational counterpart of Newton's second law; we will use the centre of gravity (point B) of the rod as our reference point. With respect to this point, the following *torques* act on the rod

1. No torque accrues from the rod's weight. This is because the weight acts precisely at the reference point.

2. The horizontal force H is at a distance of $\tfrac{L}{2}\cos\theta$ from the reference point (point B) and the torque tends to turn the rod counterclockwise, so the associated torque is $T_H = -H\tfrac{L}{2}\cos\theta$.

3. The vertical force N is at a distance of $\tfrac{L}{2}\sin\theta$ from the reference point (point B) and the torque tends to turn the rod clockwise, so the associated torque is $T_N = +N\tfrac{L}{2}\sin\theta$.

so the total torque with respect to point B is $T_N + T_H = \tfrac{L}{2}(N\sin\theta - H\cos\theta)$, therefore,

$$\tfrac{L}{2}(N\sin\theta - H\cos\theta) = I\ddot{\theta}, \qquad (2.49)$$

where $I = \tfrac{1}{12}mL^2$ is the moment of inertial of the rod about its centre of gravity, that is

$$\tfrac{\cancel{L}}{2}(N\sin\theta - H\cos\theta) = \tfrac{1}{12}mL^{\cancel{2}}\ddot{\theta},$$
$$\Rightarrow N\sin\theta - H\cos\theta = \tfrac{1}{6}mL\ddot{\theta}. \qquad (2.50)$$

Car dynamics. The car dynamics is purely translational, and the car does not move along the y axis, therefore, $R - N - Mg = 0$ by Newton's first law of motion. By Newton's second law of motion along the x axis, we have

$$M\ddot{x} = F - H. \qquad (2.51)$$

Overall, Equations (2.48a), (2.48b), (2.50) and (2.51) describe the dynamics of the system.

Chapter 2. Modelling

Remark: We can simplify the model we derived in Example 2.11. We will present a **non-rigorous** approach here, and we will revisit this example in Chapter 3. As long as the rod is close to the upright position, it is reasonable to assume that $\sin\theta \approx \theta$, $\cos\theta \approx 1$ and $\theta\dot\theta \approx 0$ (this is the non-rigorous part; see Chapter 3 for details), so Equation (2.48a) becomes[4]

$$(\text{approximate}) \quad m\ddot{x} + m\frac{L}{2}\ddot\theta = H, \qquad (2.52)$$

and, similarly, Equation (2.48b) can be approximated by

$$0 = N - mg. \qquad (2.53)$$

By combining the above equations, we can derive the following simplified **approximate** dynamics for the system of Example 2.11:

$$(M+m)\ddot{x} + m\frac{L}{2}\ddot\theta = F, \qquad (2.54\text{a})$$

$$\tfrac{2}{3}L\ddot\theta + \ddot{x} = g\theta. \qquad (2.54\text{b})$$

•

Exercise 2.8 (State space representation) Write the pair of Equations (2.54) in the form of a state space representation.

✂ ....

Hint. You will need to introduce four state variables: $x_1 = x$, $x_2 = \dot{x}$, $x_3 = \theta$, $x_4 = \dot\theta$.

In Example 2.11 we used the following versatile methodological approach:

1. We stated the problem and explained how the different components (car and rod) are linked and how they interact; we also chose a frame of reference.

2. We sketched the free body diagram and used Newton's third law on the pairs of action and reaction that develop between the interacting objects.

3. We employed Newton's second law of motion to describe all translational motions.

4. For all rotational motions, we use the rotational counterparts of Newton's laws.

---

[4]The problem with this approach is that although the approximations $\sin\theta \approx \theta$, $\cos\theta \approx 1$ and $\theta\dot\theta \approx 0$ may look intuitive — and indeed are often used in the literature — they are *non sequitur*. In Chapter 3 we will present a systematic approach to perform such simplifications known as *linearisation*. Linearisation is based on Taylor's theorem.

Often the rotational motion of an object is linked to its translational motion, or the translational motion of some other object. Typical examples are the wheels of a car, or pulleys. In such cases, the following result is very useful

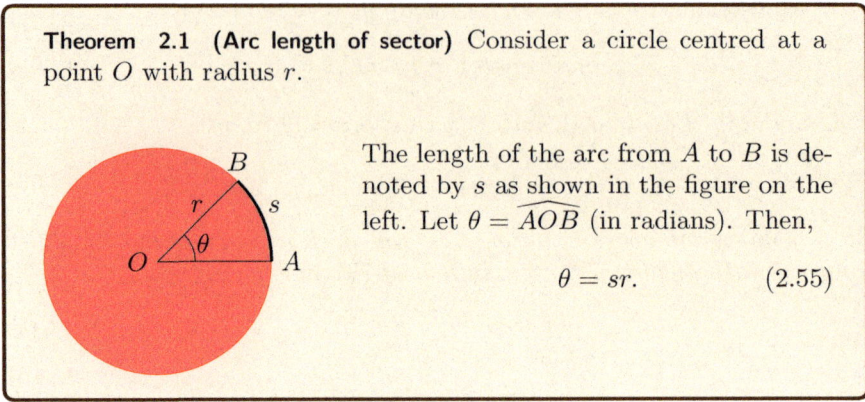

Theorem 2.1 (Arc length of sector) Consider a circle centred at a point O with radius r.

The length of the arc from A to B is denoted by s as shown in the figure on the left. Let $\theta = \widehat{AOB}$ (in radians). Then,

$$\theta = sr. \qquad (2.55)$$

Remark: *Although we have stated it as a theorem, Theorem 2.1 is the definition of an angle in radians. Note that although degrees are typically used in practice to measure angles, the radian is the SI unit for angles.* •

The above theorem is very useful for studying the motion of rolling wheels. Suppose that a wheel moves on a plane without slipping having an angular velocity $\omega = \dot{\theta}$ and its centre point has a linear velocity v as shown in the following figure.

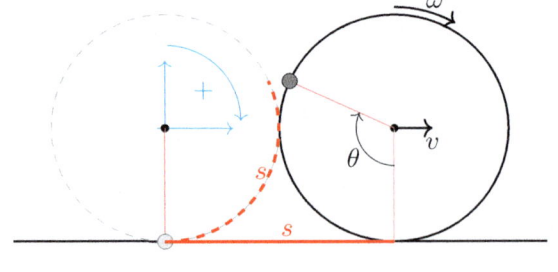

Figure 2.10: *Wheel moving to the right. Its initial position at time $t_0 = 0$ is shown with dashed circle. At a future time $t_1 > 0$ the wheel is at a distance s from its initial position.*

As the wheel moves forward, the length or the arc that has touched the ground (thick dashed red arc) is equal to its horizontal displacement. By Theorem 2.1, this is proportional to the angle θ that corresponds to this arc. By differentiating both sides of Equation (2.55) we obtain

$$\dot{\theta} = r\dot{s} \Leftrightarrow \omega = rv, \qquad (2.56)$$

where $v = \dot{s}$ is the linear velocity of the object. By differentiating once

again we obtain
$$\ddot{\theta} = r\ddot{s} \Leftrightarrow \dot{\omega} = ra, \tag{2.57}$$
where $a = \ddot{s} = \dot{v}$ is the linear acceleration of the object.

It is important to highlight that the signs in Equations (2.56) and (2.57) depend on the frame of reference introduced in Figure 2.10. Indeed, from Equation (2.56) we can infer that ω and v have the *same sign* — if $\omega > 0$ (i.e., if the wheel spins *clockwise*), then $v > 0$ (i.e., it moves *forwards*).

For example, consider the same system, but with a different frame of reference where clockwise rotations are positive as shown in the figure below where counterclockwise rotations are positive. When the wheel moves to the right, i.e., $v > 0$, it rotates clockwise, that is, $\omega < 0$, therefore we have $\omega = -rv$, and, what is the same $\dot{\omega} = -ra$.

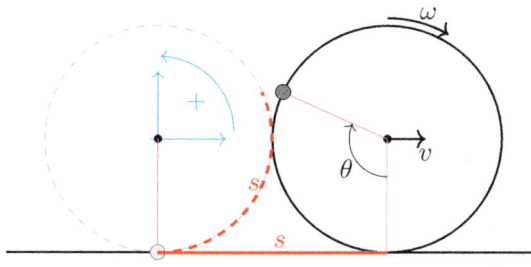

Figure 2.11: *Wheel rolling on a horizontal plane. Here the frame of reference is different from the one of Figure 2.10: counterclockwise rotations are positive.*

In the following example we derive the dynamics of a ball rolling on an inclined plane.

Example 2.12 (Rolling ball on inclined plane) Consider a solid ball on an inclined plane as in the Figure 2.12. The mass of the ball is m and its radius is r. We suppose that the ball cannot slip on the plane (there is high static friction), so if $\phi \neq 0$, it will *roll*. The rolling is a combination of translational and rotational motion.

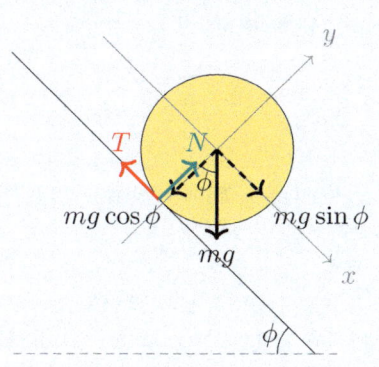

Figure 2.12: *A ball of mass m on an inclined plane at an angle ϕ. Three forces act on the ball: its weight, mg, the supporting force from the plane, N, and the static friction, T. The normal force and $mg \cos \phi$ are drawn a little away from the point of contact between the ball and the plane so that all arrows are clearly visible.*

We start by choosing the reference frame (x, y) shown below. For convenience, we use the convention that clockwise rotations are positive. We then sketch the forces that are exerted on the ball.

The ball does not move along the y axis, therefore by Newton's first law of motion, the total force must be equal to zero, that is,

$$N = mg \cos \phi. \tag{2.58}$$

Along the x axis, Newton's second law of motion gives

$$mg \sin \phi - T = m\ddot{x}, \tag{2.59}$$

where T is the static friction between the ball and the plane. The friction creates a torque on the ball with respect to its centre of mass which is equal to Tr. Essentially, it is the friction that makes the ball spin. By Newton's law for rotational motion we have

$$Tr = I\ddot{\theta}, \tag{2.60}$$

where $\ddot{\theta}$ is the angular acceleration of the ball. By virtue of Theorem 2.1, the angular acceleration of the rolling ball is related to its linear acceleration, $a = \ddot{x}$, by

$$a = \ddot{\theta} r. \tag{2.61}$$

By substituting (2.61) into (2.60) we obtain

$$T = \frac{I\ddot{\theta}}{r} = \frac{Ia}{r^2}, \tag{2.62}$$

therefore, by Equation (2.59), the acceleration along the x axis is

$$mg \sin \phi - \frac{Ia}{r^2} = ma \Rightarrow a = \frac{mg \sin \phi}{m + \frac{I}{r^2}}. \tag{2.63}$$

The moment of inertia of a ball with respect to an axis running through its centre is $I = \frac{2}{5}mr^2$, so the acceleration becomes

$$a = \frac{mg \sin \phi}{m + \frac{\frac{2}{5}mr^2}{r^2}} = \frac{\cancel{m}g \sin \phi}{\frac{7}{5}\cancel{m}} = \tfrac{5}{7}g \sin \phi, \tag{2.64}$$

and the differential equation that describes the dynamics of this system is

$$\ddot{x} = \tfrac{5}{7}g \sin \phi. \tag{2.65}$$

It is worth noting that the motion of the ball on the x-axis depends **neither** on the size nor the mass of the ball. Two balls of different size and different mass will perform identical motions on the x axis.

Remark: *The ball in Example 2.12 rolls and does not slide if the friction is below a maximum value which, as we discussed in Section 2.2.2, is given by $T_{\max} = \mu_s N = \mu_s mg \cos\phi$, where μ_s is called the **static friction coefficient**. So, for the ball not to slip we must have $T < \mu_s mg \cos\phi$.* •

Example 2.13 (Yo-Yo dynamics) A Yo-Yo of mass m has an axle of radius r and a spool of radius $R > r$ as in Figure 2.13. The string is of negligible mass and thickness. The moment of inertia of the Yo-Yo around its centre of mass is $I = \frac{1}{2}mR^2$ and its geometric centre coincides with its centre of mass. A vertical external force F is exercised on the Yo-Yo at its centre of mass.

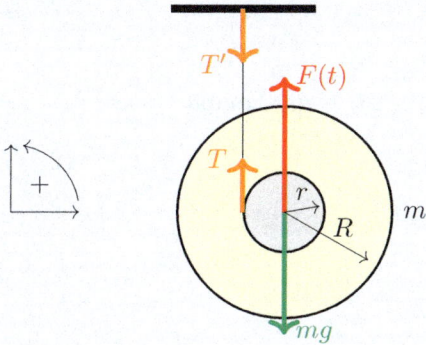

Figure 2.13: *Forces exerted on Yo-Yo: vertical force F, weight mg, and the supporting force T.*

We want to derive a dynamical model, in the form of a differential equation, that describes how the external force, F, affects the vertical position of the Yo-Yo. We use the inertial frame of reference shown in Figure 2.13.

The forces exerted on the spool are (i) the external force F, (ii) the weight, mg, and (iii) the supporting force from string:, T. There is also the reaction of T, $T' = -T$ (on ceiling). These are shown in Figure 2.13.

By Newton's second law of motion on the spool

$$F + T - mg = m\ddot{y}. \tag{2.66}$$

Force T causes the torque $-Tr$ with respect to the centre of mass of the Yo-Yo (mg and F do not cause any torques with respect to the centre of the spool). By the rotational counterpart of Newton's second law of motion

$$-Tr = I\ddot{\theta} \iff -Tr = \tfrac{1}{2}mR^2\ddot{\theta}. \tag{2.67}$$

But by Theorem 2.1 we also have that $\ddot{y} = r\ddot{\theta}$; equivalently $\ddot{\theta} = \tfrac{1}{r}\ddot{y}$ and Equation (2.67) becomes

$$-Tr = \tfrac{1}{2}mR^2\tfrac{1}{r}\ddot{y} \iff T = -\tfrac{1}{2}m\left(\tfrac{R}{r}\right)^2 \ddot{y}. \tag{2.68}$$

We will now plug (2.68) into (2.66)

$$F - \tfrac{1}{2}m\left(\tfrac{R}{r}\right)^2 \ddot{y} - mg = m\ddot{y} \tag{2.69}$$

$$\iff F - mg = \underbrace{m\left[1 + \tfrac{1}{2}\left(\tfrac{R}{r}\right)^2\right]}_{\alpha}\ddot{y}, \tag{2.70}$$

which is the desired ODE. We can write this compactly as

$$F - mg = \alpha\ddot{y}, \tag{2.71}$$

where α is a constant defined in Equation (2.70).

2.3 Circuits

2.3.1 Resistance, capacitance, inductance

According to **Ohm's law** the current I that runs through a conductor is proportional to the voltage V across its two end points. This is described by

$$V = IR, \qquad (2.72)$$

and the constant R is called the **resistance**.

The current that runs through a **capacitor** is related to the voltage across the capacitor via

$$I(t) = C\dot{V}(t). \qquad (2.73)$$

Remark: Equation (2.73) can be written as $dV(t) = \frac{1}{C}I(t)dt$ and in integral form as

$$V(t) = V(0) + \frac{1}{C}\int_0^t I(\tau)d\tau. \qquad (2.74)$$

By differentiating both sides of Equation (2.74) and applying the fundamental theorem of calculus, we obtain the expression of Equation (2.73). At the same time, the voltage across a capacitor is proportional to the **charge**, $Q(t)$, stored in it. In particular, $V(t) = \frac{Q(t)}{C}$, therefore, the charge at time t is

$$Q(t) = CV(0) + \int_0^t I(\tau)d\tau. \qquad (2.75)$$

•

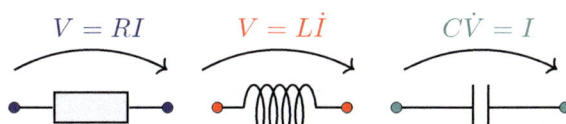

Figure 2.14: The three basic elements of an electric circuit and their dynamics.

Electric conductors tend to oppose changes in the electric current that runs through them. This effect is known as **inductance**. Inductors create voltage drops which are proportional to the rate of change of the current through them, that is,

$$V = L\dot{I}. \qquad (2.76)$$

Coefficient L is called **inductance** and its unit in the SI system is the Henry (H), which is defined as the amount of inductance which causes a voltage of 1 volt when the current is changing at a rate of one ampere per second.

2.3.2 Kirchhoff's laws

Kirchhoff's laws govern the distribution of currents through the various circuit components and the development of voltages across them. Kirchhoff's current law (KCL) is an equation that describes the distribution of current at junctions of a circuit. Roughly speaking, it states that the total current that enters the junction must be equal to the total current that exits. KCL states that:

> **Theorem 2.2 (KCL)** The algebraic **sum of currents** in a network of conductors meeting at a point is **zero**.

In order to better understand this, let us have a look at the following example

Example 2.14 (KCL example) In the following figure we see four currents meetings at a junction

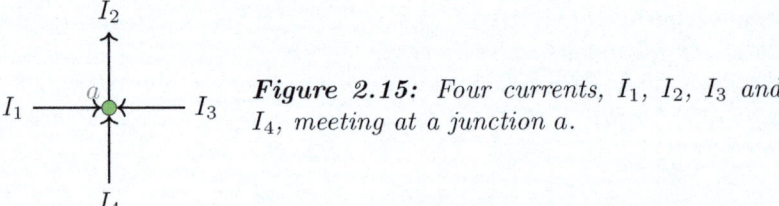

Figure 2.15: Four currents, I_1, I_2, I_3 and I_4, meeting at a junction a.

Currents I_1, I_3 and I_4 are **inbound**, therefore, they have a **positive sign**, whereas, I_2 is **outbound**, so it has a negative sign. By KCL we have that

$$I_1 + I_3 + I_4 - I_2 = 0. \tag{2.77}$$

A second important result is Kirchhoff's voltage law (KVL) which is stated below

> **Theorem 2.3 (KVL)** The directed sum of the voltages around any closed loop is zero.

We will use KVL in the following section to derive dynamical equations for circuits.

2.3.3 Dynamics of RLC circuits

Let us give a few modelling examples for circuits involving resistors, capacitors and inductances connected in series and in parallel. We will be using Kirchhoff's laws of currents and voltages.

Example 2.15 (RC circuit) In the circuit shown below a resistor and a capacitor are connected in series. We can manipulate the input voltage $V_{in}(t)$. By doing so, a current $I(t)$ will run through the circuit. We are interested in modelling the effect that V_{in} will have on the output voltage, $V_{out} = V_{bc}$, across the capacitor. The initial condition of the capacitor is zero ($V_{bc}(0) = 0$).

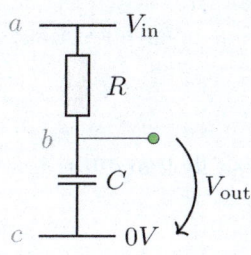

The voltage across the resistor is given by Ohm's law

$$V_{ab} = RI. \tag{2.78}$$

The voltage across the capacitor is governed by

$$\dot{V}_{bc} = \frac{1}{C}I, \tag{2.79}$$

so $\dot{V}_{out} = \frac{1}{C}I$.

By Kirchhoff's voltage law,

$$V_{in} = V_{ab} + V_{out} = RI + V_{out}. \tag{2.80}$$

Solving for I yields

$$I = \frac{1}{R}(V_{in} - V_{out}), \tag{2.81}$$

and by virtue of (2.79), we obtain

$$\dot{V}_{out} = \frac{1}{CR}(V_{in} - V_{out}). \tag{2.82}$$

This is a differential equation which describes the above dynamical system with input V_{in} and state V_{out}.

Exercise 2.9 (RL circuit) Derive a dynamical model for the following circuit that describes the effect of the manipulated variable V_{in} (input voltage) on the system state, $I(t)$ (the current through the circuit).

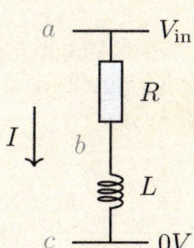

Then derive a model that describes the effect of the manipulated variable V_{in} (input voltage) on I and V_{bc} (that is, the voltage across the inductance).

Answer. The effect of V_{in} on I is described by $LI = V_{in} - IR$. Then $V_{bc} = V_{in} - IR$.

Example 2.16 (Series RLC circuit) Consider the circuit shown below where an input voltage signal V_{in} generates an output voltage signal V_{out}. All initial conditions are considered to be zero. We will derive the dynamical model of this system.

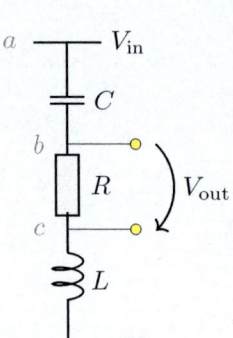

In this circuit, the input variable is the voltage $V_{in}(t)$ and the state of the system is described by $V_{out}(t)$, the output voltage. We consider zero initial conditions for all involved voltages and currents.

The input voltage is the voltage between points a and d. The voltage across the capacitor is

$$\dot{V}_{ab} = \frac{1}{C}I. \tag{2.83}$$

The voltage across the resistor is given by Ohm's law

$$V_{bc} = V_{out} = RI, \tag{2.84}$$

and the voltage across the inductor is given by

$$V_{cd} = L\dot{I}. \tag{2.85}$$

According to Kirchhoff's voltage law,

$$V_{in} = V_{ab} + V_{bc} + V_{cd}$$

$$\Rightarrow V_{in} = \frac{1}{C}\int_0^t I(\tau)d\tau + RI + L\dot{I}$$

$$\Rightarrow V_{in} = \frac{1}{RC}\int_0^t V_{out}(\tau)d\tau + V_{out} + \frac{L}{R}\dot{V}_{out}, \tag{2.86}$$

where in the last equation we simply solved (2.84) for I and substituted. This is an integro-differential equation which describes the system dynamics.

Let us now have a look at an RLC circuit where some elements are connected in parallel. In this example we will be using Kirchhoff's current law.

Example 2.17 (Parallel RLC circuit) We are interested in deriving a dynamical model for the RLC circuit shown below (we do not assume zero initial conditions):

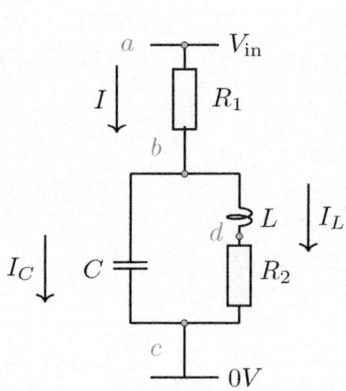

We shall follow two equivalent approaches leading to (i) a system of integro-differential equations and (ii) and system of differential equations.

Approach I. We need to model how the voltage V_{in} applied across the two ends of the circuit affects the currents I_C and I_L that run through the capacitor and the inductor respectively.

We start with Kirchhoff's current law at node b, where we have one *incoming* current, I, and two *outgoing* currents, I_C and I_L. KCL gives

$$I - I_C - I_L = 0. \tag{2.87}$$

By Kirchhoff's voltage law, we have that along the path $a \to b \to c$,

$$V_{ab} + V_{bc} = V_{\text{in}}, \tag{2.88}$$

equivalently,

$$R_1 I + V_{bc}(0) + \frac{1}{C}\int_0^t I_C(\tau)d\tau = V_{\text{in}}$$

$$\Rightarrow R_1(I_L + I_C) + V_{bc}(0) + \frac{1}{C}\int_0^t I_C(\tau)d\tau = V_{\text{in}}. \tag{2.89a}$$

Note that we used the fact that the voltage between points b and c is the voltage across the capacitor, which is given by Equation (2.73).

By Kirchhoff's voltage law for the loop $b \to d \to c$,

$$V_{bd} + V_{dc} + V_{cb} = 0$$

$$\Rightarrow L\dot{I}_L + R_2 I_L - V_{bc}(0) - \frac{1}{C}\int_0^t I_C(\tau)d\tau = 0. \tag{2.89b}$$

Here we used the fact that $V_{bc} = -V_{cb}$. Equations (2.89a) and (2.89b) are a dynamical model for the above system. The voltage V_{in} is the input signal to out system and I_L and I_C are the system states.

Approach II. In this approach, we treat V_{bc} as a system state which satisfies the differential equation

$$\dot{V}_{bc} = \frac{1}{C}I_C. \tag{2.90}$$

In addition, Equation (2.88) yields

$$R_1(I_L + I_C) + V_{bc} = V_{\text{in}} \Rightarrow I_C = \frac{1}{R_1}(V_{\text{in}} - V_{bc}) - I_L, \quad (2.91)$$

therefore, Equation (2.90) becomes

$$\dot{V}_{bc} = \frac{1}{CR_1}(V_{\text{in}} - V_{bc}) - \frac{1}{C}I_L. \quad (2.92)$$

KVL along the loop $b \to d \to c$, as we discussed above, gives $V_{bd} + V_{dc} + V_{cb} = 0$, i.e.,

$$L\dot{I}_L + R_2 I_L - V_{bc} = 0 \Leftrightarrow \dot{I}_L = \frac{1}{L}V_{bc} - \frac{R_2}{L}I_L. \quad (2.93)$$

Overall, the system dynamics can be described by the following state space equation with state vector $x = (V_{bc}, I_L)$ and input V_{in}

$$\underbrace{\begin{bmatrix} \dot{V}_{bc} \\ \dot{I}_L \end{bmatrix}}_{\dot{x}} = \underbrace{\begin{bmatrix} \frac{1}{CR_1}(V_{\text{in}} - V_{bc}) - \frac{1}{C}I_L \\ \frac{1}{L}V_{bc} - \frac{R_2}{L}I_L \end{bmatrix}}_{f(x, V_{\text{in}})}. \quad (2.94)$$

2.4 Mass transfer

In a system that involves flow of mass, such as a water tank, the catalytic converter of a car, a water cooling system for a CPU and so on, the **mass balance equation** simply states that

$$\left\{ \begin{array}{c} \text{Accumulation rate} \\ \text{of mass} \end{array} \right\} = \left\{ \begin{array}{c} \text{Rate of} \\ \text{mass inflow} \end{array} \right\} - \left\{ \begin{array}{c} \text{Rate of} \\ \text{mass outflow} \end{array} \right\}. \quad (2.95)$$

Remark: *The mass balance equation hinges on the preservation of mass. Similar balances can be derived from other similar preservation principles. For example, the preservation of charge leads to Kirchhoff's current law. The preservation of money leads to useful money flow equations in economics. The "preservation of cars" leads to traffic flow balances that read*

$$\left\{ \begin{array}{c} \text{Accumulation rate} \\ \text{of vehicles in street} \end{array} \right\} = \left\{ \begin{array}{c} \text{Inflow rate} \\ \text{of cars} \end{array} \right\} - \left\{ \begin{array}{c} \text{Outflow rate} \\ \text{of cars} \end{array} \right\}. \quad (2.96)$$

We will see later that such models are useful to control the accumulation and avoid traffic jams, overflows, bankruptcy, overcurrent and so on. •

2.4.1 Independent cylindrical tanks

Consider a tank with an inflowing stream at rate F_{in} (in kg/s) and an outflowing stream with flow rate F_{out} as shown in the figure below.

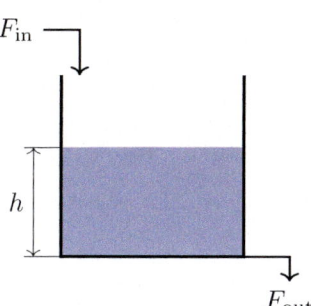

Figure 2.16: *Tank with single inflow and single outflow. The difference between the inflow and the outflow determines the rate of accumulation of mass in the tank.*

Let us denote the mass of liquid in the tank by m. The rate of accumulation of mass in the tank is $\dot{m}$. Then, according to the mass balance equation (see Equation (2.95)), we have

$$\dot{m} = F_{in} - F_{out}. \tag{2.97}$$

If the cross-section of the tank is constant, i.e., if the tank is cylindrical, and the *density* of the liquid is ρ, then, by definition

$$m = \rho V, \tag{2.98}$$

where V is the *volume* of the liquid in the tank. If the cross-section area of the tank is A and the level of the liquid in the tank is h, then

$$V = Ah, \tag{2.99}$$

therefore, Equation (2.98) becomes

$$m = \rho A h. \tag{2.100}$$

We can now write the mass balance equation in terms of the level, h, of the liquid. This is[5]

$$\frac{d}{dt}(\rho A h) = F_{in} - F_{out}$$
$$\Rightarrow \rho A \dot{h} = F_{in} - F_{out}. \tag{2.101}$$

The higher the level, h, the larger the static pressure at the bottom of the tank and, naturally, we expect the speed of the out-flowing liquid to be larger. This is, indeed, supported by **Torricelli's law** which we state below:

---
[5]Since ρ and A are constant in time we move them outside the derivative.

Theorem 2.4 (Torricelli's law) Let h be the level of liquid in a tank. Suppose that

1. The liquid is incompressible,

2. The viscosity of the liquid is negligible,

3. The flow is steady, i.e., at every point, the velocity of the liquid is constant,

4. The liquid outflows through an orifice of small area (compared to the cross-section area of the tank).

Then, the speed of the liquid exiting the tank will be

$$v = \sqrt{2gh}, \tag{2.102}$$

where g is the acceleration of gravity.

Proof. Torricelli's law is a consequence of Bernoulli's principle applied between a point at the surface of the tank and a point at the orifice. □

Torricelli's law requires that the viscosity of the liquid be negligible. This means that the liquid should be inviscid (thin). We cannot apply Torricelli's law when dealing with honey or shampoo for example.

Secondly, Torricelli's law is stated under the assumption the flow is steady; This is a reasonable approximation if (i) the liquid has been running for a while — Torricelli's law cannot model transient flows that happen, for example, once we open a valve — and (ii) the liquid flows out through an orifice without a tube connected to it (or the length of the tube is negligible). We cover the case of unsteady flow in Section 2.4.2. If the hole is not at the bottom of the tank, but at a depth h' from the surface, then Torricelli's law gives $v = \sqrt{2gh'}$ instead.

The speed of the out-flowing liquid determines the mass flow rate. In particular, if the area of the hole is A_{hole}, then the volumetric flow rate (m^3/s), that is, the volume of liquid that runs through the hole will be

$$U_{\text{out}} = A_{\text{hole}} v, \tag{2.103}$$

and the mass flow will be $F_{\text{out}} = \rho U_{\text{out}} = \rho A_{\text{hole}} v$, that is,

$$F_{\text{out}} = \rho A_{\text{hole}} \sqrt{2gh}, \tag{2.104}$$

so the overall model becomes

$$\rho A \dot{h} = F_{\text{in}} - \rho A_{\text{hole}} \sqrt{2gh}. \tag{2.105}$$

Example 2.18 (Single tank) We will derive a model that describes the dynamic correspondence between the inflow rate, F_{in} and the level of liquid h in the tank of Figure 2.16 if we know that the tank has cylindrical shape with a circular base of radius $R = 6\,\text{m}$, the liquid exits the tank from a hole of diameter $D_{hole} = 10\,\text{cm}$ at the bottom of the tank and the contains water which is incompressible and has negligible viscosity. The density of water is $\rho = 997.05\,\text{kg/m}^3$ (at $25°$). The gravitational acceleration in Belfast is $g = 9.8147\,\text{m/s}^2$.

The cross-section area of the tank is

$$A = \pi R^2 = 113.0973\,\text{m}^2, \tag{2.106}$$

and the area of the hole is

$$A_{hole} = \pi \left(\tfrac{D_{hole}}{2}\right)^2 = \pi(0.05\,\text{m})^2 = 0.0079\,\text{m}^2. \tag{2.107}$$

Indeed, the area of the hole is negligible compared to that of the tank, so all conditions of Torricelli's law are satisfied.

By Equation (2.105),

$$\rho A \dot{h} = F_{in} - \rho A_{hole} \sqrt{2gh}$$
$$\Rightarrow 997.05 \cdot 113.0973 \cdot \dot{h} = F_{in} - 997.05 \cdot 0.0079 \sqrt{2 \cdot 9.8147 \cdot h}$$
$$\Rightarrow 112763.7 \cdot \dot{h} = F_{in} - 34.898\sqrt{h}. \tag{2.108}$$

This is an ODE that describes the system dynamics. Note that we have used SI units in all derivations.

Remark: *We may derive an approximation of Torricelli's equation if we assume that the level of liquid in the tank is approximately equal to some value h^e. The approximation will be valid only for h close to h^e and will have the form $F_{out} = h/R$. This is strongly reminiscent of Ohm's law where the flow corresponds to the current, the level of liquid acts as the "voltage across the two ends of the tank" and R is a resistance term. In Chapter 3 we will discuss how such an approximation can be derived.* •

Exercise 2.10 (Tandem tanks) Derive a dynamical model for the following system of tandem tanks that explains the dynamic correspondence between the inflow F_{in} and the two levels h_1 and h_2.

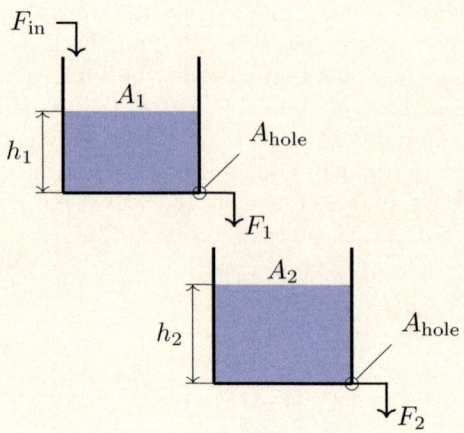

Figure 2.17: *Two tanks in tandem: the output of the first tank flows into the second one.*

The two tanks have cross-section areas A_1 and A_2 respectively and the levels of the liquid are denoted by h_1 and h_2 respectively. In both tanks, the liquid outflows through a small orifice of area A_{hole}. Both tanks contain the same incompressible liquid of negligible viscosity and constant density ρ.

✂ ···

Hint: Use Torricelli's law in each tank and state all assumptions.

2.4.2 Unsteady flow*

We have so far assumed that the flow is steady. This has allowed us to use Torricelli's law. However, if the flow is unsteady, that is, if the velocity the liquid changes in time, we need to resort to Euler's formula. Again, we assume that the liquid is inviscid and incompressible.

Euler's formula applies on a *streamline* that connects two points A and B. Let us denote the speed of the fluid on that line at a point s (between A and B) at time t by $v_s(t)$. The P_A and P_B be the pressure that the ends of the streamline. We denote by v_A and v_B the velocity at the beginning and the end of the line and h_A and h_B stand for the level at the points. Then, Euler's formula states that

$$\rho \int_A^B \frac{\partial v_s}{\partial t} ds + \left[P + \frac{1}{2}\rho v^2 + \rho g h \right]_A^B = 0, \qquad (2.109)$$

that is,

$$\rho \int_A^B \frac{\partial v_s}{\partial t} ds + P_B + \frac{1}{2}\rho v_B^2 + \rho g h_B = P_A + \frac{1}{2}\rho v_A^2 + \rho g h_A. \qquad (2.110)$$

The difficulty in using this equation is that the velocity profiles along a streamline, $v_s(t)$, is not known.

Let us apply Euler's formula to the single tank system of Section 2.4.1, where the liquid outflows through a tube of length L as illustrated in Figure 2.18

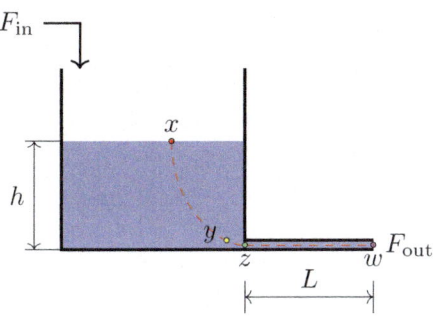

Figure 2.18: *Tank with single inflow and single outflow through a tube of length L.*

Euler's equation from point x to point w gives

$$\rho \int_x^w \frac{\partial v_s}{\partial t} ds + P_w + \frac{1}{2}\rho v_w^2 + \rho g h_w = P_x + \frac{1}{2}\rho v_x^2 + \rho g h_x. \qquad (2.111)$$

Since $P_x = P_w = P_{\text{atm}}$, $h_x = h$ and $h_w = 0$, Euler's equation becomes

$$\rho \int_x^w \frac{\partial v_s}{\partial t} ds + \frac{1}{2}\rho v_w^2 + \rho g h = \frac{1}{2}\rho v_x^2. \qquad (2.112)$$

We cannot compute $\int_x^w \frac{\partial v_s}{\partial t} ds$, but we can approximate it by breaking it down into segments as follows

$$\int_x^w \frac{\partial v_s}{\partial t} ds = \int_x^y \frac{\partial v_s}{\partial t} ds + \int_y^z \frac{\partial v_s}{\partial t} ds + \int_z^w \frac{\partial v_s}{\partial t} ds, \qquad (2.113)$$

where y is a point inside the bulk of the liquid, but not very close to the hole (with $h_y \approx 0$) and z is at the entrance of the tube (with $h_z = 0$). From point x to point y the velocity of the liquid is very small and does not change much, so we can approximate $\int_x^y \frac{\partial v_s}{\partial t} ds \approx 0^6$. The length from y to z is very small compared to the length of the tube, so it can be neglected too. Lastly, we will assume that $\frac{\partial v_s}{\partial t}$ is constant at all points along the tube, so

$$\int_x^w \frac{\partial v_s}{\partial t} ds \approx \int_z^w \frac{\partial v_s}{\partial t} ds \approx L \frac{\partial v_w}{\partial t}. \qquad (2.114)$$

Now Equation (2.111) can be approximated by

$$\rho L \dot{v}_w + \tfrac{1}{2}\rho v_w^2 = \rho g h. \qquad (2.115)$$

---

[6]We can do this assuming that the tank is large enough so that its level does not change significantly in time. This is true for large tanks with tubes of small diameter. To be more specific, we have assumed that $v_x = \dot{h}$ is negligible compared to v_z.

By the mass balance equation in the tank, we have

$$\rho A \dot{h} = F_{in} - \rho A_{hole} v_w. \tag{2.116}$$

Equations (2.115) and (2.116) describe the system dynamics. Note that if $L = 0$ we recover the model of Equation (2.105).

2.4.3 Independent tanks with non-constant cross-section area

In some applications, tanks have a non-cylindrical cross section like the one shown in the following figure.

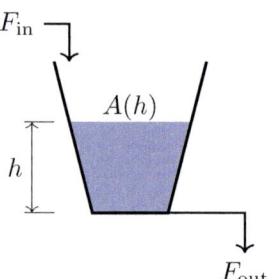

Figure 2.19: Tank with non-constant cross-section area, $A(h)$.

In such cases, the cross-section area, A, changes at different levels — it is $A = A(h)$, therefore, the mass balance becomes (see Equation (2.97))

$$\dot{m} = F_{in} - F_{out}$$
$$\Rightarrow \frac{d}{dt}(\rho V) = F_{in} - F_{out}$$
$$\Rightarrow \rho \frac{dV}{dt} = F_{in} - F_{out}, \tag{2.117}$$

but $dV = A(h)dh$ (the total volume of liquid is $V = \int_0^h A(\eta)d\eta$), so[7]

$$\Rightarrow \rho A(h) \frac{dh}{dt} = F_{in} - F_{out}. \tag{2.118}$$

The outflow is not affected by the shape of the tank; it only depends on the level of the liquid inside it, so $F_{out} = \rho A_{hole} \sqrt{h}$ as in Equation (2.104). The mass balance becomes

$$\rho A(h) \frac{dh}{dt} = F_{in} - \rho A_{hole} \sqrt{h}. \tag{2.119}$$

---

[7] A more detailed discussion of this point: We have $dV = A(h)dh$, therefore $V(h) = \int_0^h A(\eta)d\eta$. Then $\frac{d}{dt} V(h(t)) = \frac{d}{dh} \int_0^h A(\eta)d\eta \cdot \frac{d}{dt} h(t) = A(h) \dot{h}$ by virtue of the Fundamental Theorem of Calculus (Theorem A.22).

Example 2.19 (Tank of non-standard cross-section) Consider the tank of Figure 2.19. The liquid outflows from a small orifice of area A_{hole}. The tank has a circular cross-section with radius given by

$$r(h) = 1 + \frac{h}{3}, \qquad (2.120)$$

where h is the level of the liquid in the tank. The cross-section of the tank is

$$A(h) = \pi r(h)^2 = \pi \left(1 + \frac{h}{3}\right)^2. \qquad (2.121)$$

The mass balance is

$$\rho A(h) \frac{\mathrm{d}h}{\mathrm{d}t} = F_{\text{in}} - \rho A_{\text{hole}} \sqrt{h}$$

$$\Rightarrow \rho \pi \left(1 + \frac{h}{3}\right)^2 \frac{\mathrm{d}h}{\mathrm{d}t} = F_{\text{in}} - \rho A_{\text{hole}} \sqrt{h}. \qquad (2.122)$$

2.4.4 Interconnected tanks

Consider the case of two communicating vessels which are connected with a horizontal tube as shown in Figure 2.20.

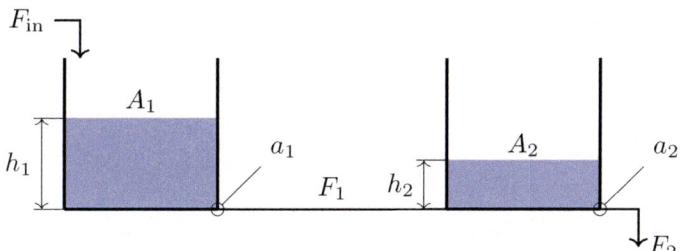

Figure 2.20: *Two communicating tanks. We denote by F_1 the mass flow from the first to the second tank.*

Under the assumption that there is no loss of pressure due to friction in the pipe that connects the two tanks and that all conditions of Torricelli's law are satisfied, the flow between the tanks is given by

$$F_1 = \begin{cases} \rho a_1 \sqrt{2g(h_1 - h_2)}, & \text{if } h_1 \geq h_2, \\ -\rho a_1 \sqrt{2g(h_2 - h_1)}, & \text{if } h_1 < h_2, \end{cases} \qquad (2.123)$$

where the minus in the second branch of the equation means that the flow will be from the second to the first tank if $h_1 < h_2$. Note that if $h_1 = h_2$, then $F_1 = 0$. Let us assume hereafter that $h_1 \geq h_2$. The mass balance of

the first tank is
$$\rho A_1 \dot{h}_1 = F_{\text{in}} - \rho a_1 \sqrt{2g(h_1 - h_2)}, \tag{2.124}$$
and the mass balance of the second tank is
$$\rho A_2 \dot{h}_2 = \rho a_1 \sqrt{2g(h_1 - h_2)} - \rho a_2 \sqrt{2gh_2}$$
$$\Rightarrow A_2 \dot{h}_2 = a_1 \sqrt{2g(h_1 - h_2)} - a_2 \sqrt{2gh_2}. \tag{2.125}$$

2.5 Simulating dynamical systems

We can use numerical methods to simulate dynamical systems given their dynamical equations in the form of an ODE and their initial conditions. Python and MATLAB provide tools to do so easily. In particular, we will focus on the following template initial value problem (see Section 2.1):

$$\frac{dy(t)}{dt} = f(t, y(t)), \tag{2.126a}$$
$$y(t_0) = y_0, \tag{2.126b}$$

where $y \in \mathbb{R}^n$. Problems involving higher-order derivatives can be transformed into this form. Take for example the dynamical system of Equation (2.31):
$$F - kx = m\ddot{x}. \tag{2.127}$$
We may define $y_1 = x$ and $y_2 = \dot{x}$. Then, the ODE of the system becomes
$$\frac{d}{dt}\underbrace{\begin{bmatrix} y_1(t) \\ y_2(t) \end{bmatrix}}_{y(t)} = \begin{bmatrix} y_2(t) \\ \frac{1}{m}(F - ky_1(t)) \end{bmatrix}. \tag{2.128}$$
In this case $y \in \mathbb{R}^2$ and
$$f(t, y) = \begin{bmatrix} y_2 \\ \frac{1}{m}(F - ky_1) \end{bmatrix}. \tag{2.129}$$

2.5.1 Python

In Python we can solve initial value problems of the form (2.126) by using `scipy.integrate.solve_ivp`. The default solver is the explicit Runge-Kutta method of order $(4, 5)$, which is adequately accurate and fact for most applications. In order to use the solver we need to specify f, the initial condition of the problem, and the time span over which we want the solver to compute the solution. Firstly, we need to import the necessary modules as follows

```python
import numpy as np
import matplotlib.pyplot as plt
import scipy.integrate as spi
```

Chapter 2. Modelling

Next we need to specify function f by implementing a Python function. For function f given in Equation (2.129) we have the following implementation

```python
def horizontal_mass_spring(t, y, mass,
                           stiffness, force):
    """Dynamics of horizontal mass-spring system

    :param t: time
    :param y: state vector where y[0] is the
              displacement of the spring and
              y[1] is its velocity
    :param mass: mass of suspended object
    :param stiffness: spring stiffness
    :param force: external force

    :return: right hand side of system dynamics
    """
    return [y[1], (force - stiffness*y[0])/mass]
```

Note that in Python, the first element of an array y is y[0] (not y[1]). We may now use `solve_ivp` to solve the initial value problem given the system parameters, an initial condition (t_0 and y_0) and the simulation time.

```python
# System parameters
m = 2           # mass (in kg)
k = 1000        # stiffness (in N/m)
F = 0           # external force
t0 = 0          # initial time
tf = 10         # final time
y0 = [0.1, 2]   # initial condition

# Call the solver
sol = spi.solve_ivp(lambda t, y:
            horizontal_mass_spring(t, y, m, k, F),
            [t0, tf], y0)
```

The above code snippet will simulate the dynamical system from t_0 to t_f with the initial condition $y(t_0) = y_0$. We have used the initial displacement $x(0) = 0.1\,\text{m}$ and initial velocity $\dot{x}(0) = 2\,\text{m/s}$. The first argument of `solve_ivp` is a **lambda expression**: an anonymous function with an explicit declaration of its free variables which corresponds to $f(t, y)$. The second argument defines the time interval over which the solver will compute an approximate solution. The last argument is the initial condition y_0.

The output of `solve_ivp` is an object whose most notable fields are (i) `t`: the discrete time points where the solution was computed, (ii) `y`: the values of the solution at the corresponding time points, (iii) `status`: is equal to zero if the computation was successful, (iv) `message`: a human-readable message of the solution status. We may now plot the obtained solution by using `matplotlib` as follows

```python
plt.figure(1)
plt.plot(sol.t, sol.y[0])
plt.plot(sol.t, sol.y[1])
plt.xlabel('Time (s)')
plt.ylabel('Solution')
plt.legend(['Position', 'Velocity'])
plt.show()
```

The solver decides the time points where the solution is computed; these are, however, often rather sparse. It is highly recommended for the users to specify time span by using the argument `t_eval` as in the following example

```python
num_time_points = 5000
t_span = np.linspace(t0, tf, num_time_points)

# Call the solver
sol = spi.solve_ivp(lambda t, y:
            horizontal_mass_spring(t, y, m, k, F),
            [t0, tf], y0, t_eval=t_span)
```

We may need to simulate the system given a time-varying force signal, for example, $F(t)$, we can directly replace `F` with a function of `t`

```python
# Call the solver
sol = spi.solve_ivp(
    lambda t, y:
    horizontal_mass_spring(t, y, m, k, np.sin(t)),
    [t0, tf], y0, t_eval=t_span)
```

2.5.2 MATLAB

MATLAB offers a toolbox of solvers for general nonlinear initial value problems of the form (2.126). The most commonly used solver is `ode45`[8], which we will present here. In order to use the solver, we need to specify (i) function f, (ii) the initial conditions t_0 and y_0, (iii) the final time, t_f, until which the solver should compute a solution.

One way to specify function f is to write a MATLAB function with the following function signature:

```matlab
function dydt = system_model(t, y, params)
%SYSTEM_MODEL is a function that describes the
%dynamics of your system.

%Implementation goes here
```

---

[8] MATLAB's `ode45` is an implementation of an explicit Runge-Kutta method of order $(4,5)$. Type `help ode45` in the MATLAB console for details.

Chapter 2. Modelling

For example, for function f in Equation (2.129), we have the following one-line implementation. We should emphasise that the documentation is an inextricable part of a software implementation. Not only is it highly unlikely that other people will find it difficult to understand what your code does, but you will have forgotten how it works a few weeks after you write it.

```
function dydt = horizontal_mass_spring(t, y, m, k, F)
%HORIZONTAL_MASS_SPRING computes the right hand side
%of the ODE for our spring-mass system
%
%
%Input Arguments:
% t       time
% y       system state y = (y1, y2), where y1 is the
%         position of the mass and y2 is its acceleration
% m       mass
% k       spring stiffness
% F       externally applied force
%
%
%Output Arguments:
%
% dydt    derivative of y at time t and state y,
%         that is, the value of f(t, y)

dydt = [y(2); (F - k*y(1))/m];
```

Let us now use `ode45` to simulate a horizontal spring-mass system with $m = 2 \, \text{kg}$, $k = 10^3 \, \text{N/m}$ when the externally applied force F is zero. The initial conditions are $x(0) = 0$ (that is, $y_1(0) = 0$) and $\dot{x}(0) = 0.5 \, \text{m/s}$. We will simulate the system in the interval $[0, t_f]$ with $t_f = 2 \, \text{s}$.

```
m = 2;              % mass
k = 1e3;            % spring stiffness
F = 0;              % external force
t0 = 0;             % initial time
tf = 2;             % final time
y0 = [0; 0.5];      % initial state, y(0)

% Simulation using ode45
[t_sol,y_sol] = ode45(...
    @(t, y)horizontal_mass_spring(t, y, m, k, F), ...
    [0, tf], y0);
```

The notation `@(t, y)` means that `horizontal_mass_spring` is treated as a function of `t` and `y`, while `m`, `k` and `F` are parameters. This is called a **function handle**.

The solver returns a numerical approximation of the solution of our initial value problem. In particular, it returns two arrays, `t_sol` and `y_sol`, so that at time `t_sol(i)` the solution is approximately equal to `y_sol(i)`. We may use `t_sol` and `y_sol` to plot the solution with

```
% Plot the solution of the initial value problem
plot(t_sol, y_sol);          % plot
xlabel('Time (s)');          % add x label to the plot
ylabel('Solution');          % add y plot
legend('Position','Speed');  % add legend
```

In the previous example, the external force was equal to a fixed value. We may, however, choose F to be a time-varying signal; for example we may want to simulate the behaviour of the system for $F(t) = 100\sin(2t)$. We then need to define F using a function handle and then pass it to the solver as in the following example

```
m  = 2;              % mass
k  = 1e3;            % spring stiffness
F  = @(t) 100*sin(2*t);  % external force
t0 = 0;              % initial time
tf = 2;              % final time
y0 = [0; 0.5];       % initial condition, y(0)

% Simulation using ode45
[t_sol,y_sol] = ode45(...
    @(t, y)horizontal_mass_spring(t, y, m, k, F(t)), ...
    [0, tf], y0);
```

Likewise, F can be a function of both t and y. In fact, in controller design the system input becomes a function of the output ($F = F(y)$). It should be easy for the reader to solve the following exercise

Exercise 2.11 (Simulations in MATLAB) Simulate the behaviour of the system given in Equation (2.128) in MATLAB using the initial condition $y_1(0) = 0.2$ m and $y_2(0) = -0.4$ m/s, mass $m = 2$ kg, spring stiffness $k = 10^3$ N/m and the external force

$$F(t, y) = -5y_2(t) + 50e^{-0.5t}\sin(2t), \qquad (2.130)$$

and plot the trajectory of the system from the initial time $t_0 = 0$ until $t_f = 10$ s. What is the maximum speed of the object and at what time is it observed?

2.6 One-minute round-up

In this chapter our job was to derive mathematical models for physical systems in terms that describe their evolution in time. These models have the form of either ordinary differential equations (ODEs) or integro-differential equations (IDEs). Overall, we learned the following

Chapter 2. Modelling

1. The time evolution of phenomena can be modelled by ordinary differential equations (ODEs) and integro-differential equations (IDEs)

2. Integro-differential equations can often be cast as ODEs

3. ODEs and IDEs can be derived from first principles of physics and science

4. Newton's laws of motions, Kirchhoff's laws of circuits and mass balance equations are examples of such first principles

5. We can use software (e.g., in Python or MATLAB) to simulate the temporal evolution of dynamical systems starting from a known initial condition

In Chapter 3 we will present a systematic way to simplify such ODEs and IDEs.

2.7 Exercises

Exercise 2.12 (Pulley) Derive a dynamical model that describes the motion of the following object of mass m which is suspended on a massless string that is wrapped around a reel of mass M and radius R. The moment of inertia of the reel is $I = \frac{1}{2}MR^2$. The reel is riveted on a wall.

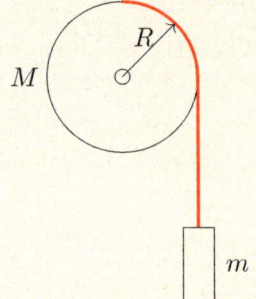

Hint. To derive a dynamical model for this system we work as in Examples 2.12 and 2.13. Firstly, introduce a frame of reference. Start by sketching the forces applied on the object and the thread that connects it to the pulley. Write down (i) Newton's second law of motion of the forces on the object, (ii) the rotational counterpart of Newton's second law for the pulley, (ii) what is the link between the angle or rotation of the pulley and the vertical position of the object?

Answer. The system dynamics is described by the ODE $mg = \left(\frac{1}{2}M + m\right)\ddot{y}$, where y increases in the downward direction and counterclockwise rotations are positive. Observe that the radius of the pulley, R, does not affect the system dynamics.

Exercise 2.13 (Another pulley) Derive a dynamical model that describes the motion of the following objects of masses m_1 and m_2 which are suspended from a massless string that is wrapped around a pulley of mass M and radius R. The moment of inertia of the pulley is $I = \frac{1}{2}MR^2$. The pulley is riveted on a wall.

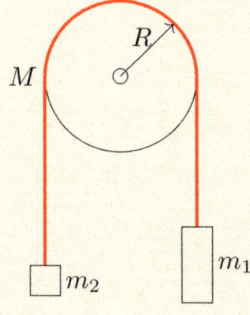

Hint. Proceed as in Exercise 2.12. As always, start by introducing a frame of reference. The additional object of mass m_2 will result to an additional torque with opposite sign to that generated by the object of mass m_1.

Answer. The system dynamics is described by the ODE $(m_1 - m_2)g = \left(\frac{1}{2}M + m_1 + m_2\right)\ddot{y}$, y is the y-coordinate of the object of mass m_1, which increases in the downward direction and counterclockwise rotations are positive.

Chapter 2. Modelling

Exercise 2.14 (Ramp and pulley) The object of mass m_1 shown below can slide on the inclined plane without friction. This is connected to the object of mass m_2 over a pulley of mass M, which is affixed to the ramp. The cord that connects the two objects is considered to be massless. Draw all forces and derive a dynamical model that describes the motion of the two objects.

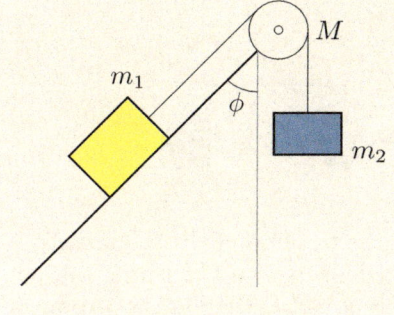

Answer. The dynamics is described by
$$m_1(g\cos\phi - \ddot{y}) - m_2(g + \ddot{y}) = \tfrac{1}{2}M\ddot{y},$$
where $\ddot{y}$ points up.

Hint. First introduce an inertial frame of reference. Write Newton's second law of motion for the two objects of masses m_1 and m_2 respectively. Write down the rotational counterpart of Newton's second law for the pulley.

Exercise 2.15 (Cylinder rolling on inclined plane) A solid cylinder of mass m is placed on an inclined plane as shown in the figure below and an elastic spring of stiffness k is attached to it.

The angle between the horizontal plane and the inclined plane is ϕ as shown in the figure. The spring is at its natural length when the cylinder is at position $x = L$ — then no restoring force is applied. The moment of inertia of the cylinder with respect to an axis running through its centre and that is perpendicular to the plane of view is given by $I = \tfrac{1}{2}mr^2$, where r is the radius of the cylinder.

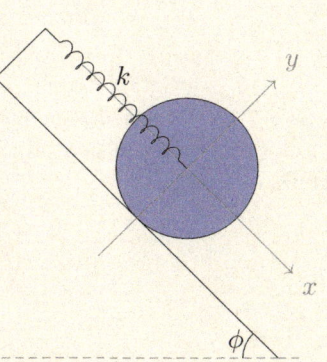

A drag force acts on the cylinder equal to $F_D = -b\dot{x}^2$. Derive the equations of motion of the system.

Answer. The system is described by the ODE $mg\sin\phi - k(x-L) - b\dot{x}^2 = \tfrac{3}{2}m\ddot{x}.$

Hint. This is similar to Example 2.12. The restoring force of the spring is $F_{\text{spring}} = -k(x-L)$ so that $F_{\text{spring}} = 0$ if $x = L$.

Exercise 2.16 (**Inverted pendulum and towing vehicle**) Derive a dynamical model of the following system that describes how the manipulated input, F, affects the angle θ between the rod and the vertical axis.

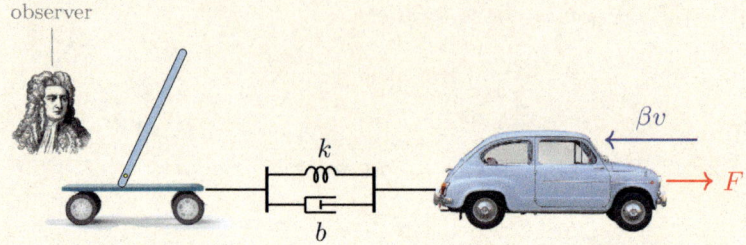

It is given that the masses of the car, the trailer and the rod are m_1, m_2 and m_3, respectively, the car is connected to the trailer via a spring of constant k and a damper of constant b, the rod is isotropic, and its length is L, an aerodynamic friction is exercised on the car as it moves, and it is given by βv, where v is the velocity of the car with respect to an external static observer. The car exercises a forward force F. The aerodynamic friction on the rod and the trailer is negligible.

Hint. This is a combination of Exercises 2.9 and 2.11.

Exercise 2.17 (**Inverted pendulum**) Derive a mathematical model for the following inverted pendulum system that describes the dynamics of angle θ.

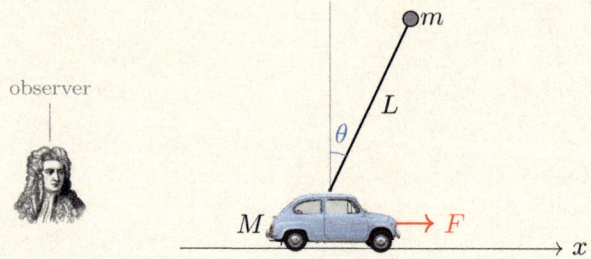

The mass of the car is M. The pendulum consists of a massless rod at the end of which there is a point particle of mass m. The length of the rod is L.

Chapter 2. Modelling

Can you simplify the model assuming that θ is small?

✱ ··

Hint. As always, we need to start by introducing a frame of reference. Sketch all the forces that are applied on the cart and the mass at the end of the pendulum. The rod is massless, so you should not consider any forces applied to it. Apply Newton's first and second law of motion on the cart and the mass of the pendulum. Equations (2.44) are still valid in this example; differentiate them twice as we did in Example 2.11. You do not need to use the rotational counterpart of Newton's second law.

✂ ··

Answer. A tension force will show up in your equations. If you eliminate it you should obtain the following model

$$\ddot{x}\cos\theta - L\ddot{\theta} + g\sin\theta = 0, \qquad (2.131a)$$
$$F + mL\dot{\theta}^2\sin\theta - mL\ddot{\theta}\cos\theta = (M+m)\ddot{x}. \qquad (2.131b)$$

When $\theta \approx 0$, then $\sin\theta \approx \theta$ and $\cos\theta \approx 1$. When θ is small and the pendulum is almost at rest at the upright position, $\dot{\theta}$ is small, therefore $\dot{\theta}^2 \approx 0$. The above equations reduce to $\ddot{x} - L\ddot{\theta} + g\theta = 0$ and $F + mL\ddot{\theta} = (M+m)\ddot{x}$.

Exercise 2.18 (Simulation of chaotic system) The **Lorenz system** is a system of ordinary differential equations known for its chaotic behaviour. It was proposed by the American mathematician and meteorologist Edward Lorenz in 1963 to describe atmospheric convection. The model is given by

$$\dot{y}_1 = \sigma(y_2 - y_1), \qquad (2.132a)$$
$$\dot{y}_2 = y_1(\rho - y_3) - y_2, \qquad (2.132b)$$
$$\dot{y}_3 = y_1 y_2 - \beta y_3. \qquad (2.132c)$$

Write a Python or MATLAB script to simulate the system starting from the initial point $y_1(0) = 1$, $y_2(0) = 1$ and $y_3(0) = 1$.

Plot the trajectories of the Lorenz system in a three-dimensional plot with axes y_1, y_2 and y_3.

The Lorenz system is very sensitive to the initial condition. Trajectories starting from neighbouring initial conditions evolve in a significantly different fashion. Simulate again the system starting from $y_1(0) = 1 + 10^{-5}$, $y_2(0) = 1$ and $y_3(0) = 1$. What do you observe?

Exercise 2.19 (Simulate a mass-spring-damper system) In Example 2.8 we derived the following system model (see Equation (2.33))

$$F + mg - b\dot{x} - kx = m\ddot{x}, \qquad (2.133)$$

where x is the displacement of the spring from its equilibrium position. Suppose that the mass of the suspended object is $m = 1500\,\text{g}$, the stiffness of the spring is $k = 853\,\text{N/m}$, the viscous damping coefficient is $b = 47.5\,\text{Ns/m}$ and the externally applied force is

$$F(t) = \begin{cases} \sin(t), & \text{for } 0 \leq t \leq \pi \\ 0, & \text{for } t > \pi \end{cases}. \qquad (2.134)$$

Write a Python or MATLAB script to simulate the system using the initial conditions $x(0) = 5\,\text{cm}$ and $\dot{x}(0) = 0\,\text{m/s}$.

✂ ....

Hint. Whether you choose to do this in Python or in MATLAB, the first thing you need to do is to determine the state space representation of this system. The state of this system will be the vector $\mathbf{x} = (x_1, x_2)'$, where $x_1 = x$ and $x_2 = \dot{x}_1$.

Exercise 2.20 (Inverted pendulum with drag force) Derive the equations of motion of the system in Example 2.11 if there is an additional drag force (friction) acting on the car, whose magnitude is given by $F_{\text{drag}} = b\dot{x}$, and its direction is the opposite of $\dot{x}$.

Advanced Exercise 2.21 (Inverted pendulum with friction at hinge) Consider the inverted pendulum of Example 2.11 and suppose that there is friction at the joint that connects the rod with the car. This friction creates a torque, with respect to point A, that resists the rotation of the rod and is given by

$$T_f = -b\dot{\theta}^2. \qquad (2.135)$$

Derive the equations of motion of this system.

Exercise 2.22 (RC circuit) Derive a mathematical model for the following circuit that explains how the input voltage, V_{in}, affects the current I that runs through the circuit and, as a result, the output voltage, V_{out}, across resistor R_2.

Chapter 2. Modelling

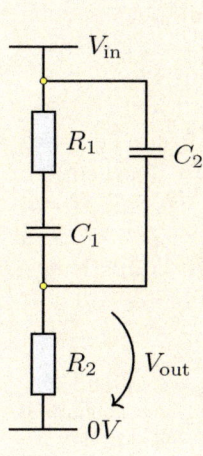

Suppose that $C_1 = 1\,\mu\text{F}$, $C_2 = 2\,\mu\text{F}$, $R_1 = 1\,\text{k}\Omega$ and $R_2 = 12\,\text{k}\Omega$, the initial charge of both capacitors is zero and the input voltage is $V_{\text{in}}(t) = \sin(100t)$ (in Volt). Write a Python or MATLAB script to simulate the system from $t_0 = 0$ till $t_f = 1\,\text{s}$. Pay attention to the units of measurement.

※

Answer. There are several equivalent formulations of the system model; for example

$$C_1 C_2 R_1 \dot{V}_{\text{out}} + \frac{C_1(R_1+R_2)+C_2 R_2}{R_2} V_{\text{out}} + \frac{1}{R_2} V_{\text{out}} = (C_1 + C_2)\dot{V}_{\text{in}} + R_1 C_1 C_2 \ddot{V}_{\text{in}}.$$

Exercise 2.23 (System of tanks) Derive a mathematical model for the following system of tanks that describes the dynamics of the level of liquid in each tank.

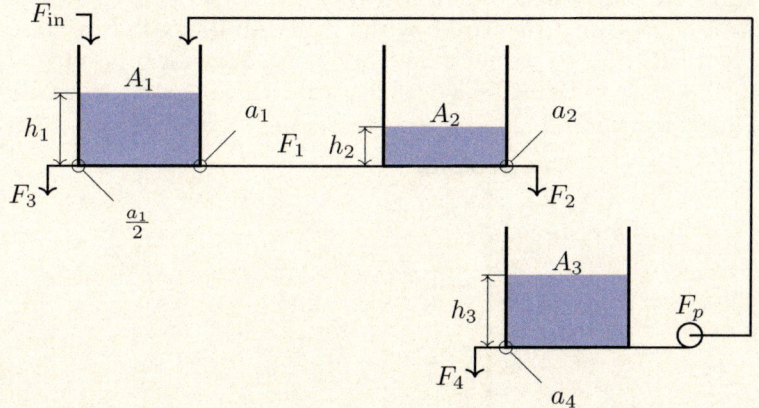

There is a reflux stream from the third to the first tank via a pump which maintains a constant flow rate F_p which is independent of the level of liquid in the third tank. Note also that there are two outflows from the first tank.

※

Answer. The system dynamics is described by the following equations (under the assumption that $h_1 > h_2$): [tank 1] $\rho A_1 \dot{h}_1 = F_{\text{in}} + F_p - \rho \frac{a_1}{2}\sqrt{2gh_1} - \rho a_1 \sqrt{2g(h_1 - h_2)} = a_1\sqrt{2g(h_1 - h_2)} - a_2\sqrt{2gh_2}$, [tank 2] $A_2 \dot{h}_2$, [tank 3] $\rho A_3 \dot{h}_3 = \rho a_2 \sqrt{2gh_2} - \rho a_4 \sqrt{2gh_3} - F_p$.

Exercise 2.24 (Pendulum with isotropic rod revisited) Let us solve Exercise 2.6 following a different approach which is more reminiscent of the way we approach the inverted pendulum problem.

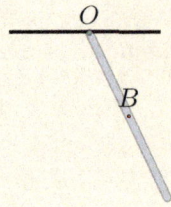

An *isotropic* rod is suspended from the ceiling as shown in the figure on the left. The centre of mass of the rod is at its geometric centre (point B). Its moment of inertia with respect to point B is $I = \frac{1}{12}mL^2$, where L is the rod's length. Derive the equations of motion of the rod.

Hint: (i) introduce an inertial frame of reference, (ii) sketch all forces acting on the rod: its weight and the reaction exerted from the ceiling to the rod, (iii) determine the total torque with respect to point B, (iv) apply Newton's second law of motion. You should obtain the same result as in Exercise 2.6.

Exercise 2.25 (YoYo moving along magnetic rail) Consider the system shown below which consists of a Yo-Yo, a spring and a damper. The Yo-Yo has a mass m, a metallic axle of radius r and a spool of radius R. All strings are of negligible mass and thickness. The moment of inertia of the Yo-Yo with respect to its axis of rotation is $I = \frac{1}{2}mR^2$ and its geometric centre coincides with its centre of mass. The axle is metallic and is attracted by a vertical magnetic rail. The axle can roll along the magnetic rail without sliding.

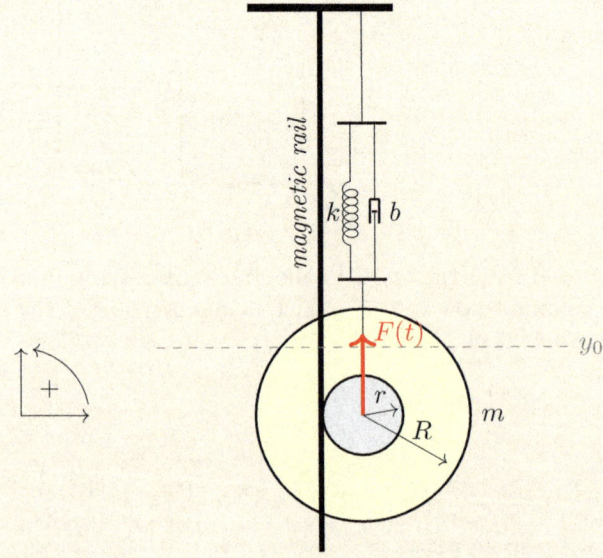

A frame of reference is provided. The spring is elastic and has stiff-

Chapter 2. Modelling

ness k, while the damper exerts the force $F_d = -bv$, where v is the linear velocity of the Yo-Yo along the vertical axis. When the centre of the Yo-Yo is at position y_0, the spring is at its natural length. An external force $F(t)$ is applied to the Yo-Yo at its centre as shown in the figure. Derive a dynamical model for this system, in the form of a differential equation, that describes how the external force affects the vertical position of the Yo-Yo.

✳ ....

Answer. The equation of motion is

$$m\left[1 + \frac{1}{2}\left(\frac{r}{R}\right)^2\right]\ddot{y} = k(y_0 - y) - b\dot{y} + F. \qquad (2.136)$$

✂ ....

Hint. Follow similar steps as in Example 2.13. Recall that according to Hooke's law (Section 2.2.3), the restoring force is proportional to the *displacement* (with respect to the natural length of the spring).

3

Linearisation

*Dynamical system are described by ODEs (see Equation (2.1)) and IDEs (see Equation (2.10)), several examples of which we encountered in Chapter 2. These can often be overly complex and too hard to study. Our objective is to approximate them by **linear** dynamical systems described by ODEs of the form $a_0 x + a_1 \dot{x} + a_2 \ddot{x} + \ldots = b_0 u + b_1 \dot{u} + \ldots$ Our basic theoretical tool will be Taylor's approximation theorem which we state in Section 3.1 for functions of one variable and in Section 3.3 for functions of several variables. The idea is that under certain smoothness conditions, we can approximate the system dynamics by a simple linear function close to an operating point of the system — an equilibrium point.*

3.1 Taylor's theorem

Systems of the very general form of Equation (2.1) are often overly complex and too difficult to analyse. For example, in the inverted pendulum case of Example 2.11, Equations (2.44) involve trigonometric functions of the system output and, similarly, Equations (2.47a) and (2.47b) involve the second derivative of the sine and cosine of the system output. Again, in Example 2.18, the output variable of the system is under a square root. Such complex dynamical systems are called **nonlinear** as opposed to simple ones which have the general form

$$a_0 x + a_1 \dot{x} + a_2 \ddot{x} + \ldots + a_n \frac{\mathrm{d}^n}{\mathrm{d}t^n} x = b_0 u + b_1 \dot{u} + \ldots + b_n \frac{\mathrm{d}^m}{\mathrm{d}t^m} u, \qquad (3.1)$$

which are called **linear**. For example, the dynamical system in Equation (2.31), which has the simple form

$$F - kx = m\ddot{x}, \qquad (3.2)$$

is a linear dynamical system with input F and output x.

We should note that, technically, the dynamical system in Equation (2.33), that is

$$F + mg - b\dot{x} - kx = m\ddot{x}, \qquad (3.3)$$

with input variable F and output x, is not linear. The reason is the presence of the constant term mg. However, admittedly, the system equation seems to be rather simple, and it would be naturally classified as simple were it not for a constant term. Such systems are called **affine** and can be easily cast as linear systems as we will discuss in what follows.

Exercise 3.1 (Linear, afffine or nonlinear?) Are the following systems linear, affine or nonlinear? Put a tick in the right box.

	Linear	Affine	Nonlinear
$\dot{x} = -3x$	☐	☐	☐
$\dot{x} = u$	☐	☐	☐
$\dot{x} = 5x + u$	☐	☐	☐
$\dot{x} = x + u + 10$	☐	☐	☐
$\dot{x} = x^2(1 + u)$	☐	☐	☐
$\dot{x} = e^u x$	☐	☐	☐
$\dot{x}^3 = x$	☐	☐	☐
$x\dot{x} = \pi x + u$	☐	☐	☐
$\dot{x} = \sqrt{2}x + \sin(4)u$	☐	☐	☐
$x\dot{x} = 3x + 5u - 1$	☐	☐	☐
$\begin{bmatrix} \dot{x}_1 = x_1 + x_2 + u \\ \dot{x}_2 = x_1 - x_2 \end{bmatrix}$	☐	☐	☐
$\begin{bmatrix} \dot{x}_1 = x_1 + 7x_2 \\ \dot{x}_2 = x_1 - x_2 - u \end{bmatrix}$	☐	☐	☐
$\begin{bmatrix} \dot{x}_1 = x_1 x_2 + u \\ \dot{x}_2 = -x_1 \end{bmatrix}$	☐	☐	☐
$\begin{bmatrix} \dot{x}_1 = -x_2 \\ \dot{x}_2 = -x_3 \\ \dot{x}_3 = x_3 + x_2 + x_1 u \end{bmatrix}$	☐	☐	☐

Chapter 3. Linearisation

	Linear	Affine	Nonlinear
$\begin{cases} \dot{x}_1 = x_3 \\ \dot{x}_2 x_3 = u \\ \dot{x}_3 = x_1^3 + x_2^3 + x_3^3 \end{cases}$	☐	☐	☐
$\dot{x} = Ax + Bu$, with $x \in \mathbb{R}^n, u \in \mathbb{R}^m$	☐	☐	☐

※ ....

Answer: L, L, L, A, N, N, N, L (the term $\dot{x}x$ is nonlinear), L (both $\sqrt{2}$ and $\sin(4)$ are constant terms), N, L, L, N ($x_1 x_2$ is nonlinear), N ($x_1 u$ is nonlinear), N, L, L.

The rest of this book focuses exclusively on linear dynamical systems. Although nonlinear systems are difficult to study, they can be approximated by linear ones, which are a lot easier to cope with. The procedure of approximating a nonlinear system by a linear one is known as **linearisation**.

Remark: *Most systems of practical interest are inherently nonlinear. The basic premise in this chapter is that if a dynamical system is operated in such a way that its states and inputs do not deviate much from a constant value, i.e., when the system is approximately at rest, its (local) behaviour can be approximated by a linear dynamical system.* •

Before we can linearise a dynamical system we need to discuss how to approximate a general (smooth) nonlinear function by a linear one. The key result is Taylor's approximation theorem, which states that smooth real functions can be approximated locally (around a given point of interest) by polynomials.

Theorem 3.1 (Taylor's approximation theorem) Let $\phi : \mathbb{R} \to \mathbb{R}$ be a function which is k times differentiable at a point $x_0 \in \mathbb{R}$. Then,

$$\phi(x) = \phi(x_0) + \phi'(x_0)(x - x_0) + \frac{\phi''(x_0)}{2!}(x - x_0)^2$$
$$+ \ldots + \frac{\phi^{(k)}(x_0)}{k!}(x - x_0)^k + R_k(x), \quad (3.4)$$

where $R_k(x)$, the **approximation error**, is $R_k(x) = o(|x - x_0|^k)$ as $x \to x_0$. Additionally, if ϕ is $k+1$ times differentiable in a neighbourhood of x_0, then $R_k(x) = \mathcal{O}(|x - x_0|^{k+1})$ as $x \to x_0$ [a].

[a] For details regarding the little-o and big-o notation, see Section A.7.

Proof. Define $P_k(x) = \phi(x_0) + \phi'(x_0)(x - x_0) + \frac{\phi''(x_0)}{2!}(x - x_0) + \ldots + \frac{\phi^{(k)}(x_0)}{k!}(x - x_0)^k$, the k-order polynomial approximation of ϕ. Define also

$$h_k(x) = \begin{cases} \frac{\phi(x) - P_k(x)}{(x - x_0)^k}, & \text{for } x \neq x_0 \\ 0, & \text{otherwise} \end{cases} \quad (3.5)$$

It suffices to show that $\lim_{x \to x_0} h_k(x) = 0$. We first observe that $\phi^{(i)}(x_0) = P_k^{(i)}(x_0)$ for all $i = 0, \ldots, k-1$, therefore, $\lim_{x \to x_0} h_k(x)$ is of the 0/0 type and the conditions of L'Hôpital's rule apply (see Theorem A.23), so

$$\lim_{x \to x_0} h_k(x) = \lim_{x \to x_0} \frac{\phi(x) - P_k(x)}{(x - x_0)^k} = \lim_{x \to x_0} \frac{\phi'(x) - P_k'(x)}{k(x - x_0)^{k-1}}.$$

By applying L'Hôpital's rule recursively, we obtain

$$\lim_{x \to x_0} h_k(x) = \lim_{x \to x_0} \frac{\phi^{(k-1)}(x) - P_k^{(k-1)}(x)}{k!} = 0. \quad (3.6)$$

We will omit the proof of the part regarding $\mathcal{O}(|x - x_0|^{k+1})$. The interested reader is referred to [M15, Sec II]. $\square$

The result of Theorem 3.1 can be written shortly as

$$\phi(x) \approx \phi(x_0) + \phi'(x_0)(x - x_0) + \frac{\phi''(x_0)}{2!}(x - x_0)^2$$
$$+ \ldots + \frac{\phi^{(k)}(x_0)}{k!}(x - x_0)^k. \quad (3.7)$$

Taylor's theorem allows us to approximate smooth functions by polynomials. Here, we are interested in first-order approximations.

> **Corollary 3.2 (Best linear approximation)** Let $\phi : \mathbb{R} \to \mathbb{R}$ be a twice differentiable function at a point $x_0 \in \mathbb{R}$. Then,
>
> $$\phi(x) = \underbrace{\phi(x_0) + \phi'(x_0)(x - x_0)}_{\text{Best Linear Approximation}} + \mathcal{O}((x - x_0)^2). \quad (3.8)$$

The result of Corollary 3.2 can be written succinctly as

$$\boxed{\phi(x) \approx \phi(x_0) + \phi'(x_0)(x - x_0).} \quad (3.9)$$

Chapter 3. Linearisation

Let us have a look at a few more examples.

Example 3.1 (Linearisation of trigonometric function) Consider the following function
$$\phi(x) = \cos(3x+1).$$
We need to find the best approximation of ϕ around $x_0 = 1$. Function ϕ is differentiable at x_0, so Taylor's Theorem applies. We have $\phi'(x) = -3\sin(3x+1)$, so
$$\phi(x) = \phi(1) + \phi'(1)(x-1) + \mathcal{O}((x-1)^2),$$
therefore, the best linear approximation of ϕ at x_0 is
$$\tilde{\phi}(x) = -0.6536 + 2.2704 \cdot (x-1) \qquad (3.10)$$
$$= 2.2704 \cdot x - 2.924. \qquad (3.11)$$

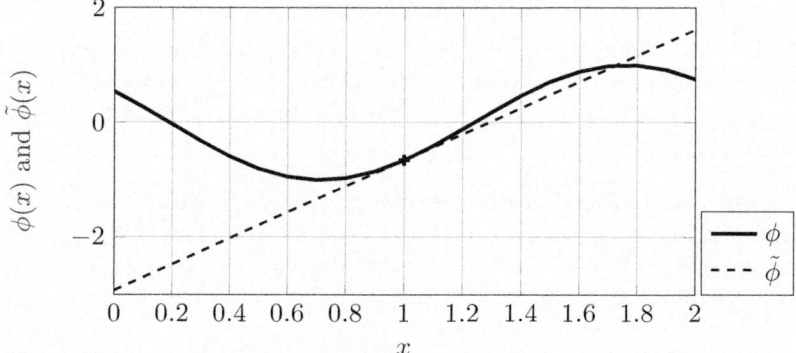

Figure 3.1: Function ϕ (solid line) and its best linear approximation, $\tilde{\phi}$ at $x_0 = 1$ (dashed line).

The two functions, ϕ and $\tilde{\phi}$, coincide at $x_0 = 1$.

It can be observed that $\tilde{\phi}$ is indeed the best linear approximation of ϕ at x_0. Note that the closer we move to x_0 (for x very close to x_0), $\tilde{\phi}(x)$ provides a good approximation of $\phi(x)$. Let us present yet another example where we follow the same procedure to derive a linear approximation of the exponential function.

Example 3.2 (Linearisation of exponential function) Consider the function $\phi(x) = e^x$. We need to find the best approximation of ϕ around x_0. Function ϕ is twice differentiable at x_0, so Corollary 3.2 applies. We have $\phi'(x) = e^x$, so
$$\phi(x) = \phi(x_0) + \phi'(x_0)(x-x_0) + \mathcal{O}((x-x_0)^2),$$

therefore, the best linear approximation of ϕ at x_0 is

$$\tilde\phi(x) = e^{x_0} + e^{x_0}(x - x_0) = e^{x_0}x + e^{x_0}(1 - x_0). \qquad (3.12)$$

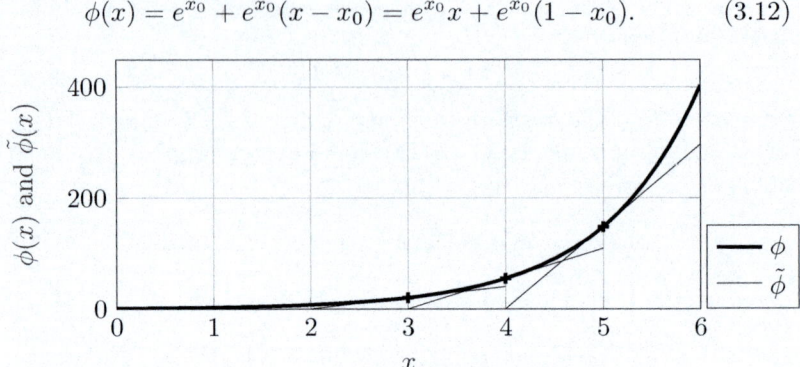

Figure 3.2: *Function ϕ and its best linear approximation, $\tilde\phi$ at three different points ($x_0 = 3$, $x_0 = 4$ and $x_0 = 5$).*

We are now able to approximate any smooth nonlinear function by a linear one around a given point x_0. The closer to x_0 we are, the better the linear approximation will be. The following exercise will convince you that this is a simple procedure.

Exercise 3.2 (**Approximation of** $\sin x$ **at 0 and** $\sqrt{x}$ **at 1**) Show that $\sin(x) \approx x$ for x close to 0. Show also that $\sqrt{x} \approx \frac{1}{2}(1 + x)$, for x close to 1. What can you say about the approximation error in each of these cases?

Exercise 3.3 (**Approximation of** $\sqrt{17}$) Use Taylor's approximation theorem of orders $k = 4$ to approximate $\sqrt{17}$. Compare you approximations to an accurate value (that you can compute using Python, MATLAB or a calculator).

❊ ┄┄┄

Answer. Define $f(x) = \sqrt{x}$ and $x_0 = 16$. We have $f'(x) = \frac{1}{2\sqrt{x}}$, $f''(x) = -\frac{1}{4}x^{-3/2}$, $f^{(3)}(x) = \frac{3}{8}x^{-5/2}$, and $f^{(4)}(x) = -\frac{15}{16}x^{-7/2}$. The fourth-order approximation of f at x_0 is

$$\sqrt{x} \approx 2 + \frac{x}{8} - \frac{(x-16)^2}{512} + \frac{(x-16)^3}{16384} - \frac{5(x-16)^4}{2097152}, \qquad (3.13)$$

from which we find that $\sqrt{17} \approx 4.1231055226$; the difference between this approximation and $\sqrt{17}$ is approximately $-9.965 \cdot 10^{-8}$.

We can leverage this linearisation methodology to linearise nonlinear differential equations. We discuss this in the following section.

3.2 Equilibrium points and deviation variables

An equilibrium point of a dynamical system is a point where the system remains at rest in the sense that under a constant input, the state of the system remains constant. We give the following definition

> **Definition 3.3 (Equilibrium point)** A state-input pair (x^e, u^e) is called an equilibrium of a dynamical system if $x(t) = x^e$ whenever $x(0) = x^e$ and $u(t) = u^e$ for all $t \geq 0$.

We can characterise and determine the equilibrium points of dynamical systems which are available in a state space representation using the following result

> **Proposition 3.4 (Characterisation of equilibrium points)** Consider a dynamical system described by the ODE
>
> $$\dot{x}(t) = f(x(t), u(t)), \tag{3.14}$$
>
> where $f(\,\cdot\,, u^e)$ is such that the initial value problem $\dot{x} = f(x, u^e)$ with $x(0) = x^0$ has a unique solution[a]. A pair (x^e, u^e) is an equilibrium point of (3.14) if and only if
>
> $$f(x^e, u^e) = 0. \tag{3.15}$$
>
> ---
> [a]The differential equations we encounter in this book have unique solutions that are defined for all $t \geq 0$. A sufficient condition for this is provided by the Picard–Lindelöf existence and uniqueness theorem, which requires that $f(\,\cdot\,, u^e)$ be Lipschitz-continuous — a notion stronger than continuity, but weaker than differentiability. We say that $f(\,\cdot\,, u^e)$ is Lipschitz-continuous if $|f(x, u^e) - f(x', u^e)| \leq L|x - x'|$ for all x, x', for some constant $L > 0$. For a detailed discussion, see [M18].

Proof. Suppose that (x^e, u^e) is an equilibrium point for the system, that is, according to Proposition 3.4 $x(t) = x^e$ whenever $x(0) = x^e$ and $u(t) = u^e$ for all $t \geq 0$. Then, $\dot{x}(t) = \frac{d}{dt}x^e = 0$, that is, $f(x(t), u(t)) = f(x^e, u^e) = 0$.

Conversely, suppose that $f(x^e, u^e) = 0$ and $x(0) = x^e$. Then $\dot{x}(0) = 0$ and a solution that satisfies $x(0) = x^e$ is $x(t) = x^e$ for all $t \geq 0$ — but this is the only solution because of the uniqueness assumption. $\square$

If a dynamical system is described by (3.14) and $x(0) = x^e$ and the input is kept constant and equal to $u(t) = u^e$ for all $t \geq 0$, then $x(t) = x^e$ for all $t \geq 0$.

Systems with higher order derivatives as well as systems described by IDEs will be discussed in the following sections.

Example 3.3 (Equilibrium points of linear system) Consider the following linear dynamical system

$$\dot{x} = 4x + u. \tag{3.16}$$

According to Proposition 3.4, a pair (x^e, u^e) is an equilibrium point of this system if

$$4x^e + u^e = 0 \Leftrightarrow u^e = -4x^e. \tag{3.17}$$

For example, the pair $(x^e, u^e) = (0,0)$ is an equilibrium point and so is $(x^e, u^e) = (1,-4)$.

Example 3.4 (Equilibrium points of nonlinear system) Suppose a dynamical system with state x and input u is described by the ODE

$$\dot{x} = x^2 u - u^2 x. \tag{3.18}$$

Let (x^e, u^e) be an equilibrium point of this system. Then,

$$(x^e)^2 u^e - (u^e)^2 x^e = 0, \tag{3.19}$$

equivalently

$$x^e u^e (x^e - u^e) = 0. \tag{3.20}$$

A pair (x^e, u^e) is an equilibrium point of this system if $x^e = 0$ (and $u^e \in \mathbb{R}$), if $u^e = 0$ and $(x^e \in \mathbb{R})$ and if $x^e = u^e$. The set of equilibrium points is illustrated below.

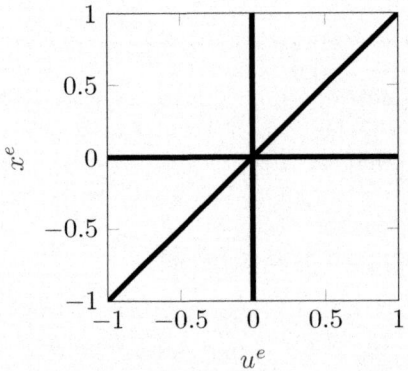

Figure 3.3: Equilibrium points of (3.18).

In Chapter 2 we described the relationship between the voltage across a conductor with the current through it by a simple linear relationship: Ohm's law (see Section 2.3.1). There exist, however, conductors which

Chapter 3. Linearisation

exhibit a nonlinear (non-Ohmic) behaviour described by relationships of the form $V = g(I)$. Let us have a look at such an example.

Example 3.5 (Non-Ohmic circuit) In the circuit shown below a non-Ohmic conductor and a capacitor are connected in series. The non-Ohmic conductor is described by the nonlinear equation (see Section A.8)

$$V = \frac{1}{\beta}\tanh(\alpha I). \tag{3.21}$$

We can manipulate the input voltage $V_i(t)$. Function tanh is called the hyperbolic tangent (see Section A.8 in the appendix for more information). By doing so, a current $I(t)$ will run through the circuit. We are interested in modelling the effect that V_i will have on the output voltage, V_o, across the capacitor. The initial condition of the capacitor are zero, i.e., $V_{bc}(0) = 0$.

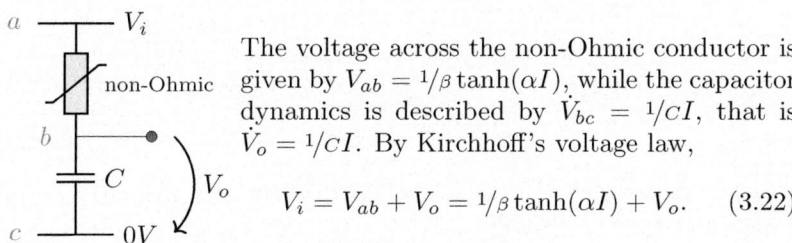

The voltage across the non-Ohmic conductor is given by $V_{ab} = 1/\beta \tanh(\alpha I)$, while the capacitor dynamics is described by $\dot{V}_{bc} = 1/cI$, that is $\dot{V}_o = 1/cI$. By Kirchhoff's voltage law,

$$V_i = V_{ab} + V_o = 1/\beta \tanh(\alpha I) + V_o. \tag{3.22}$$

Solving this for I yields $I = \frac{1}{\alpha}\tanh^{-1}(\beta(V_i - V_o))$, provided that $\beta(V_i - V_o) \in (-1, 1)$, so

$$\dot{V}_o = \frac{1}{\alpha C}\tanh^{-1}(\beta(V_i - V_o)). \tag{3.23}$$

According to the definition, a pair (V_i^e, V_o^e) is an equilibrium point for this system if $\frac{1}{\alpha C}\tanh^{-1}(\beta(V_i^e - V_o^e)) = 0$, therefore, $V_i^e = V_o^e$.

In order to linearise the above nonlinear model, we define the function

$$\phi(x) = \tanh^{-1}(\beta x). \tag{3.24}$$

This function is differentiable at $x_0 = 0$ with $\phi'(x) = \beta/1-x^2$ and $\phi'(0) = \beta$ (see Section A.8). The best linear approximation of ϕ at $x_0 = 0$ is given by

$$\phi(x) = \beta x + \mathcal{O}(x^2). \tag{3.25}$$

In other words, $\tanh^{-1}(\beta x) \approx \beta x$ for x close to 0. The linearised dynamics becomes

$$\dot{V}_0 = \frac{\beta}{\alpha C}(V_i - V_o). \tag{3.26}$$

This approximation holds for V_i close to V_i^e and for V_o close to V_o^e with $V_i^e = V_o^e$.

The result we obtained in Example 3.5 is reminiscent of that in Example 2.15. In fact, it is the same with $R = a/\beta$. What we achieved by linearising the dynamics of the above RC circuit was to find the Ohmic (linear) conductor which best approximates the given non-Ohmic (nonlinear) one.

However, the above procedure may, and typically will, yield an *affine system* instead of a linear one. Let us give an example that will motivate the definition of *deviation variables*.

Example 3.6 (Linearisation produces affine system) Consider the following nonlinear dynamical system

$$\dot{x} = x^2 + u. \tag{3.27}$$

Of course, the source of the nonlinearity is the term x^2. The equilibrium points of the system are those pairs (x^e, u^e) that satisfy

$$(x^e)^2 + u^e = 0. \tag{3.28}$$

For example the pair $(x^e, u^e) = (1, -1)$ is such an equilibrium pair. Let us linearise $\phi(x) = x^2$ at $x^e = 1$. The reader can easily verify that

$$\phi(x) \approx 2x - 1, \tag{3.29}$$

for x close to $x^e = 1$. If we substitute this approximation into Equation (3.27) we obtain

$$\dot{x} = 2x - 1 + u, \tag{3.30}$$

which is an affine system. Next we will introduce the notion of deviation variables that will help us remove such constant terms.

Definition 3.5 (Deviation variables) Consider a dynamical system described by the ODE (3.14). Let (x^e, u^e) be an equilibrium point of this system. The variables $\bar{x} = x - x^e$, and $\bar{u} = u - u^e$, are called **deviation variables** of the state and input respectively.

In practice, physical and engineering systems often operate close to an equilibrium point and fluctuate around it. Deviation variables, therefore, define the deviation from that operating point. The use of deviation variables can simplify models described by ODEs or IDEs which contain **constant** terms.

Before we proceed to an example, note that $\bar{x}$ and x are functions of t, while x^e is a constant, so $\dot{\bar{x}} = \dot{x}$ and $\dot{\bar{u}} = \dot{u}$.

Example 3.7 (Using deviation variables) The system shown below is a tank with a manipulated inflow of water, $F_{\text{in}}(t)$ (kg/s), a *constant* inflow, F (kg/s), and an outflow, $F_{\text{out}}(t)$ (kg/s). The density of water is denoted by ρ. The tank's cross-section area is denoted by A (m^2).

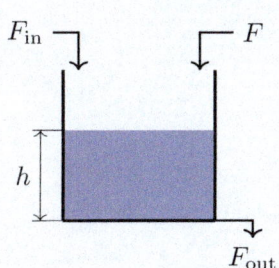

Here, the manipulated variable is the inflow F_{in} and the system state is the elevation h.

The mass balance equations yield

$$\rho A \dot{h} = F_{\text{in}} + F - F_{\text{out}}, \qquad (3.31)$$

which is an affine dynamical equation because of the constant term F.

Assume that the outflow is described by the simple linear model

$$F_{\text{out}} = \beta h, \qquad (3.32)$$

for some parameter $\beta > 0$. Equation (3.31) now becomes

$$\rho A \dot{h} = F_{\text{in}} + F - \beta h. \qquad (3.33)$$

A pair $(F_{\text{in}}^e, \beta h^e)$ is an equilibrium point of the system if

$$0 = F_{\text{in}}^e + F - \beta h^e. \qquad (3.34)$$

We now define the deviation variables

$$\bar{F}_{\text{in}} = F_{\text{in}} - F_{\text{in}}^e, \qquad (3.35a)$$
$$\bar{h} = h - h^e. \qquad (3.35b)$$

We subtract (3.33) and (3.34) by parts to obtain

$$(F_{\text{in}} - F_{\text{in}}^e) + \cancel{F} - \cancel{F} - \beta(h - h^e) = \rho A \dot{h}, \qquad (3.36)$$

and using the fact that $\dot{h} = \dot{\bar{h}}$, we have

$$\bar{F}_{\text{in}} - \beta \bar{h} = \rho A \dot{\bar{h}}. \qquad (3.37)$$

This is now a simple linear dynamical system without constant terms. Note that F does not show up in the system model now.

Let us give an example where we will combine the use of Taylor's theorem with deviation variables to linearise a nonlinear dynamical model.

Example 3.8 (Linearisation and deviation variables) Consider the same system as in the previous example. Here we shall assume that the outflow is described by Torricelli's law which states that

$$F_{\text{out}} = \gamma \sqrt{h}, \qquad (3.38)$$

where $\gamma > 0$ is a constant parameter.

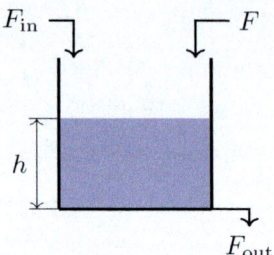

The mass balance equations now read

$$\rho A \dot{h} = F_{\text{in}} + F - \gamma \sqrt{h}. \qquad (3.39)$$

This is a nonlinear system because of the square root. We need to take the following steps:

Step 1. We determine the **equilibrium points** of (3.39). A pair (F_{in}, h^e) is an equilibrium point of (3.39) if

$$0 = F_{\text{in}} + F - \gamma \sqrt{h^e}. \qquad (3.40)$$

We will be concerned only with equilibrium points with $h^e > 0$.

Step 2. We subtract (3.39) and (3.40) by parts

$$\rho A \dot{h} = (F_{\text{in}} - F_{\text{in}}^e) - \gamma(\sqrt{h} - \sqrt{h^e}). \qquad (3.41)$$

Step 3. We linearise $\sqrt{h}$ at equilibrium, i.e., at h^e. Function $\phi(h) = \sqrt{h}$ is differentiable at all $h > 0$ and $\phi'(h) = \frac{1}{2\sqrt{h}}$. We have

$$\sqrt{h} = \sqrt{h^e} + \frac{1}{2\sqrt{h^e}}(h - h^e) + \mathcal{O}((h - h^e)^2)$$
$$\Rightarrow \sqrt{h} - \sqrt{h^e} = \frac{1}{2\sqrt{h^e}}(h - h^e) + \mathcal{O}((h - h^e)^2). \qquad (3.42)$$

Step 4. We define the deviation variables $\bar{F}_{\text{in}} = F_{\text{in}} - F_{\text{in}}^e$ and $\bar{h} = h - h^e$. We substitute (3.42) into (3.41), and we obtain

$$\rho A \dot{\bar{h}} = \bar{F}_{\text{in}} - \frac{\gamma}{2\sqrt{h^e}} \bar{h}. \qquad (3.43)$$

Equation (3.43) is a linear model for the given system. As discussed above, we used the fact that $\dot{h} = \dot{\bar{h}}$.

3.3 Linearisation in several variables

We will now turn our attention to functions of several variables, that is, ones of the form $\phi : \mathbb{R}^n \to \mathbb{R}$. We will not state Taylor's theorem for such functions. Instead, we state a result which gives us the **best linear approximation** for sufficiently smooth functions $\phi : \mathbb{R}^n \to \mathbb{R}$.

The **gradient** of a function $\phi : \mathbb{R}^n \to \mathbb{R}$ at a point $x^0 = (x_1^0, x_2^0, \ldots, x_n^0) \in \mathbb{R}^n$ is defined as the function $\nabla \phi : \mathbb{R}^n \to \mathbb{R}^n$ with

$$\nabla \phi(x^0) = \begin{bmatrix} \frac{\partial \phi}{\partial x_1}(x^0) \\ \frac{\partial \phi}{\partial x_2}(x^0) \\ \vdots \\ \frac{\partial \phi}{\partial x_n}(x^0) \end{bmatrix}, \tag{3.44}$$

provided that the partial derivatives are defined at x^0. Let us give an example.

Example 3.9 (Gradient of a function $\phi : \mathbb{R}^n \to \mathbb{R}$) Consider the function $\phi : \mathbb{R}^2 \ni x = (x_1, x_2) \mapsto \phi(x) \in \mathbb{R}$ given by

$$\phi(x) = x_1^2 + x_1 x_2^2. \tag{3.45}$$

The partial derivatives of ϕ with respect to x_1 and x_2 are

$$\frac{\partial \phi(x)}{\partial x_1} = \frac{\partial}{\partial x_1}\left(x_1^2 + x_1 x_2^2\right) = 2x_1 + x_2^2, \tag{3.46a}$$

$$\frac{\partial \phi(x)}{\partial x_2} = \frac{\partial}{\partial x_2}\left(x_1^2 + x_1 x_2^2\right) = 2x_1 x_2, \tag{3.46b}$$

so the gradient of the function is

$$\nabla \phi(x) = \begin{bmatrix} \frac{\partial \phi(x)}{\partial x_1} \\ \frac{\partial \phi(x)}{\partial x_2} \end{bmatrix} = \begin{bmatrix} 2x_1 + x_2^2 \\ 2x_1 x_2 \end{bmatrix}. \tag{3.47}$$

Note that the gradient takes a vector $x \in \mathbb{R}^2$ and returns a vector $\nabla \phi(x) \in \mathbb{R}^2$.

Next, recall that for two vectors, $x, y \in \mathbb{R}^n$, the scalar $x^\intercal y = x_1 y_1 + x_2 y_2 + \ldots + x_n y_n$ is known as their *inner product*. Moreover, for a vector $x \in \mathbb{R}^n$, its Euclidean norm is defined as $\|x\| = \sqrt{x^\intercal x}$.

We shall now state the following result, which is a generalisation of Corollary 3.2 in several variables. The result states that if a function $\phi : \mathbb{R}^n \to \mathbb{R}$ is sufficiently smooth, then we can approximate it by a linear function.

Theorem 3.6 (Best linear approximation[a]) Suppose that $\phi : \mathbb{R}^n \to \mathbb{R}$ be differentiable at a point $x^0 \in \mathbb{R}^n$. Then, the **best linear approximation** of ϕ at x^0 is

$$\tilde{\phi}_{x^0}(x) = \phi(x^0) + \nabla\phi(x^0)^\mathsf{T}(x - x^0), \quad (3.48)$$

in the sense that

$$\phi(x) = \tilde{\phi}_{x^0}(x) + o(\|x - x^0\|), \text{ as } x \to x^0. \quad (3.49)$$

---
[a]For a version of this theorem involving higher order derivatives, see [M19, Chapter 5].

Proof. The proof (for the more general case of a higher-order approximation) can be found in [M19, Chapter 5] and [M15, Section II-7]. The main idea is to define the function $g(t) = \phi(x^0 + t(x - x^0))$ of the real variable t and apply Taylor's theorem (Theorem 3.1). A more rigorous exposition of Taylor's theorem in several variables can be found in Henri Cartan's book [M22, Thm. 5.6.3]. □

As a result of Equations (3.48) and (3.49) we can write

$$\phi(x) \approx \phi(x^0) + \nabla\phi(x^0)^\mathsf{T}(x - x^0), \text{ for } x \approx x^0. \quad (3.50)$$

Note that the best linear approximation, $\tilde{\phi}_{x^0}$, can be written as

$$\tilde{\phi}_{x^0}(x) = \phi(x^0) + \frac{\partial \phi(x^0)}{\partial x_1}(x_1 - x_1^0) + \ldots + \frac{\partial \phi(x^0)}{\partial x_n}(x_n - x_n^0). \quad (3.51)$$

In a nutshell, Theorem 3.6 says that $\phi(x) \approx \tilde{\phi}_{x^0}(x)$ for x close to x^0. Let us give a few examples in order to understand how to use Theorem 3.6.

Example 3.10 (Linearisation of function of two variables #1) We will find the best linear approximation of $\phi(x_1, x_2) = e^{x_1 + 2x_2}$ at the point $x^0 = (x_1^0, x_2^0) = (1, 2)$.

Function $\phi : \mathbb{R}^2 \to \mathbb{R}$ has the following partial derivatives

$$\frac{\partial \phi}{\partial x_1}(x) = \frac{\partial e^{x_1 + 2x_2}}{\partial x_1}(x) = e^{x_1 + 2x_2}, \quad (3.52a)$$

$$\frac{\partial \phi}{\partial x_2}(x) = \frac{\partial e^{x_1 + 2x_2}}{\partial x_2}(x) = 2e^{x_1 + 2x_2}. \quad (3.52b)$$

We can now apply Theorem 3.6. The gradient of ϕ at x^0 is

$$\nabla\phi(x^0) = \begin{bmatrix} e^{x_1^0 + 2x_2^0} & 2e^{x_1^0 + 2x_2^0} \end{bmatrix}^\mathsf{T} = \begin{bmatrix} e^5 \\ 2e^5 \end{bmatrix}, \quad (3.53)$$

and $\phi(x^0)$ is
$$\phi(1,2) = e^{x_1^0 + 2x_2^0} = e^5, \qquad (3.54)$$
so the best linear approximation of ϕ at x^0 is
$$\tilde{\phi}_{x^0}(x_1, x_2) = e^5 + \begin{bmatrix} e^5 \\ 2e^5 \end{bmatrix}^\mathsf{T} (x - x^0)$$
$$= e^5 + e^5(x_1 - x_1^0) + 2e^5(x_2 - x_2^0)$$
$$\approx 148.41 x_1 + 296.83 x_2 - 593.65, \qquad (3.55)$$
which means that $\phi(x_1, x_2) = \tilde{\phi}_{x^0}(x_1, x_2) + o(\|x - x^0\|)$, or $e^{x_1 + 2x_2} \approx 148.41 x_1 + 296.83 x_2 - 593.65$, for (x_1, x_2) close to $(1, 2)$.

Let us have a look at yet another example:

Example 3.11 (Linearisation of function of two variables #2) In this example we shall compute the best linear approximation of
$$\phi(x, y) = \sin(5x)(1 + \sin^2(y)) + (x - 1)^2(y - 1)^2, \qquad (3.56)$$
at $x^0 = (0.9, 1)$. First we need to compute the gradient of ϕ, which is
$$\frac{\partial \phi(x, y)}{\partial x} = 5\cos(5x)(1 + \sin^2(y)) + 2(x - 1)(y - 1)^2, \qquad (3.57a)$$
$$\frac{\partial \phi(x, y)}{\partial y} = \sin(5x)\sin(2y) + 2(y - 1)(x - 1)^2, \qquad (3.57b)$$
as the reader can verify. These are continuous in a neighbourhood of $x^0 = (0.9, 1)$ and $\phi(x^0) \approx -1.6697$, $\frac{\partial \phi(x^0, y^0)}{\partial x} \approx -1.8003$ and $\frac{\partial \phi(x^0, y^0)}{\partial y} \approx -0.8889$.

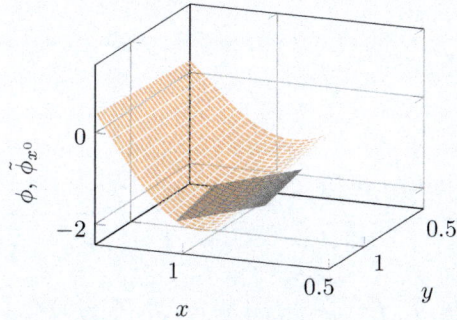

Figure 3.4: Plot of $\phi(x, y) = \sin(5x)(1 + \sin^2(y)) + (x - 1)^2(y - 1)^2$ and its best linear approximation at $x^0 = (0.9, 1)$. Function ϕ is the orange surface and $\tilde{\phi}_{x^0}$ is the grey plane.

The best linear approximation of ϕ at x^0 is given by
$$\tilde{\phi}_{x^0}(x) = -1.6697 - 1.8003(x - 0.9) - 0.8889(y - 1), \qquad (3.58)$$
according to Theorem 3.6.

We can now use the above theorem to derive approximations for dynamical systems with nonlinearities involving several variables.

Example 3.12 (Linearisation of nuclear reactor dynamics) The dynamics of a nuclear reactor is described by an ODE of the form

$$\dot{x} = ax + bxy + cu, \tag{3.59a}$$
$$\dot{y} = fy + h. \tag{3.59b}$$

This is nonlinear because of the presence of the product xy. We need to approximate it close to the equilibrium point (x^e, y^e, u^e) by a linear system. We will work as in Example 3.8.

Step 1. We need to determine the equilibrium points of the system. For (x^e, y^e, u^e) to be an equilibrium point, the following needs to be satisfied

$$0 = ax^e + bx^e y^e + cu^e, \tag{3.60a}$$
$$0 = fy^e + h. \tag{3.60b}$$

Step 2. We now subtract (3.60) and (3.59) by parts

$$\dot{x} = a(x - x^e) + b(xy - x^e y^e) + c(u - u^e), \tag{3.61a}$$
$$\dot{y} = f(y - y^e). \tag{3.61b}$$

Step 3. We now consider the function

$$\phi(x, y) = xy. \tag{3.62}$$

Its partial derivatives are $\partial \phi/\partial x(x,y) = y$ and $\partial \phi/\partial y(x,y) = x$. The conditions of the best approximation theorem (Theorem 3.6) are satisfied, therefore, the best linear approximation of ϕ is

$$\tilde{\phi}_{x^e, y^e}(x, y) = x^e y^e + y^e(x - x^e) + x^e(y - y^e)$$
$$\Rightarrow xy - x^e y^e \approx y^e(x - x^e) + x^e(y - y^e). \tag{3.63}$$

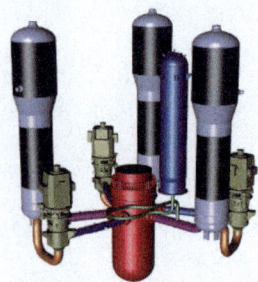

Figure 3.5: Primary coolant system of the Hualong One pressurized water reactor. Image source: Wikipedia page "Nuclear reactor," accessed on 4 Apr., 2019. Image taken from Ji Xing, Daiyong Song, Yuxiang Wu, HPR1000: Advanced Pressurized Water Reactor with Active and Passive Safety, Engineering 2(1):79-87, 2016. (CC-BY-4.0).

Chapter 3. Linearisation

Step 4. We introduce the deviation variables $\bar{x} = x - x^e$, $\bar{y} = y - y^e$ and $\bar{u} = u - u^e$. Since $\dot{\bar{x}} = \dot{x}$ and $\dot{\bar{y}} = \dot{y}$, Equation (3.61) becomes

$$\dot{\bar{x}} = a\bar{x} + b[y^e \bar{x} + x^e \bar{y}] + c\bar{u}, \tag{3.64a}$$
$$\dot{\bar{y}} = f\bar{y}. \tag{3.64b}$$

This can be written compactly as follows

$$\frac{d}{dt}\begin{bmatrix} \bar{x} \\ \bar{y} \end{bmatrix} = \begin{bmatrix} a + by^e & bx^e \\ 0 & f \end{bmatrix} \begin{bmatrix} \bar{x} \\ \bar{y} \end{bmatrix} + \begin{bmatrix} c \\ 0 \end{bmatrix} \bar{u}. \tag{3.65}$$

Example 3.13 (System of interconnected tanks) In Section 2.4.4 we derived a nonlinear model for a system of two interconnected tanks.

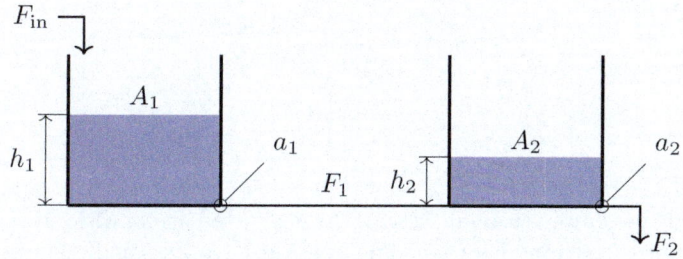

The dynamics of h_1 and h_2 is described by the equations (for $h_1 > h_2$)

$$\rho A_1 \dot{h}_1 = F_{\text{in}} - \rho a_1 \sqrt{2g(h_1 - h_2)}, \tag{3.66a}$$
$$A_2 \dot{h}_2 = a_1 \sqrt{2g(h_1 - h_2)} - a_2 \sqrt{2gh_2}. \tag{3.66b}$$

Here, we will find the equilibrium points of this system, we will linearise around an equilibrium point and will introduce deviation variables to derive a linear model.

Step 1. The triplet $(F_{\text{in}}^e, h_1^e, h_2^e)$ is an equilibrium point of the system if

$$0 = F_{\text{in}}^e - \rho a_1 \sqrt{2g(h_1^e - h_2^e)}, \tag{3.67a}$$
$$0 = a_1 \sqrt{2g(h_1^e - h_2^e)} - a_2 \sqrt{2gh_2^e}. \tag{3.67b}$$

Note: We mentioned above that Equations (3.66) hold so long as $h_1 > h_2$. We shall leave it to the reader to verify that indeed $h_1^e > h_2^e$. Additionally, the reader can show that if at time $t = 0$ we have $h_1(0) \geq h_2(0)$, then $h_1(t) \geq h_2(t)$ for all $t \geq 0$ for all possible input signals F_{in}.

Step 2. We subtract Equations (3.66) and (3.67) by parts to obtain

$$\rho A_1 \dot{h}_1 = F_{\text{in}} - F_{\text{in}}^e - \rho a_1 \sqrt{2g}\left(\sqrt{h_1 - h_2} - \sqrt{h_1^e - h_2^e}\right), \quad (3.68a)$$

$$A_2 \dot{h}_2 = a_1 \sqrt{2g}\left(\sqrt{h_1 - h_2} - \sqrt{h_1^e - h_2^e}\right) - a_2 \sqrt{2g}\left(\sqrt{h_2} - \sqrt{h_2^e}\right). \quad (3.68b)$$

Step 3-i. We consider the function

$$\phi(h_1, h_2) = \sqrt{h_1 - h_2}. \quad (3.69)$$

Its partial derivatives are

$$\frac{\partial \phi}{\partial h_1}(h_1, h_2) = \frac{1}{2\sqrt{h_1 - h_2}},$$

and

$$\frac{\partial \phi}{\partial h_2}(h_1, h_2) = -\frac{1}{2\sqrt{h_1 - h_2}}.$$

From Equation (3.67b) we can see that $h_1^e > h_2^e$, so the conditions of the best approximation theorem (Theorem 3.6) are satisfied, therefore, the best linear approximation of ϕ around (h_1^e, h_2^e) is

$$\tilde{\phi}_{h_1^e, h_2^e}(h_1, h_2) = \sqrt{h_1^e - h_2^e} + \frac{1}{2\sqrt{h_1^e - h_2^e}}(h_1 - h_1^e)$$

$$- \frac{1}{2\sqrt{h_1^e - h_2^e}}(h_2 - h_2^e). \quad (3.70)$$

For notational convenience, let us denote $c = \frac{1}{2\sqrt{h_1^e - h_2^e}}$. Essentially, the above equation says that

$$\sqrt{h_1 - h_2} \approx \sqrt{h_1^e - h_2^e} + c(h_1 - h_1^e) - c(h_2 - h_2^e), \quad (3.71)$$

when h_1 is close to h_1^e and h_2 is close to h_2^e.

Step 3-ii. Similarly, we need to linearise the function $\psi(h_2) = \sqrt{h_2}$ which appears in Equation (3.66b). This is

$$\sqrt{h_2} \approx \sqrt{h_2^e} + \frac{1}{2\sqrt{h_2^e}}(h_2 - h_2^e), \quad (3.72)$$

when h_2 is close to h_2^e.

Step 4. We introduce the deviation variables $\bar{F}_{\text{in}} = F_{\text{in}} - F_{\text{in}}^e$, $\bar{h}_1 = h_1 - h_1^e$ and $\bar{h}_2 = h_2 - h_2^e$, so the system dynamics in Equation (3.68) becomes

$$\rho A_1 \dot{\bar{h}}_1 = \bar{F}_{\text{in}} - \rho a_1 \sqrt{2g}\left(c\bar{h}_1 - c\bar{h}_2\right), \tag{3.73a}$$

$$A_2 \dot{\bar{h}}_2 = a_1 \sqrt{2g}\left(c\bar{h}_1 - c\bar{h}_2\right) - a_2 \sqrt{2g}\frac{1}{2\sqrt{h_2^e}}\bar{h}_2. \tag{3.73b}$$

To simplify the heavy notation, let $\gamma = a_1\sqrt{2g}c$, $\beta = \rho\gamma$ and $\eta = a_2\sqrt{2g}\frac{1}{2\sqrt{h_2^e}}$; then, the linearised system dynamics becomes

$$\rho A_1 \dot{\bar{h}}_1 = \bar{F}_{\text{in}} - \beta\bar{h}_1 + \beta\bar{h}_2, \tag{3.74a}$$

$$A_2 \dot{\bar{h}}_2 = \gamma\bar{h}_1 - (\gamma + \eta)\bar{h}_2. \tag{3.74b}$$

3.4 Higher-order systems

Higher-order systems can often be cast as first-order systems by a simple variable substitution. Suppose a system involves the state variable x and its derivatives up to order n, that is, $\dot{x}, \ddot{x}, \ldots, x^{(n)}$, and no derivatives on the input. Then, as we discussed in Section 2.1.1 we can introduce n state variables, namely $x_1 = x$, $x_2 = \dot{x}_1$, ..., $x_n = \dot{x}_{n-1}$, which allows us to write the system in a state space form, determine/characterise its equilibrium points and linearise it about them using the methodology we discussed previously. Let us give an example.

> **Example 3.14 (Spring-mass system, deviation variables)** Let us revisit Example 2.8. There, we derived the differential equation
>
> $$F + mg - b\dot{x} - kx = m\ddot{x}, \tag{3.75}$$
>
> where x is the system state and F is the input. The system equation involves the second-order derivative of x (the term $\ddot{x}$). We define $x_1 = x$ and $x_2 = \dot{x}_1$. Then, the system becomes
>
> $$m\dot{x}_2 = F + mg - bx_2 - kx_1, \tag{3.76a}$$
> $$\dot{x}_1 = x_2. \tag{3.76b}$$
>
> In particular, x_2 is the speed of the object. Let (x_1^e, x_2^e, F^e) be an equilibrium point of the system. Then,
>
> $$0 = F^e + mg - bx_2^e - kx_1^e, \tag{3.77a}$$
> $$0 = x_2^e, \tag{3.77b}$$

that is,
$$x_2^e = 0, \text{ and } F^e + mg - kx_1^e = 0. \tag{3.78}$$

We subtract (3.77) and (3.76) by parts, and we introduce the deviation variables $\bar{F} = F - F^e$, $\bar{x}_1 = x_1 - x_1^e$ and $\bar{x}_2 = x_2$ (since $x_2^e = 0$), to obtain

$$\dot{\bar{x}}_1 = \bar{x}_2, \tag{3.79a}$$
$$m\dot{\bar{x}}_2 = \bar{F} - b\bar{x}_2 - k\bar{x}_1. \tag{3.79b}$$

This can be compactly written as

$$\frac{\mathrm{d}}{\mathrm{d}t}\begin{bmatrix} \bar{x}_1 \\ \bar{x}_2 \end{bmatrix} = \begin{bmatrix} 0 & 1 \\ -k/m & -b/m \end{bmatrix}\begin{bmatrix} \bar{x}_1 \\ \bar{x}_2 \end{bmatrix} + \begin{bmatrix} 0 \\ 1/m \end{bmatrix}\bar{F}. \tag{3.80}$$

Let us give another example where the variables are somewhat more convolved. We will follow exactly the same procedure as above.

Example 3.15 (Yet another linearisation example) Consider a dynamical system with input u and output variables x and y which is described by the following differential equations

$$\tfrac{1}{\alpha}\ddot{x} + \beta \sin y \cos y - x\dot{y}^4 = 0, \tag{3.81a}$$
$$\beta \ddot{y} + x \cos y = u. \tag{3.81b}$$

Since the system equations involve second-order derivatives, we need to introduce the variables $x_1 = x$, $x_2 = \dot{x}$, $x_3 = y$, and $x_4 = \dot{y}$. Therefore, $\ddot{x} = \dot{x}_2$ and $\ddot{y} = \dot{x}_4$. Then, Equations (3.81) become

$$\tfrac{1}{\alpha}\dot{x}_2 + \beta \sin x_3 \cos x_3 - x_1 x_4^4 = 0, \tag{3.82a}$$
$$\beta \dot{x}_4 + x_1 \cos x_3 = u. \tag{3.82b}$$

We may now solve for $\dot{x}_2$ and $\dot{x}_4$ to obtain

$$\dot{x}_2 = \alpha(x_1 x_4^4 - \beta \sin x_3 \cos x_3), \tag{3.83a}$$
$$\dot{x}_4 = \tfrac{1}{\beta}(u - x_1 \cos x_3). \tag{3.83b}$$

In particular, the dynamical system is described by the following nonlinear differential equations (which involve derivatives of first order only)

$$\dot{x}_1 = x_2, \tag{3.84a}$$
$$\dot{x}_2 = \alpha(x_1 x_4^4 - \beta \sin x_3 \cos x_3), \tag{3.84b}$$
$$\dot{x}_3 = x_4, \tag{3.84c}$$
$$\dot{x}_4 = \tfrac{1}{\beta}(u - x_1 \cos x_3). \tag{3.84d}$$

Note that the differential equations that describe x_2 and x_4 are nonlinear. Next, we will characterise the equilibrium points of this system. A tuple $(x_1^e, x_2^e, x_3^e, x_4^e, u^e)$ is an equilibrium point of the system if

$$0 = x_2^e, \tag{3.85a}$$
$$0 = \alpha(x_1^e(x_4^e)^4 - \beta \sin x_3^e \cos x_3^e), \tag{3.85b}$$
$$0 = x_4^e, \tag{3.85c}$$
$$0 = \tfrac{1}{\beta}(u^e - x_1^e \cos x_3^e). \tag{3.85d}$$

We now subtract (3.84) and (3.85) by parts to obtain

$$\dot{x}_1 = x_2 - x_2^e, \tag{3.86a}$$
$$\dot{x}_2 = \alpha(x_1 x_4^4 - \beta \sin x_3 \cos x_3) - \alpha(x_1^e(x_4^e)^4 - \beta \sin x_3^e \cos x_3^e), \tag{3.86b}$$
$$\dot{x}_3 = x_4 - x_4^e, \tag{3.86c}$$
$$\dot{x}_4 = \tfrac{1}{\beta}(u - x_1 \cos x_3) - \tfrac{1}{\beta}(u^e - x_1^e \cos x_3^e). \tag{3.86d}$$

Define the functions

$$\phi(x_1, x_3, x_4) = \alpha(x_1 x_4^4 - \beta \sin x_3 \cos x_3), \tag{3.87a}$$
$$\psi(u, x_1, x_3) = \tfrac{1}{\beta}(u - x_1 \cos x_3). \tag{3.87b}$$

Let us linearise ϕ at an equilibrium (x_1^e, x_3^e, x_4^e). Its partial derivatives with respect to x_1, x_3 and x_4 at the equilibrium point are

$$\left.\frac{\partial \phi}{\partial x_1}\right|_{x_1^e, x_3^e, x_4^e} = \alpha(x_4^e)^4 = 0, \tag{3.88a}$$
$$\left.\frac{\partial \phi}{\partial x_3}\right|_{x_1^e, x_3^e, x_4^e} = \alpha\beta(\sin^2 x_3^e - \cos^2 x_3^e), \tag{3.88b}$$
$$\left.\frac{\partial \phi}{\partial x_4}\right|_{x_1^e, x_3^e, x_4^e} = 4\alpha x_1^e (x_4^e)^3 = 0, \tag{3.88c}$$

therefore, according to Corollary 3.2, $\phi(x_1, x_3, x_4)$ can be approximated by

$$\phi(x_1, x_3, x_4) \approx \phi(x_1^e, x_3^e, x_4^e) + \alpha\beta(\sin^2 x_3^e - \cos^2 x_3^e)(x_3 - x_3^e), \tag{3.89}$$

for (x_1, x_3, x_4) close to (x_1^e, x_3^e, x_4^e). Similarly, by means of Corollary 3.2 we find that

$$\psi(u, x_1, x_3) \approx \underbrace{\tfrac{1}{\beta}(u^e - x_1^e \cos x_3^e)}_{\psi(u^e, x_1^e, x_3^e)} + \frac{1}{\beta}(u - u^e)$$

$$-\frac{\cos(x_3^e)}{\beta}(x_1 - x_1^e) + x_1^e \frac{\sin x_3^e}{\beta}(x_3 - x_3^e) \quad (3.90)$$

We now introduce the deviation variables $\bar{u} = u - u^e$, $\bar{x}_1 = x_1 - x_1^e$, $\bar{x}_2 = x_2 - x_2^e$, $\bar{x}_3 = x_3 - x_3^e$ and $\bar{x}_4 = x_4 - x_4^e$. Then, the linearised system becomes

$$\dot{\bar{x}}_1 = \bar{x}_2, \quad (3.91a)$$

$$\dot{\bar{x}}_2 = \alpha\beta(\sin^2 x_3^e - \cos^2 x_3^e)\bar{x}_3, \quad (3.91b)$$

$$\dot{\bar{x}}_3 = \bar{x}_4, \quad (3.91c)$$

$$\dot{\bar{x}}_4 = \frac{1}{\beta}\bar{u} - \frac{\cos(x_3^e)}{\beta}\bar{x}_1 + x_1^e \frac{\sin x_3^e}{\beta}\bar{x}_3. \quad (3.91d)$$

Example 3.16 (Inverted pendulum) We will now revisit the dynamics of the inverted pendulum that we derived in Example 2.11. Recall that the system equations are given by

$$m\ddot{x} + m\ell\ddot{\theta}\cos\theta - m\ell\dot{\theta}^2\sin\theta = H, \quad (3.92a)$$

$$-m\ell\dot{\theta}^2\cos\theta - m\ell\ddot{\theta}\sin\theta = N - mg, \quad (3.92b)$$

$$N\sin\theta - H\cos\theta = \tfrac{1}{3}m\ell\ddot{\theta}, \quad (3.92c)$$

$$M\ddot{x} = F - H, \quad (3.92d)$$

where we have used the notation $\ell = L/2$ for convenience. Our objective is to **eliminate** the H and N and end up with a model which involves only x and θ — the two variables of interest. The objective of our model after all is to obtain a model that explains the effect that F has on x and θ. In this example we will follow the following procedure:

1. We will derive ODEs that describe x and θ by eliminating H and N. The resulting model will of course be nonlinear.

2. We can already expect second-order derivatives in the resulting model. We will introduce additional variables to have only first-order derivatives

3. The equilibrium point of interest is the upright position $\theta = 0$. We will verify that this is indeed an equilibrium point of our model.

4. We will identify the sources of nonlinearity, and we will use Taylor's theorem to linearise them

5. Lastly, we linearise the model.

Chapter 3. Linearisation

I. Elimination of H and N. From Equation (3.92d) we have $H = F - M\ddot{x}$, so (3.92a) becomes

$$m\ddot{x} + m\ell\ddot{\theta}\cos\theta - m\ell\dot{\theta}^2\sin\theta = F - M\ddot{x} \qquad (3.93)$$

$$\Leftrightarrow (M+m)\ddot{x} + m\ell\ddot{\theta}\cos\theta - m\ell\dot{\theta}^2\sin\theta = F. \qquad (3.94)$$

This is the first equation we were looking for.

Next, from (3.92c) we have

$$N\sin\theta = H\cos\theta + \tfrac{1}{3}m\ell\ddot{\theta}. \qquad (3.95)$$

We multiply both sides of (3.92b) by $\sin\theta$, and we have

$$-m\ell\dot{\theta}^2\sin\theta\cos\theta - m\ell\ddot{\theta}\sin^2\theta = N\sin\theta - mg\sin\theta, \qquad (3.96)$$

and we replace the term $N\sin\theta$ from Equation (3.95),

$$-m\ell\dot{\theta}^2\sin\theta\cos\theta - m\ell\ddot{\theta}\sin^2\theta = H\cos\theta + \tfrac{1}{3}m\ell\ddot{\theta} - mg\sin\theta. \qquad (3.97)$$

now recall that according to Equation (3.92a) it is $H = m\ddot{x} + m\ell\ddot{\theta}\cos\theta - m\ell\dot{\theta}^2\sin\theta$, so

$$-m\ell\dot{\theta}^2\sin\theta\cos\theta - m\ell\ddot{\theta}\sin^2\theta$$
$$= (m\ddot{x} + m\ell\ddot{\theta}\cos\theta - m\ell\dot{\theta}^2\sin\theta)\cos\theta - \tfrac{1}{3}m\ell\ddot{\theta} - mg\sin\theta$$
$$\Leftrightarrow \cancel{-m\ell\dot{\theta}^2\sin\theta\cos\theta} - m\ell\ddot{\theta}\sin^2\theta = m\ddot{x}\cos\theta + m\ell\ddot{\theta}\cos^2\theta$$
$$\cancel{-m\ell\dot{\theta}^2\sin\theta\cos\theta} - \tfrac{1}{3}m\ell\ddot{\theta} - mg\sin\theta.$$
$$\Leftrightarrow 0 = m\ell\ddot{\theta}(\underbrace{\sin^2\theta + \cos^2\theta}) + m\ddot{x}\cos\theta - \tfrac{1}{3}m\ell\ddot{\theta} - mg\sin\theta$$
$$\Leftrightarrow 0 = \tfrac{4}{3}\ell\ddot{\theta} + \ddot{x}\cos\theta - g\sin\theta. \qquad (3.98)$$

This is the second equation we were looking for. Equations (3.94) and (3.98) are the model of the inverted pendulum after we eliminated N and H.

II. First-order derivatives only. In Equations (3.94) and (3.98) we have second-order derivatives. We need to introduce additional variables as follows

$$x_1 = x, \; x_2 = \dot{x}_1 = \dot{x}, \; x_3 = \theta, \; x_4 = \dot{x}_3 = \dot{\theta}. \qquad (3.99)$$

Then, Equations (3.94) and (3.98) can be written in terms of x_1, x_2, x_3 and x_4 as follows

$$(M+m)\dot{x}_2 + m\ell\cos x_3\,\dot{x}_4 = F + m\ell x_4^2\sin x_3, \qquad (3.100a)$$
$$\cos x_3\,\dot{x}_2 \qquad + \tfrac{4}{3}\ell\,\dot{x}_4 = \qquad g\sin x_3. \qquad (3.100b)$$

We now need to solve this system of equations for $\dot{x}_2$ and $\dot{x}_4$. The reader can verify that the solution is

$$\dot{x}_2 = \frac{4m\ell x_4^2 \sin x_3 + 4F - 3mg \sin x_3 \cos x_3}{4(M+m) - 3m \cos^2 x_3}, \quad (3.101a)$$

$$\dot{x}_4 = -3\frac{m\ell x_4^2 \sin x_3 \cos x_3 + F \cos x_3 - (M+m)g \sin x_3}{[4(M+m) - 3m \cos^2 x_3]\ell}, \quad (3.101b)$$

and note that the denominator, $4(M+m) - 3m \cos^2 x_3$, cannot be equal to zero.

To sum up, the overall (nonlinear) model of the system is given by

$$\dot{x}_1 = x_2, \quad (3.102a)$$

$$\dot{x}_2 = \frac{4m\ell x_4^2 \sin x_3 + 4F - 3mg \sin x_3 \cos x_3}{4(M+m) - 3m \cos^2 x_3}, \quad (3.102b)$$

$$\dot{x}_3 = x_4, \quad (3.102c)$$

$$\dot{x}_4 = -3\frac{m\ell x_4^2 \sin x_3 \cos x_3 + F \cos x_3 - (M+m)g \sin x_3}{[4(M+m) - 3m \cos^2 x_3]\ell}. \quad (3.102d)$$

III. Equilibrium points. A tuple $(x_1^e, x_2^e, x_3^e, x_4^e, F^e)$ is an equilibrium point for this system if (by definition)

$$0 = x_2^e, \quad (3.103a)$$

$$0 = \frac{4m\ell (x_4^e)^2 \sin x_3^e + 4F^e - 3mg \sin x_3^e \cos x_3^e}{4(M+m) - 3m \cos^2 x_3^e}, \quad (3.103b)$$

$$0 = x_4^e, \quad (3.103c)$$

$$0 = -3\frac{m\ell (x_4^e)^2 \sin x_3^e \cos x_3^e + F^e \cos x_3^e - (M+m)g \sin x_3^e}{[4(M+m) - 3m \cos^2 x_3^e]\ell}. \quad (3.103d)$$

Clearly, the tuple $(x_1^e, x_2^e, x_3^e, x_4^e, F^e) = (0, 0, 0, 0, 0)$, which corresponds to the upright position with zero linear and angular velocities, is an equilibrium point.

IV. Linearisation. We will linearise (3.102) around the equilibrium point $(x_1^e, x_2^e, x_3^e, x_4^e, F^e) = (0, 0, 0, 0, 0)$. Let $\phi(F, x_3, x_4)$ and $\psi(F, x_3, x_4)$ denote the right-hand sides of equations (3.102b) and (3.102d) respectively, that is, the nonlinear system in (3.102) can be written as

$$\dot{x}_1 = x_2, \quad (3.104a)$$

Chapter 3. Linearisation

$$\dot{x}_2 = \phi(F, x_3, x_4), \tag{3.104b}$$
$$\dot{x}_3 = x_4, \tag{3.104c}$$
$$\dot{x}_4 = \psi(F, x_3, x_4). \tag{3.104d}$$

We can now compute the partial derivatives of ϕ and ψ with respect to F, x_3 and x_4 at the equilibrium point $F^e = 0$, $x_3^e = 0$ and $x_4^e = 0$. We may in fact do that in Python or MATLAB as we discuss in Section 3.7. The partial derivatives are

$$\left.\frac{\partial \phi(F, x_3, x_4)}{\partial F}\right|_{F^e, x_3^e, x_4^e} = \frac{4}{4M+m}, \tag{3.105a}$$

$$\left.\frac{\partial \phi(F, x_3, x_4)}{\partial x_3}\right|_{F^e, x_3^e, x_4^e} = -\frac{3mg}{4M+m}, \tag{3.105b}$$

$$\left.\frac{\partial \phi(F, x_3, x_4)}{\partial x_4}\right|_{F^e, x_3^e, x_4^e} = 0, \tag{3.105c}$$

and

$$\left.\frac{\partial \psi(F, x_3, x_4)}{\partial F}\right|_{F^e, x_3^e, x_4^e} = -\frac{3}{\ell(4M+m)}, \tag{3.105d}$$

$$\left.\frac{\partial \psi(F, x_3, x_4)}{\partial x_3}\right|_{F^e, x_3^e, x_4^e} = -\frac{3(M+m)g}{\ell(4M+m)}, \tag{3.105e}$$

$$\left.\frac{\partial \psi(F, x_3, x_4)}{\partial x_4}\right|_{F^e, x_3^e, x_4^e} = 0. \tag{3.105f}$$

By virtue of Theorem 3.6, the linearised dynamical system (at $(x_1^e, x_2^e, x_3^e, x_4^e, F^e) = (0,0,0,0,0)$) is given by

$$\dot{x}_1 = x_2, \tag{3.106a}$$
$$\dot{x}_2 = \frac{4}{4M+m}F - \frac{3mg}{4M+m}x_3, \tag{3.106b}$$
$$\dot{x}_3 = x_4, \tag{3.106c}$$
$$\dot{x}_4 = -\frac{3}{\ell(4M+m)}F - \frac{3(M+m)g}{\ell(4M+m)}x_3. \tag{3.106d}$$

Exercise 3.4 (Nonlinear spring-mass-damper system) Consider the spring-mass-damper system of Example 2.8 and assume that the spring does not follow Hooke's law. Instead, the restoring force exerted by the spring is given by

$$F_{\text{spring}} = -kx - cx^3, \tag{3.107}$$

where k and c are positive constants.

1. Derive the (nonlinear) dynamical model of this system
2. Characterise its equilibrium points
3. Linearise the system around an equilibrium point and introduce appropriate deviation variables; write the linearised system in terms of these deviation variables.

❋ ....

Answer. The linearised model is $\ddot{\bar{x}} = \bar{F} - b\dot{\bar{x}} - (k + 3c(x_e)^2)\bar{x} = m\ddot{\bar{x}}$, where $\bar{F}$ and $\bar{x}$ are deviation variables.

3.5 Integro-differential systems

The case of integro-differential equations bears a lot of similarities to that of ODEs with higher order derivatives. Again, the idea is to introduce additional variables. Essentially, we proceed as we did in Example 2.3.

If the system involves functions of the form $\int_0^t x(\tau)\mathrm{d}\tau$, we define $x_1(t) = x(t)$ and $x_2(t) = \int_0^t x(\tau)\mathrm{d}\tau$. Note that, by definition, $x_2(0) = 0$. Then, we have that $\dot{x}_2(t) = x_2(0) + x_1(t) = x_1(t)$. We will need, however, to impose the initial condition $x_2(0) = 0$. If the IDE involves multiple integrals, we work similarly. Let us give an example:

Example 3.17 (Linearisation of IDE) Consider the following integro-differential equation

$$\dot{x} - 3 \int_0^t x(\tau)\mathrm{d}\tau = x + u + 1, \qquad (3.108)$$

with state variable x and input u. As discussed above, let us define $x_1(t) = x(t)$ and $x_2(t) = \int_0^t x(\tau)\mathrm{d}\tau$ with the initial condition $x_2(0) = 0$. Then, the above IDE can be written as

$$\dot{x}_1 - 3x_2 = x_1 + u + 1, \qquad (3.109\mathrm{a})$$
$$\dot{x}_2 = x_1, \qquad (3.109\mathrm{b})$$

or, what is the same,

$$\dot{x}_1 = x_1 + 3x_2 + u + 1, \qquad (3.110\mathrm{a})$$
$$\dot{x}_2 = x_1. \qquad (3.110\mathrm{b})$$

As in all previous examples, an *equilibrium point* is a tuple

(x_1^e, x_2^e, u^e) with $x_2^e = 0$ and

$$0 = x_1^e + 3x_2^{e^{\cancel{0}}} + u^e + 1, \qquad (3.111a)$$
$$0 = x_1^e. \qquad (3.111b)$$

We now introduce the *deviation variables* $\bar{x}_1 = x_1 - x_1^e$, $\bar{x}_2 = x_2 - x_2^e$ and $\bar{u} = u - u^e$. By subtracting (3.111) from (3.110) we arrive at

$$\dot{\bar{x}}_1 = \bar{x}_1 + 3\bar{x}_2 + \bar{u}, \qquad (3.112a)$$
$$\dot{\bar{x}}_2 = \bar{x}_1, \qquad (3.112b)$$

which is a linear dynamical system with zero initial conditions.

Exercise 3.5 (RLC circuit, nonlinear IDE) Consider a series RLC circuit involving a resistance R, inductance L and capacitance C. At time $t = 0$, the capacitor contains an initial change $Q(0) = Q_0$.

1. Verify that the Voltage V across the ends of the circuit and the current I that runs through it satisfy

$$V = RI + L\dot{I} + \tfrac{1}{C} \int_0^t I(\tau) \mathrm{d}\tau. \qquad (3.113)$$

2. Determine the equilibrium points for the system of Equation (3.113), introduce appropriate deviation variables and cast the system in the form of a linear dynamical system as in Example 3.17.

3. Suppose that the circuit involves a nonlinear (non-Ohmic) resistor and the voltage across it is $V_{\mathrm{res}} = \alpha I + \beta I^2$, so the dynamical system becomes

$$V = \alpha I + \beta I^2 + L\dot{I} + \tfrac{1}{C} \int_0^t I(\tau) \mathrm{d}\tau. \qquad (3.114)$$

Again, determine the equilibrium points for the system of Equation (3.114), linearise the system at an equilibrium point, introduce appropriate deviation variables and cast the system in the form of a linear dynamical system as in Example 3.17.

3.6 Quality of linearisation

It is reasonable to ask what the effect of the linearisation is. How close is the linearised system to the original nonlinear system? Suppose we have an initial value problem (IVP) of the form

$$\dot{x} = f(x), \tag{3.115a}$$
$$x(0) = x_0, \tag{3.115b}$$

where f is continuously differentiable, $f(0) = 0$ (that is, $x^{\text{eq}} = 0$ is an equilibrium point of the system). For simplicity, we shall assume that the system does not have an input. Suppose that this IVP admits a unique solution, which we denote by $x(t; x_0)$. Although it is not always easy to determine the solution of such a nonlinear IVP, we can use software to determine approximate solutions numerically (see Section 2.5).

Now suppose that $f(x) \approx f'(x_0)x$ for x close to 0. Then, the linearised version of the initial value problem of Equation (3.115) is

$$\dot{x} = f'(x_0)x, \tag{3.116a}$$
$$x(0) = x_0. \tag{3.116b}$$

Let us denote the unique solution of this IVP as $x_{\text{lin}}(t; x_0)$. In fact, since the dynamical system in Equation (3.116) is linear, it is not difficult to see that

$$x_{\text{lin}}(t; x_0) = x_0 e^{f'(0)t}, \tag{3.117}$$

for $t \geq 0$.

It is reasonable to ask how $x_{\text{lin}}(t; x_0)$ compares to $x(t; x_0)$. Let us give an example.

Example 3.18 (Quality of linearisation) Consider the IVP

$$\dot{x} = -x + ax^2, \tag{3.118a}$$
$$x(0) = x_0. \tag{3.118b}$$

This is a nonlinear IVP of the *Bernoulli* type. Without going into details regarding how it can be solved analytically, this is one of the cases where we can determine a solution in closed form, which is

$$x(t; x_0) = \frac{1}{a + \left(\frac{1}{x_0} - a\right)e^t}, \tag{3.119}$$

for $t \geq 0$.

The linearised IVP is

$$\dot{x} = -x, \tag{3.120a}$$
$$x(0) = x_0, \tag{3.120b}$$

and its solution is $x_{\text{lin}}(t; x_0) = x_0 e^{-t}$. Note that the linearised IVP does not depend on a.

Let us firstly see how the two solutions compare when the initial condition is "close to" the equilibrium point; let us fix $a = 0.5$ and choose $x_0 = 0.1$.

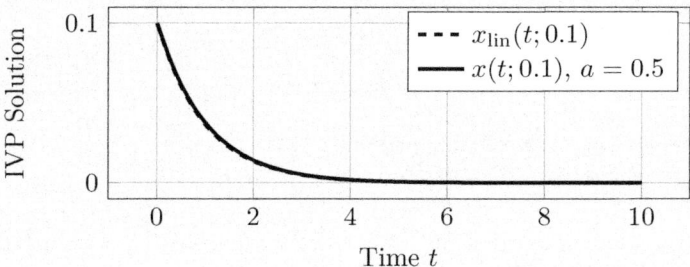

Figure 3.6: *Comparison of the linearisation-based solution (dashed line) to the actual solution of the nonlinear IVP (solid line).*

We see that the linearisation-based solution is very close to the actual solution. Let us now move further away from the equilibrium point and use the initial condition $x(0) = 0.8$.

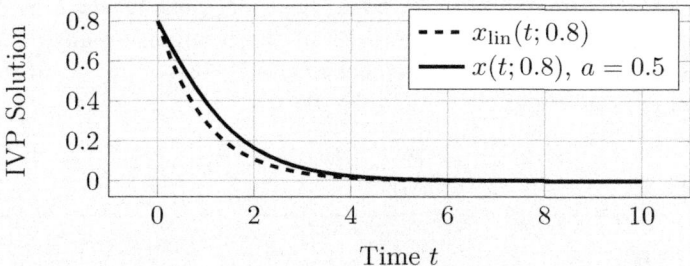

Figure 3.7: *Comparison starting from $x(0) = 0.8$.*

In this case we see that the linearisation-based solution is not so accurate. Lastly, let us compare the approximate and accurate solutions starting from $x_0 = 3$.

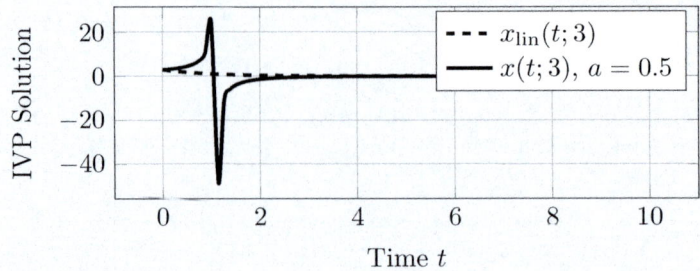

Figure 3.8: *Comparison starting from $x(0) = 3$.*

In this case we see that the linearisation does not offer a good approximation of the actual solution.

The take-home message is that the linearisation gives a good approximation of the solution only for x_0 close to the equilibrium point at which we have linearised the system <u>and</u> as long as $x(t; x_0)$ remains close to the origin.

An interesting property of the linearisation is that if $f'(0) < 0$, there is an $r > 0$ such that $|x_0| < r$, that is, if x_0 is sufficiently close to zero, the trajectory of the *nonlinear* system will converge to zero as $t \to \infty$. Moreover, if $f'(0) > 0$ then the trajectories of the nonlinear system will not converge to zero unless $x_0 = 0$. We see that the linearisation can be used to (i) approximate the trajectories of the original nonlinear IVP and (ii) understand certain qualitative characteristics of the nonlinear IVP without having — or being able — to solve it.

3.7 Symbolic differentiation

3.7.1 Python

Symbols and symbolic computations

In Python, we may use the symbolic library `sympy`, which can be installed using `pip` as follows

```
pip install sympy
```

We may need to prepend `sudo` on some Linux systems.

```
import sympy as sym
x = sym.symbols('x')    # Define symbolic variable x
```

We can then define derived symbols and symbolic expressions; for example in order to construct the symbolic expression `f=sin(cos(x))` we can write

```
import sympy as sym
x = sym.symbols('x')       # symbol x
```

Chapter 3. Linearisation

```
f = sym.sin(sym.cos(x))    # symbolic expression
```

Often, we need to substitute a symbol with another. For example, let us substitute x with $y - \pi$ in the above symbol

```
x, y = sym.symbols('x, y')    # define two symbols
f = sym.sin(sym.cos(x))
f = f.subs(x, y - sym.pi)
```

The updated value of `f` will be `-sin(cos(y))`. We can substitute multiple symbols (e.g., $x = t^2 + 1$, $y = \sqrt{t}$, etc) as in the following example

```
x, y, t = sym.symbols('x, y, t')
f = sym.sin(y * sym.cos(x*y))**2
f_subs = f.subs([(x, t**2+1), (y, sym.sqrt(t))])
```

We can of course substitute a symbol with a numeric value, but the result will be again symbolic. For example, the following code

```
f.subs(y, 1)
```

will return `sin(1)`; in order to obtain the numeric value from a symbolic expression we need to invoke `evalf` as in the following example

```
x = sym.symbols('x')
f = sym.sqrt(x + 1)
f1 = f.subs(x, 1)    # returns sqrt(2)
f1_val = f1.evalf()  # returns 1.4142135...
```

Simplification of expressions

We can simplify complicated and long expressions using sympy's `simplify`. Here is a straightforward example

```
f = sym.sin(x)**2 + sym.cos(x)**2
    + sym.sin(x)*sym.cos(x)
f = sym.simplify(f)    # returns sin(2*x)/2 + 1
```

Relevant methods are `expand` that expands polynomial expressions and `factor` which factorises polynomials. For example `sym.factor(x**3 - x**2 + x - 1)` returns `(x - 1)*(x**2 + 1)` and `sym.expand((x-1)*(x-2)**5)` gives `x**6 - 11*x**5 + 50*x**4 - 120*x**3 + 160*x**2 - 112*x + 32`.

Derivatives and gradients

Derivatives of symbolic expressions with respect to given arguments can be computed with `diff` as in the following example where we compute the partial derivatives of $f(x, y) = e^{x^2 \sin y}(x+1)(y+2)^2$

```
x, y = sym.symbols('x, y')
f = sym.exp(x**2*sym.sin(y))*(x+1)*(y+2)**2

# Compute partial derivatives
fx = sym.diff(f, x)       # partial deriv. wrt x
fy = sym.diff(f, y)       # partial deriv. wrt y

# Compute the values of the partial derivatives
# of f at x=0, y=1
fx_val = sym.subs(fx, [(x, 0), (y, 1)]).evalf()
fy_val = sym.subs(fx, [(x, 0), (y, 1)]).evalf()
```

Linear systems

One will often need to solve systems of algebraic (linear or nonlinear) equations symbolically. Suppose first we have a system of linear equations of the form

$$a_{1,1}x_1 + a_{1,2}x_2 + \ldots + a_{1,n}x_n = b_1,$$
$$a_{2,1}x_1 + a_{2,2}x_2 + \ldots + a_{2,n}x_n = b_2,$$
$$\vdots$$
$$a_{n,1}x_1 + a_{n,2}x_2 + \ldots + a_{n,n}x_n = b_n,$$

where all coefficients and right-hand sides can be symbols. Then, we first need to write the system using matrix notation, that is,

$$\underbrace{\begin{bmatrix} a_{1,1} & a_{1,2} & \cdots & a_{1,n} \\ a_{2,1} & a_{2,2} & \cdots & a_{2,n} \\ \vdots & \vdots & \ddots & \vdots \\ a_{n,1} & a_{n,2} & \cdots & a_{n,n} \end{bmatrix}}_{A} \underbrace{\begin{bmatrix} x_1 \\ x_2 \\ \vdots \\ x_n \end{bmatrix}}_{x} = \underbrace{\begin{bmatrix} b_1 \\ b_2 \\ \vdots \\ b_n \end{bmatrix}}_{b}. \qquad (3.121)$$

Then, we can use `linsolve`. As an example, we can solve the linear system we encountered in Example 3.16 (see Equation (3.100)):

$$(M+m)\ddot{x}_2 + m\ell \cos x_3 \, \ddot{x}_4 = F + m\ell x_4^2 \sin x_3, \qquad (3.122a)$$
$$\cos x_3 \, \ddot{x}_2 \qquad + \tfrac{4}{3}\ell \, \ddot{x}_4 = \qquad g \sin x_3. \qquad (3.122b)$$

We can write these equations in matrix form as follows

$$\begin{bmatrix} M+m & m\ell \cos x_3 \\ \cos x_3 & \tfrac{4}{3}\ell \end{bmatrix} \begin{bmatrix} \ddot{x}_2 \\ \ddot{x}_4 \end{bmatrix} = \begin{bmatrix} F + m\ell x_4^2 \sin x_3 \\ g \sin x_3 \end{bmatrix} \qquad (3.123)$$

We can now write a Python script to compute a symbolic solution of this system.

Chapter 3. Linearisation

```python
# Define all necessary symbols
F, M, m, ell, x3, x4, g = \
    sym.symbols('F, M, m, ell, x3, x4, g')

# Define matrices A and b (here: rhs) of the system
A = sym.Matrix(
    [[M + m,          m * ell * sym.cos(x3)],
     [sym.cos(x3),    4 * ell / 3]])
rhs = sym.Matrix([[F + m * ell * x4**2 * sym.sin(x3)],
                  [g * sym.sin(x3)]])

# Call the symbolic solver
sol, = sym.linsolve((A, rhs))
```

3.7.2 MATLAB

Symbols and symbolic computations

In MATLAB, using the Symbolic Math Toolbox, we can define **symbols** and construct expression with them. There are two ways to define symbols: either `syms x y` or `x = sym('x');, y = sym('y');` and so on. We can then construct symbolic expressions using these symbols. Let us for example define two symbols, x and y, and define a new symbol $f = x^2 y \sin(y+x)$

```matlab
syms x y                        % define symbols
f = x^2*y*sin(y + x);           % construct new symbol
```

We can then substitute certain symbols with numeric values, for example, say we have the function

$$f(x,y) = \frac{\sin(ax)^2 \cos(yx) + yx^2 \sin(2(a+1)xy^2)}{1 - a\tan^3(x + 2xy - y^2)}, \qquad (3.124)$$

where a is a constant, and we need to obtain the symbolic expression of $f(x,0)$, that is, we need to set $y = 0$ in the above expression. Then, we need to use the command **subs** as in the following example

```matlab
syms x y a                           % define symbols
f = sin(a*x)^2*cos(y*x) + ...
    y*x^2*sin(2*(a+1)*x*y^2);        % construct f
f = f / (1 - a*tan(x+2*x*y-y^2)^3);
f0 = subs(f, {y}, {0});              % substitute y=0
fprintf('f(x, 0) = %s\n', f0);       % display f(x, 0)
```

The result it `-sin(a*x)^2/(a*tan(x)^3 - 1)`. We can ask MATLAB to print the result in a prettier way with **pretty**

```matlab
pretty(f0);
```

We can also substitute symbols with other symbols. For example, suppose we want to substitute $x = ct - 1$ and $y = 1 - t^2$ in the symbolic expression of Equation (3.124).

```
syms c t                                % define symbols
ft = subs(f, {x,y}, {c*t-1, 1-t^2});    % substitute
fprintf('f(x(t), y(t)) = %s\n', ft);    % display f
```

Simplification of expressions

MATLAB can **simplify** complex symbolic expressions using `simplify`. For example, if we want to simplify the analytic expression $\sin(x)\sin(2x) + \cos(x)\cos(2x)$ we can write the following MATLAB script.

```
syms x                           % define symbol x
f = sin(x)*sin(2*x) + ...
    cos(x)*cos(2*x);             % symbolic expression
f_simple = simplify(f);          % simplify expression
fprintf('f = %s\n', f_simple);   % display result
```

This will print `f = cos(x)`.

Derivatives

MATLAB's Symbolic Math Toolbox allows us to compute derivatives with respect to a specified variable as well as gradients using `diff`. For example, suppose we need to compute the derivative of $f(x) = x\sin(x)$

```
syms x                  % Define symbol x
f = x * sin(x);         % Define expression f
df = diff(f, x);        % Compute derivative wrt x
                        % We could have written diff(f)
                        % because here f depends only on x

% Print the result...
fprintf('df = %s\n', df);
```

The above script will print `df = sin(x) + x*cos(x)`. We can compute higher-order derivatives as well. For example, to compute the second derivative of $f(x) = x\sin(x)$ we can use `diff` with the order of differentiation passed as the last argument, as in the following example

```
syms x                   % Define symbol x
f = x * sin(x);          % Define expression f
df2 = diff(f, x, 2);     % Second-order derivative
df3 = diff(f, x, 3);     % Third-order derivative
```

Let us give one more example where we compute a derivative with respect to a given variable, while the function involves additional parameters. Consider the function $f(x) = \sin(ax^2 + bx + c)$, where $a, b, c \in \mathbb{R}$. We shall compute the first, third and fifth derivative of f with respect to x.

Chapter 3. Linearisation

```
syms x a b c                % Define symbol x
f = sin(a*x^2 + b*x + c);   % Construct symbolic expr.
df  = diff(f, x);           % First derivative
df3 = diff(f, x, 3);        % Third derivative
df5 = diff(f, x, 5);        % Fifth derivative
```

We find, for example, that the third derivative of f is

$$\frac{\mathrm{d}^3 f(x)}{\mathrm{d}x^3} = -(b+2ax)^3 \cos(ax^2+x+c) - 6a(b+2ax)\sin(ax^2+x+c). \quad (3.125)$$

We can now use what we have learned so far to compute the linearisation of a function that would otherwise require length and tedious computations on paper. Consider the function $f : \mathbb{R} \to \mathbb{R}$ given by

$$f(x) = \frac{\alpha \sin x + \beta x^2 \cos x + \gamma(1-x^2)\sin x \cos x}{1 + \sin(3x)^3}. \quad (3.126)$$

We want to linearise f at $x^0 = 0$. By Corollary 3.2,

$$f(x) \approx f(0) + f'(0)x, \quad (3.127)$$

for x close to 0, but the computation of $f(0)$ and, especially, $f'(0)$ is not so straightforward. We can, however, determine $f(0)$ and $f'(0)$ in a few lines of MATLAB code

```
% LINEARISATION of function f at x^0 = 0
syms x alp bet gam
f = alp*sin(x)+bet*x^2*cos(x) ...
    + gam*(1-x^2)*sin(x)*cos(x);
f = f/(1+sin(3*x)^3);

df = diff(f,x);             % Compute the gradient

x0 = 0;                     % Given point x^0
f0 = subs(f, {x}, {x0});    % Value of f(0)
df0 = subs(df, {x}, {x0});  % Derivative f'(0)
```

and we find that $f(0) = 0$ and $f'(0) = \alpha + \gamma$, therefore,

$$f(x) \approx (\alpha + \gamma)x, \quad (3.128)$$

for x close to 0.

Gradients

In Section 3.3 we defined the gradient of a function $f : \mathbb{R}^n \to \mathbb{R}$. The gradient is a vector containing all partial derivatives of the function, so we can compute the gradient using `diff`, but MATLAB provides `gradient` that allows us to do so easily.

We have so far defined symbols which are scalar-valued. We can, however, define vector-valued symbols, or, simply, vectors of symbols. Think of a function with ten arguments, $x_1, x_2, \ldots, x_{10}$. Instead of having to define a symbol for each one by writing

```
syms x1 x2 x3 x4 x5 x6 x7 x8 x9 x10
```

we can write

```
x = sym('x', [10, 1]);
```

Then, we can define symbolic expressions using this vector symbol x. For example, consider the function $f : \mathbb{R}^{20} \to \mathbb{R}$ given by

$$f(x) = x_1 \sum_{i=1}^{20} \sin^2(x_2 x_i). \tag{3.129}$$

We can define it as follows

```
x = sym('x', [20, 1]);
f = x(1)*sum(sin(x(2)*x).^2);
```

In order to compute the **gradient** of f with respect to x, ∇f we can write

```
df = gradient(f, x);
```

Let us, for example, compute the value of $\nabla f(x^0)$ at $x^0 = [1 \ 1 \ \ldots \ 1]^\mathsf{T}$ for the above function. Here is a complete code snippet:

```
n  = 20;
x  = sym('x', [n, 1]);
f  = x(1)*sum(sin(x(2)*x).^2);
df = gradient(f, x);
x0 = ones(n, 1);
df_x0 = subs(df, x, x0);
```

Note that in this example, since x and x0 are arrays, we should not use curly brackets.

Linear systems

Suppose we need to solve a linear system of the form (3.121) which has a unique solution. Then we can solve the system using the backslash operator, A\b. For example, consider the following linear system in the variables x and y with parameters p, q and c

$$pcx + (p^2 + q^2)y = \sin^2 p,$$
$$-qx + 3\frac{p^2 + q^2}{\sqrt{p^2 + q^2}}y = pc + 1.$$

This can be written in matrix notation as follows

$$\underbrace{\begin{bmatrix} pc & p^2+q^2 \\ -q & 3\frac{p^2+q^2}{\sqrt{p^2+q^2}} \end{bmatrix}}_{A} \begin{bmatrix} x \\ y \end{bmatrix} = \underbrace{\begin{bmatrix} p \\ pc+1 \end{bmatrix}}_{b}. \qquad (3.130)$$

We can then write the following script that will give a symbolic solution of the above system and will return x and y in terms of p, q and c.

```
syms x y p q c              % define symbols
z = p^2+q^2;                % define auxiliary variable
A = [p*c         z
     -q     3*z/sqrt(z)];   % define A
b = [p
     p*c+1];                % define b
sol = A\b;                  % sol = [x; y]
```

As a practical example, in Example 3.16 we had to solve a linear system of equations (Equation (3.100)). We can use MATLAB's Symbolic Math Toolbox to determine the solution:

```
syms F M m x3 x4 ell g           % define symbols
A = [M + m           m * ell * cos(x3)
     cos(x3)       4 * ell / 3];     % matrix A
rhs = [F + m * ell * x4^2 * sin(x3)
       g * sin(x3)];               % vector b
sol = A\rhs;                       % solve system
x2_dot = sol(1);                   % x2_dot
y2_dot = sol(2);                   % y2_dot
```

3.8 One-minute round-up

We learned lots of things in this chapter. First we introduced the concepts of linear, affine and nonlinear systems. We then introduced **Taylor's theorem**, which can be used to approximate general smooth functions by polynomials. Of particular importance is Corollary 3.2 which provides the **best linear approximation** of twice differentiable functions. Next, we introduced the concept of an **equilibrium point** of a dynamical system — a pair (x^e, u^e) at which the system remains at rest. We defined the **deviation variables** $(\bar{x}, \bar{u})$ with respect to an equilibrium pair (x^e, u^e) and gave several examples on how to linearise various dynamical systems. Lastly, we introduced the concept of **symbolic computations** with examples in Python and MATLAB.

In the following chapter we will introduce the Laplace transform; a mathematical construct of central importance in control engineering that facilitates the solution of ODEs/IDEs, the analysis of dynamical systems and the design of control systems.

Remark: As a final remark, the linearisation of a dynamical system can — under certain assumptions — be used to design **local** stabilising controllers for the original systems in the sense that stability can be proven in a neighbourhood of the system's equilibrium point. The necessary conditions are provided by the Hartman-Grobman Theorem, but this is beyond the scope of this course. The interested reader can refer to [M7, Section 2.6] for details. •

Chapter 3. Linearisation

3.9 Exercises

Exercise 3.6 (Apply Taylor's theorem) Using Taylor's theorem, approximate the following functions by affine functions

1. $f_1(x) = \sin x$ at $x = \pi$
2. $f_2(x) = x^2 \cos x^2$ at $x = 0$
3. $f_3(x) = a_0 + a_1 x + a_2 x^2$ at $x = x_0$
4. $f_4(x) = \tan x$ at $x = \pi/4$

What can you say about the error of your linear approximations?

✻ ....

Answer. i. $f_1(x) \approx \pi - x + O((x-\pi)^2)$, for x close to π. ii. $f_2(x) \approx O(x^2)$ for x close to 0. iii. $f_3(x) \approx (a_1 + 2a_2 x_0)x + a_0 - a_2 x_0^2 + O((x-x_0)^2)$ for x close to x_0. iv. $f_4(x) \approx 2x - 1 + \pi/2 + O((x-\pi/4)^2)$ for x close to $\pi/4$.

Exercise 3.7 (Compute cosine without a calculator) Consider the function $f(x) = \cos(x)$. Find (i) the linear approximation of f at $x_0 = 0$, (ii) a fourth-order approximation of f at $x_0 = 0$ using Taylor's Theorem (up to $k = 4$). Use these two approximations to compute $\cos(0.1)$ and compare it with the accurate value you obtain using a calculator or software.

✻ ....

Answer. The linearisation of f at $f(x)$ is $f(x) \approx 1$. The fourth-order approximation is $f(x) \approx 1 - \frac{x^2}{2} + \frac{x^4}{24}$. Using the linearisation, $\cos 0.1 \approx 1$. Using the fourth order approximation, $\cos 0.1 \approx 0.9950041666667$. A more accurate approximation is $\cos 0.1 \approx 0.9950041652780$.

Exercise 3.8 (Find equilibrium points) A dynamical system with input $u \in \mathbb{R}$ and state $x \in \mathbb{R}$ is described by the following differential equation

$$\dot{x} = c^3 + \alpha x + \alpha' \sqrt{x} + \beta u, \tag{3.131}$$

where c, α, α' and β are constant parameters. The system is operated at an equilibrium point (x^e, u^e) with $x^e = 1$. Find u^e and linearise the system around (x^e, u^e). Use deviation variables to derive a linear system.

Exercise 3.9 (Deviation variables) A dynamical system with input $u \in \mathbb{R}$ and state $x \in \mathbb{R}$ is described by the following differential

equation
$$\dot{x} = \alpha x + \beta u + c, \tag{3.132}$$

with $\alpha, \beta \neq 0$. Is this model linear? Use deviations variables to simplify it.

❊

Answer. It is $\dot{x} = \alpha x + \beta u$, where $\bar{x} = x - x_e$ and $\bar{u} = u - u_e$, where $\alpha x_e + \beta u_e + c = 0$.

Exercise 3.10 (Car dynamics) Let y denote the position of a car with respect to the centre of the lane. The car moves with constant linear velocity, v. We denote by ψ the heading angle of the vehicle. The vehicle's lateral dynamics is described by the following simplified model

$$\dot{y} = v \sin(\psi + \beta), \tag{3.133a}$$
$$\dot{\psi} = v/L \sin(\beta), \tag{3.133b}$$

where
$$\beta = \arctan\left(\tfrac{1}{2} \tan \theta\right), \tag{3.134}$$

where L is the distance between the front axle and the centre of gravity and θ is the steering angle as illustrated in Figure 3.9. The control action is the steering angle θ.

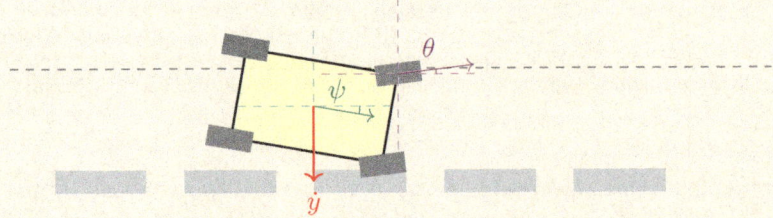

Figure 3.9: *Illustration of the vehicle with its heading angle ψ, the steering angle θ and its lateral velocity $\dot{y}$.*

Determine the equilibrium points of this system. Linearise the model around its equilibria.

Note. This is a simplified version of the *kinematic bicycle model*. You may find more information about it in [C15].

※ ....

and β_e and ψ_e satisfy $\sin(\psi_e + \beta_e) = 0$, $\sin \beta_e = 0$ and $\beta_e = \arctan(\frac{1}{2}\tan \theta_e)$.

$$a = \frac{3\cos(2\theta_e) + 5}{4}, \qquad (3.136)$$

where

$$\tilde{y} = \cos(\tilde{\psi} + \beta_e)(\tilde{\psi} + a\tilde{\theta}), \qquad (3.135a)$$

$$\tilde{\psi} = a\frac{T}{n}\cos \beta_e \tilde{\theta}, \qquad (3.135b)$$

Answer. The linearisation is

Exercise 3.11 (Linearisation) Use deviation variables to simplify the dynamical model of Example 2.9. If $b_1 = b_2 = 0$ the system equations can be simplified further.

Exercise 3.12 (Linearisation of simple pendulum equations) Linearise the model of the simple pendulum we derived in Example 2.10 around the equilibrium point which corresponds to its equilibrium position that corresponds to $\theta = 0$.

Exercise 3.13 (Linearisation of nonlinear RL circuit) Consider the following circuit where the resistance R depends on the temperature, T, of the resistor. This dependence is described by

$$R = aT^{1.2}, \qquad (3.137)$$

where $a = 0.075\,\Omega\mathrm{K}^{-1.2}$, where R and T are measured in Ω and K respectively.

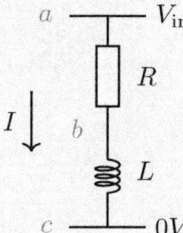

The temperature is generated by the flow of current through the resistor and is dissipated according to the model

$$\dot{T} = 15 - 0.5T + 0.6I^2. \qquad (3.138)$$

The inductance is $L = 2.5\,\mathrm{mH}$. The input variable of the system is the input voltage, V_{in}.

Questions:

1. Derive a model that describes how the input voltage, V_{in}, affects the current, I, that runs through the circuit. Is it linear or nonlinear? Why?

2. Verify that the triplet $(V_{\mathrm{in}}^e, I^e, T^e)$ with $V_{\mathrm{in}}^e = 1891.219\,\mathrm{V}$, $I^e = 18\,\mathrm{A}$ and $T^e = 145.65°C$ is an equilibrium point of the system

3. Linearise the system around the above equilibrium point

※ ....

Answer: The equations of the system are $V_{in} = a_1 T^{1/2} + LI$ and Equation (3.138); this is a nonlinear system of equations with state $x = (I, T)$ and input (I, T). The equilibrium point (V_{in}^e, I^e, T^e) must satisfy $0 = 15 - 0.5\sqrt{T^e} + 0.06(I^e)^2$ and $V_{in}^e = a_1(T^e)^{1/2}$ (note that $T^e = 418.8$ K). The linearised system is $\dot{T}_l = 21.61 - 0.5T_l$ and $\dot{V}_{in}^{in} = -4.2 \cdot 10^4 I_l - 2167.59 T_l + 400 V_{in}^{in}$, where T_l, I_l and V_{in}^{in} are the deviation variables that correspond to T, I and V_{in}, respectively.

Exercise 3.14 (How good is the linearisation?) In Example 3.13 we linearised the following dynamics (see Equation (3.66))

$$\rho A_1 \dot{h}_1 = F_{in} - \rho a_1 \sqrt{2g(h_1 - h_2)}, \qquad (3.139a)$$
$$A_2 \dot{h}_2 = a_1 \sqrt{2g(h_1 - h_2)} - a_2 \sqrt{h_2}, \qquad (3.139b)$$

and derived the following linearised dynamics

$$\rho A_1 \dot{\bar{h}}_1 = \bar{F}_{in} - \beta \bar{h}_1 + \beta \bar{h}_2, \qquad (3.140a)$$
$$A_2 \dot{\bar{h}}_2 = \gamma \bar{h}_1 - (\gamma + \eta) \bar{h}_2, \qquad (3.140b)$$

in terms of the deviation variables $\bar{F}_{in} = F_{in} - F_{in}^e$, $\bar{h}_1 = h_1 - h_1^e$ and $\bar{h}_2 = h_2 - h_2^e$.

Suppose that $A_1 = A_2 = 1\,\text{m}^2$, $a_1 = 0.010\,\text{m}^2$, $a_1 = 0.015\,\text{m}^2$, $\rho = 972\,\text{kg/m}^3$. Determine the equilibrium elevations, h_1^e and h_2^e that correspond to the inflow $F_{in}^e = 0.05\,\text{m}^3/\text{s}$.

Write a Python or MATLAB script to simulate both the nonlinear dynamical system and its linearisation with initial conditions $h_1(0) = h_1^e$, $h_2(0) = h_2^e$ and $F_{in}(t) = F_{in}^e \left(1 + \frac{1}{20} t^2 e^{-t/3}\right)$.

Compare the behaviour of the original dynamical system with that of its linearisation at the equilibrium point.

Exercise 3.15 (Another linearisation exercise) Consider the following nonlinear dynamical system

$$\dot{x} = x^2 - u + \lambda^2 x^3 u + 1 - \lambda, \qquad (3.141)$$

where λ is a positive constant parameter.

1. If the initial state of the system is $x(0) = 1$, and we apply a

Chapter 3. Linearisation

constant input signal $u(t) = 2$, for all $t \geq 0$, then the system state will satisfy $x(t) = 1$, for all $t \geq 0$. Show that $\lambda = 0.5$.

2. Suppose $\lambda = 0.5$. Let (x^e, u^e) be an equilibrium point of this system. Linearise this dynamical system at (x^e, u^e).

※ ....

Answer. 1. $\lambda = 0.5$ (hint: $(x^e, u^e) = (1, 2)$ is an equilibrium point of the system; use the fact that $\lambda > 0$). 2. the linearisation is $\dot{\tilde{x}} \approx (2x^e - 0.75(x^e)^2)\tilde{x} + 0.25(x^e)^3 - 1)\tilde{u}$, where $\tilde{x} = x - x^e$ and $\tilde{u} = u - u^e$.

Exercise 3.16 (Linearisation from graph) A dynamical system is described by the nonlinear ODE

$$\dot{x} = f(x), \tag{3.142}$$

where f is a function whose graph is shown below.

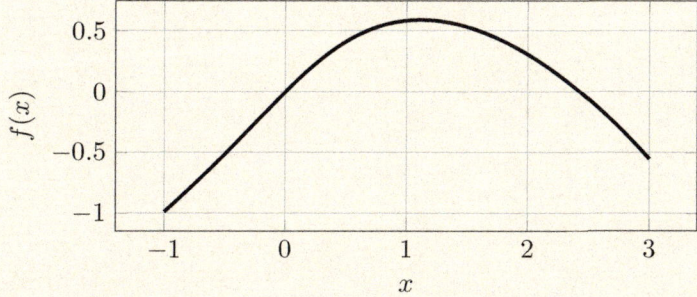

1. Determine the equilibrium points of this system
2. Linearise the system at each equilibrium point (determine the necessary derivatives approximately using the above graph)

※ ....

Answer. The function shown above (or below, depending on your perspective) is $f(x) = \arctan x - 0.2x^2$. You can determine its equilibrium points, linearise it, and compare with the results you obtained using the graph.

4

Laplace transform

In this chapter, we introduce the concept of the Laplace transform, we give the Laplace transforms of some elementary functions and study some essential properties of this new construct. The Laplace transform offers a systematic approach for solving linear differential and integro-differential equations and lays the theoretical foundations for a structured study of linear dynamical systems.

4.1 Definition

We will focus on real-valued (and complex-valued) functions of time, $t \geq 0$, as in the following definition

Definition 4.1 (Time-domain function) A function $f : [0, \infty) \to \mathbb{R}$ or $f : [0, \infty) \to \mathbb{C}$ is called a t-domain, or **time-domain** function.

We introduce the Laplace transform of a time-domain function:

Definition 4.2 (Laplace transform) The *Laplace transform* of $f : [0, \infty) \to \mathbb{R}$ (or, more generally $f : [0, \infty) \to \mathbb{C}$) is a function $\mathscr{L}f : D \to \mathbb{C}$, where $D \subseteq \mathbb{C}$ given by

$$(\mathscr{L}f)(s) = \int_0^\infty e^{-s\tau} f(\tau) \mathrm{d}\tau, \tag{4.1}$$

provided the integral exists for all $s \in D$. The Laplace transform of f, is also denoted by F instead of $\mathscr{L}f$.

The Laplace transform of a function f is a function in the domain of the **complex frequency** s, which is referred to as the s-domain.

Remark: Function $e^{-s\tau}f(\tau)$ might have a vertical asymptote as $\tau \to 0^+$ (i.e., $\lim_{t \to 0^+} f(t) = \infty$ or $\lim_{t \to 0^+} f(t) = -\infty$). An example of such a function is $f(t) = 1/\sqrt{t}$, for $t > 0$. Then, the definition we will be using hereafter becomes

$$(\mathscr{L}f)(s) = \int_{0^+}^{\infty} e^{-s\tau} f(\tau) d\tau. \tag{4.2}$$

Note that the Laplace transform is defined via an improper integral (see Section A.6.3). •

The above definition is *hardly ever used* to compute the Laplace transform of a given function (the same way the definition of the derivative[1] is hardly ever used to find the derivative of a function). Instead, there is a set of simple calculus rules we can use in practice.

A common condition imposed on f for the existence of its Laplace transform is the following:

> **Definition 4.3** **(Exponential order γ)** We say that a function $f : [0, \infty) \to \mathbb{R}$ is of exponential order $\gamma \geq 0$ if there exist positive constants $M > 0$ and $t_0 > 0$ such that
>
> $$|f(t)| \leq Me^{\gamma t}, \tag{4.3}$$
>
> for all $t \geq t_0$.

Example 4.1 (Functions of exponential order) All polynomials are of exponential order. For example, take $f(t) = t$. We know that $t < e^t$ for all $t \geq 0$, therefore, f is of exponential order $\gamma = 1$.

All functions of the form $f(t) = t^\nu$, with $\nu \in \mathbb{N}$, are of exponential order $\gamma = 1$ because it is known that[a] $t^\nu < \nu! e^t$ for all $t \geq 0$.

By definition, function $f(t) = e^{at}$ with $a \in \mathbb{R}$ is of exponential order $\gamma = a$.

---

[a]This is a direct consequence of the Taylor expansion of e^x at $x_0 = 0$, that is, $e^x = 1 + x + \frac{x^2}{2!} + \frac{x^3}{3!} + \ldots$.

A direct consequence of a function f having exponential order γ is that $\lim_{t \to \infty} e^{-st}|f(t)|$ exists for $\mathbf{re}(s) > \gamma$. This is a result that will be useful in what follows. We state this as a lemma:

---

[1]Recall: the derivative of a function $f : [0, \infty) \to \mathbb{R}$ at t is given by $f'(t) = \lim_{h \to 0} \frac{1}{h}(f(t+h) - f(t))$, provided that the limit exists.

Chapter 4. Laplace transform

> **Lemma 4.4 (Existence of limit)** Let $f : [0, \infty) \to \mathbb{R}$ have exponential order γ. Then, for all $s \in \mathbb{C}$ with $\operatorname{re}(s) > \gamma$, the limit $\lim_{t \to \infty} e^{-st} f(t)$ exists and
> $$\lim_{t \to \infty} e^{-st} f(t) = 0.$$
> Moreover, if $\lim_{t \to \infty} e^{-\gamma t} f(t)$ exists and is equal to 0, then f is of exponential order γ.

Proof. Suppose that f is of exponential order $\gamma > 0$. Then, we have

$$\begin{aligned}
0 \leq \lim_{t \to \infty} |e^{-st} f(t)| &= \lim_{t \to \infty} |e^{-(\operatorname{re}(s) + j\operatorname{im}(s))t} f(t)| \\
&= \lim_{t \to \infty} |e^{-\operatorname{re}(s)t} e^{-j\operatorname{im}(s))t} f(t)| \\
&= \lim_{t \to \infty} |e^{-\operatorname{re}(s)t} f(t)| \\
&\leq \lim_{t \to \infty} |M e^{-\operatorname{re}(s)t} e^{\gamma t}| \\
&\leq \lim_{t \to \infty} |M e^{(\gamma - \operatorname{re}(s))t}|,
\end{aligned}$$

therefore, if $\gamma - \operatorname{re}(s) < 0$ then the above upper bound is equal to 0. The second part of the lemma follows from the definition of the limit as $t \to \infty$. This completes the proof. $\square$

Let us have a look at examples and counterexamples of functions of exponential order.

Example 4.2 (Exponential order criterion #1) Consider the function $g(t) = t \sin t$; in order to test whether it is of exponential order, we compute the following limit for some constant $\gamma > 0$

$$\lim_{t \to \infty} \frac{t \sin t}{e^{\gamma t}} = 0, \tag{4.4}$$

For all γ, the above limit exists, and it is finite, therefore according to Lemma 4.4 g is of exponential order γ for any $\gamma > 0$.

Example 4.3 (Exponential order criterion #2) Function $h(t) = e^{t^2}$ is **not** of exponential order. Indeed, we see that

$$\lim_{t \to \infty} \frac{h(t)}{e^{\gamma t}} = \lim_{t \to \infty} \frac{e^{t^2}}{e^{\gamma t}} = \lim_{t \to \infty} e^{t^2 - \gamma t} = \infty,$$

for all $\gamma > 0$.

Example 4.4 (Exponential order criterion #3) We want to check whether function $q(t) = \sin(e^t)$ is of exponential order. Again, we will try to see whether there exists a $\gamma > 0$ so that the limit $\lim_{t\to\infty} \frac{q(t)}{e^{\gamma t}}$ is finite. The limit is

$$\lim_{t\to\infty} \frac{q(t)}{e^{\gamma t}} = \lim_{t\to\infty} \frac{\sin(e^t)}{e^{\gamma t}} = 0,$$

for all $\gamma > 0$. So, q is of exponential order γ for all $\gamma > 0$.

Theorem 4.5 (Existence of Laplace transform) Suppose $f : [0, \infty) \to \mathbb{R}$ is of exponential order γ and piecewise continuous (see Definition A.14). Then, its Laplace transform exists for all $s \in \mathbb{C}$ with $\mathbf{re}(s) > \gamma$.

Proof. Since f is piecewise continuous, function $t \mapsto e^{-st}f(t)$ is piecewise continuous as well, therefore, the integral

$$\int_0^T e^{-s\tau} f(\tau) \mathrm{d}\tau,$$

exists for all $T \geq 0$ (it can be written as a sum of (proper) integrals over which $e^{-st}f(t)$ is continuous in t). We need to show that this integral exists when we take the limit as $T \to \infty$. Recall that according to the definition of exponential order (see Definition 4.3), there are $M > 0$ and $t_0 > 0$ such that $|f(t)| \leq Me^{\gamma t}$ for all $t \geq t_0$. We have

$$\left| \int_0^T e^{-s\tau} f(\tau) \mathrm{d}\tau \right| \leq \int_0^T |e^{-s\tau} f(\tau)| \mathrm{d}\tau$$

$$\leq \int_0^{t_0} e^{-s\tau}|f(\tau)|\mathrm{d}\tau + \int_{t_0}^T e^{-s\tau}|f(\tau)|\mathrm{d}\tau,$$

and by defining $A = \sup_{t\in[0,t_0]} |f(t)|$ (which is finite because f is piecewise continuous), we have

$$\leq A \int_0^{t_0} e^{-s\tau} \mathrm{d}\tau + \int_{t_0}^T e^{-s\tau} Me^{\gamma\tau} \mathrm{d}\tau$$

$$= \frac{A}{s}(1 - e^{t_0 s}) + M \int_{t_0}^T e^{(\gamma-s)\tau} \mathrm{d}\tau$$

$$= \frac{A}{s}(1 - e^{t_0 s}) + \frac{M}{s-\gamma}(e^{-(s-\gamma)t_0} - e^{-(s-\gamma)T}), \quad (4.5)$$

We conclude that $\lim_{T\to\infty} \int_0^T e^{-s\tau} f(\tau) d\tau$ exists so long as $\mathbf{re}(s) - \gamma > 0$ and

$$\begin{aligned} |(\mathscr{L}f)(s)| &= \lim_{T\to\infty} \left| \int_0^T e^{-s\tau} f(\tau) d\tau \right| \\ &\leq \frac{A}{s}(1 - e^{t_0 s}) + \frac{M}{s-\gamma} e^{-(s-\gamma)t_0}, \end{aligned} \qquad (4.6)$$

which completes the proof. □

Remark: *The above theorem gives only sufficient conditions for the existence of the Laplace transform. In other words, there are functions in the time domain which do not satisfy the conditions of Theorem 4.5, yet have a Laplace transform (see Exercises 4.29 and 4.30).* •

Finally, we give the following theorem without a complete proof:

> **Theorem 4.6 (Further properties of Laplace transform)** Let f be a piecewise continuous function in the time domain with exponential order γ. Then, (i) F is infinitely many times differentiable for all $s \in \mathbb{C}$ with $\mathbf{re}(s) \geq \gamma$, and (ii) $\lim_{s\to\infty} F(s) = 0$ with $s \in \mathbb{R}$.

Proof. We shall prove the first assertion later (see Section 4.4.2). The fact that $\lim_{s\to\infty} F(s) = 0$ with $s \in \mathbb{R}$ follows from inequality (4.6). □

4.2 Laplace transforms of elementary functions

In this section we shall give the Laplace transforms of some elementary functions.

4.2.1 Constant function

The Laplace transform of the constant function $f(t) = c$, $t \geq 0$, for some real (or complex) constant x, is

$$\begin{aligned} \mathscr{L}\{c\}(s) &= \int_0^\infty c e^{-s\tau} d\tau = c \int_0^\infty e^{-s\tau} d\tau \\ &= -c \frac{1}{s} e^{-s\tau} \Big|_{\tau=0}^{\tau\to\infty} = -\frac{c}{s} \left(\lim_{t\to\infty} e^{-st} - 1 \right). \end{aligned} \qquad (4.7)$$

We need to determine the limit $\lim_{t\to\infty} e^{-st}$; since $s \in \mathbb{C}$ we can write $s = a + jb$ for some $a, b \in \mathbb{R}$. Then

$$\lim_{t\to\infty} e^{-st} = \lim_{t\to\infty} e^{-(a+jb)t} = \lim_{t\to\infty} e^{-at} e^{-jbt}. \qquad (4.8)$$

If $a > 0$, the above limit is equal to 0. Here is why:

$$0 \le \left| \lim_{t \to \infty} e^{-at} e^{-jbt} \right| = \lim_{t \to \infty} \left| e^{-at} \right| \cdot \underbrace{\left| e^{-jbt} \right|}_{1} = 0. \tag{4.9}$$

If $a = 0$, it can be seen that $e^{-st} = e^{-jbt}$ and $\arg e^{-jbt} = -bt$, which diverges. For $a < 0$, we have that $e^{-at} \to \infty$ as $t \to \infty$, and it can be seen that the modulus of $e^{-st} = e^{-at} e^{-jbt}$ diverges to infinity. We have shown the following result.

> **Theorem 4.7** (Laplace transform of constant function) For any $c \in \mathbb{R}$,
>
> $$\mathscr{L}\{c\}(s) = \frac{c}{s}, \tag{4.10}$$
>
> defined for all $s \in \mathbb{C}$ with $\mathbf{re}(s) > 0$.

4.2.2 Heaviside function

The Heaviside function is defined as follows:

$$H_a(t) = \begin{cases} 0, & \text{if } t \le a \\ 1, & \text{otherwise''} \end{cases} \tag{4.11}$$

Its Laplace transform is given in the following theorem.

> **Theorem 4.8** (Laplace of Heaviside function) The Laplace transform of H_a exists and is
>
> $$(\mathscr{L} H_a)(s) = \frac{e^{-as}}{s}. \tag{4.12}$$
>
> This function is defined for $s \in \mathbb{C}$ with $\mathbf{re}(s) > 0$.

Proof. This result is very easy to verify using the definition (exercise). □

Since all functions in the time domain are defined for $t \ge 0$, it holds that the Laplace transform of the constant function, $f(t) = 1$, is equal to the Laplace transform of $H_0(t)$, that is, $F(s) = 1/s$. The Laplace transform is an **integral transform**. We can easily see that the following function

$$\tilde{H}_a(t) = \begin{cases} 0, & \text{if } t < a \\ 1, & \text{otherwise} \end{cases} \tag{4.13}$$

has the same Laplace transform as the Heaviside function, $H_a(t)$, that is, $\mathscr{L} H_a = \mathscr{L} \tilde{H}_a$.

Chapter 4. Laplace transform

More generally, if two t-domain functions, f and g, are equal everywhere except for a finite number of points, then their Laplace transforms are equal. That is, if there are points $t_1, \ldots, t_n$ so that

$$f(t) = g(t), \text{ for all } t \in [0, \infty) \setminus \{t_1, \ldots, t_n\}, \tag{4.14}$$

then

$$F(s) = G(s). \tag{4.15}$$

4.2.3 Exponential function

The Laplace transform of the exponential function $f(t) = e^{at}$ with $a \in \mathbb{R}$, is given in the following theorem

> **Theorem 4.9 (Laplace transform of exponential)** The Laplace transform of $f(t) = e^{at}$, with $a \in \mathbb{R}$, exists and is given by
>
> $$F(s) = \mathscr{L}\{e^{at}\}(s) = \frac{1}{s-a}, \tag{4.16}$$
>
> defined for $s \in \mathbb{C}$ with $\mathbf{re}(s) > a$.

Proof. The Laplace transform of f exists because it is piecewise continuous and of exponential order a. Using the definition of the Laplace transform,

$$\begin{aligned} F(s) &= \int_0^\infty e^{-s\tau} e^{a\tau} d\tau = \left. \frac{e^{(a-s)t}}{a-s} \right|_0^\infty \\ &= \lim_{t \to \infty} \frac{e^{(a-s)t}}{a-s} - \frac{1}{a-s} \\ &= \frac{1}{s-a} \lim_{t \to \infty} e^{(a-s)t} - \frac{1}{a-s}. \end{aligned} \tag{4.17}$$

Following the same procedure as in Section 4.2.1 we can see that

$$\lim_{t \to \infty} e^{(a-s)t} = 0, \tag{4.18}$$

if $\mathbf{re}(s) > a$, while the limit diverges for $\mathbf{re}(s) \leq a$. $\square$

> **Exercise 4.1 (Laplace of complex-valued exponential function)** Show that the Laplace transform of the time-domain signal $f : [0, \infty) \to \mathbb{C}$ given by $f(t) = e^{at}$, with $a \in \mathbb{C}$, is given by
>
> $$\mathscr{L}\{e^{at}\} = \frac{1}{s-a}, \tag{4.19}$$
>
> and is defined for all $s \in \mathbb{C}$ with $\mathbf{re}(s) > \mathbf{re}(a)$.

Likewise, the Laplace transform of $g(t) = e^{-at}$ is

$$G(s) = \frac{1}{s+a}. \tag{4.20}$$

These results can be very easily verified using the definition of the Laplace transform (try this as an exercise).

4.2.4 Powers

We will use the definition to find the Laplace transform of $f(t) = t$. It is

$$F(s) = \int_0^\infty e^{-s\tau} \tau \, d\tau. \tag{4.21}$$

Recall the **integration by parts** trick. Let f, g be two continuously differentiable functions. Then,

$$\int_a^b f(x) g'(x) \, dx = f(x) g(x) \big|_a^b - \int_a^b f'(x) g(x) \, dx. \tag{4.22}$$

We will determine the Laplace transform of $f(t) = t$ which is given by $\mathscr{L}\{t\}(s) = \int_0^\infty \tau e^{-s\tau} d\tau$. Using integration by parts with $f(\tau) = \tau$ and $g(\tau) = -\frac{1}{s} e^{-s\tau}$, we have[2]

$$F(s) = \int_0^\infty e^{-s\tau} \tau \, d\tau = -\frac{1}{s} e^{-s\tau} \tau \Big|_0^\infty - \int_0^\infty -\frac{1}{s} e^{-s\tau} d\tau = \frac{1}{s^2}, \tag{4.23}$$

provided that $\mathbf{re}(s) > 0$ (see Lemma 4.4). We showed that

$$\mathscr{L}\{t\}(s) = \frac{1}{s^2}, \tag{4.24}$$

for all $s \in \mathbb{C}$ with $\mathbf{re}(s) > 0$. The reader can show that

---

Theorem 4.10 (Laplace transform of integer powers) The Laplace transform of $f(t) = t^\nu$, for $\nu = 1, 2, \ldots$ and $t \geq 0$ is

$$F(s) = \frac{\nu!}{s^{\nu+1}}, \tag{4.25}$$

defined for $s \in \mathbb{C}$ with $\mathbf{re}(s) > 0$.

Proof. The proof is left to the reader as an exercise (see Exercise 4.25). □

The above result can be generalised for real powers. Let us first define the **gamma function**:

---

[2] We have used the fact that $\lim_{t \to \infty} \tau e^{-s\tau} = 0$ for all $s \in \mathbb{C}$ with $\mathbf{re}(s) > 0$. This can be proven as follows: (i) write $s = a + jb$ for $a, b \in \mathbb{R}$, (ii) as in Section 4.2.1, use the fact that $|e^{-jbt}| = 1$ to show that the limit converges if and only if $\lim_{t \to \infty} t e^{-at} = 0$, (iii) show that $\lim_{t \to \infty} t e^{-at} = 0$ if and only if $a > 0$.

Chapter 4. Laplace transform

Definition 4.11 (Euler's gamma function) The gamma function, or Euler's gamma (Γ) function, is defined as

$$\Gamma(z) = \int_0^\infty x^{z-1} e^{-x} \, dx, \qquad (4.26)$$

for $z \in \mathbb{R} \setminus \{0, -1, -2, \ldots\}$.

The gamma function cannot be expressed in terms of other widely used functions such as powers, exponentials and/or trigonometric functions. One of the exotic properties of the gamma function is that it interpolates the factorial, that is,

$$\Gamma(n) = (n-1)!, \qquad (4.27)$$

for all $n \in \mathbb{N}$. There exist other functions with this property, but this is the most popular one. Moreover, Euler's gamma function satisfies the properties

$$\Gamma(1) = 1, \qquad (4.28a)$$
$$\Gamma(x+1) = x\Gamma(x), \qquad (4.28b)$$

for all $x > 0$. The gamma function allows us to state the following theorem.

Theorem 4.12 (Laplace transform of real powers) The Laplace transform of $f(t) = t^a$, $a > -1$ is given by

$$F(s) = \frac{\Gamma(a+1)}{s^{a+1}}, \qquad (4.29)$$

defined for $s \in \mathbb{C}$ with $\mathbf{re}(s) > 0$.

Proof. The proof is left to the reader as an exercise. Hint: use the definition of the Laplace transform and the change of variables $\xi = s\tau$. □

4.2.5 Trigonometric functions

The Laplace transforms of $\sin(\omega t)$ and $\cos(\omega t)$ can be derived from the Laplace transform of the exponential as shown in the following theorem:

Theorem 4.13 (Laplace transform of sine/cosine) The Laplace transform of the sine and cosine functions is given by

$$\mathscr{L}\{\sin(\omega t)\}(s) = \frac{\omega}{s^2 + \omega^2}, \qquad (4.30a)$$
$$\mathscr{L}\{\cos(\omega t)\}(s) = \frac{s}{s^2 + \omega^2}, \qquad (4.30b)$$

defined for all $s \in \mathbb{C}$ with $\mathbf{re}(s) > 0$.

Proof. The proof relies on the fact that we know that the Laplace transform of the exponential is (Section 4.2.3) $\mathscr{L}\{e^{at}\} = \frac{1}{s-a}$ and the exponential is intimately related to sin and cos. Recall that for all $z \in \mathbb{C}$

$$e^{jz} = \cos z + j \sin z. \tag{4.31}$$

Then, by substituting $z = \omega t$ we have

$$e^{j\omega t} = \cos(\omega t) + j \sin(\omega t), \tag{4.32}$$

and taking the Laplace transform (see Exercise 4.1)

$$\frac{1}{s - j\omega} = \mathscr{L}\{\cos(\omega t)\} + j\mathscr{L}\{\sin(\omega t)\}$$

$$\Rightarrow \frac{s + j\omega}{s^2 + \omega^2} = \mathscr{L}\{\cos(\omega t)\} + j\mathscr{L}\{\sin(\omega t)\}$$

$$\Rightarrow \frac{s}{s^2 + \omega^2} + j\frac{\omega}{s^2 + \omega^2} = \mathscr{L}\{\cos(\omega t)\} + j\mathscr{L}\{\sin(\omega t)\}. \tag{4.33}$$

Note that the Laplace transform of $e^{j\omega t}$ is defined for all $s \in \mathbb{C}$ with $\mathbf{re}(s) > \mathbf{re}(j\omega) = 0$ (see Exercise 4.1). We have an equality of complex numbers, so comparing their real and imaginary parts we complete the proof (see Section A.1.1). □

Exercise 4.2 (Trigonometric functions) (i) What is the Laplace transform of $f(t) = \sin(\omega t + \phi_0)$? (ii) What is the Laplace transform of $f(t) = 2 \sin t \cos t$?

✳ ....

Answer. It is $\mathscr{L}\{\sin(\omega t + \phi_0)\} = \frac{\cos(\phi_0)\omega + s\sin(\phi_0)}{s^2 + \omega^2}$, for $s \in \mathbb{C}$ with $\mathbf{re}(s) > 0$, and $\mathscr{L}\{2 \sin t \cos t\} = \frac{2}{s^2 + 4}$, for $s \in \mathbb{C}$ with $\mathbf{re}(s) > 0$.

✂ ....

Hint. i. You can use the trigonometric identity $\sin(a + b) = \sin(a) \cos(b) + \cos(a) \sin(b)$, ii. Use a trigonometric identity for $2 \sin t \cos t$.

4.3 Properties of the Laplace transform

4.3.1 Linearity

Firstly, the Laplace transform is **linear** as stated in the following theorem

> **Theorem 4.14 (Linearity)** Let f and g be functions in the t-domain such that their Laplace transforms, F and G respectively, exist. Then,
> $$(\mathscr{L}\{af + bg\})(s) = aF(s) + bG(s), \qquad (4.34)$$
> for all $a, b \in \mathbb{R}$.

Proof. The proof is left to the reader as an exercise. $\square$

Theorem 4.14 can be employed to compute Laplace transforms of functions such as the following:

$$\begin{aligned}
\mathscr{L}\left\{\frac{\sin t + \cos t}{2}\right\} &= \mathscr{L}\{\tfrac{1}{2}\sin t + \tfrac{1}{2}\cos t\} \\
&= \tfrac{1}{2}\frac{1}{s^2 + 1} + \tfrac{1}{2}\frac{s}{s^2 + 1} \\
&= \frac{s+1}{2(s^2+1)}. \qquad (4.35)
\end{aligned}$$

Theorem 4.14 can also be used to compute the Laplace transform of polynomial functions. For instance, let

$$p(t) = a_0 + a_1 t + a_2 t^2 + \ldots + a_n t^n. \qquad (4.36)$$

Then, the Laplace transform of p is

$$P(s) = \frac{a_0}{s} + \frac{a_1}{s^2} + \frac{2a_2}{s^3} + \ldots + \frac{n! a_n}{s^{n+1}}. \qquad (4.37)$$

4.3.2 Functions defined piecewise

We may compute the Laplace transform of functions which are defined piecewise, for example

$$h(t) = \begin{cases} f(t), & \text{if } 0 \leq t \leq t_0 \\ g(t), & \text{if } t > t_0 \end{cases} \qquad (4.38)$$

This function can be written in terms of Heaviside functions as

$$h(t) = f(t) + (g(t) - f(t))H_{t_0}(t), \qquad (4.39)$$

for $t \geq 0$. Now in order to determine the Laplace transform of h we need to state the following property

Theorem 4.15 (Shifted (delayed) functions) Let f be a function in the time domain with Laplace transform defined for $s \in \mathbb{C}$ with $\mathbf{re}(s) > \gamma$. For $a > 0$, define the following t-shifted version of f:

$$\hat{f}_a(t) = \begin{cases} 0, & \text{for } 0 \leq t < a \\ f(t-a), & \text{for } t \geq a \end{cases} \qquad (4.40)$$

Then, $\hat{f}_a$ has the following Laplace transform

$$\mathscr{L}\{\hat{f}_a(t)\} = e^{-as} F(s), \qquad (4.41)$$

which is defined for $s \in \mathbb{C}$ with $\mathbf{re}(s) > \gamma$.

Remark: *It is convenient to state Theorem 4.15 in the following way, with some abuse of notation:*

$$\mathscr{L}\{f(t-a) H_a(t)\}(s) = e^{-as} F(s). \qquad (4.42)$$

The reason this is not perfectly correct is that f is defined for nonnegative times $t \geq 0$, therefore, $f(t-a)$ is not well defined for $t \in [0, a)$. Nevertheless, $f(t-a)$ is multiplied by $H_a(t)$, therefore, whatever value we may assign to $f(t-a)$ for $t \in [0, a)$, $f(t-a) H_a(t)$ will be equal to zero — cf Equation (4.40). An example is shown in Figure 4.1 below.

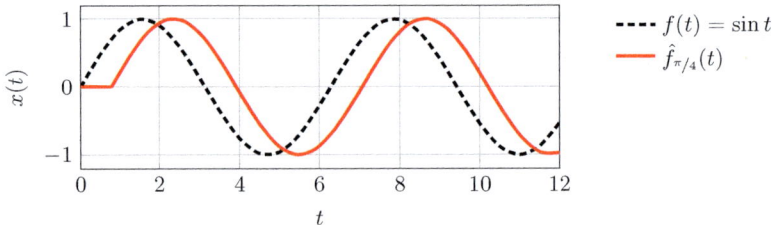

Figure 4.1: *Function $f(t) = \sin(t)$ and its shifted version, $\hat{f}_{\pi/4}(t)$*

Hereafter we will use the notation $\hat{f}_a(t) = f(t-a) H_a(t)$, for $a > 0$ and $t \geq 0$, being conscious of the fact that the correct way to write this would be as in Equation (4.40). •

Proof. By Definition 4.2, we have

$$\mathscr{L}\{f(t-a) H_a(t)\}(s) = \int_0^\infty e^{-s\tau} f(\tau - a) H_a(\tau) d\tau. \qquad (4.43)$$

We now introduce the change of variables $\xi = \tau - a$ and using the fact that $H_a(\tau + a) = H_0(\tau)$, we have

$$\mathscr{L}\{f(t-a) H_a(t)\}(s) = \int_0^\infty e^{-s(a+\xi)} f(\xi) H_a(\xi + a) d\xi$$

$$= e^{-as} \int_0^\infty e^{-s\xi} f(\xi) H_0(\xi) \mathrm{d}\xi$$
$$= e^{-as} \int_0^\infty e^{-s\xi} f(\xi) \mathrm{d}\xi = e^{-as} F(s). \qquad (4.44)$$

This completes the proof. $\square$

We cannot apply Theorem 4.15 directly to Equation (4.39) because the term $w = g - f$ is not, explicitly, a function of $t - t_0$. We need to determine a function $\tilde{w}$ so that

$$w(t) = g(t) - f(t) = \tilde{w}(t - t_0),$$

There are two main cases:

Case I. $w = g - f$ is polynomial

Suppose that w is polynomial of degree n, that is,

$$w(t) = a_0 + a_1 t + a_2 t^2 + \ldots + a_n t^n. \qquad (4.45)$$

Then, by Taylor's theorem,

$$w(t) = w(t_0) + (t - t_0) w'(t_0) + \frac{(t - t_0)^2}{2!} w''(t_0) +$$
$$\ldots + \frac{(t - t_0)^n}{n!} w^{(n)}(t_0). \qquad (4.46)$$

We denote by $w^{(n)}$ the n-th order derivative of w. Equation (4.46) is a finite sum because all derivatives of w of order higher than n are equal to 0. Let us give an example: take

$$w(t) = 1 + t - 3t^2. \qquad (4.47)$$

Then,

$$w'(t) = 1 - 6t, \qquad (4.48\text{a})$$
$$w''(t) = -6, \qquad (4.48\text{b})$$

therefore, for $t_0 = 1$,

$$w(t) = -1 - 5(t - 1) - 3(t - 1)^2. \qquad (4.49)$$

Case II. $w = g - f$ involves exponentials

Suppose that w is exponential, that is

$$w(t) = e^{at}. \qquad (4.50)$$

Then, it is straightforward to write

$$w(t) = e^{at} = e^{a(t - t_0 + t_0)} = e^{a t_0} e^{a(t - t_0)}. \qquad (4.51)$$

Using Theorem 4.15 we have the following result

> **Proposition 4.16 (Functions defined piecewise)** Consider the function
> $$h(t) = \begin{cases} f(t), & \text{if } 0 < t \leq t_0 \\ g(t), & \text{if } t > t_0 \end{cases}. \tag{4.52}$$
>
> Define $w(t) = f(t) - g(t)$ and suppose there is a time-domain function $\tilde{w}$ so that $w(t) = \tilde{w}(t - t_0)$. Then $h(t) = f(t) + \mathring{w}(t - t_0)H_{t_0}(t)$, for all $t \geq 0$. Let $\tilde{W}$ be the Laplace transform of $\tilde{w}$. Then, the Laplace transform of is given by
> $$(\mathscr{L}h)(s) = F(s) + e^{-t_0 s}\tilde{W}(s). \tag{4.53}$$

Proof. The proof relies on Theorem 4.15 and the linearity of the Laplace transform. □

Using the above tricks we can derive the Laplace transform of functions defined piecewise with two pieces. Let us give an example:

Example 4.5 (Function defined piecewise) We will compute the Laplace transform of the following function

$$h(t) = \begin{cases} 3t^2 + 1, & \text{for } 0 \leq t < 5 \\ t^2 + t + 4, & \text{for } t \geq 5 \end{cases} \tag{4.54}$$

Note here that this function is not of the same form as the one in Equation (4.38), however,

$$h(t) = (3t^2 + 1) + (t^2 + t + 4 - (3t^2 + 1))H_5(t),$$

for all but finitely many $t \geq 0$. We then simplify h,

$$h(t) = (3t^2 + 1) + (-2t^2 + t + 3)H_5(t). \tag{4.55}$$

Here we have
$$w(t) = -2t^2 + t + 3. \tag{4.56}$$

We need to find a $\tilde{w}$ so that $w(t) = \tilde{w}(t - 5)$. We proceed as above. First, we determine the derivatives of w, that is

$$w'(t) = 1 - 4t \tag{4.57a}$$
$$w''(t) = -4. \tag{4.57b}$$

Then, we apply Taylor's theorem to w with $t_0 = 5$

$$w(t) = w(5) + (t-5)w'(5) + \frac{(t-5)^2}{2}w''(5)$$
$$= -42 - 19(t-5) - 2(t-5)^2, \tag{4.58}$$

so $\tilde{w}(t) = -42 - 19t - 2t^2$. We now have
$$h(t) = (3t^2 + 1) + (-42 - 19(t-5) - 2(t-5)^2)H_5(t). \quad (4.59)$$

The Laplace transform of h is

$$(\mathscr{L}h)(s) = \left(3\frac{2!}{s^3} + \frac{1}{s}\right) + \left(-42\frac{1}{s} - 19\frac{1}{s^2} - 2\frac{2!}{s^3}\right)e^{-5s}$$
$$= \left(\frac{6}{s^3} + \frac{1}{s}\right) - \left(\frac{42}{s} + \frac{19}{s^2} + \frac{4}{s^3}\right)e^{-5s}, \quad (4.60)$$

for $s \in \mathbb{C}$ with $\mathbf{re}(s) > 0$.

Exercise 4.3 (Case I) Determine the Laplace transform of

$$h(t) = \begin{cases} t^2 + 1, & \text{for } 0 \le t < 2 \\ t^3 + 2t - 1, & \text{for } t \ge 2 \end{cases} \quad (4.61)$$

※ ....

Answer. It is

$$(\mathscr{L}h)(s) = \frac{s^3}{s^2+2} + \frac{e^{-2s}}{s^4}(6s^3 + 10s^2 + 10s + 6). \quad (4.62)$$

Exercise 4.4 (Case II) Find the Laplace transform of the following function

$$h(t) = \begin{cases} 1 + t, & \text{for } 0 \le t < 5 \\ (2 - \tfrac{1}{10}t^3 + \tfrac{1}{6}t)e^{-t}, & \text{for } t > 5 \end{cases} \quad (4.63)$$

※ ....

Answer. It is

$$(\mathscr{L}h)(s) = \frac{s^2}{s+1} - e^{-5s}\left[\frac{s^2}{s+1} - \frac{5s+1}{1} + \frac{s}{(s+1)} - \frac{17e^{-5}}{(s+1)^2} - \frac{6e^{-5}}{9(s+1)^2}\right.$$
$$\left. + \frac{10(s+1)^4}{e^{-5}(1255s^3 + 4505s^2 + 5555s + 236)}\right] \quad (4.64)$$

4.3.3 Functions with multiple branches

The above methodology can be used to compute the Laplace transform of functions defined piecewise with multiple branches, that is, functions of the form

$$f(t) = \begin{cases} f_0(t), & \text{if } 0 \leq t < t_0 \\ f_1(t), & \text{if } t_0 \leq t < t_1 \\ \vdots & \\ f_n(t), & \text{if } t \geq t_{n-1} \end{cases} \quad (4.65)$$

We can write f as follows:

$$f(t) = f_0(t) + (f_1(t) - f_0(t))H_{t_0}(t) + \ldots + (f_n(t) - f_{n-1}(t))H_{t_{n-1}}(t), \quad (4.66)$$

for all $t \geq 0$ except for finitely many.

Example 4.6 (Function with three branches) Consider the following function in the time domain

$$f(t) = \begin{cases} t^2, & \text{if } 0 \leq t < 1 \\ 1-t, & \text{if } 1 \leq t < 2 \\ e^{-3t}, & \text{if } t \geq 2 \end{cases} \quad (4.67)$$

Following Equation (4.66), f can be written as

$$f(t) = t^2 + \underbrace{(1 - t - t^2)}_{w_1(t)} H_1(t) + \underbrace{(e^{-3t} - (1-t))}_{w_2(t)} H_2(t). \quad (4.68)$$

We define $w_1(t) = 1 - t - t^2$ and $w_2(t) = e^{-3t} - (1-t)$. Using the Taylor expansion of w_1 at $t_0 = 1$ (up to order $k = 2$ as in Theorem 3.1) we have

$$w_1(t) = -\left[1 + 3(t-1) + (t-1)^2\right]. \quad (4.69)$$

For w_2 we have

$$\begin{aligned} w_2(t) &= e^{-3t} - 1 + t \\ &= e^{-3(t-2+2)} + t - 2 + 2 - 1 \\ &= e^{-6}e^{-3(t-2)} + (t-2) + 1, \end{aligned} \quad (4.70)$$

therefore,

$$f(t) = t^2 - (1 + 3(t-1) + (t-1)^2)H_1(t) + (e^{-6}e^{-3(t-2)} + (t-2) + 1)H_2(t),$$

so, its Laplace transform is

$$F(s) = \frac{2}{s^3} - e^{-s}\left(\frac{1}{s} + \frac{3}{s^2} + \frac{2}{s^3}\right) \\ + e^{-2s}\left(e^{-6}\frac{1}{s+3} + \frac{1}{s^2} + \frac{1}{s}\right), \quad (4.71)$$

for $s \in \mathbb{C}$ with $\text{re}(s) > 0$.

Chapter 4. Laplace transform

4.3.4 Time scaling

The Laplace transform has the following property:

> **Theorem 4.17 (Time scaling)** Let f be a function in the time domain with Laplace transform F. Then, for $a > 0$,
>
> $$\mathscr{L}\{f(at)\} = 1/a\, F(s/a), \qquad (4.72)$$
>
> provided that s/a is in the domain of F.

Proof. You can prove this theorem as an exercise (hint: use the definition of the Laplace transform). □

> **Exercise 4.5 (Application of time scaling property)** Using only Theorem 4.17 and the fact that
>
> $$\mathscr{L}\{\sin t\}(s) = \frac{1}{s^2+1}, \qquad (4.73)$$
>
> show that for all $\omega > 0$,
>
> $$\mathscr{L}\{\sin(\omega t)\}(s) = \frac{\omega}{s^2+\omega^2}. \qquad (4.74)$$

4.3.5 Translation in the s-domain

The following result allows us to find the Laplace transform of products of exponential functions with functions whose Laplace transform is known. The Laplace transform of products of functions is not always easy to compute[3].

> **Theorem 4.18 (Translation in s domain)** Let f be a function in the time domain with Laplace transform F. Then,
>
> $$\mathscr{L}\{e^{-at}f(t)\} = F(s+a), \qquad (4.75)$$
>
> for $s \in \mathbb{C}$ such that $s+a$ is in the domain of F.

Proof. This can be proven as an exercise using the definition of the Laplace transform. □

---

[3]The Laplace transform of the product of two functions, or even the square of a function, can be extremely difficult to compute. There are certain exceptions: (i) products with exponential functions (Theorem 4.18), (ii) products with t (Theorem 4.26), (iii) products with t^n (Exercise 4.10), (iv) products with $1/t$ (Theorem 4.27) or $1/t^\nu$.

Let us give an example of how we can use Theorem 4.18. Consider the time-domain function $f(t) = e^{-t}\sin 3t$. Then,

$$\mathscr{L}\{e^{-t}\sin 3t\} = \frac{3}{(s+1)^2 + 3^2} = \frac{3}{(s+1)^2 + 9}. \tag{4.76}$$

4.3.6 First-order derivatives

We shall now state the most important result in the chapter so far:

> **Theorem 4.19 (Laplace transform of derivative)** Let f be a time domain function with derivative $f'{}^{a}$. Suppose that f' is piecewise continuous and f is of exponential order γ. Then the Laplace transform of f' exists and is given by
>
> $$(\mathscr{L}\{f'\})(s) = sF(s) - f(0^+), \tag{4.77}$$
>
> for $s \in \mathbb{C}$ with $\mathbf{re}(s) > \gamma$.
>
> ---
> [a] We can in fact assume that f' is defined over $[0, \infty)$ with the exception of a set of countably many points.

Proof. By definition

$$\mathscr{L}\{f'\}(s) = \int_0^\infty e^{-s\tau} f'(\tau) \mathrm{d}\tau,$$

for all $s \in \mathbb{C}$ with $\mathbf{re}(s) > \gamma$. Using integration by parts we have

$$\mathscr{L}\{f'\}(s) = e^{-s\tau} f(\tau)\Big|_0^\infty - \int_0^\infty -se^{-s\tau} f(\tau)\mathrm{d}\tau, \tag{4.78}$$

where the first term is meant in the sense[4]

$$e^{-s\tau} f(\tau)\Big|_0^\infty = \lim_{\tau \to \infty} e^{-s\tau} f(\tau) - \lim_{\tau \to 0^+} e^{-s\tau} f(\tau) = -f(0^+).$$

The existence of the first limit, $\lim_{\tau \to \infty} e^{-s\tau} f(\tau)$, is a consequence of Lemma 4.4 (this is why we need the condition $\mathbf{re}(s) > \gamma$). As a result, from Equation (4.78) we have

$$\mathscr{L}\{f'\}(s) = s\int_0^\infty e^{-s\tau} f(\tau)\mathrm{d}\tau - f(0^+) = sF(s) - f(0^+).$$

This completes the proof. $\square$

---

[4] The integration by parts formula, namely $\int_a^b fg\,\mathrm{d}\tau = fg\big|_a^b - \int_a^b fg'\mathrm{d}\tau$, requires that both f and g be *continuous* over the closed interval $[a, b]$. This may not be the case here — function $e^{-s\tau} f(\tau)$ might be discontinuous (it is only assumed to be piecewise continuous). This is what motivates taking the limit as $\tau \to 0^+$.

Chapter 4. Laplace transform

Example 4.7 (ODEs to algebraic equations) The Laplace transform of the derivative is a powerful tool, and we will be using it heavily throughout this book. Let us have a look at an example. The kinematics of a horizontal spring-mass system is described by

$$\dot{v}(t) = g + \tfrac{1}{m}f(t) - \tfrac{k}{m}x(t), \tag{4.79a}$$
$$\dot{x}(t) = v(t), \tag{4.79b}$$

where x is the displacement of the mass from its equilibrium position, v is its velocity, f is an externally applied force, k is the spring constant and g is the acceleration of gravity.

Let X, V, F be the Laplace transforms of x, v and f respectively. By applying the Laplace transform on both sides of (4.79), we obtain

$$sV(s) - v(0) = \tfrac{g}{s} + \tfrac{1}{m}F(s) - \tfrac{k}{m}X(s), \tag{4.80a}$$
$$sX(s) - x(0) = V(s). \tag{4.80b}$$

For now, we will proceed no further. It is worth noting that we have transformed the **differential equations** in (4.79) to **algebraic equations** in (4.80)! The latter will be **much easier** to work with. We start to understand the reason why we introduced the Laplace transform in the first place.

4.3.7 Higher-order derivatives

The results we stated above for the first-order derivative can be generalised to higher-order derivatives. We state the following result:

> **Theorem 4.20 (Second-order derivative)** Let f be a twice differentiable time domain function. Suppose that f'' is piecewise continuous and f' is of exponential order γ. Then the Laplace transform of f'' exists, and it is given by
>
> $$(\mathscr{L}f'')(s) = s^2 F(s) - sf(0^+) - f'(0^+), \tag{4.81}$$
>
> which is defined for $s \in \mathbb{C}$ with $\mathbf{re}(s) > \gamma$.

Proof. We simply need to apply Theorem 4.19 to f'. □

Example 4.8 (Higher-order system) The kinematics in Equation (4.79) can be written as

$$\ddot{x}(t) = g + \tfrac{1}{m}f(t) - \tfrac{k}{m}x(t), \tag{4.82}$$

where $\ddot{x}(t)$ is the second derivative of the displacement at time t, that is, the acceleration of the mass. We apply the Laplace transform on both sides and apply Theorem 4.20

$$s^2 X(s) - sx(0) - \dot{x}(0) = \tfrac{g}{s} + \tfrac{1}{m}F(s) - \tfrac{k}{m}X(s). \tag{4.83}$$

Again, this is an algebraic equation that links the control action, F, with the system response, X, in the s-domain.

The above result can be generalised to higher-order derivatives under analogous assumptions and can be proven by induction.

Theorem 4.21 (Higher-order derivatives) Let f be an n times differentiable function in the time domain. Suppose that $f^{(n)}$ is piecewise continuous over $[0, T]$ for all $T > 0$ and that $f^{(n-1)}$ is of exponential order γ. Then the Laplace transform of $f^{(n)}$ exists and is given by

$$(\mathscr{L} f^{(n)})(s) = s^n F(s) - s^{n-1} f(0^+) - \\ \ldots - s f^{(n-2)}(0^+) - f^{(n-1)}(0^+), \tag{4.84}$$

which is defined for $s \in \mathbb{C}$ with $\mathbf{re}(s) > \gamma$.

Proof. The proof is left to the reader as an exercise. □

4.3.8 Time integrals

Example 4.7 has offered some understanding of what we are up to with the Laplace transform: it allows us to write differential equations as algebraic ones. We will see later that several properties of interest of the original differential equations can be studied by means of their Laplace transforms — their algebraic counterparts.

But what about integro-differential equations? We saw in Chapter 2 that certain physical systems give rise to integro-differential equations. Similarly to the derivative of a function, the integral has also an easily computable integral.

Chapter 4. Laplace transform

> **Theorem 4.22 (Laplace transform of integral)** Let f be a piecewise continuous time domain function of exponential order γ and Laplace transform $F(s) = (\mathscr{L}f)(s)$. Define $I(t) = \int_0^t f(\tau)\mathrm{d}\tau$. Then, I has a Laplace transform and
>
> $$(\mathscr{L}I)(s) = \frac{F(s)}{s}, \qquad (4.85)$$
>
> with $s \neq 0$ and $\mathrm{re}(s) > \gamma$.

Proof. By definition, I is a continuous function in t. As an exercise, show that I has exponential order γ. As a result, by Theorem 4.5, I has a Laplace transform which is

$$\begin{aligned}(\mathscr{L}I)(s) &= \int_0^\infty e^{-s\tau} \int_0^\tau f(\xi)\mathrm{d}\xi \mathrm{d}\tau \\ &= \int_0^\infty (-\tfrac{1}{s}e^{-s\tau})' \int_0^\tau f(\xi)\mathrm{d}\xi \mathrm{d}\tau \\ &= -\tfrac{1}{s}e^{-s\tau} \int_0^\tau f(\xi)\mathrm{d}\xi \Big|_0^\infty - \int_0^\infty -\tfrac{1}{s}e^{-s\tau} f(\tau) \mathrm{d}\tau \qquad (4.86)\end{aligned}$$

Using the fact that I is sub-exponential, yields $(\mathscr{L}I)(s) = F(s)/s$. $\square$

Example 4.9 (Laplace transform of integral) Let us find the Laplace transform

$$\mathscr{L}\left\{\int_0^t e^{-a\tau} \cos(\omega\tau)\mathrm{d}\tau\right\}. \qquad (4.87)$$

This is the Laplace transform of an integral, so we should be looking at Theorem 4.22. Function $f(t) = e^{-at}\cos(\omega t)$ satisfies the assumptions of the theorem (you can verify this as an exercise). Then,

$$\begin{aligned}\mathscr{L}\left\{\int_0^t e^{-a\tau} \cos(\omega\tau)\mathrm{d}\tau\right\}(s) &= \frac{1}{s}\mathscr{L}\{e^{-at}\cos(\omega t)\}(s) \\ &= \frac{1}{s}\frac{s+a}{(s+a)^2 + \omega^2} = \frac{s+a}{s[(s+a)^2 + \omega^2]}\end{aligned}$$

In the second equality we used the Laplace transform of cosine (see Exercise 4.2) and Theorem 4.18. The above Laplace transform is defined for $\mathrm{re}(s) > -a$ (see Theorem 4.22).

Exercise 4.6 (Double integral) Let f be a piecewise continuous function in the time domain which has exponential order γ and Laplace transform $F(s) = (\mathscr{L}f)(s)$. Define

$$J(t) = \int_0^t \int_0^\tau f(\xi) d\xi d\tau. \tag{4.88}$$

Show that J has a Laplace transform given by $(\mathscr{L}J)(s) = \frac{1}{s^2}F(s)$ for all $s \in \mathbb{C}$ with $\mathbf{re}(s) > \gamma$.

✂ ....

Hint: Apply Theorem 4.22 to the function $\int_0^t f(\xi) d\xi$; pay attention to the fact that you have to show that the domain of $\mathscr{L}J$ is all $s \in \mathbb{C}$ with $\mathbf{re}(s) > \gamma$. See also Exercise 4.27.

4.4 Finer properties of the Laplace transform

4.4.1 Periodic functions

Let us recall the definition of a periodic function

> **Definition 4.23** (**Periodic function**) We say that a function $f : [0, \infty) \to \mathbb{R}$ is periodic with period $T > 0$ if
>
> $$f(t + T) = f(t), \tag{4.89}$$
>
> for all $t \in \mathbb{R}$.

Typical examples of periodic functions are: (i) the trigonometric functions $\sin t$ and $\cos t$, with period $T = 2\pi$, (ii) the square wave with period T defined as[5] $x(t) = (-1)^{\lfloor 2t/T \rfloor}$, and (iii) the triangle wave with period $T = 2$ given by $x(t) = \frac{2}{\pi} \sin^{-1}(\sin(\pi t))$. The Laplace transform of periodic functions can be computed using the following result

> **Theorem 4.24** (**Laplace transform of periodic functions**) Let $f : [0, \infty) \to \mathbb{R}$ be a piecewise continuous function and periodic with period $T > 0$. Then, if its Laplace transform exists, it is given by
>
> $$F(s) = \frac{1}{1 - e^{-sT}} \int_0^T e^{-s\tau} f(\tau) d\tau. \tag{4.90}$$

Proof. Function f satisfies the requirements of Theorem 4.5, so its Laplace

---

[5] The *floor function* of a real number x is denoted as $\lfloor x \rfloor$, and it is the largest integer that is no larger than x. For example, $\lfloor 2.1 \rfloor = 2$, and $\lfloor 3.9 \rfloor = 3$.

transform exists. By invocation of the dominated convergence theorem

$$(\mathscr{L}f)(s) = \int_0^\infty e^{-s\tau} f(\tau) d\tau = \sum_{k=0}^\infty \int_{kT}^{(k+1)T} e^{-s\tau} f(\tau) d\tau. \tag{4.91}$$

We use the change of variables $\tau = \xi + kT$:

$$\int_{kT}^{(k+1)T} e^{-s\tau} f(\tau) d\tau = \int_0^T e^{-s(\xi+nT)} f(\xi + nT) d\xi$$

$$= e^{-skT} \int_0^T e^{-s\tau} f(\tau) d\tau, \tag{4.92}$$

where the last equality is because of the periodicity of f. Therefore,

$$(\mathscr{L}f)(s) = \left(\sum_{k=0}^\infty e^{-skT}\right) \int_0^T e^{-s\tau} f(\tau) d\tau = \frac{1}{1 - e^{-sT}} \int_0^T e^{-s\tau} f(\tau) d\tau.$$

This completes the proof. □

Example 4.10 (Application of Theorem 4.24) We already know that $\mathscr{L}\{\sin t\}(s) = \frac{1}{s^2+1}$. Function $f(t) = \sin t$ is periodic with period $T = 2\pi$. Let us (re)derive its Laplace transform using Theorem 4.24. We have

$$\mathscr{L}\{\sin t\}(s) = \frac{1}{1 - e^{-2\pi s}} \int_0^{2\pi} e^{-s\tau} \sin \tau d\tau$$

$$= \frac{1}{1 - e^{-2\pi s}} \frac{1 - e^{-2\pi s}}{s^2 + 1}$$

$$= \frac{1}{s^2 + 1}. \tag{4.93}$$

We leave it to the reader to verify that $\int_0^{2\pi} e^{-s\tau} \sin \tau d\tau = \frac{1-e^{-2\pi s}}{s^2+1}$.

Let us now have a look at a more interesting example: the Laplace transform of a square wave.

Example 4.11 (Laplace transform of square wave) We shall determine the Laplace transform of

$$f(t) = \begin{cases} 1, & \text{for } 2nT \le t < (2n+1)T \\ -1, & \text{for } (2n+1)T \le t < (2n+2)T \end{cases} \tag{4.94}$$

with $n \in \mathbb{N}$. Function f is periodic with period $2T$. By Theorem

4.24, the Laplace transform of f exists and is given by

$$F(s) = \frac{1}{1-e^{-2Ts}} \int_0^{2T} e^{-s\tau} f(\tau) d\tau$$

$$= \frac{1}{1-e^{-2Ts}} \left(\int_0^T e^{-s\tau} d\tau - \int_T^{2T} e^{-s\tau} d\tau \right)$$

$$= \frac{1}{1-e^{-2Ts}} \left(-\frac{1}{s} e^{-s\tau} \Big|_0^T + \frac{1}{s} e^{-s\tau} \Big|_T^{2T} \right)$$

$$= \frac{1}{1-e^{-2Ts}} \frac{1}{s} (1 - 2e^{-Ts} + 2e^{-2Ts})$$

$$= \frac{1}{1-e^{-2Ts}} \frac{1}{s} (1 - e^{-Ts})^2$$

$$= \frac{1}{(1-e^{-Ts})(1+e^{-Ts})} \frac{1}{s} (1 - e^{-Ts})^2$$

$$= \frac{1}{s} \frac{e^{Ts} - 1}{e^{Ts} + 1} = \frac{1}{s} \tanh(Ts/2). \qquad (4.95)$$

We shall now restate Theorem 4.24 in an equivalent way which, however, will turn out to be more convenient

Theorem 4.25 (Laplace transform of periodic functions, v2) Let f and T be as in Theorem 4.24. Then,

$$F(s) = \frac{\mathscr{L}\{f_0(t)\}}{1 - e^{-sT}}, \qquad (4.96)$$

where $f_0(t) = f(t) \cdot (1 - H_T(t))$.

Proof. From Theorem 4.24, we have

$$F(s) = \frac{1}{1-e^{-sT}} \int_0^T e^{-s\tau} f(\tau) d\tau$$

$$= \frac{1}{1-e^{-sT}} \int_0^\infty e^{-s\tau} f_0(\tau) d\tau = \frac{\mathscr{L}\{f_0(t)\}}{1-e^{-sT}}, \qquad (4.97)$$

which proves the assertion. □

This theorem allows us to determine the Laplace transform of a T-periodic function f if we can determine the Laplace transform of $f_0(t)$. Note that $f_0(t)$ is equal to $f(t)$ over $[0,T]$ and is equal to zero for $t > T$, that is

$$f_0(t) = \begin{cases} f(t), & \text{for } 0 \leq t \leq T, \\ 0, & \text{for } t > T \end{cases} \qquad (4.98)$$

Let us now solve again Example 4.11 using this second version of the theorem.

Example 4.12 (**Example 4.11, second approach**) The square wave with period $2T$ is equal to $1 - 2H_T(t)$ for $t \in [0, 2T]$. Then,

$$\begin{aligned}
f_0(t) &= (1 - 2H_T(t))(1 - H_{2T}(t)) \\
&= 1 - H_{2T}(t) - 2H_T(t) + 2H_T(t)H_{2T}(t) \\
&= 1 - H_{2T}(t) - 2H_T(t) + 2H_{2T}(t) \\
&= 1 + H_{2T}(t) - 2H_T(t).
\end{aligned} \tag{4.99}$$

The Laplace transform of $f_0(t)$ is

$$\begin{aligned}
(\mathscr{L}f_0)(s) &= \frac{1}{s} + \frac{e^{-2Ts}}{s} - 2\frac{e^{-Ts}}{s} \\
&= \frac{1 + e^{-2Ts} - 2e^{-Ts}}{s} \\
&= \frac{(1 - e^{-Ts})^2}{s},
\end{aligned} \tag{4.100}$$

so from Theorem 4.25,

$$\begin{aligned}
(\mathscr{L}f)(s) &= \frac{1}{1 - e^{-2Ts}} \frac{(1 - e^{-Ts})^2}{s} \\
&= \frac{1}{(1 - e^{-Ts})(1 + e^{-Ts})} \frac{(1 - e^{-Ts})^2}{s} \\
&= \frac{1}{s} \tanh(Ts/2),
\end{aligned} \tag{4.101}$$

which is the same as the result we obtained previously in Example 4.11.

One example is never enough. Here is another one where, again, there is no need to determine any integrals:

Example 4.13 (**Laplace transform of rectified sine wave**) We want to determine the Laplace transform of

$$f(t) = |\sin(t)|, \tag{4.102}$$

defined for $t \geq 0$. This is a periodic function with period $T = \pi$ (this is left to the reader to verify). Then,

$$\begin{aligned}
f_0(t) &= |\sin(t)|(1 - H_\pi(t)) \\
&= \sin(t)(1 - H_\pi(t)) \\
&= \sin(t) - \sin(t)H_\pi(t).
\end{aligned} \tag{4.103}$$

In order to determine the Laplace transform of f_0 we employ Theorem 4.15, from which

$$\begin{aligned}(\mathscr{L}f_0)(s) &= \mathscr{L}\{\sin(t)\} - \mathscr{L}\{\sin(t)H_\pi(t)\} \\ &= \mathscr{L}\{\sin(t)\} + \mathscr{L}\{\sin(t-\pi)H_\pi(t)\} \\ &= \frac{1}{s^2+1} + \frac{e^{-\pi s}}{s^2+1} \\ &= \frac{1+e^{-\pi s}}{s^2+1}.\end{aligned} \qquad (4.104)$$

We also used the fact that $\sin\theta = -\sin(\theta-\pi)$ for all θ. Finally, from Theorem 4.25

$$(\mathscr{L}|\sin t|)(s) = \frac{1}{1-e^{-\pi s}} \frac{1+e^{-\pi s}}{s^2+1}. \qquad (4.105)$$

If you like, you can write this result in terms of the hyperbolic tangent or hyperbolic cotangent function.

Exercise 4.7 (**Laplace transform of rectified cosine wave**) Find the Laplace transform of $f(t) = |\cos t|$, $t \geq 0$.

✻ ....

Hint: **i.** Note that $f(t) = |\cos t|$ is π-periodic and define $f_0(t) = |\cos t|(1 - H_\pi(t))$. **ii.** Note that f_0 can be written as $f_0(t) = \cos(t)(1 - H_{\pi/2}(t)) - 2H_{\pi/2}(t) + H_\pi(t))$. **iii.** Use the trigonometric identities $\cos t = -\sin(t-\pi/2)$ and $\cos t = -\cos(t-\pi)$. **iv.** Use Equation (4.42) to determine the Laplace transform of f_0. **v.** Apply Theorem 4.25 to determine the Laplace transform of f.

Answer: It is

$$(\mathscr{L}f)(s) = \frac{1}{1-e^{-\pi s}} \cdot \frac{s(1-e^{-\pi s}) + 2e^{-\frac{\pi}{2}s}}{s^2+1}.\qquad(4.106)$$

This formula can be simplified using the definition of the hyperbolic cosecant function, $\operatorname{csch} x = \frac{2}{e^x - e^{-x}}$. We then have

$$(\mathscr{L}f)(s) = \frac{s + \operatorname{csch}(\pi s/2)}{s^2+1}. \qquad (4.107)$$

✂ ....

Exercise 4.8 (**Laplace transform of sawtooth wave**) The sawtooth wave is given by

$$f(t) = t - \lfloor t \rfloor, \qquad (4.108)$$

for $t \geq 0$. Show that f is periodic. What is the period of f? Determine its Laplace transform. Lastly, determine the Laplace transform of a sawtooth waveform with period T.

✳ ....

Answer. The sawtooth function, f, is 1-periodic. Its Laplace transform is

$$(\mathscr{L}f)(s) = \frac{1 - e^{-s} - se^{-s}}{s^2(1 - e^{-s})}. \qquad (4.109)$$

✂ ....

Hint. Note that the *floor function*, $\lfloor t \rfloor$, gives the largest integer that does not exceed t. For example $\lfloor 2.1 \rfloor = 2$, and $\lfloor 2.999 \rfloor = 2$.

Exercise 4.9 (**Laplace transform of sine half-wave**) Find the Laplace transform of $f(t) = \max\{0, \sin(\omega t)\}$.

✳ ....

Answer. It is

$$(\mathscr{L}f)(s) = \frac{\omega}{(s^2 + \omega^2)(1 + e^{-\frac{\pi}{\omega}s})}. \qquad (4.110)$$

4.4.2 Derivative and integral of Laplace transform

Earlier we mentioned that the Laplace transform is an infinitely many times differentiable function. The following theorem gives the derivative of the Laplace transform and at the same time allows us to determine the Laplace transform of functions of the form $tf(t)$.

Theorem 4.26 (**Derivative of Laplace transform**) Let $f : [0, \infty) \to \mathbb{R}$ be a piecewise continuous function of exponential order γ. Suppose that f has a Laplace transform $F(s) = (\mathscr{L}f)(s)$. Then, F is differentiable, $tf(t)$ has a Laplace transform and

$$F'(s) = \mathscr{L}\{-tf(t)\}, \qquad (4.111)$$

holds for all $s \in \mathbb{C}$ with $\mathbf{re}(s) > \gamma$.

Proof. We should firstly note that under the assumption that f is piecewise continuous and of exponential order γ, function $tf(t)$ is piecewise continuous and of exponential order $\gamma + c$ for any $c > 0$. For simplicity and without loss of generality, we shall hereafter assume that $tf(t)$ is continuous. Using the definition of the Laplace transform (Definition 4.2), the derivative of F is

$$\frac{\mathrm{d}}{\mathrm{d}s}F(s) = \frac{\mathrm{d}}{\mathrm{d}s}\int_0^\infty e^{-s\tau}f(\tau)\mathrm{d}\tau. \qquad (4.112)$$

Note that we cannot apply Leibniz's rule (Theorem A.17) to interchange the derivative and the integral because the integral is an improper one. We should rather write

$$\frac{d}{ds}F(s) = \frac{d}{ds}\lim_{\nu\to\infty}\int_0^\nu e^{-s\tau}f(\tau)d\tau. \qquad (4.113)$$

In order to interchange the derivative and the limit we use the following fact without a proof[6]

> Let $F_\nu : A \to \mathbb{C}$, where $A \subseteq \mathbb{C}$ is an open set, be a sequence of holomorphic functions that converges locally uniformly (see Definition A.12) to a function F. Then F is holomorphic and F'_n converges locally uniformly to F' as $\nu \to \infty$.

To use the above fact, let us define $F_\nu(s) = \int_0^\nu e^{-s\tau}f(\tau)d\tau$. We may use Leibniz's rule[7] to determine the derivative of F_ν because $e^{-s\tau}f(\tau)$ is continuous in (s,τ) and its derivative, $\frac{d}{ds}e^{-s\tau}f(\tau) = -\tau e^{-s\tau}f(\tau)$, is continuous in (s,τ). By virtue of Theorem A.17,

$$F'_\nu(s) = \int_0^\nu -\tau e^{-s\tau}f(\tau)d\tau. \qquad (4.114)$$

We need to show that $F'_\nu \to \int_0^\infty -\tau e^{-s\tau}f(\tau)d\tau$ *uniformly* as $\nu \to \infty$ over $s \in \mathbb{C}$ with $\mathbf{re}(s) > \gamma$. According to the definition of uniform convergence, we need to show that for every $\epsilon > 0$, there is a $T_0 > 0$ such that for all $T > T_0$ and s

$$\left|\int_T^\infty \frac{\partial}{\partial s}e^{-s\tau}f(\tau)d\tau\right| \equiv \left|\int_T^\infty \tau e^{-s\tau}f(\tau)d\tau\right| < \epsilon. \qquad (4.115)$$

The reader can verify that this holds for $s \in \mathbb{C}$ with $\mathbf{re}(s) \geq \gamma + c$ for any $c > 0$ (using the exponential order of $tf(t)$). Then,

$$\begin{aligned}\frac{d}{ds}F(s) &= \frac{d}{ds}\lim_\nu F_\nu(s) = \lim_\nu \frac{d}{ds}F_\nu(s) \\ &= \lim_\nu \int_0^\nu e^{-s\tau}(-\tau)f(\tau)d\tau \\ &= \int_0^\infty e^{-s\tau}(-\tau)f(\tau)d\tau \\ &= \mathscr{L}\{(-t)f(t)\}(s), \qquad (4.116)\end{aligned}$$

for $s \in \mathbb{C}$ with $\mathbf{re}(s) > \gamma$. This completes the proof. □

---

[6] The proof can be found on ProofWiki, entry "Derivative of Uniform Limit of Analytic Functions" as well as in [M14, Theorem 1.2].

[7] In fact, we use Leibniz's rule for the complex function F_ν defined for $s \in \mathbb{C}$ with $\mathbf{re}(s) > \gamma$. This is Lemma II.3.3 in [M11].

Chapter 4. Laplace transform

Exercise 4.10 (Higher-order derivatives of Laplace transform) Let f be a function in the time domain with Laplace transform $F(s) = (\mathscr{L}f)(s)$. Find the second derivative of the Laplace transform. Find the n-th order derivative of F.

※ ....

Answer. It is $F^{(n)}(s) = \mathscr{L}\{(-t)^n f(t)\}(s)$.

✂ ....

Hint. (i) Check that the conditions of the above theorem are satisfied, i.e., show that if f has a Laplace transform, then $tf(t), t^2 f(t), t^3 f(t), \ldots$ also have a Laplace transform, and (ii) Apply Theorem 4.26 recursively.

Example 4.14 (Laplace transform of $tf(t)$) Let us compute the Laplace transform of
$$f(t) = t\cos(\omega t). \tag{4.117}$$

According to Theorem 4.26,

$$\begin{aligned}\mathscr{L}\{-t\cos(\omega t)\} &= \tfrac{\mathrm{d}}{\mathrm{d}s}\mathscr{L}\{\cos(\omega t)\}(s) \\ &= \tfrac{\mathrm{d}}{\mathrm{d}s}\frac{s}{s^2+\omega^2} \\ &= \frac{\omega^2 - s^2}{(s^2+\omega^2)^2}. \end{aligned} \tag{4.118}$$

Because of the linearity of the Laplace transform (Theorem 4.14), $\mathscr{L}\{-t\cos(\omega t)\} = -\mathscr{L}\{t\cos(\omega t)\}$, therefore,

$$\mathscr{L}\{t\cos(\omega t)\} = \frac{s^2 - \omega^2}{(s^2+\omega^2)^2}. \tag{4.119}$$

Theorem 4.26 can be used to find the Laplace transform of functions of the form $f(t)/t$ when the Laplace transform of f is known. The result is stated below:

Theorem 4.27 (Integral of Laplace transform) Let $f : [0, \infty) \to \mathbb{R}$ have a Laplace transform $F(s) = (\mathscr{L}f)(s)$ and suppose that $f(t)/t$ has a Laplace transform[a]. Then,

$$\mathscr{L}\{\tfrac{1}{t}f(t)\}(s) = \int_s^\infty F(\xi)\mathrm{d}\xi. \tag{4.120}$$

[a] For example, this is the case if f is piecewise continuous and $\lim_{t\to 0+} f(t)/t$ exists and is finite.

Proof. Let us define $h(t) = f(t)/t$ and apply Theorem 4.26. We have

$$F(s) = (\mathscr{L}f)(s) = (\mathscr{L}\{t\tfrac{f(t)}{t}\})(s) = (\mathscr{L}\{th(t)\}) = -\frac{\mathrm{d}H(s)}{\mathrm{d}s},$$

so we have the differential equation $\frac{\mathrm{d}}{\mathrm{d}s}H(s) = -G(s)$, whose integral representation is

$$H(s) = \eta - \int_c^s F(\xi)\mathrm{d}\xi, \qquad (4.121)$$

for some constants η and $c > 0$. In order to determine η and c we use condition (ii) in Theorem 4.6, that is H must satisfy $\lim_{s \to \infty} H(s) = 0$, therefore

$$\eta - \lim_{s \to \infty} \int_c^s F(\xi)\mathrm{d}\xi = 0$$

$$\Leftrightarrow \eta - \int_c^\infty F(\xi)\mathrm{d}\xi = 0 \Leftrightarrow \eta = \int_c^\infty F(\xi)\mathrm{d}\xi, \qquad (4.122)$$

Therefore,

$$H(s) = \int_c^\infty F(\xi)\mathrm{d}\xi - \int_c^s F(\xi)\mathrm{d}\xi = \int_s^\infty F(\xi)\mathrm{d}\xi. \qquad (4.123)$$

This completes the proof. $\square$

Example 4.15 (Laplace transform of $\sin(\omega t)/t$) Let us determine the Laplace transform of $f(t) = \sin(\omega t)/t$ with $\omega > 0$. We know that (see Section 4.2.5) $\mathscr{L}\{\sin(\omega t)\}(s) = \frac{\omega}{s^2+\omega^2}$ for all $s \in \mathbb{C}$ with $\mathbf{re}(s) > 0$. By virtue of Theorem 4.26,

$$\mathscr{L}\left\{\frac{\sin(\omega t)}{t}\right\}(s) = \int_s^\infty \frac{\omega}{\xi^2 + \omega^2}\mathrm{d}\xi$$

$$= \int_s^\infty \frac{\frac{1}{\omega}}{(\frac{\xi}{\omega})^2 + 1}\mathrm{d}\xi$$

$$= \frac{\pi}{2} - \arctan\left(\frac{s}{\omega}\right)$$

$$= \arctan\left(\frac{\omega}{s}\right). \qquad (4.124)$$

Note. Recall that $\int \frac{1}{x^2+1}\mathrm{d}x = \arctan(x)$, $\lim_{x \to +\infty} \arctan(x) = \pi/2$ and that $\frac{\pi}{2} - \arctan\phi = \arctan\frac{1}{\phi}$.

Exercise 4.11 (Laplace transform of sine integral) Find the Laplace transform of the following function which is known as the **sine in-**

tegral

$$\mathrm{Si}(t) = \int_0^t \frac{\sin(\tau)}{\tau}\,d\tau. \tag{4.125}$$

Hint. Use the result obtained in Example 4.15.

4.4.3 Asymptotic analysis

The initial and final value theorems allow us to compute $\lim_{t \to 0^+} f(t)$ and $\lim_{t \to +\infty} f(t)$ using the Laplace transform of f. The final value theorem will prove a valuable tool for the analysis of control systems as we will often need to find these limits without knowing f explicitly.

> **Theorem 4.28 (Initial value theorem)** Let f be a function in the time domain with Laplace transform $F(s) = (\mathscr{L}f)(s)$. Suppose that f is piecewise continuous and of exponential order γ. Then, $\lim_{t \to 0^+} f(t)$ exists and is given by
>
> $$\lim_{t \to 0^+} f(t) = \lim_{s \to \infty} sF(s). \tag{4.126}$$
>
> The limit $\lim_{s \to \infty} sF(s)$ is taken with $s \in \mathbb{R}$.

Proof. The existence of the Laplace transform of f is guaranteed by Theorem 4.5. We observe that

$$sF(s) = s\int_0^\infty f(\tau)e^{-s\tau}\,d\tau = \int_0^\infty sf(\tau)e^{-s\tau}\,d\tau, \tag{4.127}$$

by introducing the variable $y = s\tau$, this becomes

$$sF(s) = \int_0^\infty f\left(\frac{y}{s}\right)e^{-y}\,dy. \tag{4.128}$$

Define the function $f_s(y) = f\left(\frac{y}{s}\right)e^{-y}$. Since f is of exponential order γ, there is an $M > 0$ and a $t_0 > 0$ such that

$$|f_s(y)| = \left|f\left(\frac{y}{s}\right)e^{-y}\right| \leq Me^{\gamma \frac{y}{s}}e^{-y} = M\exp\left(y(\tfrac{\gamma}{s}-1)\right), \tag{4.129}$$

for all $y \geq t_0$. Choose an $\alpha > 0$; then for all $s > \frac{\gamma}{1-\alpha}$, $|f_s(y)| \leq M\exp(-\alpha y)$, so the conditions of the dominated convergence theorem apply (see Theorem A.20). We apply the dominated convergence theorem and take $s \to \infty$:

$$\lim_{s \to \infty} sF(s) = \lim_{s \to \infty} \int_0^\infty f\left(\frac{y}{s}\right)e^{-y}\,dy$$

$$= \int_0^\infty \lim_{s \to \infty} f\left(\frac{y}{s}\right) e^{-y} dy$$
$$= \int_0^\infty f(0^+) e^{-y} dy = f(0^+). \tag{4.130}$$

This completes the proof. □

Exercise 4.12 (Alternative IVT) We can prove the initial value theorem under a different set of assumptions. Let f be a function in the time domain with Laplace transform $F(s) = (\mathscr{L}f)(s)$. Suppose that the following conditions are satisfied:

1. the conditions of Theorem 4.19 are satisfied

2. $\lim_{t \to 0^+} f(t)$ exists and is finite,

then,
$$\lim_{t \to 0^+} f(t) = \lim_{s \to \infty} sF(s), \tag{4.131}$$

where the limit $\lim_{s \to \infty} sF(s)$ is taken with $s \in \mathbb{R}$.

✂ ....

Hint. Consider following these steps:

1. *Apply Theorem 4.19.*
2. *Take the limit as $s \to \infty$ on the equation you derived in the previous step*
3. *Use Theorem 4.6(ii).*

It is tempting to apply the initial value theorem in cases where we have at our disposal the Laplace transform of a time-domain function, but not the function itself. Indeed, it is possible to use the theorem to determine $\lim_{t \to 0^+} f(t)$ using $F(s)$. However, there are a few caveats we need to keep in mind:

1. If $\lim_{s \to \infty} sF(s)$ is not finite, we should reject the result; we should not conclude that $\lim_{t \to 0^+} f(t)$ is infinite (see Example 4.17),

2. If $sF(s)$ is a proper rational function (i.e., a rational function whose denominator has a larger or equal degree than the numerator) then we can apply the initial value theorem. This is the case when F is a strictly proper rational function (the degree of its denominator is strictly larger than the degree of the numerator).

3. If $sF(s)$ is the product of a proper rational function with an exponential function, for example $F(s) = \frac{s+1}{s^2+4s+1}e^{-5s}$, then we can apply the initial value theorem.

The reason why the initial value theorem can be applied when F is a proper rational function is because in that case f satisfies the conditions of Theorem 4.28. This will become clear in the following examples and in the next chapter.

Example 4.16 (Application of IVT) Let f be a function in the time domain with Laplace transform

$$F(s) = \frac{1}{s+1}. \tag{4.132}$$

Note that this is strictly proper rational function. We compute

$$\lim_{s \to \infty} sF(s) = \lim_{s \to \infty} \frac{s}{1+s} = 1. \tag{4.133}$$

The limit exists, therefore,

$$\lim_{t \to 0^+} f(t) = 1. \tag{4.134}$$

We can see that the said function f is in fact $f(t) = e^{-t}$, and we can verify that $\lim_{t \to 0^+} f(t) = 1$.

Exercise 4.13 (Verify IVT) Verify the initial value theorem for the function $f(t) = A\cos(t)$ with $A \in \mathbb{R}$: find its Laplace transform, verify that the conditions of the theorem are satisfied, verify that $\lim_{t \to 0^+} f(t) = \lim_{s \to \infty} sF(s)$.

Example 4.17 (Conditions of IVT: a counterexample) We need to be careful when applying the initial and final value theorems as their conditions must be satisfied, otherwise we may be led to erroneous conclusions. For example, let f be a function in the time domain with Laplace transform

$$F(s) = \frac{s}{s+1}. \tag{4.135}$$

We compute

$$\lim_{s \to \infty} sF(s) = \lim_{s \to \infty} \frac{s^2}{s+1} = \infty. \tag{4.136}$$

The limit is not finite, therefore, we cannot accept the result. Note also that the function $\frac{s^2}{s+1}$ is not proper and F is not strictly proper.

The initial value theorem can also be employed to find the initial value of the derivative of the state of a system (the initial velocity of the system). An example follows.

Example 4.18 (Initial value of derivative) A dynamical system with state x and input u is described by the following algebraic equation in the s-domain:

$$X(s) = \frac{3s+2}{s^2+5s+3}U(s), \tag{4.137}$$

where $X(s) = (\mathscr{L}x)(s)$ and $U(s) = (\mathscr{L}u)(s)$. The system is excited with a step pulse, that is,

$$u(t) = \begin{cases} 0, & \text{for } t = 0 \\ 1, & \text{for } t > 0 \end{cases} \tag{4.138}$$

for which $U(s) = 1/s$ for all $s \in \mathbb{C}$ with $\operatorname{re}(s) > 0$. We are interested in determining its *initial velocity*, that is $\dot{x}(0^+)$. To that end, we will use the initial value theorem. We will assume that $x(0^+) = 0$. The Laplace transform of the derivative of x is

$$(\mathscr{L}\dot{x})(s) = sX(s). \tag{4.139}$$

According to the initial value theorem,

$$\dot{x}(0^+) = \lim_{t \to 0^+} \dot{x}(t) = \lim_{s \to \infty} s(\mathscr{L}\dot{x})(s)$$

$$= \lim_{s \to \infty} s^2 X(s) = \lim_{s \to \infty} \frac{\not{s^2}(3s+2)}{\not{s}(s^2+5s+3)}$$

$$= \lim_{s \to \infty} \frac{3+\frac{2}{s}}{1+\frac{5}{s}+\frac{3}{s^2}} = 3. \tag{4.140}$$

We will now state a result that describes the behaviour of a function f at infinity in terms of its Laplace transform; this is known as the final value Theorem (FVT).

Theorem 4.29 (Final value theorem) Let f be a function in the time domain with Laplace transform $F(s) = (\mathscr{L}f)(s)$. Suppose that the following conditions are satisfied:

1. f is bounded,

2. $\lim_{t \to \infty} f(t)$ exists and is finite,

then,

$$\lim_{t \to \infty} f(t) = \lim_{s \to 0^+} sF(s), \tag{4.141}$$

where the limit is taken over $s \in \mathbb{R}$.

Proof. Similar to the proof of the initial value theorem (see Theorem 4.28),

$$sF(s) = \int_0^\infty f\left(\frac{y}{s}\right) e^{-y} dy. \tag{4.142}$$

Define the function $f_s(y) = f\left(\frac{y}{s}\right) e^{-y}$. Since f is assumed to be bounded, there is an $M > 0$ such that $|f(t)| \leq M$ for all $t \geq 0$, therefore, $|f_s(y)| \leq Me^{-y}$. As a result, we can apply the dominated convergence theorem (Theorem A.20):

$$\lim_{s \to 0^+} sF(s) = \lim_{s \to 0^+} \int_0^\infty f\left(\frac{y}{s}\right) e^{-y} dy$$

$$\stackrel{\text{dct}}{=} \int_0^\infty \lim_{s \to 0^+} f\left(\frac{y}{s}\right) e^{-y} dy = \lim_{t \to \infty} f(t). \tag{4.143}$$

This completes the proof. □

The converse of FVT is not true, i.e., if $\lim_{s \to 0^+} sF(s)$ exists and is finite, this does not mean that $\lim_{t \to \infty} f(t)$ is finite (see Example 4.20).

The final value theorem requires that the limit $\lim_{t \to \infty} f(t)dt$ exists. This is, however, hard to know if we have the Laplace transform of F. We may of course determine f from F using the inverse Laplace transform, but then we would not need the final value theorem in the first place. We may, however, tell whether $\lim_{t \to \infty} f(t)dt$ exists by examining F alone.

In particular, if F is a *rational* function (the quotient of two polynomial functions) or a rational function multiplied by an exponential, that is

$$F(s) = e^{-as} \frac{P(s)}{Q(s)} \tag{4.144}$$

and

1. the (complex) non-zero roots of Q (called the **poles** of F), say $s_1, \ldots, s_n$, are *all* in the open left half-plane (i.e., $\mathbf{re}(s_i) < 0$ for all $s_i \neq 0$), and

2. F has at most one pole at the origin,

then the final value theorem applies (under these assumptions, all requirements of the theorem are satisfied).

Exercise 4.14 (Alternative FVT) We can prove the final value theorem under a different set of assumptions. Let f be a function in the time domain with Laplace transform $F(s) = (\mathscr{L}f)(s)$. Suppose that the following conditions are satisfied:

1. the conditions of Theorem 4.19 are satisfied

2. f' is absolutely integrable, that is, $\int_0^\infty |f'(\tau)| d\tau$ is finite

3. $\lim_{t \to \infty} f(t)$ exists and is finite,

then,
$$\lim_{t\to\infty} f(t) = \lim_{s\to 0^+} sF(s), \qquad (4.145)$$
where the limit is taken over $s \in \mathbb{R}$.

✂ ....

Hint. Guidance:

1. Under Assumption 1 we can use Theorem 4.19.
2. Take the limit as $s \to 0^+$ on the equation you derived in the previous step
3. By means of Assumption 2 and the dominated convergence theorem (see Theorem A.20), interchange the limit and the integral in the equation you derived in the previous step

Example 4.19 (FVT example) Consider the *s*-domain function
$$F(s) = \frac{1}{s(s^2 + s + 1)}. \qquad (4.146)$$

Indeed, F satisfies the above conditions: (i) it is a rational function, (ii) it has exactly one pole at the origin, (iii) the reader can verify that its other two poles have a negative real part. Therefore,
$$\lim_{t\to\infty} f(t) = \lim_{s\to 0^+} sF(s) = \lim_{s\to 0^+} \frac{1}{s^2 + s + 1} = 1. \qquad (4.147)$$

The final value theorem cannot be applied, for example, in the following case:

Example 4.20 (FVT counterexample) Take
$$F(s) = \frac{1}{s^2 + 4}. \qquad (4.148)$$

We know already that this is the Laplace transform of $f(t) = \sin(2t)$ which does not have a limit at infinity. The conditions for applying the final value theorem do not apply: the poles of F are $\pm 2j$ (whose real part is zero). Nevertheless, if we computed the limit
$$\lim_{s\to 0^+} sF(s) = \lim_{s\to 0^+} \frac{s}{s^2 + 4} = 0, \qquad (4.149)$$
we would find a wrong "finite" value: this does not mean that $\lim_{t\to\infty} f(t) = 0$.

Exercise 4.15 (Assumptions of FVT are not satisfied) Verify that the final value theorem is not applicable in the following cases:

1. $F(s) = \frac{s+1}{s-2}$

2. $F(s) = \frac{2}{s^2(s^2+s+1)}$

3. $F(s) = \frac{se^{-s}}{s^2+1}$

The following example demonstrates the value of the final value theorem (pun intended): it allows us to study the asymptotic behaviour of a dynamical system without solving the associated differential equations.

Example 4.21 (FVT in action) A physical system with input u and state x is described, in the time domain, by the differential equation

$$x(t) + \tau \dot{x}(t) = Ku(t), \qquad (4.150)$$

where K and τ are positive constants. Suppose that $x(0^+) = 0$. We are interested in determining the asymptotic behaviour of the response of the system as $t \to \infty$ when it is excited with a constant input signal, $u(t) = 1$ for all $t \geq 0$.

We apply the Laplace transform on both sides of (4.150) and use the fact that $u(t) = 1$, so $U(s) = 1/s$

$$X(s) + \tau s X(s) = K \Rightarrow X(s) = \frac{K}{s(1+\tau s)}. \qquad (4.151)$$

The conditions of the final value theorem are satisfied for $X(s)$ (indeed, X has two poles: $s_1 = 0$ and $s_2 = -1/\tau < 0$), so

$$\lim_{t \to \infty} x(t) = \lim_{s \to 0^+} sX(s) = \lim_{s \to 0^+} \not{s}\frac{K}{\not{s}(1+\tau s)} = K. \qquad (4.152)$$

Note that FVT allowed us to compute $\lim_{t \to \infty} x(t)$ without the need to compute $x(t)$.

4.4.4 Dirac's delta

Physical, mechanical, biological and cyber systems can be exposed to impulses: abrupt changes in the system's input which have a very high magnitude and very short duration. In the following chapters we will study the effect that such pulses have on the said dynamical systems. Such *impulsive* phenomena give rise to the definition of the **rectangle pulse function** δ_ϵ defined as

> **Definition 4.30 (Pulse function)** The pulse function $\delta_\epsilon : [0, \infty) \to \mathbb{R}$ with parameter $\epsilon > 0$ is defined as
> $$\delta_\epsilon(t) = \begin{cases} \frac{1}{\epsilon}, & \text{if } 0 \leq t < \epsilon \\ 0, & \text{otherwise} \end{cases} \qquad (4.153)$$

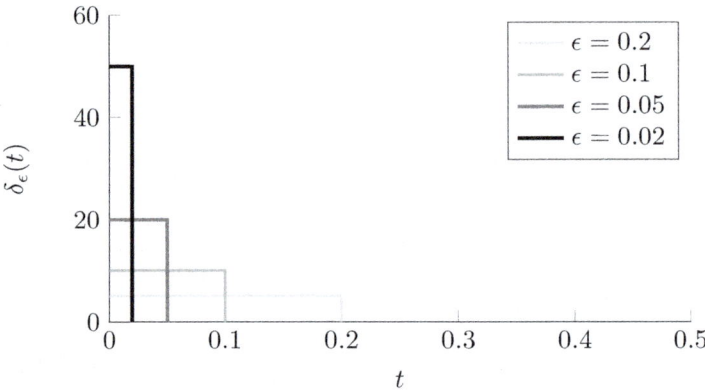

Figure 4.2: Definition of Dirac's delta function as the limit of δ_ϵ as $\epsilon \to 0^+$.

The Laplace transform of δ_ϵ can be computed easily using the definition

$$(\mathscr{L}\delta_\epsilon)(s) = \int_0^\infty e^{-s\tau}\delta_\epsilon(\tau)\mathrm{d}\tau = \int_{-\epsilon}^\epsilon \frac{1}{\epsilon}e^{-s\tau}\mathrm{d}\tau$$
$$= -\frac{1}{\epsilon s}e^{-s\tau}\bigg|_0^\epsilon = \frac{1}{\epsilon s}(1 - e^{-\epsilon s}). \qquad (4.154)$$

The time integral of δ_ϵ over $[0, \infty)$ is

$$I_\epsilon = \int_0^\infty \delta_\epsilon(\tau)\mathrm{d}\tau = \int_0^\epsilon \frac{1}{\epsilon}\mathrm{d}\tau = 1. \qquad (4.155)$$

The integral, I_ϵ, is equal to 1 and is independent of ϵ, therefore, trivially $\lim_{\epsilon \to 0^+} I_\epsilon = 1$, while δ_ϵ behaves as follows

$$\lim_{\epsilon \to 0^+} \delta_\epsilon(t) = \begin{cases} +\infty, & \text{if } t = 0 \\ 0, & \text{if } t \neq 0 \end{cases} \qquad (4.156)$$

Chapter 4. Laplace transform

The concept of this limit is illustrated in Figure 4.2 and leads to the definition of a construct with very exotic properties. Although a proper definition of the Dirac delta requires either a little measure theory or an exposition of theory of Schwartz distributions or generalised functions — which goes beyond the scope of this course — we give the following non-rigorous definition (rather, description).

> **Definition 4.31** (**Dirac delta**) The limit of δ_ϵ as $\epsilon \to 0^+$ defines the Dirac delta "function[a]", δ, which has the following properties
>
> $$\delta(t) = 0, \text{ for } t > 0, \quad (4.157a)$$
>
> $$\text{``}\int_0^\infty \delta(\tau)d\tau = 1\text{''}. \quad (4.157b)$$
>
> ---
> [a] The Dirac delta is not a function $\delta : [0, \infty) \to \mathbb{R}$. It is easy to see that the limit of δ_ϵ does not converge to a function over $[0, \infty)$. Besides, a function $\delta : [0, \infty) \to \mathbb{R}$ cannot take the value $+\infty$ and if a function is everywhere equal to zero except at a single point, then its integral is zero, not 1.

Remark: *The interested reader can refer to [M13, Chapter 6] for a rigorous definition of the Dirac delta function. We should note that the above definition is not rigorous. If δ were an ordinary function, since it is equal to zero everywhere but at a single point, its integral over $[0, \infty)$ would be zero.* •

We may *define* the Laplace transform of δ as the limit of the Laplace transforms of δ_ϵ as $\epsilon \to 0^+$, so

$$\mathscr{L}\delta(s) := \lim_{\epsilon \to 0^+} \mathscr{L}\delta_\epsilon(s) = \lim_{\epsilon \to 0^+} \frac{1}{\epsilon s}\frac{1 - e^{-\epsilon s}}{\epsilon s} = 1, \quad (4.158)$$

where the last limit is computed using L' Hôpital's rule (see Theorem A.23). Note that from Theorem 4.6 we know that a Laplace transform must satisfy $\lim_{s \to \infty} \mathscr{L}\delta(s) = 0$, which is not the case here. The reason is that δ is *not* a function.

Remark: *The Dirac delta function has the following property:*

$$\int_0^{+\infty} f(\tau)\delta(\tau)d\tau = f(0), \quad (4.159)$$

for all smooth compactly-supported functions f (we will not go into details here). One rigorous definition of δ makes use of Equation (4.159) [M8]: it defines δ to be a functional, $\int_0^{+\infty} \cdots \delta(\tau)d\tau$, which maps a smooth compactly-supported function f to $f(0)$ according to Equation (4.159). We can write, $\delta(f) = f(0)$. That said, in the left-hand side of Equation (4.159), the integral is to be understood as $\delta(f)$. In other words, rigorously speaking, in this context there is no such thing as $\delta(t)$ — only $\delta(f)$ — while the integral in Equation (4.159) is not a regular Riemann–Stieltjes integral. •

Example 4.22 (Kicking a ball) At time $t = 0$, a football player kicks a ball of mass m which is initially at rest. We assume that the player kicks the ball horizontally, there is no friction and the ball does not spin. The ball's motion is described by Newton's second law:

$$m\dot{v}(t) = F(t), \tag{4.160}$$

where $F(t)$ is the force exercised by the kick and $v(t)$ is the ball's velocity at time t. The kick can be modelled by

$$F(t) = A\delta(t), \tag{4.161}$$

where A is the *magnitude* of the kick. By applying the Laplace transform on both sides of the above kinematic equation we obtain

$$msV(s) = A \Leftrightarrow V(s) = \frac{A}{ms} \tag{4.162}$$

But wait! We've seen this before (see Section 4.2.2). This is the Laplace transform of

$$v(t) = \frac{A}{m}. \tag{4.163}$$

This means that the ball will move at a constant speed.

In Example 4.22 we used the **inverse Laplace transform**: given the function A/ms in the s-domain, we found the function A/m in the t-domain so that $\mathscr{L}\{A/m\} = A/ms$. This is the topic of the following chapter.

Remark: *The Dirac delta can be seen as a **measure**. We will try to give a brief presentation of the concept of a measure without the heavy mathematical formalism that comes with measure theory.*

In mathematics, measures generalise the notion of the length of a line segment, the area of a surface or the volume of a 3D object. One way to measure intervals on the real line, $[a,b]$, is to define the measure $\mu([a,b]) = b - a$. Note that $\mu(\{0\}) = 0$ (a single point has measure zero).

We can generalise this definition to subsets of the set of real numbers, $\mathbb{R}$, that are not intervals — for example $A = [1,2] \cup [3,4]$ — as follows: (i) if $A \subseteq \mathbb{R}$ can be written as $A = \bigcup_{i=1}^{\infty} [a_i, b_i]$, where the sets $[a_i, b_i]$ are mutually disjoint, we can define

$$\mu(A) = \sum_{i=1}^{\infty} \mu([a_i, b_i]), \tag{4.164}$$

while μ has another two notable properties: (ii) $\mu(\emptyset) = 0$ and (iii) it is nonnegative-valued. That said, $\mu([1,2] \cup [3,4]) = \mu([1,2]) + \mu([3,4]) = 2$. Any function that satisfies these three properties is a measure (see [M8, Chapter 1] for a proper definition).

Chapter 4. Laplace transform

Given a measure m we can define the integral of a function f over a set A with respect to m, which is denoted by $\int_A f \, dm$. This integral has the property $\int_A dm = m(A)$. We can use this property to determine the integral of constant functions, $\int_A c \, dm = cm(A)$, and, similarly, piecewise constant functions. We can then extend this to be able to determine the integrals with respect to m of a larger family of functions which can be written as limits of piecewise constant ones.

We may now define the **Dirac measure**, δ, which has the property $\delta(\{0\}) = 1$. In general,

$$\delta([a,b]) = \begin{cases} 1, & \text{if } 0 \in [a,b] \\ 0, & \text{otherwise} \end{cases} \tag{4.165}$$

The Dirac measure has the property $\int_0^\infty f \, d\delta = f(0)$ — cf. Equation (4.159). Essentially, "$\delta(\tau) d\tau$" in Equation (4.159) corresponds to integration with respect to the measure δ.

We may now extend the definition of the Laplace transform to include measures. We define the Laplace transform of a measure m over $\mathbb{R}$ to be $(\mathscr{L}m)(s) = \int_0^\infty e^{-s\tau} dm$. That said, the Laplace transform of the Dirac delta is

$$(\mathscr{L}\delta)(s) = \int_0^\infty e^{-s\tau} d\delta = e^{-s\tau}\big|_{\tau=0} = 1. \tag{4.166}$$

Lastly, the Laplace transform of a time-domain function, $f : [0, \infty) \to \mathbb{R}$, can be seen as the Laplace transform of the measure $f d\mu$ (which can also be denoted as $f dx$). The measure $f dx$ is defined via $(f d\mu)(A) = \int_A f d\mu$.

•

4.5 Symbolic computations

4.5.1 Symbolic computations in Python

In Python, we may use the symbolic library SymPy, which can be installed using pip as follows

```
pip install sympy
```

We may then compute Laplace transforms symbolically as in the following example where we compute $\mathscr{L}\{\sin(t)\}$

```
import sympy as sym
s,t = sym.symbols('s,t')   # Define symbols s, t
x_signal = sym.sin(t)       # Signal, x(t)
x_laplace = sym.laplace_transform(x_signal, t, s)
```

Here we first defined two symbolic variables, s and t, we then constructed a symbolic function, x_signal, which is a function of t, and, lastly, we computed a symbolic Laplace transform of this function.

In the last line in the above example, function laplace_transform returns a tuple (F, a, cond), where F is the Laplace transform of x_signal (which is a function of s) and a is a real number that defines the domain of the Laplace transform in the form $\{s \in \mathbb{C} : \mathbf{re}(s) > a\}$. If we only need the Laplace transform, we should call laplace_transform with the optional argument noconds=True.

Consider the following example

```
x_signal = sym.sin(t)*sym.cos(t)**3
x_laplace = sym.laplace_transform(x_signal, t, s)
```

This will return the 3-tuple ((s**2 + 10)/((s**2 + 4)*(s**2 + 16)), 0, True). This means that the Laplace transform is

$$\mathscr{L}\{\sin(t)\cos(t)^3\} = \frac{s^2 + 10}{(s^2 + 4)(s^2 + 16)}, \tag{4.167}$$

defined for $s \in C$ with $\mathbf{re}(s) > 0$.

4.5.2 Symbolic computations in MATLAB

We may, similarly, compute symbolic Laplace transforms in MATLAB using MATLAB's Symbolic Math Toolbox. Here is an example

```
t = sym('t');                        % Define symbol `t`
x_signal = sin(t)*cos(t)^3;          % Signal in t-domain
x_laplace = laplace(x_signal);       % Laplace transform
```

The result, x_laplace is 1/(2*(s^2 + 4)) + 1/(2*(s^2 + 16)); the reader can verify that this is equal to the Laplace transform we computed above using SymPy in Python. MATLAB, however, does not compute the domain of the result.

Chapter 4. Laplace transform

4.6 One-minute round-up

In this chapter we learned the following

1. What the Laplace transform is, we studied some of its properties, and learned how to compute it for various functions

2. Important facts about the Laplace transform: it translates derivatives and integrals in the time domain into simple functions in the s-domain. If we apply the Laplace transform on an ODE/IDE (Chapter 2), we obtain and algebraic equation.

3. The initial and final value of a function $f(t)$ can be computed by knowing the Laplace transform of that function alone (Initial and final value theorems)

4. We introduced the Dirac function – a construct with exotic properties. The Dirac function models impulsive effects.

We **did not learn** how to use the Laplace transform to simulate the response of dynamical systems. All following chapters will be about understanding dynamical systems in terms of the Laplace transforms of their ODE/IDE.

4.7 Exercises

Exercise 4.16 (Pair the functions) Pair the following functions with their Laplace transforms (we have omitted the domains of the Laplace transforms here)

1. $f(t) = e^t$
2. $f(t) = e^{-t}$
3. $f(t) = t$
4. $f(t) = te^{-t}$
5. $f(t) = \sin(2t)$
6. $f(t) = \cos(2t)$
7. $f(t) = e^{-t}\cos t$
8. $f(t) = t\cos(t)$
9. $f(t) = e^{-2t}\sin t$
10. $f(t) = e^{-2t}\cos t$
11. $f(t) = \sqrt{t}$
12. $f(t) = \delta(t)$

a. $F(s) = \frac{1}{s+1}$
b. $F(s) = \frac{1}{s-1}$
c. $F(s) = \frac{\sqrt{\pi}}{2s^{1.5}}$
d. $F(s) = \frac{2}{s^2+4}$
e. $F(s) = \frac{1}{(s+1)^2}$
f. $F(s) = 1/s^2$
g. $F(s) = \frac{s}{s^2+4}$
h. $F(s) = \frac{s+1}{(s+1)^2+1}$
i. $F(s) = \frac{s^2-1}{(s^2+1)^2}$
j. $F(s) = \frac{1}{(s+2)^2+1}$
k. $F(s) = 1$
l. $F(s) = \frac{s+2}{(s+2)^2+1}$

Write your answers below:

1. _____   2. _____
3. _____   4. _____
5. _____   6. _____
7. _____   8. _____
9. _____   10. _____
11. _____ 12. _____

Answer: 1-b, 2-a, 3-f, 4-e, 5-
d, 6-g, 7-h, 8-i, 9-j, 10-l, 11-c,
12-k

Exercise 4.17 (Find the Laplace transform) Find the Laplace transforms of the following functions:

1. $f_1(t) = 1 - t + 5t^2$
2. $f_2(t) = e^{-\pi t}(5\cos(3t) + 4\sin(3t))$
3. $f_3(t) = 6\sin t + t\sin t$
4. $f_4(t) = \sqrt{t} + 5t^{5/2}$
5. $f_5(t) = e^{-at^2}$, $a \in \mathbb{R}$, $a < 0$
6. $f_6(t) = -\frac{\sin(2t)}{t}$
7. $f_7(t) = t^2\cos(\omega t)$, $\omega \in \mathbb{R}$
8. $f_8(t) = \cos^2(\omega t)$, $\omega \in \mathbb{R}$
9. $f_9(t) = \frac{\sin^2(t)}{t}$ for $t > 0$
10. $f_{10}(t) = \sin^3(\beta t)$, $\beta \in \mathbb{R}$
11. $f_{11}(t) = t\sinh t$
12. $f_{12}(t) = \cosh t\cos t$
13. $f_{13}(t) = e^{-t}\frac{d^{10}}{dt^{10}}(t^{10})$
14. $f_{14}(t) = H_1(t)e^{-2t}(1+t^3)$

Chapter 4. Laplace transform

Recall that $\sinh t = \frac{1}{2}(e^t - e^{-t})$ and $\cosh t = \frac{1}{2}(e^t + e^{-t})$.

※ ....

Answer. $(\mathscr{L}f_1)(s) = \frac{s^2}{s^2-10}$, $(\mathscr{L}f_2)(s) = \frac{5s+12s+9}{s(s+n)+9}$, $(\mathscr{L}f_3)(s) = \frac{s^2+1}{6} +$ $\frac{s^2+1}{2s}$. For f_4 use the fact that $\Gamma(n+\frac{1}{2}) = \frac{(2n)!}{4^n n!}\sqrt{\pi}$ for all $n \in \mathbb{N}$ to find that $(\mathscr{L}f_4)(s) = \sqrt{\pi}\left(\frac{25\sqrt{3}/2}{s^{7/2}} + \frac{88\sqrt{7}/2}{75\, s^{9/2}}\right)$, $(\mathscr{L}f_5)(s) = \frac{2}{(s+a)^3}$, $(\mathscr{L}f_6)(s) = -\arctan 2/s$, $(\mathscr{L}f_7)(s) = \frac{2s(s^2-3m^2)}{(s^2+m^2)^3}$, $(\mathscr{L}f_8)(s) = \frac{s(s^2+2m^2)}{s^2+4m^2}$ (for f_8 use the trigonometric identity $\cos^2(t) = \frac{1}{2}(1-\cos(2t))$), $(\mathscr{L}f_9)(s) = \frac{1}{4}\ln\left(1 + \frac{s^2}{4}\right)$, $(\mathscr{L}f_{10})(s) = \frac{6\beta^3}{(s^2+\beta^2)(s^2+9\beta^2)}$ (use the trigonometric identity $\sin^3 x = \frac{1}{4}(3\sin x - \sin(3x))$ for all $x \in \mathbb{R}$), $(\mathscr{L}f_{11})(s) = \frac{2s}{(s^2-1)^2}$, $(\mathscr{L}f_{12})(s) = \frac{s^2+4}{s^3}$, $(\mathscr{L}f_{13})(s) = \frac{s+1}{10\,!}$, $(\mathscr{L}f_{14})(s) = e^{-s-2} \cdot 2\,\frac{s^3+15s^2+45s+46}{(s+2)^4}$.

Exercise 4.18 (Simple exercise) Let $F(s) = \mathscr{L}\{e^{-t}\sin t\}(s)$. Compute $F(a+jb)$ for $a, b \in \mathbb{R}$ (determine the real and imaginary parts of $F(a+bj)$), if possible.

※ ....

$$F(a+jb) = \frac{-b^2 + (a+1)^2 + 1}{4b^2(a+1)^2 + \left(-b^2+(a+1)^2+1\right)^2}$$
$$- j\cdot\frac{2b(a+1)}{4b^2(a+1)^2 + \left(-b^2+(a+1)^2+1\right)^2}$$

Answer. Following Theorem 4.18, $F(a+bj)$ is defined only if $a + 1 > 0$. Then,

Exercise 4.19 (Function defined piecewise) Write the following function $f:[0,\infty) \to \mathbb{R}$ in terms of Heaviside functions

$$f(t) = \begin{cases} 1, & \text{for } 0 \le t < 1 \\ t, & \text{for } t \ge 1 \end{cases} \tag{4.168}$$

and determine its Laplace transform.

※ ....

$$(\mathscr{L}f)(s) = \frac{e^{-s}}{s^2} + \frac{1}{s}, \tag{4.169}$$

for $s \in \mathbb{C}$ with $\mathrm{re}(s) > 0$.

Answer. It is $f(t) = 1 + (t-1)H_1(t)$ and its Laplace transform is

Exercise 4.20 (Initial and final values) What are the values of $\lim_{t\to 0^+} x(t)$, $\lim_{t\to 0^+} \dot{x}(t)$ and $\lim_{t\to\infty} x(t)$ of the time-domain signal $x(t)$ if its Laplace transform is given by

$$X(s) = \frac{2s^2 + s + 1}{20s^4 + 72s^3 + 77s^2 + 27s + 2} e^{-0.01s}. \qquad (4.170)$$

✂ .... ✻ ....

Hint. Employ the initial and final value theorems. In order to use the final value theorems. In order to use the final value theorem, you need to check whether all poles of X have negative real part. You and can confirm that -2 is a pole of X. Use -0.5, -0.1, $x(0^+) = 0$ and $\lim_{t\to\infty} x(t) = 0$. For the derivative it Horner's scheme to determine the remaining three poles.

is $x(0^+) = 0$.

Exercise 4.21 (Positive triangle wave) The triangle wave of amplitude $A > 0$ and half-period $a > 0$ is the periodic signal shown in Figure 4.3.

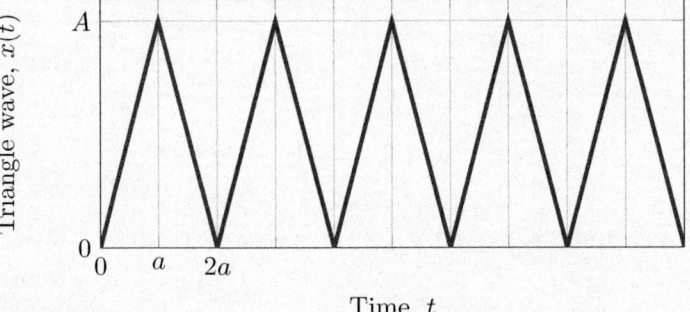

Figure 4.3: Triangle wave with period $2a$ and amplitude A.

(i) Determine the Laplace transform of the triangle wave using Theorem 4.24 or Theorem 4.25, (ii) Show that this signal can be written as the time-integral of a square wave (see Example 4.11), (iii) Use this fact to derive the Laplace transform of the triangle wave from that of the square wave using Theorem 4.22.

✻ ....

Answer. (i) The Laplace transform of the triangle wave shown in Figure 4.3 is $\frac{A}{s^2}\tanh(2as)$, where tanh is the hyperbolic tangent function (see Section A.8). (ii) Note that the above signal, $f(t)$, for $t \in [0, 2a]$, can be written as the integral of $g(t) = A(1 - 2H_a(t))$, for $t \in [0, 2a]$. This can be extended to a $2a$-periodic square wave.

Chapter 4. Laplace transform

Exercise 4.22 ($\mathscr{L}$ cos **from** $\mathscr{L}$ sin) Using the fact that $\mathscr{L}\{\sin(\omega t)\}(s) = \frac{\omega}{s^2+\omega^2}$, and Theorem 4.19, show that $\mathscr{L}\{\cos(\omega t)\}(s) = \frac{s}{s^2+\omega^2}$.

✂ ....

Hint. Use the fact that $\cos t = (\sin t)'$.

Exercise 4.23 (**Properties of solution of linear ODE**) Consider the dynamical system

$$5x(t) + 2\dot{x}(t) = \begin{cases} a, & \text{for } 0 \leq t < 1 \\ a+1, & \text{for } 1 \leq t < 2 \\ a+2, & \text{for } 2 \leq t \end{cases}. \tag{4.171}$$

If $x(0^+) = b$, determine $\lim_{t \to \infty} x(t)$ and $\lim_{t \to 0^+} \dot{x}(t)$.

✻ ....

Answer. The conditions of the final value theorem are satisfied and $\lim_{t \to \infty} x(t) = \lim_{s \to 0^+} sX(s) = \frac{a+2}{5}$. By the initial value theorem we have $\lim_{t \to 0^+} \dot{x}(t) = \lim_{s \to \infty} s(sX(s) - x(0^+)) = \frac{1}{2}(a - 5b)$.

✂ ....

Hint. Define $X(s) = \mathscr{L}\{x(t)\}(s)$ and apply the Laplace transform on both sides of (4.171). The right-hand side of the above (or below) — that depends on your perspective — equation, is a function defined piecewise (see Section 4.3.2). Solve for $X(s)$ and apply the final value theorem (after you check that its conditions are satisfied).

Exercise 4.24 (**True or false?**) Warning: tricky questions ahead!

	True	False
Function f has a Laplace transform if and only if it is of exponential order	☐	☐
Function f has a Laplace transform if it is of exponential order	☐	☐
All polynomials have a Laplace transform	☐	☐
$\mathscr{L}\{e^{at}\}(s) = \frac{1}{s-a}$ for all $a \in \mathbb{C}$ with $\mathbf{re}(s) > a$	☐	☐
$\mathscr{L}\{t^2\}(s) = \frac{2}{s^3}$ for all $s \in \mathbb{C}$ with $\mathbf{re}(s) > 0$	☐	☐
$\mathscr{L}\{t^3\}(s) = \frac{3}{s^4}$ for all $s \in \mathbb{C}$ with $\mathbf{re}(s) > 0$	☐	☐
If f and g have a Laplace transform, then $\mathscr{L}\{f+g\} = \mathscr{L}f + \mathscr{L}g$	☐	☐

	True	False
If f and g have a Laplace transform, then $\mathscr{L}\{f - g\} = \mathscr{L}f - \mathscr{L}g$	☐	☐
If f has a Laplace transform, then $\mathscr{L}\{5f\} = 5\mathscr{L}f$	☐	☐
If f and g have a Laplace transform, then $\mathscr{L}\{fg\} = \mathscr{L}f \cdot \mathscr{L}g$	☐	☐
We can apply the FVT to determine $\lim_{t \to \infty} f(t)$ if $F(s) = \frac{1}{s^2+1}$	☐	☐
We can apply the IVT to determine $\lim_{t \to 0^+} f(t)$ if $F(s) = \frac{1}{s^2+1}$	☐	☐
The Laplace transform of the derivative is $\mathscr{L}\{f'\}(s) = sF(s)$	☐	☐
The functions $f_1(t) = \sin(t)$ and $f_2(t) = \cos(t)$ are periodic	☐	☐
Function $f(t) = |\sin(t)|$ is periodic	☐	☐
If f is smooth and has a Laplace transform and $f(0^+) = 0$, then $\mathscr{L}\{f'\}(s) = sF(s)$	☐	☐
The Laplace transform of a piecewise continuous periodic function f is $F(s) = \frac{1}{1-e^{-sT}} \int_0^T e^{-s\tau} f(\tau) \mathrm{d}\tau$	☐	☐
The Laplace transform of $f(t) = e^t$ is $F(s) = \frac{1}{s-1}$ for $s \in \mathbb{C}$	☐	☐

※ ....

Answer: F, F, F, F, T, T, F, F, T, T, F (be careful: the domain of this Laplace transform is $s \in \mathbb{C}$ with $\mathrm{re}(s) > 1$)

Exercise 4.25 (Integer powers) Prove Theorem 4.10 by following the following procedure: (i) what is the Laplace transform $\mathscr{L}\{1\}$? (ii) use Theorem 4.19 to show that

$$\mathscr{L}\{f(t)\} = \frac{1}{s}\left(\mathscr{L}\{f'(t)\} - f(0^+)\right), \qquad (4.172)$$

provided f satisfies the assumptions of the theorem, (iii) apply Equation (4.172) to $f(t) = t$, (iv) use (4.172) and apply mathematical induction to complete the proof.

Chapter 4. Laplace transform

Exercise 4.26 (Verify Theorem 4.6) For each t-domain function in questions 1–5 in Exercise 4.17 verify that the Laplace transforms you determined satisfy the conditions of Theorem 4.6.

Exercise 4.27 (Exponential order of integral) Show that if f is of exponential order γ, then the integral $I(t) = \int_0^t f(\xi)\mathrm{d}\xi$ is also of exponential order γ.

Exercise 4.28 (Saturated sinusoidal function) Determine the Laplace transform of the following saturated sinusoidal signal

$$f(t) = \min\left\{|\sin t|, \sqrt{3}/2\right\}. \tag{4.173}$$

✂ ....

Answer. It is

$$\mathscr{L}\{f\}(s) = \frac{2s - (\sqrt{3} + s)e^{(\sqrt{3}+s)\frac{\pi}{6}} + (\sqrt{3} - s)e^{(\sqrt{3}-s)\frac{\pi}{6}} + 2se^{\pi s}}{2s(s^2 + 1)(e^{\pi s} - 1)} \tag{4.174}$$

✂ ....

Hint. First show that f is periodic with period $T = \pi$. For $t \in [0, \pi]$, f can be written as a function defined piecewise. It will be useful to plot f.

Exercise 4.29 (Laplace transform of $1/\sqrt{t}$) Show that the function $f(t) = \frac{1}{\sqrt{t}}$, for $t > 0$, has the Laplace transform

$$(\mathscr{L}\{f(t)\})(s) = \frac{\sqrt{\pi}}{\sqrt{s}}, \tag{4.175}$$

defined for $s \in \mathbb{C}$, $\mathrm{re}(s) > 0$.

✂ ....

Hint. (i) In order to compute $(\mathscr{L}\{f(t)\})(s)$ you need to use the definition of the Laplace transform, (ii) in order to compute the integral of the Laplace transform, try to introduce a convenient change of variables, (iii) it is known that

$$\int_0^\infty e^{-\xi^2}\mathrm{d}\xi = \frac{\sqrt{\pi}}{2}. \tag{4.176}$$

This is known as the Gaussian integral. Alternatively, we can use Theorem 4.19. Note that $(2\sqrt{t})' = 1/\sqrt{t}$, for $t > 0$.

In Theorem 4.5 we showed that if a time-domain function is of exponential order and piecewise continuous in the sense of Definition A.14, then it has a Laplace transform. These are sufficient but *not necessary* conditions. In the following exercise you are asked to show that a given function has a Laplace transform, although it is not of exponential order.

Exercise 4.30 (Exponential order) Show that the following function

$$f(t) = 2te^{t^2}\cos(e^{t^2}), \qquad (4.177)$$

is *not* of exponential order, but has a Laplace transform.

✂ ....

Hint. You do not need to compute the Laplace transform of f. You only need to show that it exists. Note that f is the derivative of a simple function g, that is, $f(t) = g'(t)$. Show that g has a Laplace transform and invoke Theorem 4.19.

Exercise 4.31 (Laplace transform of $t^2 \cos t$) Use the result of Exercise 4.10 to show that

$$\mathscr{L}\{t^2 \cos t\} = \frac{2s^3 - 6s}{(s^2 + 1)^3}. \qquad (4.178)$$

✂ ....

Hint. Observe that $\frac{d^2}{ds^2}\mathscr{L}\{\cos t\} = \mathscr{L}\{t^2 \cos t\}$.

Advanced Exercise 4.32 (Taylor expansion of $\sin t$) Certain time-domain functions can be written as their Taylor expansion, which is a power series. For example, $f(t) = \sin t$ can be written as follows

$$f(t) = \sum_{n=0}^{\infty} \frac{(-1)^n t^{2n+1}}{(2n+1)!}, \qquad (4.179)$$

defined for all $t \geq 0$. Use this expansion to compute the Laplace transform of f.

✂ ....

Hint. (i) You may use the fact that

$$\frac{1}{1-x} = \sum_{n=0}^{\infty} x^n, \qquad (4.180)$$

(ii) keep in mind that the integral of a series is not necessarily equal to the series of integrals, (iii) determine the domain of $F(s)$ (for what values of s is $F(s)$ well defined?)

Chapter 4. Laplace transform   181

Exercise 4.33 (**Interesting Laplace transform**) What is the Laplace transform of

$$f(t) = \frac{\cos\sqrt{t}}{\sqrt{t}}, \qquad (4.181)$$

defined for $t > 0$? Use the definition of the Laplace transform.

✂ .... ※ ....

Hint. Use the definition of the Laplace transform (Definition 4.2). In order to evaluate the integral, introduce the change of variables $y = \sqrt{t}$. Recall the Gaussian integral formula:

$$\int_{-\infty}^{0} e^{-z^2}\,dz = \frac{\sqrt{\pi}}{2}.$$

Answer. The result is

$$(\mathcal{L}f)(s) = \sqrt{\frac{\pi}{s}}\, e^{-\frac{1}{4s}},$$

for $s \in \mathbb{C}$ with $\mathrm{re}(s) > 0$.

Exercise 4.34 (**Exercise 4.33 revisited**) Solve Exercise 4.33 again, but now we will use Theorem 4.19. Determine an antiderivative of function f in Equation (4.181), that is, determine a function g such that $g' = f$ for all $t > 0$. Apply Theorem 4.19 to g.

✂ ....

Hint. Choose $g(t) = 2\sin\sqrt{t}$.

Advanced Exercise 4.35 (**Existence of Laplace transform of integral**) Find an example to show that the fact that a function f has a Laplace transform does not imply that function $h(t) := \int_0^t f(\tau)d\tau$ has a Laplace transform.

✂ ....

Hint. (i) In light of Theorem 4.22, we are looking for a function f which does not satisfy its assumptions, i.e., is not piecewise continuous and/or is not of exponential order, (ii) try $f(t) = \ln|1 - t|$ for $t \geq 0$, $t \neq 1$ and $f(1) = 0$. (iii) interpret the integral of f as $\int_0^t f(\tau)d\tau = \int_0^{1-} f(\tau)d\tau + \int_{1+}^{t} f(\tau)d\tau$, whenever $t > 1$.

Advanced Exercise 4.36 (**Laplace transform of a Taylor expansion**) Suppose that $f : [0, \infty) \to \mathbb{R}$ is n times differentiable over $(0, \infty)$ and its n-th order derivative is piecewise continuous, f is of exponential order, the limits $f(0^+), f'(0^+), \ldots, f^{(n)}(0^+)$ exist and are all equal to zero, that is

$$f(0^+) = f'(0^+) = \ldots = f^{(n)}(0^+) = 0. \qquad (4.182)$$

Let us denote the Laplace transform of f by F. Show that

$$F(s) = o\left(1/s^{n+1}\right), \text{ as } s \to \infty, s \in \mathbb{R}, \tag{4.183}$$

where o is the little-O notation (see Definition A.27).

✂ ··

Hint: Use Taylor's approximation Theorem (Theorem 3.1). You need to convince yourself that you can have a Taylor expansion up to order n taking one-sided derivatives at 0^+.

Advanced Exercise 4.37 (**Generalisation of final value theorem**) Let $f :$ $[0, \infty) \to \mathbb{R}$ be a continuous, bounded function and let F denote its Laplace transform. Suppose that

$$\lim_{T \to \infty} \tfrac{1}{T} \int_0^T f(\tau) \mathrm{d}\tau =: L, \tag{4.184}$$

exists and is finite. Then,

$$\lim_{s \to 0^+} sF(s) = L, \tag{4.185}$$

where the limit is taken over $s \in \mathbb{R}$.

The above exercise is a generalisation of the final value theorem (Theorem 4.29). Note that if $\lim_{t \to \infty} f(t)$ exists and is finite, then the limit in Equation (4.184) also exists and is finite. However, the converse is not true. For example, for $f(t) = \sin t$, the limit $\lim_{t \to \infty} f(t)$ does not exist, but $L = 0$. This is a somewhat special case of [C25, Theorem 1]. Another generalisation is given in [C26] which covers the case where the limit is infinite.

Advanced Exercise 4.38 (**Limit-$\mathscr{L}$ interchange counterexample!**) Let $(f_n)_n$ be a sequence of time-domain functions that converges pointwise to f (see Definition A.11). It is always true that

$$\lim_{n \to \infty} (\mathscr{L} f_n)(s) = \mathscr{L}\{f\}(s)? \tag{4.186}$$

✂ ··

Hint: This is not always true. Here are two counterexamples: (i) Define f_n to be such that $f_n(0) = 0$, and $f'_n(t) = 1 - H_1(t)/n$ for $t > 0$. Then $f_n(t) \to 0$ pointwise for all $t \geq 0$ as $n \to \infty$, so $\lim_n \mathscr{L}\{f_n\} = 0$, but $\lim_n \mathscr{L}\{f_n\}(s) = 1$. (ii) take $f_n(t) = e^{t^2} (H_n(t) - H_{n^2}(t))$. Then $f_n \to 0$ pointwise as $n \to \infty$, so $\mathscr{L}\{\lim_n f_n\}(s) = 0$, but for every s, $\lim_n \mathscr{L}\{f_n\}(s) = \infty$.

Chapter 4. Laplace transform

Exercise 4.39 (**Almost everywhere equal to 1**) What is the Laplace transform of

$$f(t) = \begin{cases} 1, & \text{if } t \geq 0, \text{ and } t \neq 2, \\ 0, & \text{if } t = 2 \end{cases} \tag{4.187}$$

✂ .... ※ ....

Answer. $(\mathscr{L}f)(s) = \frac{1}{s}$, for $s \in \mathbb{C}$ with *Hint*. You may want to use the definition of the Laplace transform. re(s) > 0.

Advanced Exercise 4.40 (**Error function**) The *error function* is defined as

$$\operatorname{erf}(t) = \frac{2}{\sqrt{\pi}} \int_0^t e^{-\tau^2} \, d\tau. \tag{4.188}$$

1. Determine $\mathscr{L} \operatorname{erf}(t)$ as a series

2. Determine $\mathscr{L} \operatorname{erf}(\sqrt{t})$ as a series and show that $\mathscr{L} \operatorname{erf}\left(\sqrt{t}\right) = \frac{1}{s\sqrt{s+1}}$.

3. Show that

$$\mathscr{L}\left\{\int_0^t \operatorname{erf}\left(\sqrt{\tau}\right) d\tau\right\} = \frac{1}{s^2\sqrt{s+1}}. \tag{4.189}$$

※ ....

$$\operatorname{erf} t = \frac{2}{\sqrt{\pi}} \sum_{\nu=0}^{\infty} \frac{(-1)^\nu \nu!}{(2\nu)_\nu!} \frac{1}{t^{2\nu+1}}, \quad \mathscr{L} \operatorname{erf} t = \frac{2}{\sqrt{\pi}} \sum_{\nu=0}^{\infty} \frac{(-1)^\nu \nu!}{(2\nu)_\nu!} \frac{1}{s^{2(\nu+1)}}. \tag{4.190}$$

Answer. 1. It is

✂ ....

question. 3. Use Theorem 4.22.
using the definition directly, but here you need to use the result of the second
Hint. 1. Use the Taylor expansion of $e^{-\tau^2}$ at 0. 2. You can determine $\mathscr{L} \operatorname{erf}(\sqrt{t})$

Advanced Exercise 4.41 (**Bessel functions of the first kind**) The Bessel function is an "oscillatory" function with many applications in physics, signal processing and probability theory. A Bessel function of the first kind of order $n \in \mathbb{N}$, $J_n(t)$, is a solution of Bessel's differential equation, which is

$$t\ddot{x} + t\dot{x} + (t^2 - n^2)x = 0. \tag{4.191}$$

1. Determine $J_n(t)$ in the form of a power series

2. Show that $\mathscr{L}\{J_0(t)\}(s) = \frac{1}{\sqrt{s^2+1}}$

3. Show that $\mathscr{L}\{J_0(\sqrt{t})\}(s) = \frac{1}{s\sqrt{s^2+1}}$

4. Using the recursive formula [M23, Section 4.9]

$$J_{n+1}(t) = -t^n \frac{d}{dt}\left(t^{-n} J_n(t)\right), \qquad (4.192)$$

determine the Laplace transform of J_n for $n = 1, 2, \ldots$

✻ ....

Answer. 1. $J_n(t) = \sum_{i=0}^{\infty} \frac{(-1)^i \left(\frac{t}{2}\right)^{2i+n}}{i!(n+i)!}$, 4. $\mathscr{L}\{J_n(t)\}(s) = \frac{(\sqrt{s^2+1}-s)^n}{\sqrt{s^2+1}}$, $n = 1, 2, \ldots$

Exercise 4.42 (**Laplace transform of double integral**) Let f be a piecewise continuous function in the time domain which has exponential order γ and Laplace transform $F(s) = (\mathscr{L}f)(s)$. Show that

$$\mathscr{L}\left\{\int_a^t \int_a^\tau f(\xi) d\xi d\tau\right\} = \frac{1}{s^2} F(s) + \frac{1}{s^2} \int_a^0 f(\tau) d\tau$$
$$+ \frac{1}{s} \int_a^0 \int_a^\tau f(\xi) d\xi d\tau. \qquad (4.193)$$

Exercise 4.43 (**Laplace transform of** $\ln(t)$) The Euler-Mascheroni constant is $\gamma := -\Gamma'(1)$, and is approximately equal to 0.5772156649. Show that the Laplace transform of $f(t) = \ln t$, $t > 0$, is

$$\mathscr{L}\{\ln t\}(s) = -\frac{\ln s + \gamma}{s}, \qquad (4.194)$$

defined for $s \in \mathbb{C}$ with $\mathbf{re}(s) > 0$.

✂ ....

Hint. You may consider one of the following approaches: (i) Use the definition of the Laplace transform (Definition 4.2), or (ii) Use the Laplace transform of t^a, $a > 0$, differentiate with respect to a, and take $a \to 0^+$.

Exercise 4.44 (**Integrals using the Laplace transform**) Show that

$$\int_0^\infty \frac{\sin t}{t} dt = \frac{\pi}{2}, \text{ and } \int_0^\infty \frac{e^{-t} \sin t}{t} dt = \frac{\pi}{4}. \qquad (4.195)$$

✂ ....

Hint. What is the Laplace transform of $\sin t/t$? Use Theorem 4.27 and go through Example 4.15.

5

Inverse Laplace transform

The Laplace transform maps time-domain representations (of signals and systems) into s-domain representations. This way, it casts differential equations as algebraic ones which are easier to work with. If we are interested in determining the system response in the time domain though, we will have to "undo" the Laplace transform and return to the domain of time. This is done by means of the inverse Laplace transform.

5.1 General properties

In Chapter 4 we saw how to derive the Laplace transform of a given time-domain function f. We also stated conditions under which the Laplace transform exists (Theorem 4.5). Here we pose the inverse problem: given a function F in the s-domain, is there a function f so that $F(s) = (\mathscr{L}f)(s)$? Such a function is called an **inverse Laplace transform** of F. We are also interested in whether the inverse Laplace transform is unique. Is it possible that for an s-domain function F there are more than one t-domain functions f_1 and f_2, not equal to one another, such that $\mathscr{L}f_1 = \mathscr{L}f_2 = F$? We give the following definition

> **Definition 5.1 (Inverse Laplace transform)** Let f be a function in the time domain with Laplace transform $F(s) = (\mathscr{L}f)(s)$. We say that f is an inverse Laplace transform of F, and we denote
>
> $$f(t) = \mathscr{L}^{-1}\{F(s)\}(t). \quad (5.1)$$

In light of Definition 5.1 it is natural to ask whether $\mathscr{L}^{-1}F$ is unique. This is *almost*[1] true in the following sense:

---
[1]Pun intended: In *measure theory*, the field of mathematics that is concerned with

> **Theorem 5.2 (Lerch)** Let f and g be two piecewise continuous functions in the time domain which have exponential order. If there is an $a \in \mathbb{R}$ so that
> $$(\mathscr{L}f)(s) = (\mathscr{L}g)(s), \qquad (5.2)$$
> for all $s \in \mathbb{C}$ with $\mathbf{re}(s) > a$, then $f(t) = g(t)$ for all $t > 0$, expect, potentially, at the discontinuity points of f and g.

Proof. The proof can be found in [M16, Theorem 2.1]. □

Lerch's theorem allows us to compile a table of Laplace transforms and establish a one-to-one correspondence between t-domain and s-domain functions.

Table 5.1: Table of Laplace transforms

Function $f(t)$	Laplace transform	Defined for	See Section
$\delta(t)$	1	$s \in \mathbb{C}$	4.4.4
$H_0(t)$	$1/s$	$\mathbf{re}(s) > 0$	4.2.2
e^{at}	$1/{s-a}$	$\mathbf{re}(s) > a$	4.2.3
$t^\nu,\ \nu \in \mathbb{N}_+$	$\nu!/s^{\nu+1}$	$\mathbf{re}(s) > 0$	4.2.4
$t^a,\ a \in \mathbb{R}_+$	$\Gamma(a+1)/s^{a+1}$	$\mathbf{re}(s) > 0$	4.2.4
$\sin(\omega t)$	$\omega/s^2+\omega^2$	$\mathbf{re}(s) > 0$	4.2.5
$\cos(\omega t)$	$s/s^2+\omega^2$	$\mathbf{re}(s) > 0$	4.2.5
$t^n g(t),\ n \in \mathbb{N}_+$	$(-1)^n G^{(n)}(s)$		4.4.2

According to Table 5.1,

$$\mathscr{L}^{-1}\left\{\frac{1}{s}\right\}(t) = H_0(t),$$

$$\mathscr{L}^{-1}\left\{\frac{1}{s-a}\right\}(t) = e^{at},$$

$$\mathscr{L}^{-1}\left\{\frac{\omega}{s^2+\omega^2}\right\}(t) = \sin(\omega t),$$

and so on.

---

measuring the size of sets, we say that functions such as f and g in Lerch's Theorem, are equal *almost* everywhere.

Chapter 5. Inverse Laplace transform

The inverse Laplace transform is linear as stated in the following theorem

> **Theorem 5.3 (Linearity)** Let f and g be two piecewise continuous functions in the time domain whose Laplace transforms are F and G respectively. Then, for all $a, b \in \mathbb{R}$:
> $$\mathscr{L}^{-1}\{aF + bG\}(t) = af(t) + bg(t), \tag{5.3}$$
> except, potentially, for those points t where f or g are discontinuous.

Proof. This is a consequence of the linearity of the Laplace transform (Theorem 4.14) and Lerch's Theorem (Theorem 5.2). $\square$

This property, alongside Table 5.1, can be used to compute some inverse Laplace transforms. Here is an example:

Example 5.1 (Function with imaginary poles) We shall find the inverse Laplace transform of
$$F(s) = \frac{2s+1}{s^2+2}. \tag{5.4}$$

In light of Theorem 5.3,

$$\begin{aligned}(\mathscr{L}^{-1}F)(t) &= \mathscr{L}^{-1}\left\{\frac{2s+1}{s^2+2}\right\} \\ &= \mathscr{L}^{-1}\left\{\frac{2s}{s^2+2} + \frac{1}{s^2+2}\right\} \\ &= 2\mathscr{L}^{-1}\left\{\frac{s}{s^2+\sqrt{2}^2}\right\} + \frac{1}{\sqrt{2}}\mathscr{L}^{-1}\left\{\frac{\sqrt{2}}{s^2+\sqrt{2}^2}\right\} \\ &= 2\cos(\sqrt{2}t) + \frac{\sqrt{2}}{2}\sin(\sqrt{2}t),\end{aligned}$$

for $t \geq 0$. A plot of this time domain function is shown below:

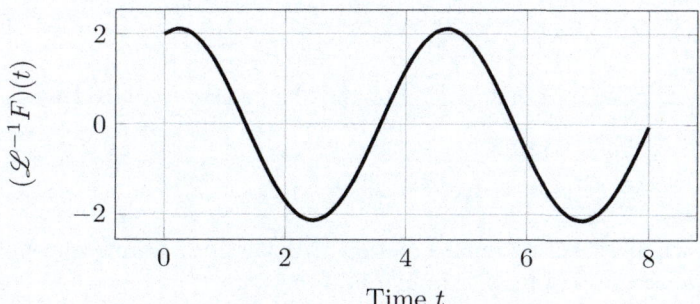

Figure 5.1: *Inverse Laplace of F in Equation (5.4).*

5.2 Rational functions

Hereafter, we focus mainly on the inverse Laplace of rational functions, that is, functions of the form

$$F(s) = \frac{P(s)}{Q(s)},$$

where P and Q are *coprime polynomials*[2] and the degree of P is smaller than the degree of Q[3]. Hereafter, we denote the degree of P by n and the degree of Q by m, that is

$$P(s) = a_0 + a_1 s + \ldots + a_n s^n,$$
$$Q(s) = b_0 + b_1 s + \ldots + b_m s^m,$$

with $a_n, b_m \neq 0$. The roots of the denominator, Q, are called the **poles** of F. The roots of P are called the **zeroes** of F.

The way we compute the inverse Laplace transform of a given function in the s domain depends on the location of its poles on the complex plane. We will use different methodologies depending on whether the function has distinct real poles, multiple real poles and single or multiple complex conjugate poles.

5.2.1 Distinct simple real poles

In this section we assume that Q has m distinct real poles, that is Q can be factorised as follows

$$Q(s) = a(s - s_1)(s - s_2) \cdot \ldots \cdot (s - s_m). \tag{5.5}$$

Our objective is to write F as a sum of simple terms — in particular, simple fractions — akin to those in Example 5.1. The trick is that we may expand F into **partial fractions**:

$$F(s) = \frac{P(s)}{Q(s)} = \frac{A_1}{s - s_1} + \frac{A_2}{s - s_2} + \ldots + \frac{A_m}{s - s_m}, \tag{5.6}$$

where $A_1, A_2, \ldots, A_m$ are real constants and the inverse Laplace transform of F will be

$$(\mathcal{L}^{-1} F)(t) = A_1 e^{s_1 t} + A_2 e^{s_2 t} + \ldots A_m e^{s_m t}. \tag{5.7}$$

Apparently, when F is a rational function with distinct simple real poles, its inverse Laplace is a sum of exponential functions.

---
[2] Two polynomials P and Q are said to be coprime if they share no roots.
[3] This is the most typical case in physical systems. The case where the P and Q have the same degree is discussed in Section 5.2.7

Example 5.2 (Two distinct real poles) We will find the inverse Laplace transform of
$$F(s) = \frac{s}{s^2 - 1}. \tag{5.8}$$
It is easy to factorise the denominator of F:
$$F(s) = \frac{s}{(s-1)(s+1)}. \tag{5.9}$$
Indeed, F has two distinct real poles, $s_1 = 1$ and $s_2 = -1$. We can, therefore, expand F into partial fractions as follows
$$F(s) = \frac{s}{(s-1)(s+1)} = \frac{A_1}{s-1} + \frac{A_2}{s+1}, \tag{5.10}$$
where A_1 and A_2 are constants to be determined. We multiply both sides of this last equation by $(s-1)(s+1)$ to obtain
$$s = A_1(s+1) + A_2(s-1), \tag{5.11}$$
and we collect terms of the same order on both sides, that is
$$1\,s + 0 = (A_1 + A_2)\,s + (A_1 - A_2). \tag{5.12}$$
The two polynomials are equal if and only if the corresponding coefficients are equal to one another, that is
$$\text{Order } s^0 : 0 = A_1 - A_2, \tag{5.13a}$$
$$\text{Order } s^1 : 1 = A_1 + A_2. \tag{5.13b}$$
Solving this linear system of equations gives $A_1 = A_2 = 1/2$, so
$$F(s) = \frac{1/2}{s-1} + \frac{1/2}{s+1}, \tag{5.14}$$
thus,
$$(\mathscr{L}^{-1} F)(t) = \tfrac{1}{2}(e^t + e^{-t}) = \cosh t, \tag{5.15}$$
defined for $t \geq 0$.

Exercise 5.1 (Distinct real poles) Find the inverse Laplace transform of the s-domain function
$$F(s) = \frac{s + \tfrac{1}{2}}{s^3 + 8s^2 + 17s + 10}. \tag{5.16}$$

※

> *Hint.* Note that $s = -1$ is a pole of F. You may use Horner's scheme (see Section A.3.2).
>
> ✂ ··
>
> *Answer.* The inverse Laplace transform of F is $(\mathscr{L}^{-1}F)(t) = \frac{1}{2}e^{-2t} - \frac{2}{3}e^{-t} - \frac{8}{3}e^{-5t}$, for $t \geq 0$.

Let us present another example:

Example 5.3 (Function with pole at zero) We shall determine the inverse Laplace transform of

$$F(s) = \frac{2s+3}{s(s+2)(2s+1)}. \tag{5.17}$$

We first observe that F has only distinct simple real poles, namely, $s_1 = 0$, $s_2 = -2$ and $s_3 = -1/2$. We expand F into partial fractions:

$$F(s) = \frac{s + 3/2}{s(s+2)(s+1/2)} = \frac{A_1}{s} + \frac{A_2}{s+2} + \frac{A_3}{s+1/2}. \tag{5.18}$$

In order to determine the real constants A_1, A_2 and A_3, we multiply both sides of Equation (5.18) by $s(s+2)(s+1/2)$.

$$\frac{s + 3/2}{s(s+2)(s+1/2)} = \frac{A_1}{s} + \frac{A_2}{s+2} + \frac{A_3}{s+1/2}$$
$$\Rightarrow s + 3/2 = A_1(s+2)(s+1/2) + A_2 s(s+1/2) + A_3 s(s+2)$$
$$\Rightarrow s + 3/2 = A_1(s^2 + 5/2 s + 1) + A_2(s^2 + 1/2 s) + A_3(s^2 + 2s)$$
$$\Rightarrow s + 3/2 = (A_1 + A_2 + A_3)s^2 + (5/2 A_1 + 1/2 A_2 + 2A_3)s + A_1$$
$$\Rightarrow 0 \cdot s^2 + 1 \cdot s + 3/2 = (A_1 + A_2 + A_3)s^2$$
$$\qquad\qquad + (5/2 A_1 + 1/2 A_2 + 2A_3)s + A_1$$

This is an equality of polynomials, therefore, by comparing the corresponding coefficients we obtain the linear system

$$A_1 + A_2 + A_3 = 0, \tag{5.19a}$$
$$5/2 A_1 + 1/2 A_2 + 2A_3 = 1, \tag{5.19b}$$
$$A_1 = 3/2. \tag{5.19c}$$

The solution of this system is

$$A_1 = 3/2, A_2 = -1/6, A_3 = -4/3, \tag{5.20}$$

so F is written as

$$F(s) = \frac{3/2}{s} - \frac{1/6}{s+2} - \frac{4/3}{s+1/2}, \tag{5.21}$$

therefore,
$$(\mathscr{L}^{-1}F)(t) = \frac{3}{2} - \frac{1}{6}e^{-2t} - \frac{4}{3}e^{-\frac{t}{2}}, \qquad (5.22)$$
for $t \geq 0$.

5.2.2 Multiple real poles

Suppose that Q has a root z of multiplicity κ, that is
$$Q(s) = (s - z)^\kappa q(s), \qquad (5.23)$$
where q is a polynomial with $q(z) \neq 0$. Then, the partial fractions expansion associated with this root is not merely $A/s-z$, but
$$\frac{A_1}{s-z} + \frac{A_2}{(s-z)^2} + \ldots + \frac{A_\kappa}{(s-z)^\kappa}. \qquad (5.24)$$
We then use the following inverse Laplace rule
$$\mathscr{L}^{-1}\left\{\frac{1}{(s-z)^\nu}\right\} = \frac{t^{\nu-1}}{(\nu-1)!}e^{zt}, \qquad (5.25)$$
for $t \geq 0$. For instance,
$$\mathscr{L}^{-1}\left\{\frac{1}{(s+3)^2}\right\} = te^{-3t}, \qquad (5.26)$$
for $t \geq 0$. Let us demonstrate this concept with an example:

Example 5.4 (Multiple real poles) We will find the inverse Laplace transform of
$$F(s) = \frac{s+3}{(s+1)^2(s-3)}. \qquad (5.27)$$
We observe that F has three poles: $s_1 = s_2 = -1$ (with multiplicity 2) and $s_3 = 3$. We expand F as a sum of partial fractions:
$$F(s) = \frac{s+3}{(s+1)^2(s-3)} = \frac{A}{s-3} + \frac{B_1}{s+1} + \frac{B_2}{(s+1)^2}. \qquad (5.28)$$
We now multiply both sides by $(s+1)^2(s-3)$ and obtain
$$s + 3 = A(s+1)^2 + B_1(s-3)(s+1) + B_2(s-3)$$
$$\Rightarrow s + 3 = A(s^2 + 2s + 1) + B_1(s^2 - 2s - 3) + B_2(s - 3)$$
$$\Rightarrow s + 3 = (A + B_1)s^2 + (2A - 2B_1 + B_2)s + (A - 3B_1 - 3B_2)$$
$$\Rightarrow 0 \cdot s^2 + 1 \cdot s + 3 = (A + B_1)s^2$$
$$\qquad + (2A - 2B_1 + B_2)s + (A - 3B_1 - 3B_2).$$

This equality of polynomials leads to the system

$$A + B_1 = 0, \qquad (5.29a)$$
$$2A - 2B_1 + B_2 = 1, \qquad (5.29b)$$
$$A - 3B_1 - 3B_2 = 3. \qquad (5.29c)$$

The solution is

$$A = 3/8,\ B_1 = -3/8,\ B_2 = -1/2, \qquad (5.30)$$

so, F is written as

$$F(s) = \frac{3/8}{s-3} - \frac{3/8}{s+1} - \frac{1/2}{(s+1)^2}. \qquad (5.31)$$

and using Equation (5.25),

$$(\mathscr{L}^{-1}F)(t) = \tfrac{3}{8}e^{3t} - \tfrac{3}{8}e^{-t} - \tfrac{1}{2}te^{-t}, \qquad (5.32)$$

for $t \geq 0$.

5.2.3 Simple complex conjugate poles

If the denominator contains factors of the form

$$(s+a)^2 + b^2, \qquad (5.33)$$

that is, if $-a \pm jb$ are simple complex conjugate poles, then the corresponding fractions in the partial fractions expansion will have the form

$$\frac{B_1 s + B_2}{(s+a)^2 + b^2}. \qquad (5.34)$$

In order to apply the inverse Laplace transform we use the identities

$$\mathscr{L}^{-1}\left\{\frac{b}{(s+a)^2 + b^2}\right\} = e^{-at}\sin(bt), \qquad (5.35a)$$

$$\mathscr{L}^{-1}\left\{\frac{s+a}{(s+a)^2 + b^2}\right\} = e^{-at}\cos(bt). \qquad (5.35b)$$

Exercise 5.2 (Proof) Prove the above identities.

These identities lead to the following result which will be used to compute the inverse Laplace transform of terms of the form (5.34).

Chapter 5. Inverse Laplace transform

Proposition 5.4 (Simple complex conjugate poles) The following holds

$$\mathscr{L}^{-1}\left\{\frac{B_1 s + B_2}{(s+a)^2 + b^2}\right\} = K e^{-at} \sin(bt + \phi_0), \tag{5.36}$$

for $t \geq 0$, where

$$K = \tfrac{1}{b}\sqrt{(bB_1)^2 + (B_2 - B_1 a)^2}, \tag{5.37}$$

and

$$\phi_0 = \operatorname{atan2}(bB_1, B_2 - B_1 a). \tag{5.38}$$

Proof. The proof is based on the inverse Laplace transforms in Equation (5.35) and the trigonometric identity in Equation (A.37). In particular,

$$\begin{aligned}
\mathscr{L}^{-1}\left\{\frac{B_1 s + B_2}{(s+a)^2 + b^2}\right\} &= \mathscr{L}^{-1}\left\{\frac{B_1(s+a) + B_2 - B_1 a}{(s+a)^2 + b^2}\right\} \\
&= \mathscr{L}^{-1}\left\{\frac{B_1(s+a) + \frac{B_2 - B_1 a}{b} b}{(s+a)^2 + b^2}\right\} \\
&= B_1 \mathscr{L}^{-1}\left\{\frac{s+a}{(s+a)^2 + b^2}\right\} \\
&\quad + \frac{B_2 - B_1 a}{b} \mathscr{L}^{-1}\left\{\frac{b}{(s+a)^2 + b^2}\right\} \\
&= B_1 e^{-at} \cos(bt) + \frac{B_2 - B_1 a}{b} e^{-at} \sin(bt) \\
&= \tfrac{1}{b} e^{-at}\left[bB_1 \cos(bt) + (B_2 - B_1 a)\sin(bt)\right] \\
&= K e^{-at} \sin(bt + \phi_0),
\end{aligned}$$

for $t \geq 0$, with

$$K = \tfrac{1}{b}\sqrt{(bB_1)^2 + (B_2 - B_1 a)^2}, \tag{5.39a}$$
$$\phi_0 = \operatorname{atan2}(bB_1, B_2 - B_1 a), \tag{5.39b}$$

by virtue of the trigonometric identity of Equation (A.37). This completes the proof. $\square$

Example 5.5 (Function with complex poles #1) We shall find the inverse Laplace transform of

$$F(s) = \frac{s+2}{s^2 + s + 1}. \tag{5.40}$$

Its poles are $s_{1,2} = -1/2 \pm j\sqrt{3}/2$, we may, therefore, write it in the

form
$$F(s) = \frac{s+2}{(s+1/2)^2 + (\sqrt{3}/2)^2}. \tag{5.41}$$

Now F is in the form of Equation (5.36) with $B_1 = 1$, $B_2 = 2$, $a = 1/2$ and $b = \sqrt{3}/2$. We compute K and ϕ_0

$$K = \tfrac{1}{b}\sqrt{(bB_1)^2 + (B_2 - B_1 a)^2} = 2, \tag{5.42}$$

and

$$\begin{aligned}\phi_0 &= \operatorname{atan2}(bB_1, B_2 - B_1 a) \\ &= \operatorname{atan2}\left(\sqrt{3}/2, 3/2\right) = \arctan \sqrt{3}/3 = \pi/6.\end{aligned} \tag{5.43}$$

Note that ϕ_0 is reported in radians, not degrees. By means of Proposition (5.4), we now have

$$f(t) = 2e^{-t/2} \sin(\sqrt{3}t/2 + \pi/6), \tag{5.44}$$

for $t \geq 0$.

Let us give one more example

Example 5.6 (Function with complex poles #2) A dynamical system is described, in the s-domain, by the following equation

$$X(s) = \frac{1}{s^2 + 2s + 5} U(s), \tag{5.45}$$

where $X(s) = (\mathscr{L}x)(s)$ is the Laplace transform of the system state and $U(s) = (\mathscr{L}u)(s)$ is the Laplace transform of its input. The system is excited with a Heaviside function (a step), $u(t) = H_0(t)$, and all initial conditions are assumed to be zero ($x(0) = \dot{x}(0) = \ddot{x}(0) = 0$). We need to find the system response, $x(t)$.

The Laplace transform of $u(t) = H_0(t)$ is $U(s) = 1/s$, so $X(s)$ becomes

$$X(s) = \frac{1}{s(s^2 + 2s + 5)}. \tag{5.46}$$

The roots of $s^2 + 2s + 5$ are $-1 \pm j2$, so, X is written as

$$X(s) = \frac{1}{s((s+1)^2 + 4)}, \tag{5.47}$$

and its partial fraction expansion will be

$$X(s) = \frac{1}{s((s+1)^2 + 4)} = \frac{A}{s} + \frac{B_1 s + B_2}{(s+1)^2 + 4}. \tag{5.48}$$

Chapter 5. Inverse Laplace transform

In order to determine A, B_1 and B_2 we multiply both sides of this last equation by $s((s+1)^2 + 4)$, and we have

$$1 = A((s+1)^2 + 4) + (B_1 s + B_2)s$$
$$\Rightarrow 1 = A(s^2 + 2s + 5) + B_1 s^2 + B_2 s$$
$$\Rightarrow 1 = As^2 + 2As + 5A + B_1 s^2 + B_2 s$$
$$\Rightarrow 0s^2 + 0s + 1 = (A + B_1)s^2 + (2A + B_2)s + 5A.$$

This is an equality of polynomials, so, by comparing the corresponding coefficients, we have the system

$$A + B_1 = 0, \qquad (5.49\text{a})$$
$$2A + B_2 = 0, \qquad (5.49\text{b})$$
$$5A = 1, \qquad (5.49\text{c})$$

which has the solution

$$A = 1/5, B_1 = -1/5, B_2 = -2/5, \qquad (5.50)$$

so,

$$X(s) = \frac{1/5}{s} - \frac{1}{5} \frac{s+2}{(s+1)^2 + 4}. \qquad (5.51)$$

The inverse Laplace of the first term is equal to $1/5$ for all $t \geq 0$ — or, what is the same, $1/5 H_0(t)$ — and the second term is of the form presented in Proposition 5.4. We compute K and ϕ_0 for

$$\mathscr{L}^{-1}\left\{\frac{s+2}{(s+1)^2 + 4}\right\}.$$

It is

$$K = \tfrac{1}{b}\sqrt{(bB_1)^2 + (B_2 - B_1 a)^2} = \sqrt{5}/2, \qquad (5.52)$$

and

$$\phi_0 = \operatorname{atan2}(1 \cdot 2, 2 - 1 \cdot 1) = \operatorname{atan2}(2, 1) = \arctan 2, \qquad (5.53)$$

Therefore,

$$x(t) = 1/5 - \sqrt{5}/10 \cdot e^{-t} \sin(2t + \arctan 2), \qquad (5.54)$$

for $t \geq 0$. The response is shown below:

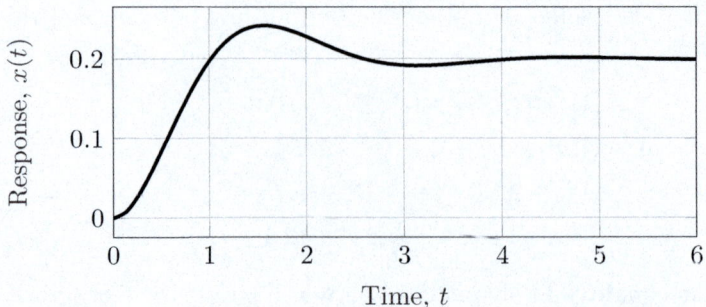

Figure 5.2: *Response of the system, described by Equation (5.54).*

Exercise 5.3 (Function with real, pure imaginary and complex poles) Find the inverse Laplace transform of the following function

$$F(s) = \frac{2s+7}{(2s+1)(s^2+1)((s+2)^2+9)}. \qquad (5.55)$$

✻ ....

Answer. The inverse Laplace transform of F is

$$(\mathscr{L}^{-1}F)(t) = \frac{75}{16}e^{-\frac{t}{2}} - \frac{40}{106}\sin(t - 1.1506) + \frac{120}{\sqrt{10}}e^{-2t}\sin(3t + 0.9653), \qquad (5.56)$$

defined for $t \geq 0$.

✂ ....

You should find that $A_1 = \frac{75}{16}$, $B_1 = -\frac{200}{47}$, $B_2 = \frac{200}{21}$, $C_1 = \frac{600}{13}$, and $C_2 = \frac{600}{53}$.

Hint. Expand F in partial fractions as follows

$$F(s) = \frac{A_1}{s + 1/2} + \frac{B_1 s + B_2}{s^2 + 1} + \frac{C_1 s + C_2}{(s+2)^2 + 3^2}. \qquad (5.57)$$

Remark: In Exercise 5.31 we give an alternative way of determining inverse Laplace transforms, which is particularly convenient when it comes to working with lengthy partial fraction expansions. It is a good idea to solve Exercise 5.31 and revisit Exercise 5.3. •

5.2.4 Simple complex conjugate poles – revisited**

It is possible to treat the case of simple complex conjugate the same way as that of distinct real poles. For example, suppose that the denominator of F has a pair $z = a + jb$, $\bar{z} = a - jb$, of complex conjugate poles. This can be associated with the following (complex) terms in the partial fraction expansion

$$\frac{A_1}{s-z} + \frac{A_2}{s-\bar{z}}. \tag{5.58}$$

The inverse Laplace transforms of these terms are $A_1 e^{zt} = A_1 e^{(a+jb)t} = A_1 e^{a}(\cos(bt) + j\sin(bt))$ and $A_1 e^{\bar{z}t}$. This will be better understood through the following example.

Example 5.7 (Function with complex poles) We will determine the inverse Laplace transform of function F in Example 5.5,

$$F(s) = \frac{s+2}{s^2+s+1} = \frac{s+2}{[s-(-1/2+j\sqrt{3}/2)][s-(-1/2-j\sqrt{3}/2)]}. \tag{5.59}$$

We define $z = a + jb$ with $a = -1/2$ and $b = \sqrt{3}/2$. We now expand this function in partial fractions

$$\frac{s+2}{(s-z)(s-\bar{z})} = \frac{A_1}{s-z} + \frac{A_2}{s-\bar{z}}$$
$$\Rightarrow s+2 = A_1(s-\bar{z}) + A_2(s-z)$$
$$\Rightarrow 1 \cdot s + 2 = (A_1+A_2)s + (-A_1\bar{z} - A_2 z), \tag{5.60}$$

which leads to the system

$$A_1 + A_2 = 1, \tag{5.61a}$$
$$-\bar{z}A_1 - zA_2 = 2, \tag{5.61b}$$

leading to $A_1 = \frac{1}{2}(1 - j\sqrt{3})$ and $A_2 = \frac{1}{2}(1 + j\sqrt{3})$. So, the inverse Laplace transform is

$$f(t) = A_1 e^{zt} + A_2 e^{\bar{z}t} = 2e^{-t/2}\sin(\pi/6 + \sqrt{3}/2\,t), \tag{5.62}$$

for $t \geq 0$.

5.2.5 Multiple complex conjugate poles

We will only cover the case of complex conjugate poles of multiplicity 2, that is, factors of the form $(s^2 + b^2)^2$, for example

$$F(s) = \frac{s+3}{(s^2 - 2s + 1)(s^2 + 3^2)^2}.$$

The partial fraction expansion terms that correspond to such factors have the form
$$\frac{A_1 s + A_2}{s^2 + b^2} + \frac{B_1 s + B_2}{(s^2 + b^2)^2}.$$

The inverse Laplace transform of the second term is given in the following proposition

> **Proposition 5.5 (Multiple complex conjugate poles)** The following hold
> $$\mathcal{L}^{-1}\left\{\frac{1}{(s^2 + \omega^2)^2}\right\} = \frac{1}{2\omega^3}(\sin(\omega t) - \omega t \cos(\omega t)), \quad (5.63a)$$
> $$\mathcal{L}^{-1}\left\{\frac{s}{(s^2 + \omega^2)^2}\right\} = \frac{t}{2\omega}\sin(\omega t). \quad (5.63b)$$

Proof. The proof is left to the reader as an exercise. $\square$

> **Exercise 5.4 (Multiple complex conjugate poles)** Compute the inverse Laplace transforms of the following functions
> $$F_1(s) = \frac{1}{((s+a)^2 + \omega^2)^2}, \quad (5.64a)$$
> $$F_2(s) = \frac{s}{((s+a)^2 + \omega^2)^2}, \quad (5.64b)$$
>
> and then combine the above results to determine the inverse Laplace transform of
> $$F_3(s) = \frac{A_1 s + A_2}{((s+a)^2 + \omega^2)^2}. \quad (5.64c)$$
>
> ✂ ....
>
> *Answer.* The inverse Laplace transform of F_3 is
> $$\mathcal{L}^{-1}\left\{\frac{A_1 s + A_2}{((s+a)^2 + \omega^2)^2}\right\} = \frac{1}{2\omega^3} e^{-at} \cdot$$
> $$[(A_1(\omega^2 - a) + A_2)\sin(\omega t) + (A_1 a - A_2)\omega t \cos(\omega t)], \quad (5.65)$$
>
> for $t \geq 0$.
>
> *Hint.* Recall that according to Theorem 4.18, $\mathscr{L}\{e^{-at}f(t)\} = F(s+a)$.
>
> ✂ ....

Let us have a look at the following example

Chapter 5. Inverse Laplace transform

Example 5.8 (Double imaginary poles) We shall compute the inverse Laplace transform of

$$F(s) = \frac{s}{(s+1)(s^2+4)^2}. \tag{5.66}$$

Note that F has a real pole, $s_1 = -1$, and two imaginary poles $s_{2,3} = \pm j2$, each with multiplicity 2. We first need to expand F in partial fractions

$$F(s) = \frac{s}{(s+1)(s^2+4)^2} = \frac{A}{s+1} + \frac{B_1 s + B_2}{s^2+4} + \frac{C_1 s + C_2}{(s^2+4)^2}. \tag{5.67}$$

We multiply both sides by $(s+1)(s^2+4)^2$

$$s = A(s^2+4)^2 + (B_1 s + B_2)(s+1)(s^2+4) + (C_1 s + C_2)(s+1)$$
$$\Rightarrow s = A(s^4 + 8s^2 + 16) + (B_1 s + B_2)(s^3 + s^2 + 4s + 4)$$
$$\quad + C_2 + C_1 s + C_2 s + C_1 s^2$$
$$\Rightarrow s = (A + B_1)s^4 + (B_1 + B_2)s^3 + (8A + 4B_1 + B_2 + C_1)s^2$$
$$\quad + (4B_1 + 4B_2 + C_1 + C_2)s + 16A + 4B_2 + C_2.$$

This is again an equality of polynomials, so, by comparing the corresponding coefficients, we obtain the linear system

$$A + B_1 = 0, \tag{5.68a}$$
$$B_1 + B_2 = 0, \tag{5.68b}$$
$$8A + 4B_1 + B_2 + C_1 = 0, \tag{5.68c}$$
$$4B_1 + 4B_2 + C_1 + C_2 = 1, \tag{5.68d}$$
$$16A + 4B_2 + C_2 = 0. \tag{5.68e}$$

Its solution is

$$A = -1/25, B_1 = 1/25, B_2 = -1/25, C_1 = 1/5, C_2 = 4/5, \tag{5.69}$$

so, F can be written as

$$F(s) = -\frac{1/25}{s+1} + \frac{1}{25}\frac{s-1}{s^2+4} + \frac{1}{5}\frac{s+4}{(s^2+4)^2}$$
$$= -\frac{1/25}{s+1} + \frac{1}{25}\frac{s-1}{s^2+4} + \frac{1}{5}\left(\frac{s}{(s^2+4)^2} + \frac{4}{(s^2+4)^2}\right). \tag{5.70}$$

Therefore, the inverse Laplace transform of F is

$$f(t) = -1/25 e^{-t} + 1/25(\cos(2t) - 1/2\sin(2t))$$
$$\quad + 1/5 \left(\tfrac{t}{4}\sin(2t) + \tfrac{1}{4}\sin(2t) - \tfrac{1}{2}t\cos(2t)\right), \tag{5.71}$$

defined for $t \geq 0$.

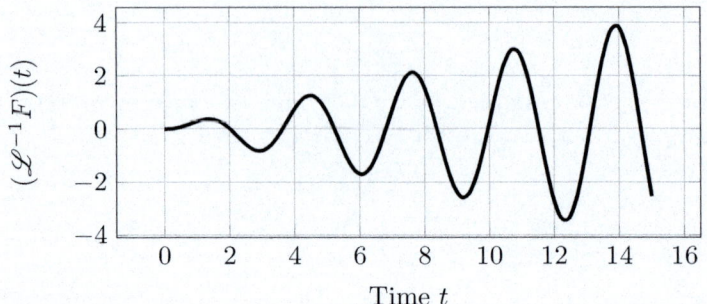

Figure 5.3: *Response of the system with multiple complex poles, described by Equation* (5.71).

Remark: *It is possible to compute inverse Laplace transforms for rational functions with complex poles with multiplicity 3 or higher. However, equations get quite involved. There exist symbolic algebra software (such as Mathematica, SymPy in Python, MATLAB's Symbolic Math Toolbox and more) we can use to derive such inverse Laplace transforms.* •

Exercise 5.5 (Multiple complex conjugate poles) Find the inverse Laplace transform of the following functions

$$F_1(s) = \frac{s+1}{(5s+1)(s^2+1)^2}, \tag{5.72a}$$

$$F_2(s) = \frac{3s^2+s+1}{(s-1/3)(s^2+s+1)^2}. \tag{5.72b}$$

Note that you should first solve Exercise 5.4. Beware that this exercise will try your patience.

※

Answer. The inverse Laplace transform of F_1 is

$$(\mathscr{L}^{-1}F_1)(t) = \frac{169}{25}e^{-\frac{t}{5}} - \frac{5}{2} + \frac{169}{25}\cos t + \frac{338}{49}\sin t - \frac{26}{3}t\cos t - \frac{13}{1}t\sin t, \tag{5.73}$$

for $t \geq 0$. The inverse Laplace transform of F_2 is

$$(\mathscr{L}^{-1}F_2)(t) = \frac{169}{135}e^{\frac{t}{3}} - \frac{169}{135}e^{-\frac{t}{2}}\left[\cos\left(\frac{\sqrt{3}}{2}t\right) + \frac{9}{5\sqrt{3}}\sin\left(\frac{\sqrt{3}}{2}t\right)\right] \\ + \frac{9}{4\sqrt{3}}e^{-\frac{t}{2}}\left[\frac{9-18t}{13}\sin\left(\frac{\sqrt{3}}{2}t\right) - \frac{3\sqrt{3}t}{13}\cos\left(\frac{\sqrt{3}}{2}t\right)\right], \tag{5.74}$$

for $t \geq 0$ (this expression could perhaps be simplified a little further).

✂

Hint. The partial fractions expansion of F_1 is

$$F_1 = \frac{\frac{1}{5} + \frac{1}{4}}{s^2+4} - \frac{\frac{1}{25}s - \frac{1}{25}}{s^2+4} - \frac{1}{25(s+1)}, \tag{5.75}$$

and the partial fractions expansion of F_2 is

$$F_2(s) = \frac{169(3s-1)}{405} + \frac{((s+1/2)^2 + 3/4)}{24} + \frac{13}{6} + \frac{\frac{169}{135}s + \frac{169}{180}}{(s+1/2)^2 + 3/4}. \tag{5.76}$$

5.2.6 One-size-fits-all approach

If F has multiple complex conjugate poles we can work as in Section 5.2.4: we will perform the partial fraction expansion using the complex poles directly.

In Sections 5.2.1 and 5.2.2 we saw that if F has multiple real poles of multiplicity κ, i.e., it involves terms of the form

$$\frac{1}{(s-z)^\kappa}, \tag{5.77}$$

then, the corresponding terms in the partial fraction expansion are

$$\frac{A_1}{s-z} + \frac{A_1}{(s-z)^2} + \ldots + \frac{A_\kappa}{(s-z)^\kappa}, \tag{5.78}$$

where $A_1, \ldots, A_\kappa \in \mathbb{R}$ are constant coefficients and

$$\mathcal{L}^{-1}\left\{\frac{1}{(s-z)^\nu}\right\} = \frac{t^{\nu-1}}{(\nu-1)!}e^{zt}. \tag{5.79}$$

The exact same methodology can be applied in the general case of complex conjugate poles with the only difference that $A_1, \ldots, A_\kappa$ will be complex coefficients.

This means that in the general case where $F(s)$ is a rational function with poles $s_1, \ldots, s_n \in \mathbb{C}$ of multiplicities $\kappa_1, \ldots, \kappa_n$, its inverse Laplace transform has the form

$$f(t) = \sum_{i=1}^{n}\sum_{k=1}^{\kappa_i} A_{i,k} t^{k-1} e^{s_i t}, \tag{5.80}$$

for some constant coefficients $A_{i,k} \in \mathbb{C}$. This is the general form of the inverse Laplace transform of rational functions.

Recall Euler's formula:

$$e^{jb} = \cos b + j\sin b. \tag{5.81}$$

We should expect an inverse Laplace of a rational function to involve terms of the form e^{ct}, $t^a e^{ct}$, $\cos(ct)$, $\sin(ct)$, $t^a \cos(ct)$ and $t^a \sin(ct)$.

5.2.7 Numerator and denominator have equal degrees

So far we have focused on rational functions where the denominator has larger degree than the numerator. The case where the degrees are equal is not too different.

Again, consider a rational function of the form

$$F(s) = \frac{P(s)}{Q(s)},$$

where P and Q are coprime polynomials of equal degree n

$$P(s) = a_0 + a_1 s + \ldots + a_n s^n,$$
$$Q(s) = b_0 + b_1 s + \ldots + b_n s^n.$$

Then, the partial fraction expansion of F is of the same form as outlined above plus the constant term a_n/b_n. Let us give an example.

Example 5.9 (Inverse Laplace transform of non-strictly proper rational function) We will determine the partial fraction expansion of the following function

$$F(s) = \frac{51s^2 - 4s + 5}{3s^2 + 18s + 15}. \tag{5.82}$$

The poles of F are $s_1 = -5$ and $s_2 = -1$ (we can determine them by solving the quadratic equation $3s^2 + 18s + 15 = 0$). The partial fraction expansion of F is

$$F(s) = \cancel{\frac{51}{3}}^{17} + \frac{A_1}{s+5} + \frac{A_2}{s+1}. \tag{5.83}$$

We can then determine A_1 and A_2 as before. We first note that $3s^2 + 18s + 15 = 3(s+1)(s+5)$. We then have

$$\frac{51s^2 - 4s + 5}{3s^2 + 18s + 15} = 17 + \frac{A_1}{s+5} + \frac{A_2}{s+1}$$

$$\Rightarrow \frac{51s^2 - 4s + 5}{\cancel{3s^2 + 18s + 15}} \cancel{3(s+1)(s+5)} = 17(s+1)(s+5)$$

$$\quad + \frac{A_1}{\cancel{s+5}}(s+1)\cancel{(s+5)} + \frac{A_2}{\cancel{s+1}}\cancel{(s+1)}(s+5)$$

$$\Rightarrow \tfrac{1}{3}(51s^2 - 4s + 5) = 17s^2 + 102s + 85 + A_1 s + A_1 + A_2 s + 5A_2$$

$$\Rightarrow 17s^2 - \tfrac{4}{3}s + \tfrac{5}{3} = 17s^2 + (A_1 + A_2 + 102)s + A_1 + 5A_2 + 85. \tag{5.84}$$

By comparing the corresponding terms we formulate the system of equations

$$A_1 + A_2 + 102 = -\tfrac{4}{3}, \tag{5.85a}$$

$$A_1 + 5A_2 + 85 = \tfrac{5}{3}, \tag{5.85b}$$

which yields

$$A_1 = -\tfrac{325}{3} \tag{5.86a}$$
$$A_2 = 5, \tag{5.86b}$$

therefore, the partial fraction expansion is

$$F(s) = \frac{51s^2 - 4s + 5}{3s^2 + 18s + 15} = 17 - \frac{325}{3(s+5)} + \frac{5}{s+1}, \tag{5.87}$$

therefore, the inverse Laplace transform of F is

$$f(t) = (\mathscr{L}^{-1}F)(t) = 17\delta(t) - \tfrac{325}{3}e^{-5t} + 5e^{-t}, \tag{5.88}$$

defined for $t \geq 0$.

5.2.8 Rational times exponential

The inverse Laplace transform of the product of a rational function with an exponential can be computed using the following result

> **Proposition 5.6 (Inverse Laplace of rational-times-exponential)** Let f be a function in the time domain with Laplace transform F. Then, for all $a \geq 0$,
> $$\mathscr{L}^{-1}\{e^{-as}F(s)\} = H_a(t)f(t-a). \tag{5.89}$$

Proof. This follows from Theorem 4.15. Let us repeat that the right-hand side of Equation (5.89) is an abuse of notation as $f(t-a)$ is not defined for $0 \leq t < a$. We have addressed this in the remark following Theorem 4.15. □

We observe the inverse Laplace transform of a function F multiplied by an exponential is a time-delayed version of $f(t) = \mathscr{L}^{-1}\{F(s)\}(t)$.

Example 5.10 (Inverse Laplace of rational-times-exponential) Take the following function in the s domain:

$$F(s) = \frac{e^{-2s}}{s+1}. \tag{5.90}$$

This is the product of the exponential function e^{-2s} with the rational function $G(s) = \frac{1}{s+1}$. The inverse Laplace of G is $g(t) = e^{-t}$,

therefore,

$$\mathscr{L}^{-1}\left\{\frac{e^{-2s}}{s+1}\right\} = \mathscr{L}^{-1}\left\{e^{-2s}G(s)\right\} = H_2(t)g(t-2) = H_2(t)e^{2-t},$$
(5.91)

defined for $t \geq 0$.

Exercise 5.6 (**Inverse Laplace transform of rational-times-exponential function**) Determine the inverse Laplace transform of the following s-domain function

$$F(s) = \frac{5e^{-3s}}{s(2s^2 + 10s + 12)}.$$
(5.92)

❋ ⋯⋯

Answer. The inverse Laplace transform of F is

$$(\mathscr{L}^{-1}F)(t) = 5H_3(t)\left(\frac{1}{2}e^{\partial}9 - \frac{1}{4}e^{-9\partial} - \frac{1}{2}\frac{1}{2}\right).$$
(5.93)

5.2.9 An alternative approach

There is an alternative approach we can follow to determine the coefficients of a partial fractions expansion. Suppose we have a strictly proper rational function $F(s) = \frac{P(s)}{Q(s)}$ and $Q(s) = a(s-s_1)(s-s_2)\ldots(s-s_n)$, $a \neq 0$. Then, the partial fractions expansion of F is

$$F(s) = \frac{P(s)}{Q(s)} = \sum_{i=1}^{n} \frac{A_i}{s-s_i}$$

$$\Leftrightarrow \frac{1}{a}P(s) = A_1 \underbrace{\prod_{i=2,\ldots,n}(s-s_i)}_{\text{Product of all terms except } i=1} + A_2 \underbrace{\prod_{i=1,3\ldots,n}(s-s_i)}_{\text{Product of all terms except } i=2} + \ldots$$

$$+ A_n \underbrace{\prod_{i=1,2\ldots,n-1}(s-s_i)}_{\text{Product of all terms except } i=n}.$$
(5.94)

In order to determine A_1 we can set $s = s_1$ in Equation (5.94) where all terms on the right-hand side, except for the first one, will vanish[4]. We have

$$\frac{1}{a}P(s_1) = A_1 \prod_{i=2,\ldots,n}(s_1-s_i) \Rightarrow A_1 = \frac{P(s_1)}{a\prod_{i=2,\ldots,n}(s_1-s_i)}.$$
(5.95)

---

[4]Since $F(s) = \frac{P(s)}{Q(s)}$, the poles of F are not in its domain: In Equation (5.94) it is assumed that $s \neq s_i$ for all $i = 1, \ldots, n$, so we are not allowed to substitute $s = s_i$. However, we can take the limit as $s \to s_i$, which produces the same result.

Likewise, we can determine all other coefficients. Let us see how this works in an example.

Example 5.11 (Alternative approach to determine the PFE coefficients) Let us determine the partial fraction expansion of

$$F(s) = \frac{s+3}{(s+1)(s+5)(s+10)}. \tag{5.96}$$

It is

$$F(s) = \frac{s+3}{(s+1)(s+5)(s+10)} = \frac{A_1}{s+1} + \frac{A_2}{s+5} + \frac{A_3}{s+10}, \tag{5.97}$$

and by multiplying by $(s+1)(s+5)(s+10)$ we obtain

$$s+3 = A_1(s+5)(s+10) + A_2(s+1)(s+10) + A_3(s+1)(s+5). \tag{5.98}$$

This is precisely Equation (5.94). Then we set $s = -1$ in Equation (5.98)

$$-1+3 = A_1(-1+5)(-1+10) \Rightarrow A_1 = \tfrac{1}{18}. \tag{5.99}$$

Then substitute $s = -5$ in Equation (5.98)

$$-5+3 = A_2(-5+1)(-5+10) \Rightarrow A_2 = \tfrac{1}{10}. \tag{5.100}$$

Lastly, substitute $s = -10$ in Equation (5.98)

$$-10+3 = A_3(-10+1)(-10+5) \Rightarrow A_3 = -\tfrac{7}{45}, \tag{5.101}$$

therefore,

$$F(s) = \frac{\tfrac{1}{18}}{s+1} + \frac{\tfrac{1}{10}}{s+5} - \frac{\tfrac{7}{45}}{s+10}, \tag{5.102}$$

from which we can determine the inverse Laplace transform of F.

The exact same procedure can be applied when we have simple complex poles. We can either use the expansion we presented in Section 5.2.3, or the one from Section 5.2.4. Here is an example

Example 5.12 (Simple complex conjugate poles) Consider the function

$$F(s) = \frac{5s+2}{(s+2)((s+1)^2+1)}. \tag{5.103}$$

We need to determine the coefficients of the following partial fractions expansion:

$$F(s) = \frac{A_1}{s+2} + \frac{B_1 s + B_2}{(s+1)^2+1}. \tag{5.104}$$

By multiplying by $(s+2)((s+1)^2+1)$ we obtain

$$5s + 2 = A_2((s+1)^2 + 1) + (B_1 s + B_2)(s+2). \tag{5.105}$$

For $s = -2$ we obtain

$$-8 = 2A_1 \Rightarrow A_1 = -4. \tag{5.106}$$

For $s = j - 1$, which is one of the poles of F, we have

$$5(j-1) + 2 = (B_1(j-1) + B_2)(j - 1 + 2)$$
$$\Leftrightarrow 5j - 3 = B_2 - 2B_1 + B_2 j. \tag{5.107}$$

By comparing the real and imaginary parts in the above equation we have

$$B_2 = 5, \tag{5.108a}$$
$$B_2 - 2B_1 = -3 \Leftrightarrow B_1 = 4. \tag{5.108b}$$

The approach presented in this section can be extended to be applicable to rational functions with multiple poles. Firstly, let us see why the method is not directly applicable.

Example 5.13 (Limitations of the alternative approach) Suppose we need to determine the partial fractions expansion of the following function

$$F(s) = \frac{1}{(s+1)(s+2)^2} = \frac{A_1}{s+1} + \frac{A_2}{s+2} + \frac{B_2}{(s+2)^2}. \tag{5.109}$$

We multiply both sides of the equation by $(s+1)(s+2)^2$ and obtain

$$1 = A_1(s+2)^2 + A_2(s+1)(s+2) + B_2(s+1). \tag{5.110}$$

We see that if we set $s = -1$ or $s = -2$, the term that involves A_2 vanishes. Therefore, it is not possible to determine A_2. Besides, by setting $s = -1$ and $s = -2$ we end up with two equations, but we have three unknowns (A_1, A_2 and B_2).

In the above example one could recommend to set s to a value that is not a pole of F; for example, $s = 0$. However, there is a more efficient way to solve such problems. The trick is that if s_0 is a root of a polynomial Q with multiplicity κ, then it is a root all derivatives of Q up to order $\kappa - 1$. For instance, in Example 5.13 $s = -2$ is a root of $Q(s) = (s+1)(s+2)^2$ with multiplicity $\kappa = 2$ — indeed it is a root of $Q'(s) = 2(s+1)(s+2) + (s+2)^2$. It will become clear how we can use this fact through the following example

Chapter 5. Inverse Laplace transform

Example 5.14 (Limitations of the alternative approach: solution) Again, we want to determine the partial fractions expansion of function F in Equation (5.109) using Equation (5.110). For $s = -1$ we have

$$1 = A_1(-1+2)^2 \Leftrightarrow A_1 = 1. \tag{5.111}$$

For $s = -2$ we have

$$1 = B_2(-2+1) \Leftrightarrow B_2 = -1. \tag{5.112}$$

By differentiating both sides of Equation (5.110) we obtain

$$0 = 2A_1(s+2) + A_2[(s+1) + (s+2)] + B_2 \tag{5.113}$$

and for $s = -2$ we have

$$0 = A_2(-2+1) - 1 \Leftrightarrow A_2 = -1. \tag{5.114}$$

Exercise 5.7 (Application of alternative methodology) Use the above alternative methodology to expand the following function into a partial fractions expansion

$$F(s) = \frac{s}{(2s+1)^3(s+1)(s+2)}. \tag{5.115}$$

Answer. It is

$$F(s) = -\frac{s+1}{1} - \frac{s+2}{27} - \frac{2s+1}{50} + \frac{(2s+1)^2}{\frac{14}{9}} - \frac{(2s+1)^3}{\frac{2}{3}}. \tag{5.116}$$

5.3 Convolution theorem

The **convolution** of two functions is one of the most important operations in systems theory and is heavily used in engineering and mathematics. It has several uses in signal and image processing, probability, statistics, computer vision and differential equations to name a few. Let us first give the definition.

> **Definition 5.7 (Convolution)** Let f and g be two functions in the time domain. Their convolution is denoted by $f * g$; it is a function in the time domain defined as
>
> $$(f * g)(t) = \int_0^t f(\tau)g(t-\tau)\mathrm{d}\tau. \qquad (5.117)$$

Often, the convolution of two time-domain functions is defined as

$$(f * g)(t) = \int_0^\infty f(\tau)g(t-\tau)\mathrm{d}\tau. \qquad (5.118)$$

Here, the integration is taken from $\tau = 0$ to $\tau = \infty$ instead of up to $\tau = t$. These two definitions are **identical**, and we will be using them interchangeably. The second definition implies that functions f and g are defined over all $t \in \mathbb{R}$, whereas, we have defined functions in the time domain to live on the half line $[0, \infty)$. However, t-functions can be **extended** to the whole real line assuming that $f(t) = 0$ for all $t < 0$. This leads to the following proposition.

> **Proposition 5.8 (Equivalence of definitions)** The two definitions of the convolution are equivalent, that is, for two functions $f : \mathbb{R} \to \mathbb{R}$ and $g : \mathbb{R} \to \mathbb{R}$, with $f(t) = 0$ and $g(t) = 0$ for $t < 0$,
>
> $$\int_0^t f(\tau)g(t-\tau)\mathrm{d}\tau = \int_0^\infty f(\tau)g(t-\tau)\mathrm{d}\tau. \qquad (5.119)$$

Proof. It suffices to show that $\int_t^\infty f(\tau)g(t-\tau)\mathrm{d}\tau = 0$. But then $t - \tau < 0$ over the integration interval, so $g(t-\tau) = 0$, which proves the assertion. $\square$

In order to familiarise with this new operator, let us compute the convolution of two functions

Example 5.15 (Convolution of two functions) Let $f(t) = \sin t$ and $g(t) = t^2$. Then,

$$(f * g)(t) = \int_0^t \sin(\tau)(t-\tau)^2 \mathrm{d}\tau = \int_0^t -(\cos(\tau))'(t-\tau)^2 \mathrm{d}\tau,$$

and sing integration by parts

$$= -\cos(\tau)(t-\tau)^2\Big|_0^t - \int_0^t -\cos(\tau)((t-\tau)^2)'d\tau$$

$$= t^2 - \int_0^t 2(t-\tau)\cos\tau d\tau = t^2 - 2\int_0^t (t-\tau)(\sin\tau)'d\tau,$$

and sing integration by parts again

$$= t^2 - 2\left((t-\tau)\sin\tau\Big|_0^t - \int_0^t \sin\tau d\tau\right)$$

$$= t^2 + 2\cos t - 2, \tag{5.120}$$

for $t \geq 0$.

Exercise 5.8 (Properties of the convolution) Prove that the convolution operator has the following properties

1. $f * 1 = \int_0^t f(\tau)d\tau$
2. $f * g = g * f$
3. $f * (g+h) = f * g + f * h$
4. $f * (g * h) = (f * g) * h$
5. $f * 0 = 0$

Assume that f and g are such that the above convolutions are well defined. In #6 it is assumed that f and g are differentiable.

Exercise 5.9 (Unity of convolution operation) As we saw in Exercise 5.8, the convolution of two functions has some properties similar to multiplication. However, we saw that $f * 1$ is not equal to f. Is there a function g such that for all f, $f * g = f$?

※

Answer: For the Dirac delta we have $f * g = f$.

Exercise 5.10 (A few convolutions) Determine the following convolutions:

1. $\sin t * \sin(2t)$
2. $t * t$
3. $\sin(\omega t) * \cos(\omega t)$
4. $t * \sin(at)$
5. $e^t * e^t$
6. $t^2 * \sqrt{t}$
7. $t * p(t)$, where p is an n-th order polynomial

Then, sketch the above convolutions (you may use Python or MATLAB to plot $f * g$):

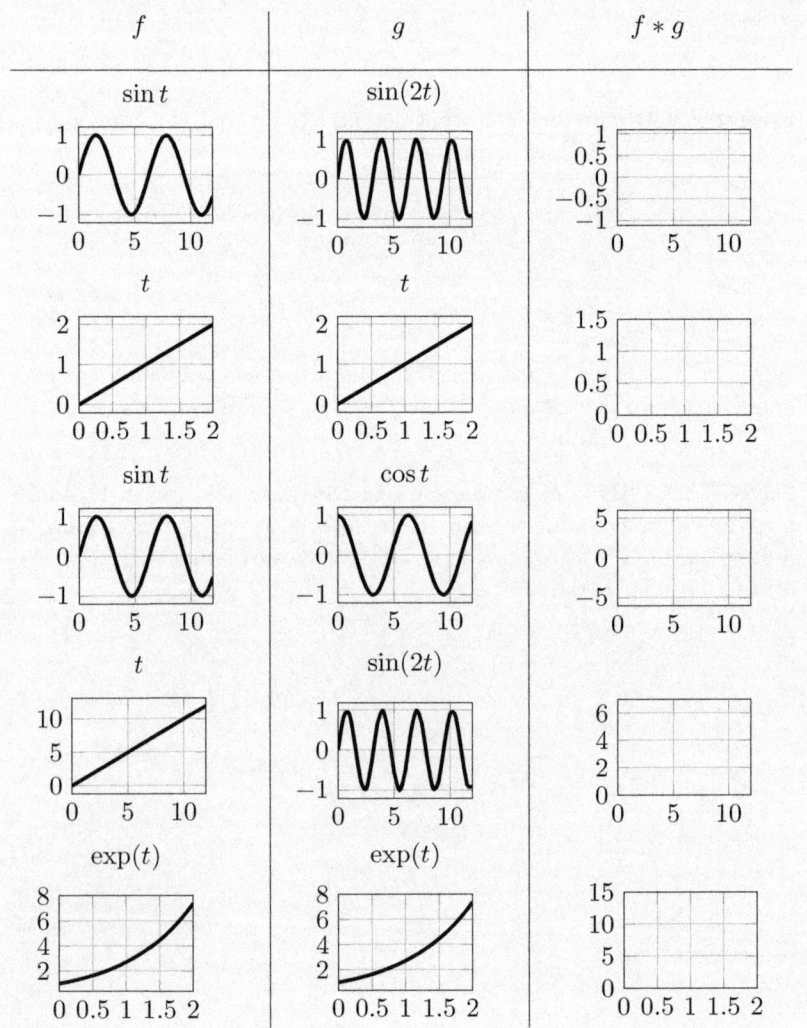

Chapter 5. Inverse Laplace transform

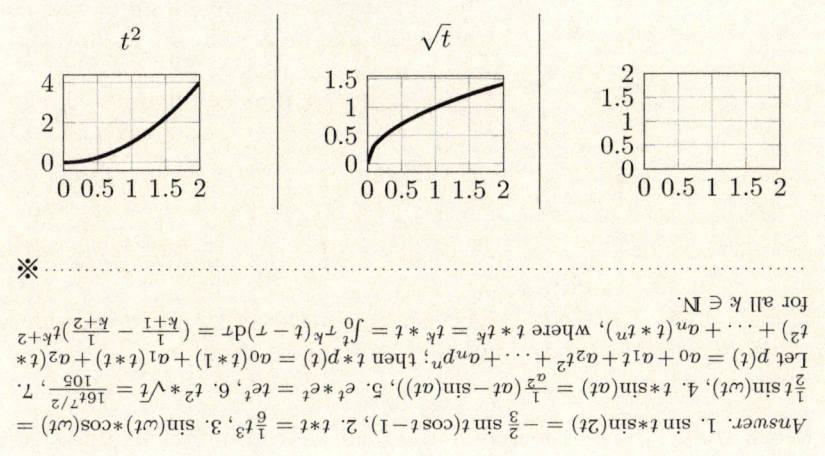

Answer. 1. $\sin t * \sin(2t)$, 2. $t * t$, $-\frac{2}{3}\sin t(\cos t - 1)$, 3. $\frac{2}{3}t^{\frac{3}{2}}$, $t*t = \frac{1}{6}t^3$, 4. $t\sin(\omega t)$, 5. $\frac{a}{2}(a t - \sin(a t))$, 6. $t * t^2 = \frac{1}{12}t^4$, 7. $\frac{16t^{7/2}}{105}$.

Let $p(t) = a_0 + a_1 t + a_2 t^2 + \cdots + a_n t^n$; then $t * p(t) = a_0(t * 1) + a_1(t * t) + a_2(t * t^2) + \cdots + a_n(t * t^n)$, where $t * t^k = \int_0^t t^k \tau (t - \tau) d\tau = \frac{t^{k+2}}{k+1} \left(\frac{t}{k+1} - \frac{t}{k+2} \right)$ for all $k \in \mathbb{N}$.

The convolution operation behaves nicely under the Laplace transform.

> **Theorem 5.9 (Laplace transform of convolution)** Let f, g be two functions in the time domain with Laplace transforms F and G respectively. Then,
> $$\mathcal{L}\{f * g\}(s) = F(s)G(s). \tag{5.121}$$

Proof. Using the definition of the Laplace transform

$$F(s)G(s) = \int_0^\infty e^{-s\xi} f(\xi) d\xi \int_0^\infty e^{-s\tau} g(\tau) d\tau$$
$$= \int_0^\infty \left(\int_0^\infty e^{-s(\xi + \tau)} f(\xi) d\xi \right) g(\tau) d\tau,$$

where we use the change of variables $t(\xi) = \xi + \tau$, therefore, $dt = d\xi$, $t(0) = \tau$ and $t(\infty) = \infty$, so

$$= \int_0^\infty \int_\tau^\infty e^{-st} f(t - \tau) dt g(\tau) d\tau$$
$$= \int_0^\infty \int_\tau^\infty e^{-st} f(t - \tau) g(\tau) dt d\tau.$$

Here we are integrating over a set D in the (t, τ) plane so that τ traverses the half-line $[0, \infty)$ and t goes from τ to ∞. It would be the same to integrate over the same set D having t move from 0 to ∞ and τ go from 0 to t.

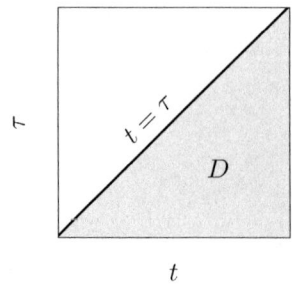

Figure 5.4: Illustration of the domain $D = \{(t, \tau) : \tau < t\}$.

This is illustrated in the figure on the left: the highlighted area is the set D over which we integrate, in other words the above integral can be written as follows

$$F(s)G(s) = \int_D e^{-st} f(t-\tau)g(\tau)(\mathrm{d}t \times \mathrm{d}\tau)$$

$$= \int_0^\infty \int_0^t e^{-st} f(t-\tau)g(\tau)\mathrm{d}\tau \mathrm{d}t$$

The change of the order of integration is allowed by Fubini's theorem.

Recall that Fubini's theorem states that we can interchange the order of integration, provided that the integral is finite. Next, we have

$$F(s)G(s) = \int_0^\infty e^{-st} \left[\int_0^t f(t-\tau)g(\tau)\mathrm{d}\tau \right] \mathrm{d}t$$

$$= \mathscr{L}\left\{ \int_0^t f(t-\tau)g(\tau)\mathrm{d}\tau \right\}(s) = \mathscr{L}\{f * g\}(s).$$

This completes the proof. □

The convolution theorem is very useful for the computation of the inverse Laplace transforms of functions which can be analysed in products whose terms have easier Laplace transforms. For example, if the inverse Laplace transform of $F(s)$ and $G(s)$ can be easily found to be $f(t)$ and $g(t)$ respectively, but the Laplace transform of their product is more difficult to determine directly, we can use the convolution theorem according to which $\mathscr{L}\{f * g\}(s) = F(s)G(s)$ or equivalently $\mathscr{L}^{-1}\{F(s)G(s)\}(t) = f(t) * g(t)$.

For instance, if we are able to compute the inverse Laplace transform of $F(s)$, and it is $\mathscr{L}^{-1}\{F(s)\} = f(t)$, then $\mathscr{L}^{-1}\{F(s)^2\} = f(t) * f(t)$. This way, the problem of computing a difficult inverse Laplace transform reduces to the easier problem of computing a convolution.

Example 5.16 (Application of the convolution theorem) In Proposition 5.5 we gave the inverse Laplace transform of

$$F(s) = \frac{s}{(s^2 + \omega^2)^2}.$$

We can easily compute this by splitting $F(s)$ into a product of functions with known inverse Laplace transform:

$$\mathscr{L}^{-1}\left\{ \frac{s}{(s^2+\omega^2)^2} \right\} = \mathscr{L}^{-1}\left\{ \frac{s}{s^2+\omega^2} \cdot \frac{1}{s^2+\omega^2} \right\},$$

Chapter 5. Inverse Laplace transform

and using the convolution theorem (Theorem 5.9)

$$= \mathscr{L}^{-1}\left\{\frac{s}{s^2+\omega^2}\right\}(t) * \mathscr{L}^{-1}\left\{\frac{1}{s^2+\omega^2}\right\}(t)$$

$$= \frac{1}{\omega}\mathscr{L}^{-1}\left\{\frac{s}{s^2+\omega^2}\right\}(t) * \mathscr{L}^{-1}\left\{\frac{\omega}{s^2+\omega^2}\right\}(t),$$

where these inverse Laplace transforms are known (see Table 5.1), so

$$= \frac{1}{\omega}\cos(\omega t) * \sin(\omega t),$$

and sing the definition of the convolution operator

$$= \frac{1}{\omega}\int_0^t \cos(\omega\tau)\sin(\omega t - \omega\tau)d\tau,$$

so using the trigonometric identity $\sin(a-b) = \sin a \cos b - \cos a \sin b$

$$= \frac{1}{\omega}\int_0^t \cos(\omega\tau)[\sin(\omega t)\cos(\omega\tau) - \cos(\omega t)\sin(\omega\tau)]d\tau$$

$$= \frac{1}{\omega}\left[\sin(\omega t)\int_0^t \cos^2(\omega\tau)d\tau - \cos(\omega t)\int_0^t \sin(\omega\tau)\cos(\omega\tau)d\tau\right]$$

$$= \frac{1}{\omega}\left[\sin(\omega t)\frac{2\omega t + \sin(2\omega t)}{4\omega} - \cos(\omega t)\frac{\sin^2 \omega t}{2\omega}\right],$$

and, finally, using the identity $\sin(2\theta) = 2\sin\theta\cos\theta$,

$$= \frac{t}{2\omega}\sin(\omega t),$$

which confirms the result we obtained in Proposition 5.5(ii).

Exercise 5.11 (Inverse Laplace using the convolution theorem) Prove that

$$\mathscr{L}^{-1}\left\{\frac{s}{(s^2+1)^3}\right\} = \frac{1}{8}\left(t\sin t - t^2 \cos t\right),$$

using the convolution theorem.

✂ ⋯⋯⋯

Hint. There are several ways one can solve this exercise. For example, define $F(s) = \frac{s}{(s^2+1)^3}$ which you can write as $F(s) = F_1(s)F_2(s)$, with $F_1(s) = \frac{s}{(s^2+1)^2}$ and $F_2(s) = \frac{1}{s^2+1}$. The inverse Laplace transforms of F_1 and F_2 are known: $(\mathscr{L}^{-1}F_1)(t) = \frac{t}{2}(\sin t - t\cos t)$ and $(\mathscr{L}^{-1}F_2)(t) = \cos t$. Then, use Theorem 5.9.

Exercise 5.12 (Another exercise using the convolution theorem) Using the convolution theorem, show that

$$\mathscr{L}^{-1}\left\{\frac{1}{(s^2+\omega^2)^2}\right\} = \frac{1}{2\omega^3}(\sin(\omega t) - \omega t \cos(\omega t)). \tag{5.122}$$

This was stated in Proposition 5.5(i).

Let us have a look at another example:

Example 5.17 (System driven by a force $f(t) = \sin^2(t)$) We exercise a force $f(t)$ on an object of mass m which is initially at the origin ($x(0) = 0$) and is at rest ($\dot{x}(0) = 0$). No other forces act on the object. The system dynamics is described by Newton's second law:

$$f(t) = m\ddot{x}(t). \tag{5.123}$$

Suppose that the force is given by

$$f(t) = \sin^2(t), \tag{5.124}$$

for $t \geq 0$. We apply the Laplace transform on both sides of (5.123)

$$F(s) = ms^2 X(s), \tag{5.125}$$

and we solve for $X(s)$

$$X(s) = \frac{1}{ms^2} F(s), \tag{5.126}$$

where $X(s) = (\mathscr{L}x)(s)$ and $F(s) = (\mathscr{L}f)(s)$ are the Laplace transforms of the input and state signals respectively. We have not encountered the Laplace transform of $\sin^2 t$ before; we can either compute it using the definition, or write $\sin^2 t$ in terms of functions whose Laplace transforms are known. By means of the identity (see Equation (A.36b))

$$\sin^2 t = \frac{1}{2}(1 - \cos(2t)), \tag{5.127}$$

we have that

$$F(s) = \frac{2}{s(s^2+4)}, \tag{5.128}$$

so, $X(s)$ becomes

$$X(s) = \frac{2}{ms^3(s^2+4)}. \tag{5.129}$$

We can compute the inverse Laplace transform of the above rational function using a partial fraction expansion. Alternatively, we can use the convolution theorem. In particular,

$$\mathscr{L}^{-1}\left\{\frac{1}{s^3(s^2+4)}\right\} = \mathscr{L}^{-1}\left\{\frac{1}{s^3}\frac{1}{(s^2+4)}\right\} = \tfrac{1}{2}t^2 * \tfrac{1}{2}\sin(2t)$$

$$= \tfrac{1}{4}\left[t^2 * \sin(2t)\right]$$

$$= \tfrac{1}{4}\int_0^t (t-\tau)^2 \sin(2\tau)\mathrm{d}\tau. \qquad (5.130)$$

We have already computed this convolution above

$$= \tfrac{1}{16}\left(2t^2 + \cos(2t) - 1\right), \qquad (5.131)$$

therefore,

$$x(t) = \frac{1}{8m}\left(2t^2 + \cos(2t) - 1\right), \qquad (5.132)$$

for $t \geq 0$.

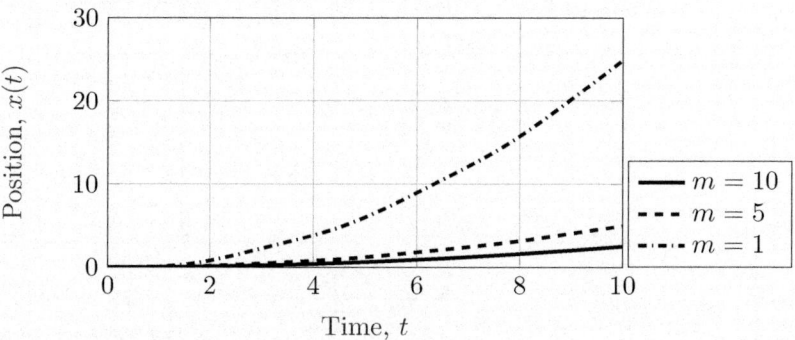

Figure 5.5: *Position of the object, $x(t)$, for three different masses.*

5.4 Solving initial value problems

The Laplace transform maps functions, their derivatives and integrals from the t-domain into the s-domain. There, ODEs and IDEs take the form of simple algebraic equations. Then, the inverse Laplace transform takes functions from the s-domain to the t-domain. By doing so we can solve linear initial value problems of the general form

$$a_0 x + a_1 \dot{x} + \ldots + a_n x^{(n)} = b_0 u + b_1 \dot{u} + \ldots + b_m u^{(m)}, \tag{5.133}$$

with $m < n$, with the initial conditions

$$x(0) = x_0, \dot{x}(0) = \dot{x}_0, \ldots, x^{(n-1)}(0) = x_0^{(n-1)}, \tag{5.134}$$

and a given input signal u. The pertinent procedure is illustrated in Figure 5.6 below.

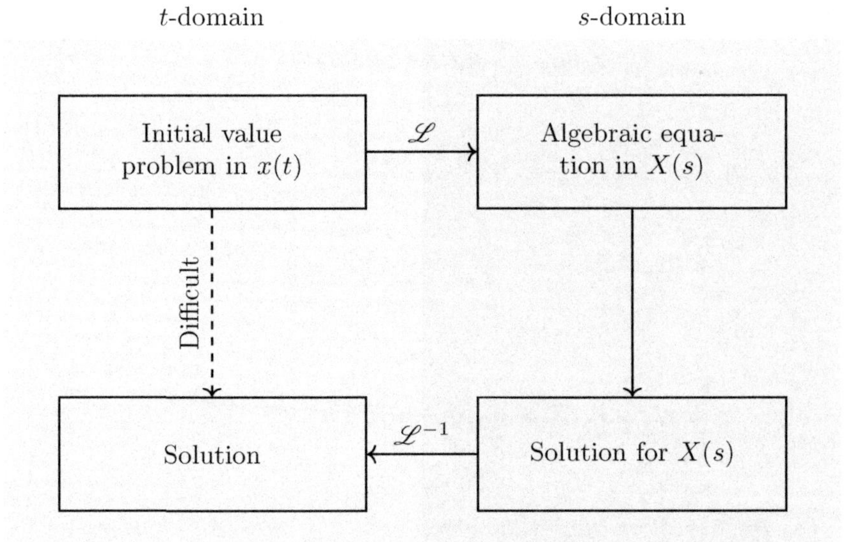

Figure 5.6: *Procedure for solving initial value problems using the Laplace transform. Solving an IVP directly can be difficult; instead, we may apply the Laplace transform on the given ODE/IDE to obtain an algebraic equation that we can solve for the system output. By applying the inverse Laplace transform we recover the solution of the IVP.*

In particular, in order to solve an initial value problem,

1. We apply the Laplace transform on the ODE/IDE to obtain an equation in the s domain
2. We solve the (algebraic) equation in the s domain
3. We apply the inverse Laplace transform to obtain a solution

Let us give an example.

Chapter 5. Inverse Laplace transform

Example 5.18 (Simple initial value problem) Take the initial value problem

$$\dot{x}(t) = x(t), \quad (5.135a)$$
$$x(0) = 1. \quad (5.135b)$$

The first step is to apply the Laplace transform on both sides of the given ODE (see Section 4.3.6):

$$sX(s) - x(0) = X(s) \Leftrightarrow sX(s) - 1 = X(s). \quad (5.136)$$

The second step is to solve for $X(s)$:

$$X(s) = \frac{1}{s-1}. \quad (5.137)$$

Lastly, we apply the inverse Laplace transform to obtain a solution, that is

$$x(t) = \mathscr{L}^{-1}\left\{\frac{1}{s-1}\right\}(t) = e^t, \quad (5.138)$$

for $t \geq 0$. We see that this solution satisfies the initial condition of Equation (5.135b).

Here is another example

Example 5.19 (Nutation control of satellite) The rotational motion of a satellite about its longitudinal axis is called *nutation*. The nutation angle, θ, is affected by the thrust, u, exerted by the thrusters of the satellite. Suppose that the system dynamics is described by

Figure 5.7: *The Terra scientific research satellite. Image taken from Wikipedia, "Terra (satellite)" page, accessed on 17 Feb 2022.*

$$5\ddot{\theta} + \dot{\theta} + 0.1\theta = u + 0.5\dot{u}. \quad (5.139)$$

We want to find the solution, $\theta(t)$ of this system when the thrust

signal is
$$u(t) = e^{-2t}\sin(t/3), \tag{5.140}$$
and the initial conditions are $\theta(0) = 0$ and $\dot{\theta}(0) = 0$.

Step 1. Again, the first step is to apply the Laplace transform
$$5s^2\Theta(s) + s\Theta(s) + 0.1\Theta(s) = U(s) + 0.5sU(s), \tag{5.141}$$
where $\Theta(s) = \mathscr{L}\{\theta(t)\}(s)$ and $U(s) = \mathscr{L}\{u(t)\}(s)$. The Laplace transform of u is
$$U(s) = \mathscr{L}\{u(t)\}(s) = \mathscr{L}\{e^{-2t}\sin(t/3)\}(s) = \frac{1/3}{(s+2)^2 + 1/9}, \tag{5.142}$$
then the system equation, in the s-domain, becomes
$$5s^2\Theta(s) + s\Theta(s) + 0.1\Theta(s) = \frac{1/3(1+s)}{(s+2)^2 + 1/9} \tag{5.143}$$

Step 2. We now solve Equation (5.143) for $\Theta(s)$,
$$\Theta(s) = \frac{1/3(1+s)}{[(s+2)^2 + 1/9](5s^2 + s + 0.1)}. \tag{5.144}$$

Step 3. We now apply the inverse Laplace transform to $\Theta(s)$ in the equation above, that is, $\theta(t) = \mathscr{L}^{-1}\{\Theta(s)\}(t)$. The reader can verify that
$$\theta(t) = 0.00139\Big[e^{-0.1t}(\cos(0.1t) + 116\sin(0.1t))$$
$$- e^{-2t}(\cos(t/3) + 40.51\sin(t/3))\Big]. \tag{5.145}$$

This is the solution of the given initial value problem. The reader can verify that the given initial conditions are satisfied.

Exercise 5.13 (Solve an IVP) Follow the procedure described above to solve the initial value problem
$$\ddot{x} - 3\dot{x} + 10x = 10\sin(2t), \tag{5.146a}$$
$$x(0) = -1, \dot{x}(0) = -2. \tag{5.146b}$$
Determine the solution in the s-domain and verify that the initial conditions are satisfied.

※

Chapter 5. Inverse Laplace transform

Answer. In the s-domain, the solution is

$$X(s) = \frac{-s^3 + s^2 - 4s + 24}{s^4 - 3s^3 + 14s^2 - 12s + 40}. \quad (5.147)$$

We can use the initial value theorem to verify that the initial conditions are indeed satisfied.

In the following example, we use the above procedure for solving initial value problems to find the trajectories of ideal oscillators with sinusoidal excitation.

Example 5.20 (Spring-mass system and resonance) An object of mass m is attached to a spring with constant k. An external force $F(t)$ is exerted on the object as shown in the figure below.

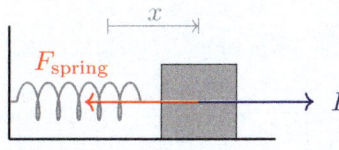

The system is described by the differential equation

$$F(t) - kx(t) = m\ddot{x}(t). \quad (5.148)$$

Suppose that the initial conditions are $x(0) = 0$ and $\dot{x}(0) = 0$.

Suppose that the exercised force is

$$F(t) = A\sin(\omega t), \quad (5.149)$$

so the ODE of the system becomes

$$A\sin(\omega t) - kx(t) = m\ddot{x}(t). \quad (5.150)$$

In order to solve this initial value problem, the first step is to apply the Laplace transform to the ODE. This yields:

$$\frac{A\omega}{s^2 + \omega^2} - kX(s) = ms^2 X(s). \quad (5.151)$$

Then, we solve for $X(s)$,

$$\frac{A\omega}{s^2 + \omega^2} - kX(s) = ms^2 X(s)$$

$$\Rightarrow (ms^2 + k) X(s) = \frac{A\omega}{s^2 + \omega^2}$$

$$\Rightarrow X(s) = \frac{A\omega}{(s^2 + \omega^2)(ms^2 + k)}$$

$$\Rightarrow X(s) = \frac{A\omega}{m(s^2 + \omega^2)(s^2 + k/m)}.$$

For convenience, we define $\beta = \sqrt{k/m}$, so

$$X(s) = \frac{A\omega}{m(s^2 + \omega^2)(s^2 + \beta^2)}. \quad (5.152)$$

The roots of the denominator — the *poles* of X — are $\pm j\omega$ and $\pm j\beta$.

Case I ($\omega \neq \beta$). Then, X has two pairs of complex conjugate poles. The inverse Laplace transform of X is computed as discussed in Section 5.2.3. In particular, X is expanded in partial fractions as follows

$$X(s) = \frac{A\omega}{m(s^2 + \omega^2)(s^2 + \beta^2)} = \frac{B_1 s + B_2}{s^2 + \omega^2} + \frac{C_1 s + C_2}{s^2 + \beta^2}. \quad (5.153)$$

The reader can confirm that the partial fraction expansion of X is

$$X(s) = \frac{A\omega}{m(\beta^2 - \omega^2)(s^2 + \omega^2)} - \frac{A\omega}{m(\beta^2 - \omega^2)(s^2 + \beta^2)}, \quad (5.154)$$

so its inverse Laplace transform is

$$x(t) = \frac{A}{\beta m(\beta^2 - \omega^2)} \left[\beta \sin(\omega t) - \omega \sin(\beta t)\right]. \quad (5.155)$$

Next, we plot $x(t)$ for a system with $\beta = 100$, $m = 1$, $A = 1$ and for different frequencies

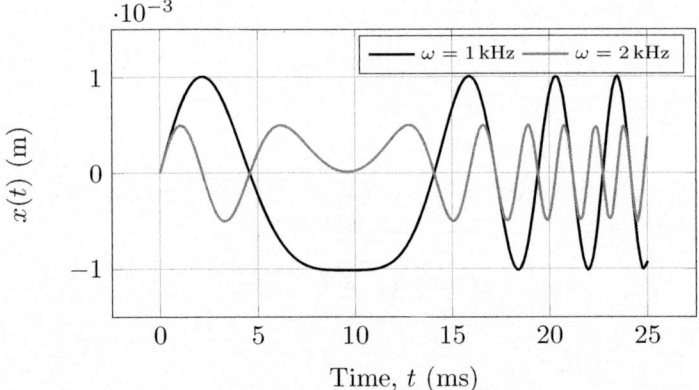

Figure 5.8: Response, $x(t)$, of a perfect oscillator to a force $F(t) = A\sin(\omega t)$ with $\omega \neq \beta$ (Case I).

Case II ($\omega = \beta$). If $\omega = \beta$, then the denominator of $X(s)$ has a pair of complex conjugate roots with multiplicity 2, that is

$$X(s) = \frac{A\omega}{m(s^2 + \omega^2)^2}. \quad (5.156)$$

Then, the inverse Laplace transform of X is

$$x(t) = \frac{A(\sin(\omega t) - \omega t \cos(\omega t))}{2m\omega^2}. \quad (5.157)$$

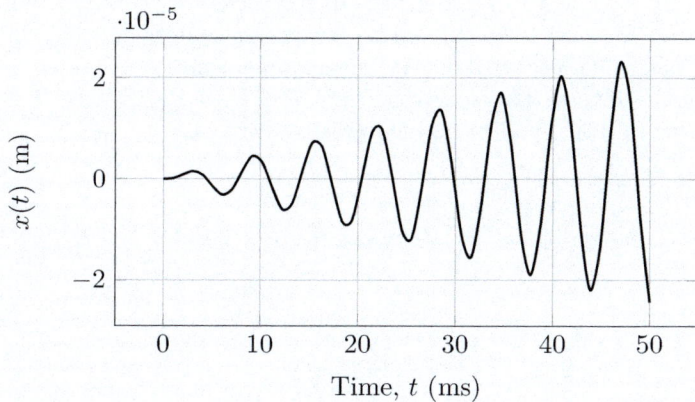

Figure 5.9: Response, $x(t)$, of a perfect oscillator to a force $F(t) = A\sin(\omega t)$ with $\omega = \beta$ (Case II).

In this example we observed a very interesting property of the ideal oscillator: for all $\omega \neq \beta$, the system response remains bounded, but for $\omega = \beta$, the response grows unbounded. This phenomenon is known as **resonance** (see Figure 5.9).

5.5 Symbolic computations

5.5.1 Symbolic computations in Python

In Python we may use the symbolic library SymPy, which can be installed using pip as follows

```
pip install sympy
```

We may then compute inverse Laplace transforms symbolically as in the following example where we compute $\mathscr{L}^{-1}\{\frac{1}{s^2+1}\}$

```
import sympy as sym
# Define the symbolic variables s and t
s,t = sym.symbols('s,t')
# Function in s-domain
x_laplace = s/(s**2+1)
# Symbolic computation of inverse Laplace transform
x_t = sym.inverse_laplace_transform(x_laplace, s, t)
```

The result will be

```
cos(t)*Heaviside(t)
```

where Heaviside(t) is to emphasise that this is a time-domain function defined for $t \geq 0$.

Note that sympy always assumes that all parameters are complex-valued. In certain cases, we need to specify that certain parameters are real-valued (or positive, negative, etc, if so). As an example, suppose that we need to determine the inverse Laplace transform $\mathscr{L}^{-1}\{\frac{\omega}{s^2+\omega^2}\}t$, where $\omega \in \mathbb{R}$, which we know that is equal to sin(ωt). The output of

```
s,t,w = sym.symbols('s,t,w')
x_laplace = w/(s**2+w**2)
x_t = sym.inverse_laplace_transform(x_laplace, s, t)
```

will be

```
I*(-I*exp(2*t*im(w))*sin(t*re(w))
  + exp(2*t*im(w))*cos(t*re(w))
  - I*sin(t*re(w)) - cos(t*re(w)))
    *exp(-t*im(w))*Heaviside(t)/2
```

where we have the imaginary unit, I, and the real and imaginary parts of ω. If, however, we use w = sym.symbols('w', real=True), the result is sin(t*w)*Heaviside(t).

Exercise 5.14 (**Inverse Laplace transform in Python**) We know that the inverse Laplace transform of $F(s) = \frac{1}{s^2+a}$, where $a > 0$, is $(\mathscr{L}^{-1}F)(t) = \omega^{-1/2}\sin(\sqrt{\omega}t)$, for $t \geq 0$. However, the following Python script

```
s,t= sym.symbols('s,t')
a = sym.symbols('a')
fs = 1/(s**2 + a)
ft = sym.inverse_laplace_transform(fs, s, t)
```

returns a different result. Can you fix the issue?

※ ....

Answer: The issue here is that sympy does not know that a is real and positive. Use a = sym.symbols('a', real=True, positive=True) to fix it.

5.5.2 Symbolic computations in MATLAB

In MATLAB we may use `ilaplace` to eschew the tedious drudgery of computing the inverse Laplace transform of an s-domain function

```
s = syms('s');         % Define symbolic variable
X = 1/(s+1);           % Function in s-domain
x = ilaplace(F);       % Computation of inverse Laplace
fplot(x, [0, 10]);     % Plot x(t) for t in [0, 10]
```

With MATLAB's Symbolic Math Toolbox we can compute lengthy inverse Laplace transforms. For example

Chapter 5. Inverse Laplace transform

```
s = syms('s');
X = (7*s+1)/((s^3+1)*(2*s+1)*s);
x = ilaplace(F);
```

will return the following result

```
(20*exp(-t/2))/7 - 2*exp(-t) - (13*exp(t/2)*(
    cos((3^(1/2)*t)/2)
- (5*3^(1/2)*sin((3^(1/2)*t)/2))/39))/7 + 1
```

which will look better if we display it using `pretty(x)` instead. Lastly, same as in Python, we may need to define certain variables as real-valued or positive using `x = sym('x','real')` or `x = sym('x','positive')`. Type `help sym` and `help syms` for details.

5.6 One-minute round-up

In this chapter we learned the following

1. The Laplace transform can be inverted

2. We can compute the inverse Laplace transform for rational functions. There exist different techniques depending on the poles of the function.

3. We can use the inverse Laplace transform of an s-domain function to simulate how a dynamical system behaves under a certain excitation

It is important to keep in mind that if $F(s)$ is a rational function with poles $s_1, s_2, \ldots, s_n \in \mathbb{C}$ of multiplicities $\kappa_1, \ldots, \kappa_m$, and the degree of its denominator is less than that of its denominator, then its inverse Laplace transform has the form

$$f(t) = \sum_{i=1}^{m} \sum_{k=1}^{\kappa_i} a_{i,k} t^{k-1} e^{s_i t}, \tag{5.158}$$

for some constant coefficients $a_{i,k} \in \mathbb{C}$. If, instead, its numerator and denominator have the same degree, then

$$f(t) = a_0 \delta(t) + \sum_{i=1}^{m} \sum_{k=1}^{\kappa_i} a_{i,k} t^{k-1} e^{s_i t}, \tag{5.159}$$

for some constant coefficients $a_{i,k} \in \mathbb{C}$ and $a_0 \in \mathbb{R}$.

We should note that the Laplace transform can only be applied to *linear* ODEs and IDEs, that is, equations of the form $a_0 x + a_1 \dot{x} + a_2 \ddot{x} + \ldots = b_0 u + b_1 \dot{u} + \ldots$. The Laplace transform cannot be used to solve nonlinear ODEs, for example, of the form $\dot{x} = x^2 + u$. In such cases, we can resort to the linearisation methodology described in Chapter 3 and then apply the Laplace transform to the approximate linearised version of the original equation.

In the following chapters we will discover a more systematic way to analyse dynamical systems. The main idea is that given an ODE/IDE, we will apply the Laplace transform to obtain an algebraic equation that describes the system dynamics.

5.7 Exercises

Exercise 5.15 (Inverse Laplace practice) Find the inverse Laplace tranform of the following functions

1. $F_1(s) = \frac{2s-1}{(s-1)(s+3)}$
2. $F_2(s) = \frac{s^2+3s-1}{(s-1)^3}$
3. $F_3(s) = \frac{1-3s}{(s-1)^2+1}$
4. $F_4(s) = \frac{1}{(s+1)(s+1)(s+3)}$
5. $F_5(s) = \frac{s+1}{s^3+4s^2+s-6}$
6. $F_6(s) = \frac{5s-3}{s^4-s^3-7s^2+s+6}$

✂ ....

Hint. Use Horner's method to determine the roots of the denominator, where necessary.

※ ....

Answer. $(\mathscr{L}^{-1}F_1)(t) = \frac{1}{4}(7e^{-3t} + e^t)$, $(\mathscr{L}^{-1}F_2)(t) = e^t(\frac{3}{2}t^2 + 5t + 1)$, $(\mathscr{L}^{-1}F_3)(t) = -\sqrt{13}e^t \sin(t + \arctan 3/2)$, $(\mathscr{L}^{-1}F_4)(t) = \frac{1}{4}e^{-t} - \frac{1}{2}te^{-t} - \frac{1}{4}e^{-3t}$, $(\mathscr{L}^{-1}F_5)(t) = \frac{3}{8}e^{-2t} - \frac{2}{5}e^{-3t} + \frac{10}{15}e^{-t}$, for $t \geq 0$.

Exercise 5.16 (More practice) Find the inverse Laplace transform of the following functions

1. $F_1(s) = \frac{s+1}{(s^2+s+1)(s+3)}$
2. $F_2(s) = \frac{s+5}{s(s+3)^2(s-5)^3}$
3. $F_3(s) = \frac{s+2}{s^2+3}$
4. $F_4(s) = \frac{se^{-7s}}{s^2+2}$
5. $F_5(s) = \frac{s-2}{((s-1)^2+16)^2}$
6. $F_6(s) = \frac{1}{(s^2+1)^4}$

※ ....

Answer. $(\mathscr{L}^{-1}F_1)(t) = \frac{2}{7}e^{-\frac{t}{2}}\left[\cos(\sqrt{3}t/2) + \frac{3}{2}\sqrt{3}\sin(\sqrt{3}t/2)\right] - \frac{2}{7}e^{-3t}$ $(\mathscr{L}^{-1}F_2)(t) = -\frac{7}{2}e^{-3t} - (0.8148) + \frac{29}{18432}e^{-3t}$ $+ \frac{5100}{147}e^{5t} + \frac{768}{7}te^{5t} - \frac{640}{1}t^2e^{5t} + \frac{64}{1}t^2e^{5t} - \frac{1}{225}$, $(\mathscr{L}^{-1}F_3)(t) = \cos(\sqrt{3}t) + \frac{2}{3}\sqrt{3}\sin(\sqrt{3}t) = \frac{\sqrt{21}}{3}\sin(\sqrt{3}t + 0.8571)$, $(\mathscr{L}^{-1}F_4)(t) = \cos(\sqrt{2}(t-7))$, $(\mathscr{L}^{-1}F_5)(t) = \left[\frac{32}{128}\cos(4t) - \frac{128}{128}\sin(4t) + \frac{1}{8}t\sin(4t)\right]e^t$, $(\mathscr{L}^{-1}F_6)(t) = \frac{1}{48}(t^3 \cos t - 15t^2 \sin t + 15t \cos t)$, for $t \geq 0$.

Exercise 5.17 (Computation of convolutions) Write a Python or MATLAB function that determines the convolution of two given functions

(i) symbolically and (ii) numerically.

Exercise 5.18 (Simple pendulum) The ODE that described the motion of a pendulum for small angles θ is (see Example 2.10 and Exercise 3.12)
$$\ddot{\theta} + \frac{g}{L}\theta = 0, \qquad (5.160)$$
where θ is the angle between the string of the pendulum and a vertical line running through its upper end, g is the gravitational acceleration and L is the length of the string.

The initial angle of the pendulum is $\theta(0) = 10°$ and the initial angular velocity is $\dot{\theta}(0) = 0$. What is the period of oscillation? What is the maximum angular velocity, and when is it attained?

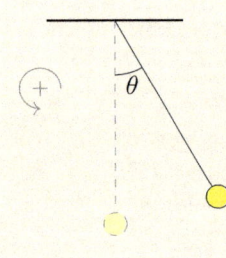

※

Answer. The answer is: $\theta(t) = \frac{18}{\pi}\cos(\omega t)$, for $t \geq 0$. This is a periodic signal with period $T = \frac{2\pi}{\omega}$, where $\omega = \sqrt{\frac{g}{L}}$. The angular velocity is $\dot{\theta}(t) = -\frac{18}{\pi}\omega\sin(\omega t)$, for $t \geq 0$. The maximum angular velocity (in absolute value) is $\dot{\theta}_{\max} = \frac{18}{\pi}\omega$ and it is attained at $t = \frac{\pi}{2L} + kT$, and $t = \frac{3\pi}{2L} + kT$, for $k \in \mathbb{Z}$.

Exercise 5.19 (Spring-mass system with friction) A spring-mass system is shown in the following figure:

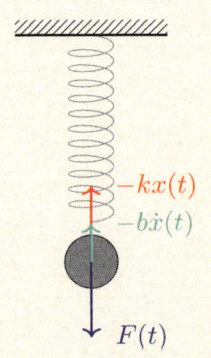

Let $x(t)$ denote the displacement of the mass from its equilibrium at time t. Let $F(t)$ be an external force. The spring applies a force which satisfies Hooke's law, $-kx(t)$. A damping force is also present, $-b\dot{x}(t)$. Suppose that the system is initially at rest ($x(0) = 0$ and $\dot{x}(0) = 0$).

1. Write the differential equations that describe the dynamics of this system

2. Apply the Laplace transform on the differential equations

3. Determine the system output, $x(t)$, for an arbitrary input $F(t)$ (Hint: Use the convolution theorem).

4. Determine the response of the system when the input signal is: (i) $F(t) = \delta(t)$, (ii) $F(t) = 1$, (iii) $F(t) = e^{-t}$.

※ ....

Answer. 1. $F - kx - b\dot{x} = m\ddot{x}$, 2. $F = (ms^2 + bs + k)X$, 3. $x(t) = \frac{1}{a}e^{-\frac{a t}{2m}}\sin\left(\frac{\alpha t}{2m}\right) * F(t)$, for $t \geq 0$, where $\alpha = \sqrt{4km - b^2}$, 4. (i) $x(t) = \frac{1}{a}e^{-\frac{a t}{2m}}\sin\left(\frac{\alpha t}{2m}\right)$, (ii) it is

$$x(t) = \frac{k}{1}\left[1 - e^{-\frac{a t}{2m}}\left(\cos\frac{\alpha t}{2m} + \frac{a}{b}\sinh\frac{\alpha t}{2m}\right)\right], \quad (5.161)$$

$t \geq 0$, (iii) it is

$$x(t) = ce^{-t} - ce^{-\frac{a t}{2m}}\left(\cosh\left(\frac{\alpha t}{2m}\right) + \frac{b - 2m}{\alpha}\sinh\left(\frac{\alpha t}{2m}\right)\right), \quad (5.162)$$

with $t \geq 0$, where $c = \frac{1}{k - b + m}$.

※ ....

Exercise 5.20 (Some IVPs) Solve the following initial value problems using the Laplace transform

1. $\dot{x} + 7x - te^{-3t} = 0$ with $x(0) = -1$

2. $\ddot{x} = \dot{x} + 2x$ with $x(0) = 1$, $\dot{x}(0) = 0$

3. $\ddot{x} + x = \sin(2t)$ with $x(0) = 0$, $\dot{x}(0) = 0$

4. $x^{(3)} - 3\ddot{x} + x = 0$ with $x(0) = 5$, $\dot{x}(0) = -2$, $\ddot{x}(0) = 0$. Hint. You may use Python or MATLAB to confirm that the polynomial $s^3 - 3s^2 + 1 = 0$ has the roots 2.879385, 0.652703 and -0.532089.

※ ....

Answer. 1. $x(t) = \frac{4}{1}te^{-3t} - \frac{16}{15}e^{-3t} - \frac{1}{1}e^{-7t} - \frac{16}{15}e^{-3t}$, $t \geq 0$, 2. $x(t) = \frac{2}{3}e^{-t} + \frac{1}{3}e^{2t}$, $t \geq 0$, 3. $x(t) = \frac{2}{3}\sin(t) - \frac{1}{3}\sin(2t)$, $t \geq 0$, 4. $x(t) = 4.073\exp(-0.53321t) + 1.124\exp(0.65271t) - 0.1968\exp(2.8794t)$, $t \geq 0$.

Exercise 5.21 (Curious logarithmic function) Compute the inverse Laplace transform

$$\mathscr{L}^{-1}\left\{\ln\frac{s+1}{s-1}\right\}. \quad (5.163)$$

※ ....

Answer. It is

$$\mathscr{L}^{-1}\left\{\ln\frac{s+1}{s-1}\right\} = \frac{2\sinh t}{t}. \quad (5.164)$$

✂··

$$\mathscr{L}^{-1}\{F'(s)\}(t) = \mathscr{L}^{-1}\{-tf(t)\}. \quad (5.165)$$

where $F(s) = \mathscr{L}\{f(t)\}(s)$. If we apply the inverse Laplace transform on both sides we obtain

$$\mathscr{L}^{-1}\{F'(s)\}(t) = -tf(t) \iff f(t) = -\frac{1}{t}\mathscr{L}^{-1}\{F'(s)\}(t). \quad (5.166)$$

Hint. By virtue of Theorem 4.26,

Exercise 5.22 (**Inverse Laplace using Horner's scheme**) Determine the inverse Laplace transform of the following function

$$F(s) = \frac{3s+2}{s^3 + 5s^2 + 7s + 3}. \quad (5.167)$$

※··

Answer. $(\mathscr{L}^{-1}F)(t) = \frac{7}{4}e^{-t} - \frac{1}{2}te^{-t} - \frac{7}{4}e^{-3t}$, for $t \geq 0$. Note that the poles of F are -3, and -1 (double).

Exercise 5.23 (**Inverse Laplace**) Compute the inverse Laplace transform of $X(s) = \arctan \frac{1}{s}$.

※··

Answer. The answer is $\frac{\sin t}{t}$, for $t \geq 0$.

✂··

Hint. (i) work as in Exercise 5.21, (ii) recall that

$$(\arctan x)' = \frac{1}{x^2 + 1}. \quad (5.168)$$

Exercise 5.24 (**Integrodifferential equation**) Determine a signal $x(t)$, $t \geq 0$, so that

$$x(t) = 1 + \int_0^t \cos(t-\tau)x(\tau)\mathrm{d}\tau. \quad (5.169)$$

※··

Answer. $x(t) = 1 + \frac{\sqrt{3}}{3}e^{\frac{t}{2}}\sin\left(\frac{\sqrt{3}}{2}t\right)$, for $t \geq 0$.

✂··

Hint. Take the Laplace transform on both sides of this integro-differential equation. Apply the convolution theorem (Theorem 5.9).

Chapter 5. Inverse Laplace transform

Exercise 5.25 (Solve an IDE) Solve the following IDE

$$x(t) = e^{-t} + \int_0^t \sin(2\xi) x(t-\xi) \mathrm{d}\xi, \qquad (5.170)$$

where $\alpha > 0$.

✳ ....

Answer. It is $x(t) = \frac{5}{3} e^{-t} - \frac{2}{3} \cos(\sqrt{2}t) + \frac{\sqrt{2}}{3} \sin(\sqrt{2}t)$, for $t \geq 0$.

✂ ....

Hint. Observe that we have a convolution on the right-hand side; apply Theorem 5.9.

Exercise 5.26 (Decomposition theorem) A system with state x and input u is described by the second-order differential equation

$$\ddot{x} + a\dot{x} + bx = u, \qquad (5.171)$$

with initial conditions $x(0) = x_0$ and $\dot{x}(0) = v_0$. Assume that u has a Laplace transform. Then, its solution can be written as

$$x(t) = x_{\mathrm{hom}}(t) + (x_{\mathrm{imp}} * u)(t), \qquad (5.172)$$

where x_{hom} is the solution of the "homogeneous" IVP

$$\ddot{x} + a\dot{x} + bx = 0, x(0) = x_0, \dot{x}(0) = v_0, \qquad (5.173)$$

and x_{imp} is the impulse response of the system, that is, the solution of the IVP

$$\ddot{x} + a\dot{x} + bx = u, x(0) = 0, \dot{x}(0) = 0, \qquad (5.174)$$

✂ ....

Hint. It suffices to verify that $x(t)$ in Equation (5.172) satisfies the differential equation in (5.171) and the given initial conditions. It is easy to see that $(x_{\mathrm{imp}} * u)(0) = 0$. This result will be generalised later in Section 6.5.2.

Exercise 5.27 (Pair the functions) Pair the following functions in the s domain with the plots of their inverse Laplace transforms

$X(s) = \dfrac{1}{(s+1)^2 + 1}$

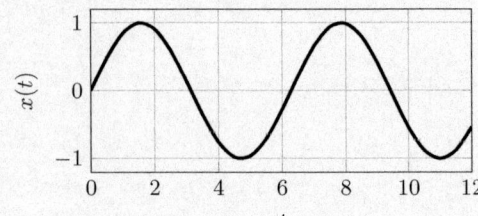

$X(s) = \dfrac{s}{s^2 + 1}$

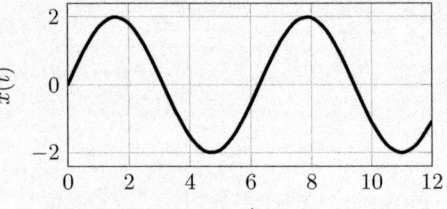

$X(s) = \dfrac{s}{(s - 1/10)^2 + 1}$

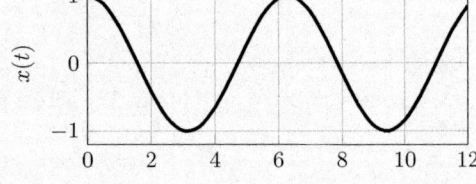

$X(s) = \dfrac{s}{(s + 1/5)^2 + 1}$

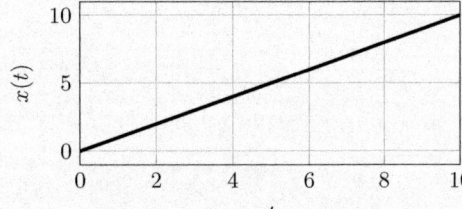

$X(s) = \dfrac{e^{-5s}}{s^2 + 1}$

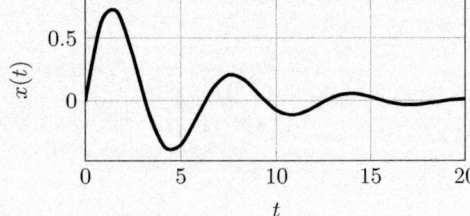

Chapter 5. Inverse Laplace transform

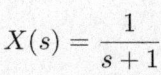

$$X(s) = \frac{1}{s+1}$$

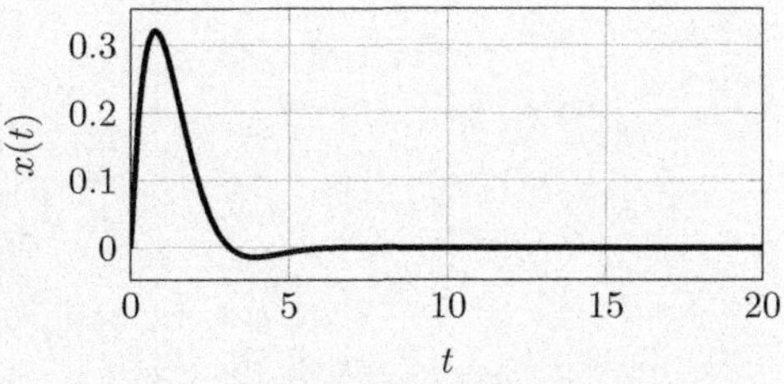

$$X(s) = \frac{1}{(s+0.2)^2 + 1}$$

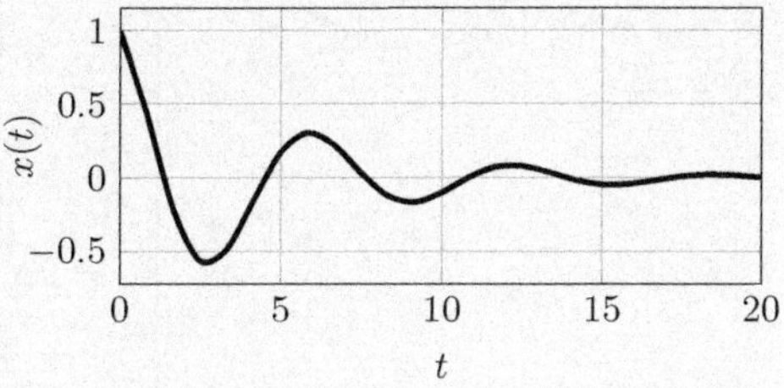

$$X(s) = \frac{1}{s^2 + 1}$$

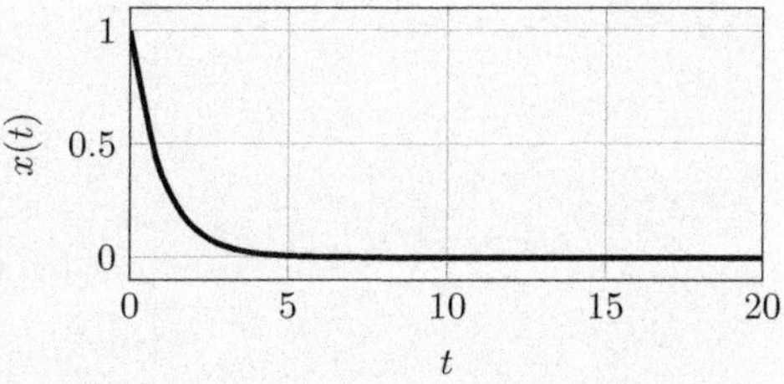

$$X(s) = \frac{2}{s^2 + 1}$$

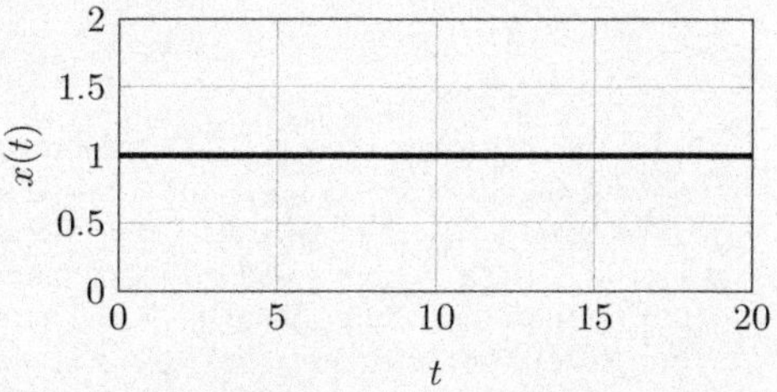

$$X(s) = \frac{1}{s^2}$$

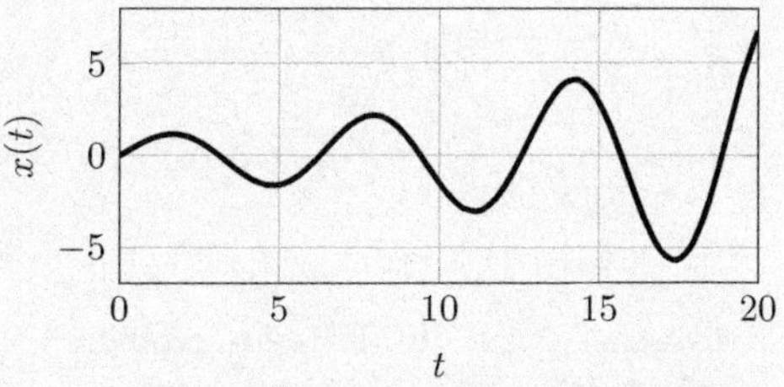

$$X(s) = \frac{1}{5s - 1}$$

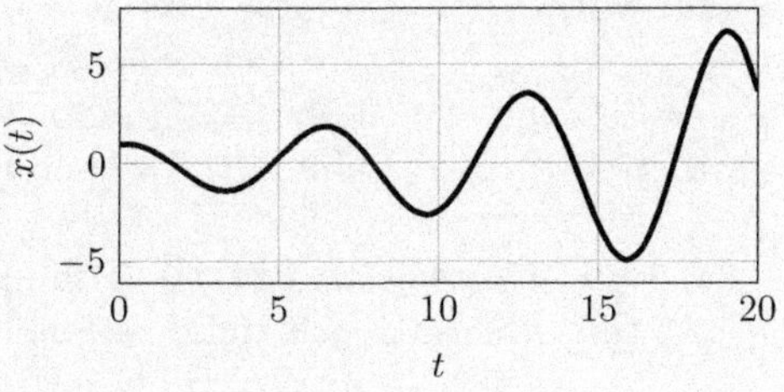

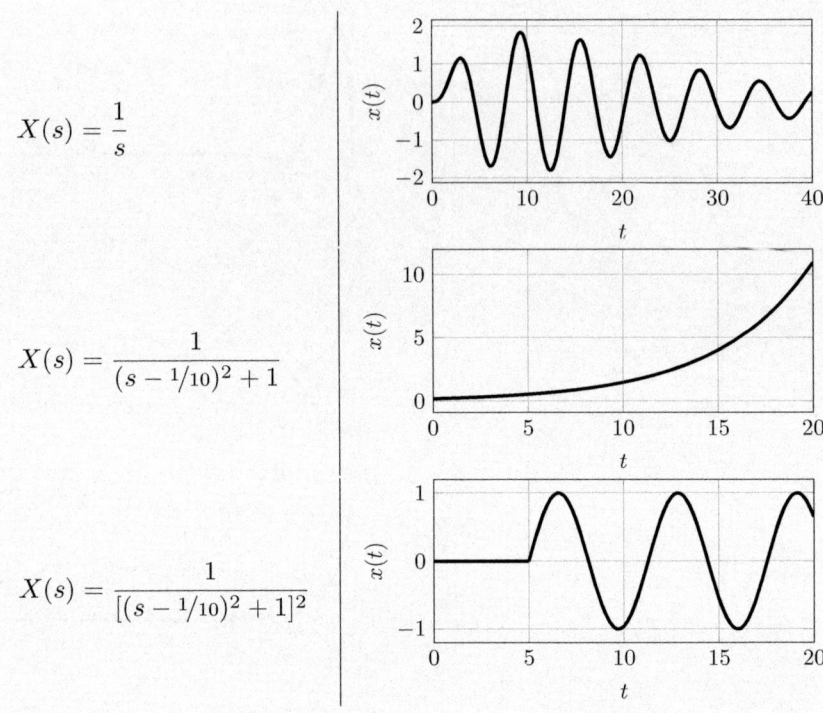

$$X(s) = \frac{1}{s}$$

$$X(s) = \frac{1}{(s - 1/10)^2 + 1}$$

$$X(s) = \frac{1}{[(s - 1/10)^2 + 1]^2}$$

Hint: (i) Start by determining the poles of each function X in the first column. The presence of non-real poles means that the corresponding time-domain function, $x(t)$, has an oscillatory component, (ii) if all poles have a strictly negative real part, $\lim_{t \to \infty} x(t) = 0$, (iii) is there is a pole with a strictly positive real part, then $|x(t)|$ grows unbounded as $t \to \infty$, (iv) use the initial value theorem, (v) you might need to use the initial value theorem for $\dot{x}$ as well, (vi) revise the results of Section 5.2.

Answer: 1-6, 2-3, 3-11, 4-7, 5-14, 6-8, 7-5, 8-1, 9-2, 10-4, 11-13, 12-9, 13-10, 14-12.

Exercise 5.28 (Initial value problem) Consider the system of Exercise 2.15 with $b = 0$ (no drag force is acting on the cylinder). Note that for $b = 0$ the system dynamics is affine, so the Laplace transform can be applied directly. At time $t = 0$ the cylinder is at $x = (1 - \alpha)L$ where $\alpha \in (0, 1)$. The initial velocity of the cylinder is zero.

(i) Determine the trajectory of the system, $x(t)$, (ii) At what time will the cylinder reach the position $\bar{x} = (1 + \frac{\alpha}{2})L$ for the first time?

You may use Python or MATLAB for your symbolic computations.

Answer. Define the constant $\mu = \frac{mg}{k}\sin\phi + L$, $\lambda = \frac{mg}{k}\sin\phi + \alpha L\left(1 + \frac{3m}{2(1-\alpha)}\right)$, and $\omega = \sqrt{\frac{1}{3} \cdot \frac{gk}{m}}$. Then
$$x(t) = \mu - \lambda\cos(\omega t),$$
for $t \geq 0$, so the time, t_*, at which x is equal to $\bar{x}$ is $t_* = \frac{1}{\omega}\arccos\frac{\mu - \bar{x}}{\lambda}$. (5.175)

※

Exercise 5.29 (Convolution: an interesting property) Suppose that f and g are differentiable functions and $f * g$ exists. Show that
$$(f * g)' = f' * g = f * g'. \tag{5.176}$$

※

Answer. Use Theorem 4.19 and the convolution theorem (Thm. 5.9). It is
$$\mathscr{L}\{(f * g)'\} = s\mathscr{L}\{f * g\} = s\mathscr{L}\{f\}\mathscr{L}\{g\} = \mathscr{L}\{f'\}\mathscr{L}\{g\} = \mathscr{L}\{f' * g\}. \tag{5.177}$$
By Lerch's Theorem (Theorem 5.2), $(f * g)' = f' * g$ almost everywhere, but $f * g$ is continuous, so the equality holds for all $t \geq 0$.

Exercise 5.30 (Recursion) Determine the partial fractions expansion and inverse Laplace transform of
$$F(s) = \frac{1}{(s+1)(s+2)\cdot \ldots \cdot (s+N)}, \tag{5.178}$$
where $N \in \mathbb{N}$ with $N \geq 1$.

※

Answer. The inverse Laplace transform of F is
$$(\mathscr{L}^{-1}F)(t) = \sum_{i=1}^{N} \frac{(-1)^{i-1}}{(i-1)!(N-i)!} e^{-it},$$
for $t \geq 0$.

✂

Hint. It is perhaps more convenient to use the alternative approach of Section 5.2.9 for this exercise.

Exercise 5.31 (A little trick!) Determine the inverse Laplace transform of the following function
$$F(s) = \frac{1}{\prod_{i=1}^{N} s^2 + b_i^2}, \tag{5.179}$$

where $b_i \neq b_\nu$ for $i \neq \nu$, for all $i, \nu = 1, \ldots, N$.

✻ ··

$$B_i = \frac{1}{\prod_{\substack{\nu=1 \\ \nu \neq i}}^{N}(b_\nu^2 - b_i^2)}, \quad (5.180)$$

for all $i = 1, \ldots, N$.

Answer. It is $A_i = 0$ for all $i = 1, \ldots, N$ and

✂ ··

Hint. The partial fractions expansion of F must have the following form

$$\frac{\prod_{i=1}^{N} 1}{s^2 + b_i^2} = \sum_{i=1}^{N} \frac{A_i s + B_i}{s^2 + b_i^2}. \quad (5.181)$$

We then multiply both parts by the denominator of the left-hand side:

$$1 = \sum_{i=1}^{N}(A_i s + B_i) \prod_{\substack{\nu=1 \\ \nu \neq i}}^{N}(s^2 + b_\nu^2). \quad (5.182)$$

Instead of applying the distributive property and comparing the corresponding coefficients we can substitute on both sides of the equation the values $s = \mp j b_1, \mp j b_2, \ldots, \mp j b_N$. All coefficients can then be determined easily.

Exercise 5.32 (Application of the above trick) Now that you know the trick of Exercise 5.31, determine the inverse Laplace transform of the function

$$F(s) = \frac{1}{\prod_{i=1}^{N}(s + a_i)^2 + b_i^2}, \quad (5.183)$$

where all pairs (a_i, b_i) are distinct, that is, for all $i, \nu = 1, \ldots, N$, $a_i \neq a_\nu$ or $b_i = b_\nu$.

Advanced Exercise 5.33 (Inverse Laplace from series) Let x be a time-domain function that is continuous for all $t \geq 0$ and of exponential order. Define $X(s) = (\mathscr{L}x)(s)$. Suppose that F is expanded in the following partial fractions series expansion:

$$X(s) = \sum_{n=0}^{\infty} \frac{a_n}{s^{n+1}}, \quad (5.184)$$

where the series converges absolutely, uniformly over $s \in \mathbb{C}$ with $\mathbf{re}(s) > c \geq 0$ for some constant c. Then,

$$(\mathscr{L}^{-1}X)(t) = \sum_{n=0}^{\infty} \frac{a_n t^{n+z}}{n!}, \quad (5.185)$$

for $t \geq 0$.

✂ ....

Hint. Recall that a series is said to converge absolutely if the series of the absolute values of its summand converges and is finite. This is Theorem 6 in [MIT, Chap. 2].

Advanced Exercise 5.34 (Generalisation of Exercise 5.33) Let x be a time-domain function that is continuous for all $t \geq 0$ and of exponential order. Define $X(s) = (\mathscr{L}x)(s)$. Suppose that F is expanded in the following partial fractions series expansion:

$$X(s) = \sum_{n=0}^{\infty} \frac{a_n}{s^{n+1+z}}, \qquad (5.186)$$

where the series converges absolutely, uniformly over $s \in \mathbb{C}$ with $\mathrm{re}(s) > c \geq 0$ for some constant c and $z \in [0, 1)$. Show that

$$(\mathscr{L}^{-1}X)(t) = \sum_{n=0}^{\infty} \frac{a_n t^{n+z}}{\Gamma(z+n+1)}, \qquad (5.187)$$

for $t \geq 0$.

Advanced Exercise 5.35 (Inverse Laplace Transform from series: an application) Use the result of Exercise 5.34 to show that

$$\mathscr{L}^{-1}\left\{\frac{1}{\sqrt{s}} e^{-1/s}\right\} = \frac{1}{\sqrt{\pi t}} \cos\left(2\sqrt{t}\right), \qquad (5.188)$$

for $t \geq 0$.

6

Transfer functions and block diagrams

The **transfer function** of a linear dynamical system allows us to simulate the output of the system for any given input (which has a Laplace transform). In this chapter we introduce the concept of the transfer function of a linear dynamical system, and we give a few examples of computing the system output to certain "standard" inputs. We define the **step**, **impulse** and **frequency response** of a linear dynamical system: its response when the input is a Heaviside (step function), a Dirac pulse and a sine respectively. Next, we describe how different systems interact by connecting the output of one to the input of another; we construct **block diagrams** that illustrate such interconnections, and we discuss multiple systems may be connected and how they interact. Lastly, we introduce the notion of **impedance**. A notion that facilitates the analysis and modelling of electric circuits.

6.1 Transfer functions

Consider a linear dynamical system in the following form

$$a_0 x + a_1 \dot{x} + \ldots + a_n x^{(n)} = b_0 u + b_1 \dot{u} + \ldots + b_m u^{(m)}, \quad (6.1)$$

with $m < n$ and with state x and input u. Assume that the initial conditions are

$$x(0) = \dot{x}(0) = \ldots = x^{(n-1)}(0) = 0, \quad (6.2a)$$
$$u(0) = \ldots = u^{(m-1)} = 0. \quad (6.2b)$$

We apply the Laplace transform on both sides of (6.1) and, in light of the assumed zero initial conditions, we have

$$a_0 X(s) + a_1 s X(s) + \ldots + a_n s^n X(s) = b_0 U(s) + \ldots + b_m s^m U(s), \quad (6.3)$$

where $X(s) = \mathscr{L}\{x(t)\}(s)$ and $U(s) = \mathscr{L}\{u(t)\}(s)$, so

$$(a_0 + a_1 s + \ldots + a_n s^n)X(s) = (b_0 + \ldots + b_m s^m)U(s), \quad (6.4)$$

and, by rearranging the terms we obtain

$$\frac{X(s)}{U(s)} = \frac{b_0 + \ldots + b_m s^m}{a_0 + a_1 s + \ldots + a_n s^n}. \quad (6.5)$$

Equation (6.5) was obtained from the original ODE of the dynamical system and describes the system dynamics. This defines the **transfer function** of the system which is

$$G(s) = \frac{b_0 + \ldots + b_m s^m}{a_0 + a_1 s + \ldots + a_n s^n}. \quad (6.6)$$

We may now give the following definition

> **Definition 6.1 (Transfer function)** Let x and u denote the input and state of a linear dynamical system and let X and U be the corresponding Laplace transforms. Assume all initial conditions are zero. The **transfer function** of the system is defined as the complex function
>
> $$G(s) = \frac{X(s)}{U(s)}. \quad (6.7)$$

The transfer function of a linear dynamical system describes its input-output correspondence. It describes the effect that the input has on the output. Note that the definition requires that the dynamical system have zero initial conditions. For that reason, we will describe all systems in terms of deviation variables as we discussed in Chapter 3, and we will assume that at time $t = 0$ the system is at equilibrium.

The transfer function of a system describes fully the system dynamics. For example, if we know the transfer function, $G(s)$, of a system we can predict its response to any input. This is how:

$$X(s) = G(s)U(s) \Rightarrow x(t) = \mathscr{L}^{-1}\{G(s)U(s)\}(t). \quad (6.8)$$

We will revisit this in Section 6.2 where we will study the response of dynamical systems to certain inputs.

Let us first give a few examples of deriving the transfer function of a dynamical system

Example 6.1 (Transfer function of simple RC circuit) In Example 2.15 we derived the following model (Equation (2.82))

$$\dot{V}_{\text{out}}(t) = \frac{1}{CR}(V_{\text{in}}(t) - V_{\text{out}}(t)), \quad (6.9)$$

where $V_{\text{in}}(t)$ is the input variable and $V_{\text{out}}(t)$. With some abuse of notation, we denote the respective Laplace transforms by $V_{\text{in}}(s)$ and $V_{\text{out}}(s)$.

We assume that the initial conditions are zero, i.e., $V_{\text{out}}(0) = 0$. We apply the Laplace transform on both sides of the system model to obtain

$$sV_{\text{out}}(s) = \frac{1}{CR}(V_{\text{in}}(s) - V_{\text{out}}(s)), \qquad (6.10)$$

or, equivalently,

$$CRsV_{\text{out}}(s) = V_{\text{in}}(s) - V_{\text{out}}(s)$$
$$\Rightarrow (CRs+1)V_{\text{out}}(s) = V_{\text{in}}(s)$$
$$\Rightarrow \frac{V_{\text{out}}(s)}{V_{\text{in}}(s)} = \frac{1}{CRs+1}, \qquad (6.11)$$

therefore, the transfer function of this system is

$$G(s) = \frac{1}{CRs+1}. \qquad (6.12)$$

Example 6.2 (Transfer function of single tank system) In Example 3.8 we derived the following ODE model for a single tank system

$$\rho A \dot{\bar{h}} = \bar{F}_{\text{in}} - \frac{\gamma}{2\sqrt{h^e}} \bar{h}, \qquad (6.13)$$

where $\bar{h}(t)$ is the deviation variable associated with the elevation of water in the tank and $\bar{F}_{\text{in}}(t)$ is the deviation variable associated with the inflow of water. We assume that at time $t = 0$ the system is at equilibrium, that is, $\bar{F}_{\text{in}}(0) = 0$ and $\bar{h}(0) = 0$. We denote the Laplace transform of $\bar{h}(t)$ by $\bar{H}(s)$ and, with some abuse of notation, we denote the Laplace transform of $\bar{F}_{\text{in}}(t)$ by $\bar{F}_{\text{in}}(s)$.

Then, applying the Laplace transform to both sides of (6.13) we obtain

$$\rho As \bar{H}(s) = \bar{F}_{\text{in}}(s) - \frac{\gamma}{2\sqrt{h^e}} \bar{H}(s)$$
$$\Rightarrow \left(\rho As + \frac{\gamma}{2\sqrt{h^e}}\right) \bar{H}(s) = \bar{F}_{\text{in}}(s)$$
$$\Rightarrow \frac{\bar{H}(s)}{\bar{F}_{\text{in}}(s)} = \frac{1}{\rho As + \frac{\gamma}{2\sqrt{h^e}}}. \qquad (6.14)$$

This is the transfer function of the system. In particular

$$G(s) = \frac{1}{\rho As + \frac{\gamma}{2\sqrt{h^e}}}. \qquad (6.15)$$

In the last two examples, the transfer functions we derived were rational functions. This is the most common case. We will now give an example of a spring-mass-damper system which is a second-order system.

Example 6.3 (Spring-mass-damper system) The dynamics of a spring-mass-damper system, as in Example 2.8, is described by the ODE

$$\bar{F}(t) - b\dot{\bar{x}}(t) - k x(t) = m\ddot{\bar{x}}(t), \qquad (6.16)$$

where $\bar{F}(t)$ is the deviation variable associated with the externally applied force $F(t)$ and $\bar{x}(t)$ is the deviation of the displacement of the spring from its equilibrium position. Assuming zero initial conditions, that is $\bar{x}(0) = 0$ and $\dot{\bar{x}}(0) = 0$, we apply the Laplace transform to both sides of the ODE to obtain

$$\bar{F}(s) - bs\bar{X}(s) - k\bar{X}(s) = ms^2 \bar{X}(s). \qquad (6.17)$$

After rearranging the terms, we obtain

$$G(s) = \frac{\bar{X}(s)}{\bar{F}(s)} = \frac{1}{ms^2 + bs + k}. \qquad (6.18)$$

The same procedure can be applied to integro-differential equations using Theorem 4.22. Let us give an example

Example 6.4 (Series RLC circuit) In Example 2.16 we derived the following model for a series RLC series

$$V_{\text{in}}(t) = \frac{1}{RC} \int_0^t V_{\text{out}}(\tau) d\tau + V_{\text{out}}(t) + \frac{L}{R} \dot{V}_{\text{out}}(t). \qquad (6.19)$$

We assume all initial conditions are zero, and we apply the Laplace transform to both sides of the IDE to obtain

$$V_{\text{in}}(s) = \frac{1}{RCs} V_{\text{out}}(s) + V_{\text{out}}(s) + \frac{L}{R} s V_{\text{out}}(s), \qquad (6.20)$$

and by rearranging the terms we derive the transfer function

$$G(s) = \frac{V_{\text{out}}(s)}{V_{\text{in}}(s)} = \frac{RCs}{LCs^2 + RCs + 1}. \qquad (6.21)$$

In Example 2.17 we derived a model for an RLC system involving two integro-differential equations involving two states (the two currents I_L and I_C) and one input (the input voltage V_{in}). Although these two equations describe adequately the system, it is not obvious how we can describe the correspondence between the input voltage V_{in} and I_C without I_L. In other words, this system has one input (V_{in}), one output (I_C) and two states(I_L

Chapter 6. Transfer functions and block diagrams

and I_C) and we want to reveal the relationship between the input and the output.

The Laplace transform transforms the ODE/IDE of a dynamical system into algebraic equations. This facilitates their manipulation. In the following example we revisit Example 2.17 and derive the transfer function that describes how the input voltage, V_{in}, affects the current I_C. In doing so, we will eliminate the state variable I_L.

Example 6.5 (**RLC circuit**) Consider the dynamical system of Example 2.17. The input signal is the input voltage V_{in} and the output is

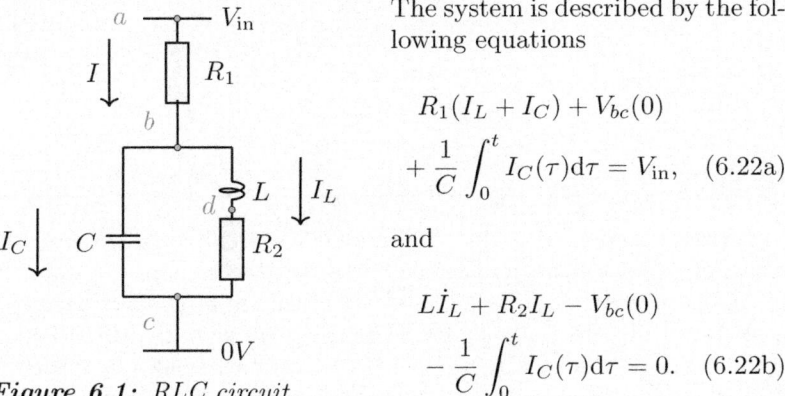

Figure 6.1: *RLC circuit*

The system is described by the following equations

$$R_1(I_L + I_C) + V_{bc}(0)$$
$$+ \frac{1}{C}\int_0^t I_C(\tau)d\tau = V_{in}, \quad (6.22a)$$

and

$$L\dot{I}_L + R_2 I_L - V_{bc}(0)$$
$$- \frac{1}{C}\int_0^t I_C(\tau)d\tau = 0. \quad (6.22b)$$

We assume zero initial conditions, and we apply the Laplace transform to both equations. We obtain

$$R_1(I_L(s) + I_C(s)) + \cancel{V_{bc}(0)} + \frac{1}{Cs}I_C(s) = V_{in}(s), \quad (6.23a)$$
$$LsI_L(s) + R_2 I_L(s) - \cancel{V_{bc}(0)} - \frac{1}{Cs}I_C(s) = 0. \quad (6.23b)$$

We need to eliminate I_L. We will now solve the first equation for I_L and substitute it into the second one. From (6.23a) we have

$$I_L(s) = \frac{V_{in}(s)}{R_1} - \left(1 + \frac{1}{R_1 Cs}\right) I_C(s). \quad (6.24)$$

We now substitute this expression into (6.23b) and obtain

$$LsI_L(s) + R_2 I_L(s) - \frac{1}{Cs}I_C(s) = 0$$
$$\Rightarrow (Ls + R_2)I_L(s) \frac{1}{Cs}I_C(s) = 0$$
$$\Rightarrow (Ls + R_2)\left[\frac{V_{in}(s)}{R_1} - \left(1 + \frac{1}{R_1 Cs}\right)\right] I_C(s) - \frac{1}{Cs}I_C(s) = 0$$

$$\Rightarrow \frac{1}{R_1}(Ls + R_2)V_{\text{in}}(s) = \left[(Ls + R_2)\left(1 + \frac{1}{R_1 Cs}\right) + \frac{1}{Cs}\right]I_C(s), \tag{6.25}$$

so the wanted input-output correspondence is

$$\frac{I_C(s)}{V_{\text{in}}(s)} = \frac{\frac{1}{R_1}(Ls + R_2)}{(Ls + R_2)\left(1 + \frac{1}{R_1 Cs}\right) + \frac{1}{Cs}}. \tag{6.26}$$

This is the transfer function of the system. We may multiply the numerator and denominator by Cs to simplify it.

$$G(s) = \frac{I_C(s)}{V_{\text{in}}(s)} = \frac{Cs(Ls + R_2)}{Cs(Ls + R_2)(R_1 Cs + 1) + 1}. \tag{6.27}$$

6.2 System response

The transfer function of a linear dynamical system — with zero initial conditions — allows us to simulate its response to a given excitation, insofar as we can compute the Laplace transform of the input signal. We may, for example, compute the output of a system when excited with a Heaviside function (a step), a Dirac function (a pulse) or a sinusoidal function.

The **impulse response** of a system with transfer function G is its response, $x(t)$, in the time domain, when excited with a pulse, i.e., when $u(t) = \delta(t)$. In that case, $U(s) = 1$, therefore, the system response is

$$x(t) = \mathscr{L}^{-1}\{G(s)U(s)\} = \mathscr{L}^{-1}\{G(s)\}. \tag{6.28}$$

The **step response** of a system is its response, $x(t)$, in the time domain, when excited with a Heaviside function, i.e., $u(t) = H_0(t)$. Then, $U(s) = 1/s$ and the corresponding response of a system with transfer function G becomes

$$x(t) = \mathscr{L}^{-1}\{G(s)U(s)\} = \mathscr{L}^{-1}\left\{\frac{G(s)}{s}\right\}. \tag{6.29}$$

The frequency response of a system with transfer function G is its response on excitation with a sinusoidal function of the form

$$u(t) = A\sin(\omega t). \tag{6.30}$$

We call A the **amplitude** of the input signal and ω its **angular frequency**. The Laplace transform of this signal is

$$U(s) = \frac{A\omega}{s^2 + \omega^2}, \tag{6.31}$$

therefore, the response of a system with transfer function G on excitation with a sinusoidal signal is

$$x(t) = \mathcal{L}^{-1}\{G(s)U(s)\} = A\omega\mathcal{L}^{-1}\left\{\frac{G(s)}{s^2+\omega^2}\right\}. \tag{6.32}$$

The following example demonstrates how we can use the transfer function of a dynamical system to determine its impulse, step and frequency response.

Example 6.6 (Velocity dynamics of vehicle) In a car, the position of the accelerator creates a moving force u (input) that affects the speed of the vehicle, v (output).

Figure 6.2: *Forces at play on a moving vehicle. The image of the car was obtained from* `pixabay.com` *and is royalty free (Free for commercial use, No attribution required).*

The velocity dynamics is governed by Newton's second law of motion

$$m\dot{v} = u - bv, \tag{6.33}$$

which is a linear differential equation. The system is at equilibrium (u^e, v^e) if $u^e - bv^e = 0$. We define the deviation variables $\bar{u} = u - u^e$ and $\bar{v} = v - v^e$. The system dynamics now becomes

$$m\dot{\bar{v}} = \bar{u} - b\bar{v}. \tag{6.34}$$

Assuming zero initial conditions, that is, $\bar{v}(0) = 0$ (equivalently, $v(0) = v^e$), and applying the Laplace transform, we obtain

$$ms\bar{V}(s) = \bar{U}(s) - b\bar{V}(s), \tag{6.35}$$

where $\bar{V}$ and $\bar{U}$ stand for the Laplace transforms of $\bar{v}$ and $\bar{u}$ respectively. The transfer function of the system is

$$G(s) = \frac{\bar{V}(s)}{\bar{U}(s)} = \frac{1}{ms+b}. \tag{6.36}$$

Now assume the car has a mass of $m = 500\,\text{kg}$ and $b = 100\,\text{Ns/m}$. We assume also that the equilibrium speed is $v^e = 20\,\text{m/s}$.

The impulse response of the car is then given by

$$\bar{v}(t) = \mathcal{L}^{-1}\{G(s)\} = \mathcal{L}^{-1}\left\{\frac{1}{500s+100}\right\} = \frac{1}{500}e^{-t/5}, \tag{6.37}$$

so,

$$v(t) = v^e + \bar{v}(t) = v^e + \frac{1}{500}e^{-t/5}. \tag{6.38}$$

The reader can verify the computation of the inverse Laplace transform as an exercise. The impulse response of the velocity dynamics is shown below

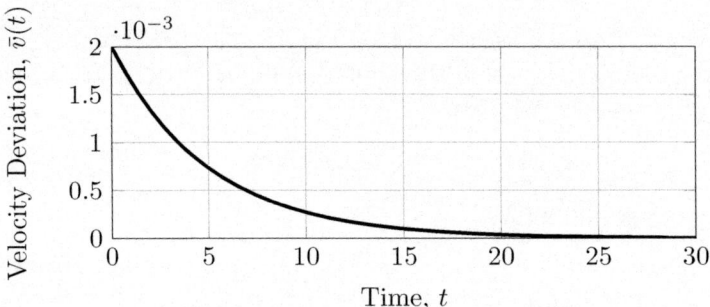

Figure 6.3: *Impulse response of the velocity dynamics. This is what happens if we give a pulse signal through the accelerator, i.e., $\bar{u}(t) = \delta(t)$.*

What happens if we provide a constant thrust input by keeping the accelerator at a constant position such that $\bar{u}(t) = 1$ for all $t \geq 0$? The step response of the velocity dynamics is

$$\begin{aligned}\bar{v}(t) &= \mathscr{L}^{-1}\left\{\frac{G(s)}{s}\right\} \\ &= \mathscr{L}^{-1}\left\{\frac{1}{s(500s + 100)}\right\} = \frac{1}{100}\left(1 - e^{-t/5}\right),\end{aligned} \tag{6.39}$$

which translates to

$$v(t) = v^e + \bar{v}(t) = v^e + \frac{1}{100}\left(1 - e^{-t/5}\right). \tag{6.40}$$

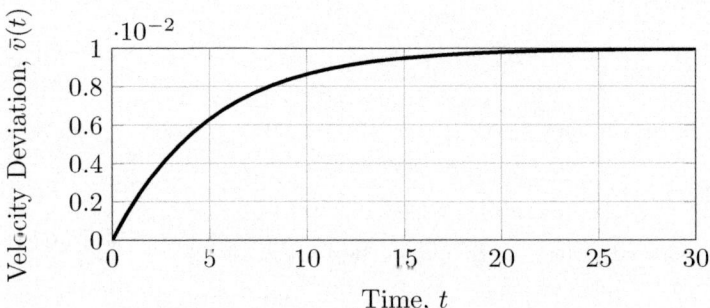

Figure 6.4: *Step response of the velocity dynamics.*

Lastly, the frequency response of the car to a sinusoidal excitation of the form
$$\bar{u} = A\sin(\omega t), \tag{6.41}$$
is given by
$$\begin{aligned}\bar{v}(t) &= A\omega \mathscr{L}^{-1}\left\{\frac{G(s)}{s^2+\omega^2}\right\} \\ &= A\omega \mathscr{L}^{-1}\left\{\frac{1}{(500s+100)(s^2+\omega^2)}\right\},\end{aligned} \tag{6.42}$$

and taking the inverse Laplace transform (see Chapter 5)
$$\bar{v}(t) = \frac{A\sin(\omega t) - 5A\omega\cos(\omega t)}{2500\omega^2+100} + \frac{A\omega e^{-t/5}}{5(100\omega^2+4)}. \tag{6.43}$$

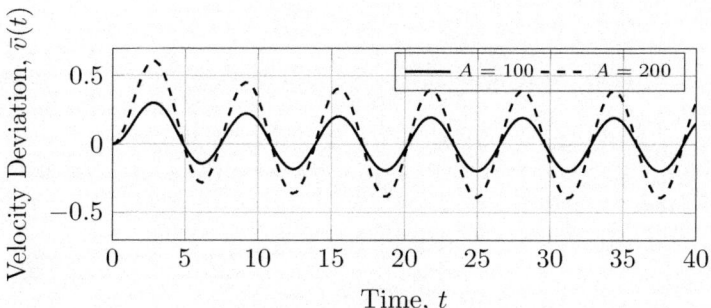

Figure 6.5: *Frequency response of the velocity dynamics for a constant angular frequency $\omega = 1$ and two different amplitudes.*

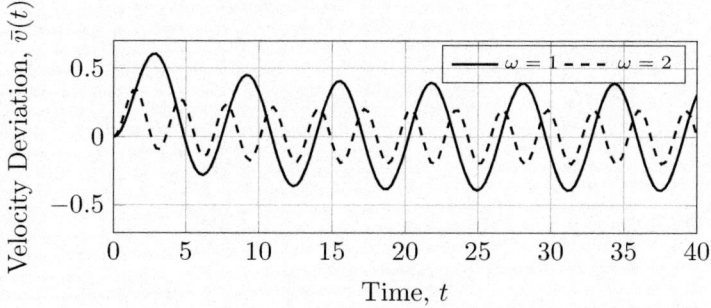

Figure 6.6: *Frequency response of the velocity dynamics for a constant amplitude $A = 200$ and two different angular frequencies.*

Note that, once again, we see that using deviation variables allows us to deal with non-zero initial conditions in the system variables (inputs and outputs). In the above example, the non-zero initial conditions $v(0) = v^e$

and $u(0) = u^e$ translate to $\bar{v}(0) = 0$ and $\bar{u}(0) = 0$.

Exercise 6.1 (**Impulse, step and frequency responses**) Determine the impulse, step and frequency responses of a linear dynamical system with transfer function

$$G(s) = \frac{s}{\tau s + 1}, \qquad (6.44)$$

with $\tau > 0$. Plot the responses of the system for $\tau = 1$.

❋ ....

Answer. The impulse response is $x_{(\text{imp})}(t) = \mathscr{L}^{-1}\varrho(t) - \tau^{-2} \exp(-t/\tau)$. The step response is $x_{\text{step}}(t) = \mathscr{L}^{-1} \exp(-t/\tau)$. The frequency response (when the input is $u(t) = \sin(\omega t)$) has the form $x_{\text{freq}}(t) = K \sin(\omega t + \phi_0)$, where $K = \sqrt{\frac{\omega^2}{\tau^2\omega^2+1}}$, $\phi_0 = \arctan \frac{1}{\tau\omega}$ and $B = \frac{\tau^2\omega^2}{\tau^2\omega^2+1}$.

Let us give an interesting practical example

Example 6.7 (**How not to fly a plane**) The roll angle of an aeroplane is controlled by its ailerons. Let u be the angle of the ailerons and x be the roll angle and let U and X be the corresponding Laplace transforms. Suppose that the input-output correspondence is described by the transfer function

$$G(s) = \frac{X(s)}{U(s)} = \frac{3s + 0.5}{2s^3 + 1.5s^2 + 2s}. \qquad (6.45)$$

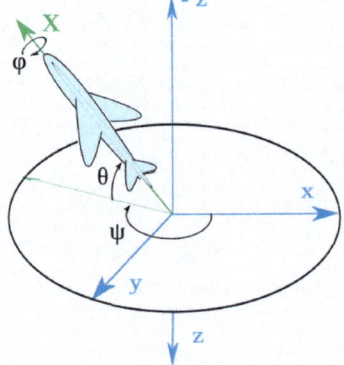

Figure 6.7: *Roll (φ), pitch (θ) and yaw (ψ) angles of an aeroplane. Image source: Wikipedia article "Orientation (geometry)," accessed on 17 February 2022 (Licence: CC-BY-3.0).*

The impulse response is

$$x(t) = \mathscr{L}^{-1}\{G(s)\} = \mathscr{L}^{-1}\left\{\frac{3s + 0.5}{2s^3 + 1.5s^2 + 2s}\right\}$$
$$= 1/4 \left(1 - e^{-0.375t}\left(\cos(0.927t) - 6.0678\sin(0.927t)\right)\right). \qquad (6.46)$$

The reader can verify this computation as an exercise. The impulse response of the system is shown below.

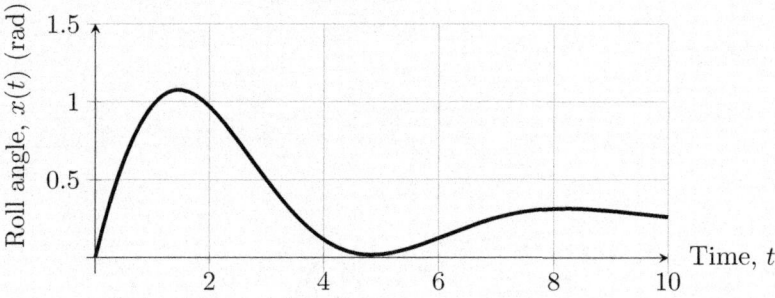

Figure 6.8: Impulse response of the roll dynamics. This is what happens if you kick the aeroplane's yoke.

The step response of the roll dynamics is given by

$$x(t) = \mathscr{L}^{-1}\left\{\frac{G(s)}{s}\right\} = \mathscr{L}^{-1}\left\{\frac{3s+0.5}{s(2s^3+1.5s^2+2s)}\right\}$$
$$= 0.25t - 1.3125\exp(-0.357t)\Big(\cos(0.927t) + 0.61\sin(0.927t)\Big) + 1.3125. \qquad (6.47)$$

The reader can verify this computation as an exercise.

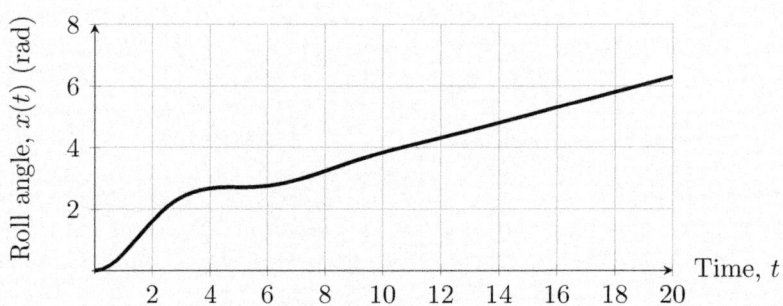

Figure 6.9: Step response of the roll dynamics. This is what will happen if one keeps the control yoke of the aircraft at a fixed angle. As we see here, the aircraft will eventually keep spinning around its roll axis at a constant rate.

Lastly, if we excite the system with a sinusoidal signal, $u(t) = \sin t$, whose Laplace transform is

$$U(s) = \frac{1}{s^2+1}. \tag{6.48}$$

Then, the system response is

$$x(t) = \mathscr{L}^{-1}\left\{\frac{G(s)}{s^2+1}\right\} = \mathscr{L}^{-1}\left\{\frac{3s+0.5}{(s^2+1)(2s^3+1.5s^2+2s)}\right\}. \tag{6.49}$$

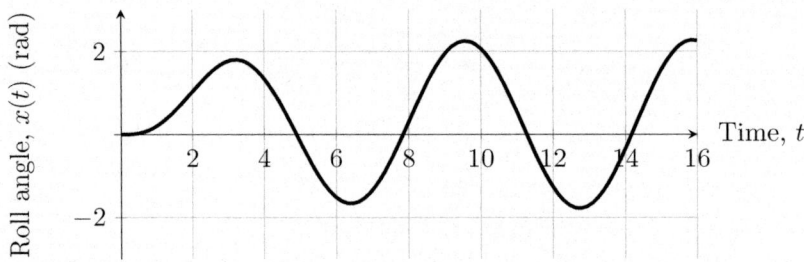

Figure 6.10: *Frequency response of the roll dynamics with excitation* $u(t) = \sin t$.

Remark: *The convolution theorem (Theorem 5.9) enables us to give an additional interpretation of the transfer function of a linear dynamical system with zero initial conditions. In particular, we know that if the input signal is $u(t)$ and its Laplace transform is a function $U(s)$, then the system response can be described, in the s-domain, by $X(s) = G(s)U(s)$, where G is the transfer function of the system. This means that*

$$x(t) = \mathscr{L}^{-1}\{G(s)U(s)\} = \mathscr{L}^{-1}\{G(s)\} * u(t). \tag{6.50}$$

The time-domain function $g(t) = \mathscr{L}^{-1}\{G(s)\}(t)$ is the impulse response of the system. This means that given the impulse response of the system, $g(t)$, we can determine the response of the system to any input signal via

$$x(t) = (g*u)(t). \tag{6.51}$$

This is an important result that we should state as a proposition. •

Chapter 6. Transfer functions and block diagrams

Proposition 6.2 (Response of linear dynamical system) If a linear dynamical system with transfer function G and zero initial conditions is excited with an input signal u (the Laplace transform of u is assumed to exist), then, the response of the system is

$$x(t) = (g * u)(t), \tag{6.52}$$

where $g(t) = \mathscr{L}^{-1}\{G(s)\}(t)$ is the impulse response of the system.

Proof. See above. □

Example 6.8 (System response in terms of convolution with input) Consider a linear dynamical system with transfer function

$$G(s) = \frac{1}{s+1}, \tag{6.53}$$

with zero initial conditions. The impulse response is

$$g(t) = \mathscr{L}^{-1}\{G(s)\}(t) = \mathscr{L}^{-1}\left\{\frac{1}{s+1}\right\} = e^{-t}, \tag{6.54}$$

$t \geq 0$. By Proposition 6.2, if the system is excited with an input signal u, the output will be

$$x(t) = e^{-t} * u(t), \tag{6.55}$$

for $t \geq 0$.

Let us give a couple of simple examples of Proposition 6.2. It should be noted that the result of Proposition 6.2 holds true even for non-strictly proper rational functions.

Example 6.9 (System response for systems with non-strictly proper transfer functions) Consider a linear dynamical system with transfer function

$$G(s) = \frac{s}{s+1}, \tag{6.56}$$

with zero initial conditions. Note that G is not strictly proper. The impulse response is

$$g(t) = \mathscr{L}^{-1}\{G(s)\}(t)$$
$$= \mathscr{L}^{-1}\left\{\frac{s}{s+1}\right\} = \mathscr{L}^{-1}\left\{1 - \frac{1}{s+1}\right\} = \delta(t) - e^{-t}, \tag{6.57}$$

$t \geq 0$. By Proposition 6.2, if the system is excited with an input signal u, the output will be

$$\begin{aligned} x(t) &= (\delta(t) - e^{-t}) * u(t) \\ &= \delta(t) * u(t) - e^{-t} * u(t) \\ &= u(0) - e^{-t} * u(t), \end{aligned} \quad (6.58)$$

for $t \geq 0$. Note that we used the property $\delta(t) * u(t) = u(0)$ (see Equation (4.158)).

6.3 Solving IVPs using transfer functions*

In Section 5.4 we used the Laplace transform to solve (linear) initial value problems. The same can be done using the transfer function of a dynamical system without the need to determine the corresponding ODE.

Suppose we have a proper rational transfer function $G(s) = P(s)/Q(s)$ with $\deg P = m$ and $\deg Q = n$ (with $m \leq n$). In particular, suppose that $P(s) = a_0 + a_1 s + \ldots + a_m s^m$ and $Q(s) = b_0 + b_1 s + \ldots + b_n s^n$.

Before we proceed, we will introduce the following convenient notation: using the coefficients of Q we define a sequence of polynomials $Q_{1:n}$, $Q_{2:n}$ up to $Q_{n:n}$ as follows

$$Q_{1:n}(s) = b_1 + b_2 s + \ldots + b_n s^{n-1}, \quad (6.59)$$
$$Q_{2:n}(s) = b_2 + b_3 s + \ldots + b_n s^{n-2}, \quad (6.60)$$
$$Q_{n:n}(s) = b_n. \quad (6.61)$$

Note that $\deg Q_{i:n} = n - i$ for $i = 1, \ldots, n$. Likewise, we define the polynomials $P_{1:m}, P_{2:m}, \ldots, P_{m:m}$.

A well-posed initial value problem is accompanied by n initial conditions of the form

$$x(0) = x_0, \ \dot{x}(0) = \dot{x}_0, \ \ldots, \ x^{(n-1)}(0) = x_0^{(n-1)}, \quad (6.62)$$

and for the input we have the initial condition

$$u(0^+) = u_0, \ \dot{u}(0^+) = \dot{u}_0, \ \ldots, \ u^{(m-1)}(0^+) = u_0^{(n-1)}. \quad (6.63)$$

Suppose that the input is $U(s)$ in the s-domain. Then, the state of the system given the above initial conditions is

$$X(s) = \frac{1}{Q(s)} \left[P(s) U(s) - \sum_{i=1}^{m} P_{i:m}(s) u_0^{(i-1)} + \sum_{i=1}^{n} Q_{n:n}(s) x^{(i-1)} \right]. \quad (6.64)$$

Let us give an example.

Example 6.10 (IVP from transfer function) Consider the transfer function
$$G(s) = \frac{5s - 1}{2s^3 + 5s^2 + 4s + 1}. \tag{6.65}$$
Here $P(s) = -1 + 5s$ ($m = 1$) and $Q(s) = 1 + 4s + 5s^2 + 2s^3$ ($n = 3$), therefore,

$$Q_{1:3}(s) = 4 + 5s + 2s^2, \tag{6.66a}$$
$$Q_{2:3}(s) = 5 + 2s, \tag{6.66b}$$
$$Q_{3:3}(s) = 2, \tag{6.66c}$$
$$P_{1:1}(s) = 5. \tag{6.66d}$$

We need to determine the output of the system given that the input is $u(t) = te^{-t}$ and given the initial conditions $x(0) = 3$, $\dot{x}(0) = 4$ and $\ddot{x}(0) = -2$.

We have that $U(s) = \mathscr{L}\{u(t)\}(s) = \frac{1}{(s+1)^2}$ and $u(0^+) = 0$. Following Equation (6.64) we have

$$X(s) = \frac{\frac{5s-1}{(s+1)^2} + 3(4 + 5s + 2s^2) + 4(5 + 2s) + 2(-2)}{2s^3 + 5s^2 + 4s + 1}$$
$$= \frac{6s^4 + 35s^3 + 80s^2 + 84s + 27}{(2s+1)(s+1)^4}. \tag{6.67}$$

Using Python or MATLAB we can determine the inverse Laplace transform of X, which is

$$x(t) = \tfrac{1}{2}e^{-t}\left(6t + 16\,e^{t/2} + 7t^2 + 2t^3 - 10\right), \tag{6.68}$$

for $t \geq 0$. The reader can verify that the given initial conditions are indeed satisfied.

6.4 Block diagrams

6.4.1 An algebra of blocks

A block diagram is a graphical representation of a transfer function, or multiple interconnected transfer function. A single transfer function is shown in Figure 6.11.

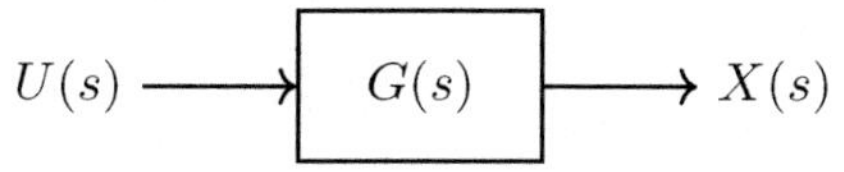

$$U(s) \longrightarrow \boxed{G(s)} \longrightarrow X(s)$$

Figure 6.11: *Block diagram of a transfer function. The input, U, and*

output, X, are shown in the diagram. The system output is given by $X(s) = G(s)U(s)$.

Block diagrams can be composed in three main ways: (i) in series, (ii) in parallel, (iii) through feedback. A **series** connection is one where the output of one block is fed to the input of a second one as shown below

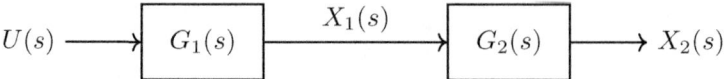

Figure 6.12: *Series connection of two blocks with transfer functions $G_1(s)$ and $G_2(s)$.*

In Figure 6.12 we see that the output of the first block is $X_1(s) = G_1(1)U(s)$. This is fed into the second block, so its output will be $X_2(s) = G_1(s)G_s(2)U(s)$. Therefore, the transfer function that links $U(s)$ with $X_2(s)$ is

$$G_{1,2}(s) = G_1(s)G_2(s). \tag{6.69}$$

A parallel series of blocks is shown below

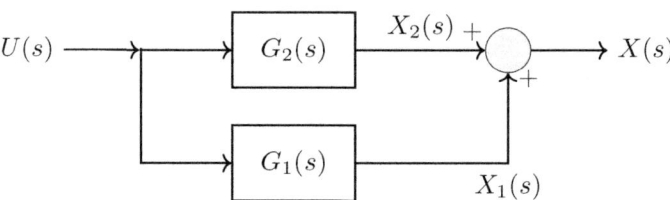

Figure 6.13: *Block diagram of a parallel connection of two transfer function, G_1 and G_2.*

The output, $X(s)$, in Figure 6.13 is the sum of the outputs of the two transfer functions $G_1(s)$ and $G_2(s)$, that is $X(s) = G_1(s)U(s) + G_2(s)U(s)$, therefore, the overall transfer function is

$$G_{1\|2}(s) = G_1(s) + G_2(s). \tag{6.70}$$

Lastly, two transfer functions can be connected through feedback as shown below

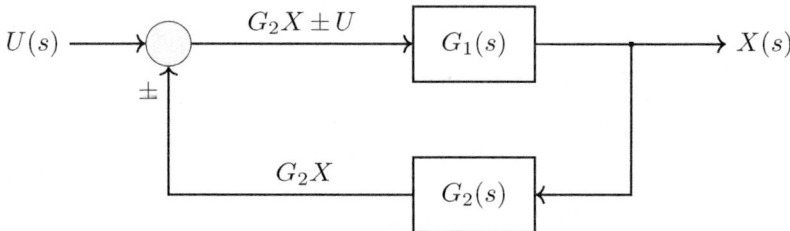

Figure 6.14: Connection of two blocks with transfer functions $G_1(s)$ and $G_2(s)$ through (positive or negative) feedback.

In Figure 6.14 we see that the output, $X(s)$ is fed back through G_2; the resulting signal G_2X is added to U and the sum, $G_2X + U$ is fed into G_1. If we focus on G_1 we have that

$$X = G_1(U \pm G_2X) \Rightarrow X = G_1U \pm G_1G_2X \Rightarrow (1 \mp G_1G_2)X = G_1U,$$

therefore, the transfer function of the system (from U to X) is

$$G_{[1,2]}(s) = \frac{G_1}{1 \mp G_1G_2}. \tag{6.71}$$

The "$\mp$" sign in the denominator means that if we have positive feedback, the sign should be negative and vice versa. Let us see how to use these rules to derive the transfer function of a more complex block diagram.

Example 6.11 (Block diagram) Consider the following block diagram:

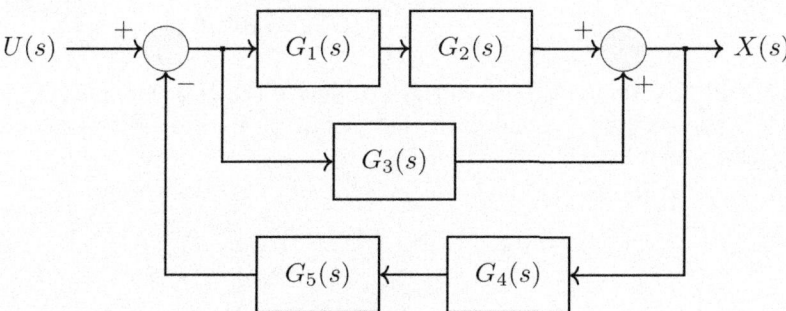

Figure 6.15: Example block diagram.

We need to derive the transfer function that associates the input $U(s)$ with the output function $X(s)$.

We start by noting that G_1 is in series with G_2 and G_4 is connected in series with G_5. The block diagram reduces to

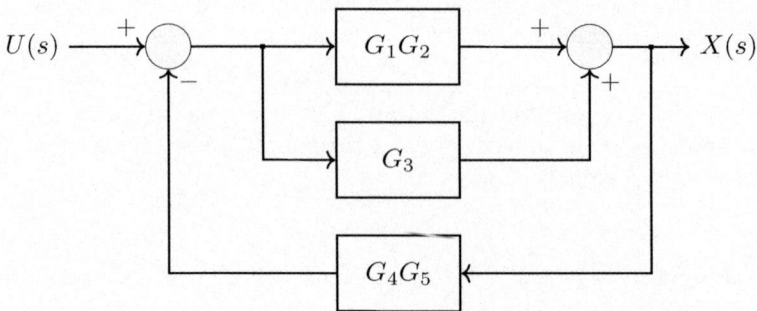

Next, we observe that G_1G_2 is connected in parallel with G_3, so the block diagram reduces to

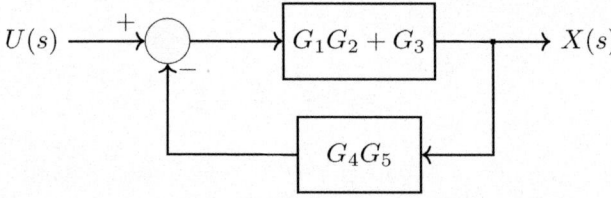

The two blocks with transfer functions $G_1G_2 + G_3$ and G_4G_5 are connected through *negative* feedback, therefore, the overall transfer function is

$$G(s) = \frac{X(s)}{U(s)} = \frac{G_1G_2 + G_3}{1 + (G_1G_2 + G_3)G_4G_5}. \qquad (6.72)$$

We will now apply the above rules to derive the transfer function that connects the set-point signal with the output of a controlled system in a feedback control loop.

Chapter 6. Transfer functions and block diagrams

Example 6.12 (Altitude control of quadcopter) A feedback control system is used to control the altitude of a quadcopter is shown below

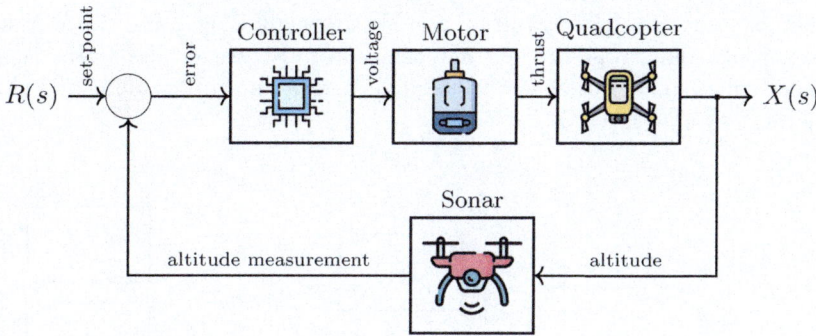

Figure 6.16: Block diagram of an altitude hold system on a quadcopter[a].

The closed loop comprises

1. the quadcopter dynamics, with transfer function $G_p(s)$, which specifies the correspondence between the thrust, generated by the spinning propellers, and the altitude $X(s)$

2. an ultrasonic sonar, with transfer function G_m, which measures the altitude of the quadcopter above the ground; this transfer function describes the dynamic correspondence between the altitude and its measurement

3. a controller, with transfer function $G_c(s)$, which commands a voltage to the motors

4. a set of four motors which are actuated with a voltage at their input and produce a thrust; this correspondence is described by a transfer function $G_a(s)$

We will find the transfer function between the set-point signal, R, and the altitude of the quadcopter, X. The controller, the motor and the quadcopter are connected in series. The loop is closed with the transfer function of the sonar. The reader can verify as an exercise (see Equations (6.69) and (6.71)) that the transfer function is

$$G(s) = \frac{X(s)}{R(s)} = \frac{G_c(s)G_a(s)G_p(s)}{1 + G_m(s)G_c(s)G_a(s)G_p(s)}. \qquad (6.73)$$

---

[a]Sources: controller icon created by @GoodWare, onar and motor icons created by @Freekpic, quadcopter icon by @Eucalyp on https://www.flaticon.com.

6.4.2 Reduction of block diagrams

Here we will give some additional rules that will allow us to simplify complex block diagrams and extract their overall transfer function. A very useful rule is that we can move a node before and after a block as shown below:

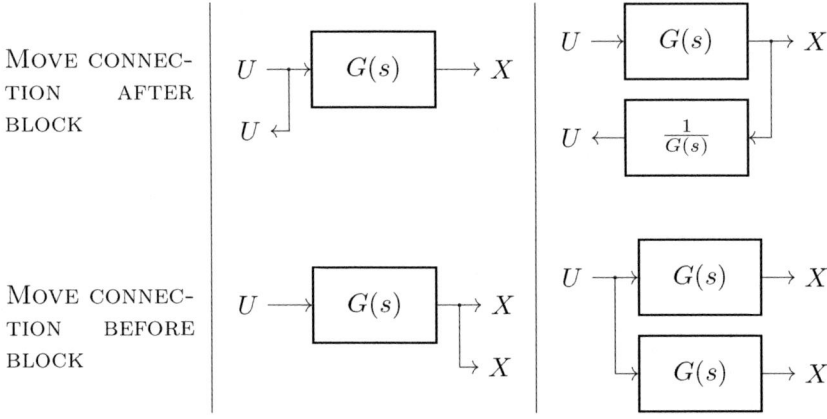

Figure 6.17: Rules for reducing block diagrams (see Example 6.13).

Example 6.13 (Reduction of complicated block diagram) We will now determine the transfer function of the following block diagram

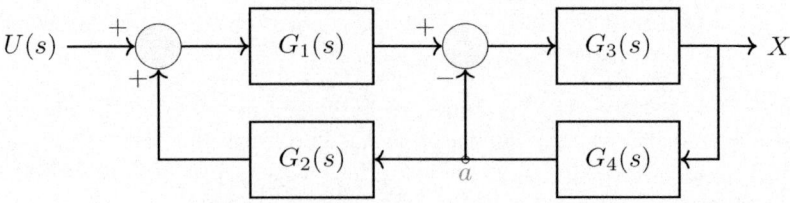

Firstly note that we cannot apply the rules for series, parallel and feedback rules we introduced in Section 6.4.1 because the two loops are combined. One thing we can do to simplify the above block diagram is to move node a before G_4 as follows:

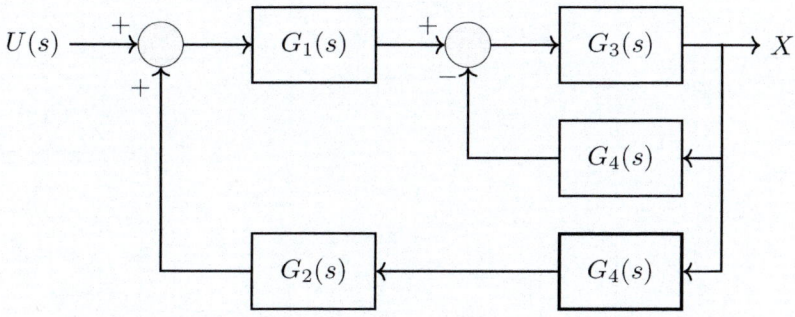

We can now combine G_3 and G_4 which are connected via feedback, into a block with transfer function $G_{[3,4]} = \frac{G_3}{1+G_3G_4}$, and it is easy to see that the overall transfer function is

$$G(s) = \frac{X(s)}{U(s)} = \frac{G_1 G_{[3,4]}}{1 - G_1 G_2 G_4 G_{[3,4]}}. \tag{6.74}$$

The reader may try herself to solve the following exercise following the procedure of Example 6.13.

Exercise 6.2 (Reduction of complicated block diagram) Determine the transfer function $G(s) = X(s)/U(s)$ given the following block diagram:

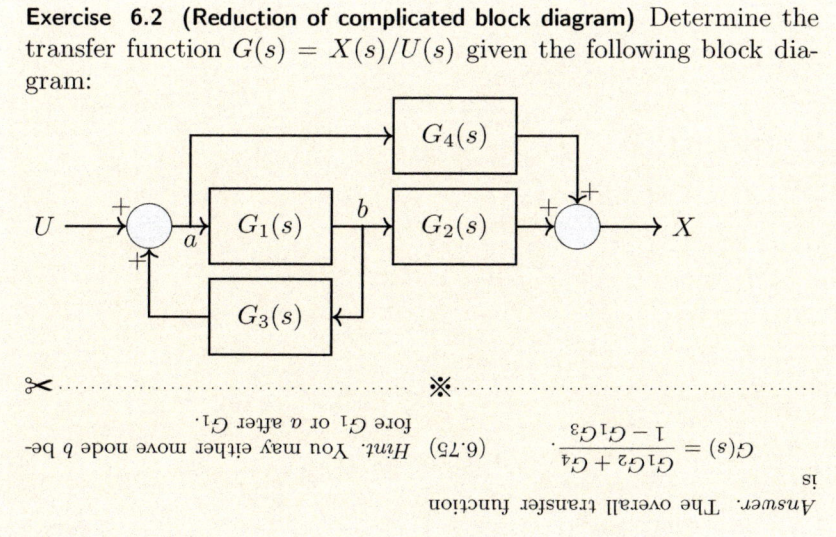

(6.75) *Hint*: You may either move node b before or after G_1.

Answer: The overall transfer function is
$$G(s) = \frac{G_1 G_2 + G_4}{1 - G_1 G_3}.$$

Another useful trick is that we can move a summation node before and after a block as shown in the following table

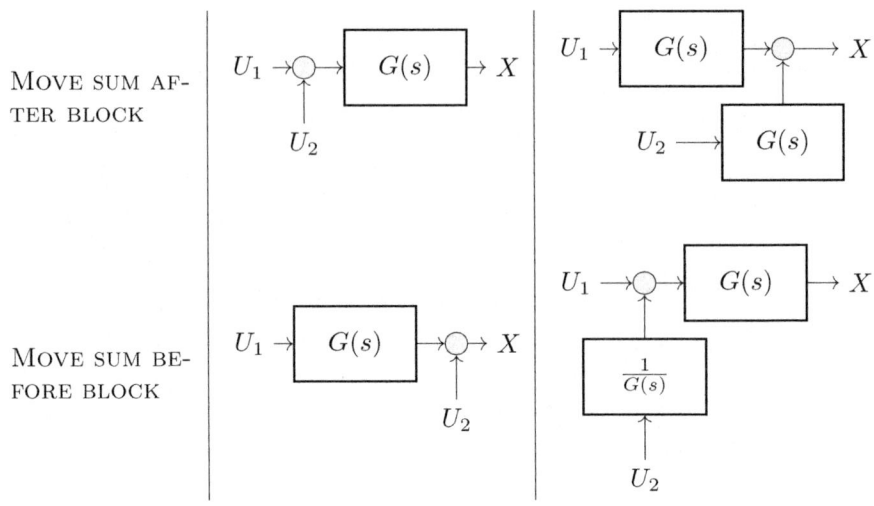

Figure 6.18: Further rules for reducing block diagrams (see Exercise 6.3).

Chapter 6. Transfer functions and block diagrams

Exercise 6.3 (**Reduction of block diagram**) Determine the transfer function $G(s) = X(s)/U(s)$ for the system shown in the following block diagram:

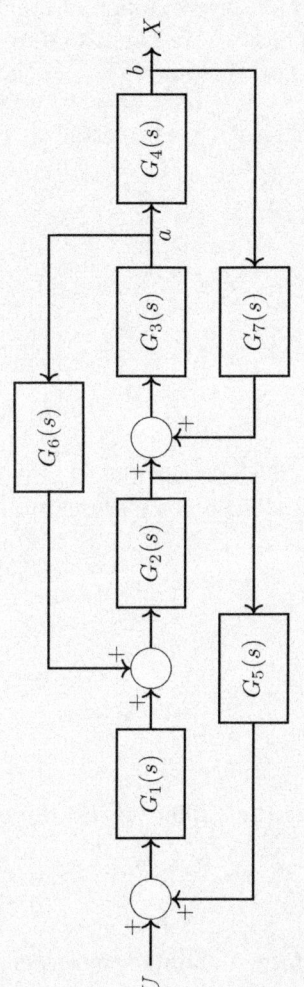

※ ....

Answer. You may move the second summation node to the left of G_1 and then interchange the first and the second sums. Then, move the node at a to the right of G_4 (at node b). You should find that the overall transfer function is

$$G(s) = \frac{X(s)}{U(s)} = \frac{G_1^2 G_2 G_3 G_4^2}{G_1 G_4 [1 - G_1 G_2 G_5)(1 - G_3 G_4 G_7) - G_2 G_3 G_6]}.$$

6.5 More transfer functions

6.5.1 Transfer functions to time domain

We have so far been deriving the transfer function of dynamical systems using their differential or integro-differential equation. We can also derive the ODE/IDE of a dynamical system from its transfer function. To that end, we introduce the *derivative operator* $D = \frac{d}{dt}$. This operator acts on differentiable functions and returns their derivative. In other words $Dx = \dot{x}$. We then define the operators [M21]

$$D^2 = DD, \qquad (6.76)$$

that is,

$$D^2 x = \ddot{x}, \qquad (6.77)$$

and, recursively, we define

$$D^{k+1} = DD^k, \qquad (6.78)$$

for all $k = 1, 2, \ldots$. As a convention, we define D^0 to be the identity operator, that is, $D^0 x = x$. These operators offer an alternative representation of differential equations. For example, the ODE

$$\ddot{x} + 3\dot{x} + 2x = u, \qquad (6.79)$$

can be written as

$$D^2 x + 3Dx + 2x = u, \qquad (6.80)$$

or equivalently[1],

$$(D^2 + 3D + 2)x = u, \qquad (6.81)$$

Assuming zero initial conditions and applying the Laplace transform on both sides of the last equation yields

$$(s^2 + 3s + 2)X(s) = U(s). \qquad (6.82)$$

We see that the application of the Laplace transform amounted to replacing the derivative operator D with s. This trick works in the opposite direction as well. For example, if a system has the following transfer function

$$H(s) = \frac{X(s)}{U(s)} = \frac{s-1}{3s^2 + 0.5s + 1}, \qquad (6.83)$$

we can replace s with D, $X(s)$ with $x(t)$ and $U(s)$ with $u(t)$ to obtain the differential equation of the system. It is

$$(3s^2 + 0.5s + 1)X(s) = (s-1)U(s)$$

---

[1]Strictly speaking, we should write $D^2 + 3D + 2I$, where I is the identity operator $Ix = x$.

$$\Rightarrow (3D^2 + 0.5D + 1)x(t) = (D - 1)u(t)$$
$$\Rightarrow 3D^2 x(t) + 0.5Dx(t) + x(t) = Du(t) - u(t)$$
$$\Rightarrow 3\ddot{x}(t) + 0.5\dot{x}(t) + x(t) = \dot{u}(t) - u(t). \tag{6.84}$$

The replacement of s with D and vice versa is valid only under the assumption of zero initial conditions and only for *linear* dynamical systems.

6.5.2 Interlude: Solving IVPs with the method of linear differential operators*

Linear differential operators

In Section 6.5.1 we defined the operators $D, D^2, D^3, \ldots$. We can then define *polynomials* in D, which have the form

$$p(D) = c_0 + c_1 D + c_2 D^2 + \ldots + c_n D^n, \tag{6.85}$$

where $c_0, c_1, \ldots, c_n \in \mathbb{C}$.

Let $x : \mathbb{R} \to \mathbb{C}$ be a smooth function. The polynomial $p(D)$ in Equation (6.85) is an operator such that $p(D)(x) = c_0 x + c_1 \dot{x} + \ldots + c_n x^{(n)}$. Such operators are referred to as *linear differential operators* (LDOs)[2].

LDOs inherit many of the properties of polynomials. The following exercise states certain useful properties of LDOs:

Exercise 6.4 (Linear differential operators) Prove the following

1. (Linearity) Let x_1 and x_2 be two sufficiently smooth functions and $c_1, c_2 \in \mathbb{C}$. Then

$$p(D)(c_1 x_1 + c_2 x_2) = c_1 p(D) x_1 + c_2 p(D) x_2. \tag{6.86}$$

2. (Sum of two operators) If p and q are two linear differential operators and x is a sufficiently smooth function, then

$$(p + q)(D)x = p(D)x + q(D)x. \tag{6.87}$$

3. (Multiplication rule) Let p and q be two linear differential operators and u be a sufficiently smooth function. Then,

$$p(D)(q(D)u) = (p(D)q(D))(u). \tag{6.88}$$

4. (Exponential functions) For every $a \in \mathbb{C}$ and any linear differential operator p,

$$p(D)e^{at} = p(a)e^{at}. \tag{6.89}$$

---

[2] Technical detail: Let $C^\infty(\mathbb{R}; \mathbb{C})$ be the set of differentiable functions, $x : \mathbb{R} \to \mathbb{C}$. Then, $p(D) : C^\infty(\mathbb{R}; \mathbb{C}) \to C(\mathbb{R}; \mathbb{C})$ is a linear operator.

Hint. The first three questions follow from the definition. Regarding the fourth question, you can start by showing that $D_v e^{at} = a_v e^{at}$, for all $v \in \mathbb{N}$.

ODEs via linear differential operators

Linear ODEs can be written compactly in the form

$$p(D)\tilde{x} = u, \qquad (6.90)$$

using an LDO $p(D)$. In particular, consider an ODE of the form

$$a_0 x(t) + a_1 \dot{x}(t) + a_2 \ddot{x}(t) + \ldots + a_n x^{(n)}(t) = u(t). \qquad (6.91)$$

By defining the LDO

$$p(D) = a_0 + a_1 D + \ldots + a_n D^n, \qquad (6.92)$$

we can write the above ODE in the form of Equation (6.90). If a function, $\tilde{x}(t)$, satisfies $p(D)\tilde{x} = u$, we say that it is a *solution* of the ODE. Note that such linear ODEs have infinitely many solutions; let us give an example that will assist our intuition.

Example 6.14 (Non-uniqueness of solutions) Consider the ODE

$$\dot{x} = \sin t \Leftrightarrow Dx(t) = \sin(t), \qquad (6.93)$$

so we have the LDO $p(D) = D$ and $u(t) = \sin(t)$. Equation (6.93) suggests that x is an antiderivative of $\sin(t)$ (see Section A.6.5), that is,

$$x(t) = \int \sin(t)\, dt = -\cos(t) + c, \qquad (6.94)$$

for $c \in \mathbb{R}$. The notation $\int \sin(t)\, dt$ refers to *an* antiderivative of $\sin t$ (for example, $-\cos t$). Any function of the form $\tilde{x}(t) = -\cos(t) + c$, with $x \in \mathbb{R}$, is a solution of the ODE in (6.93).

Solutions of $p(D)x = u$

To solve an IVP which is given in the form of Equation (6.90) (alongside some initial conditions), we recall that $p(D)$ is a linear operator. At this point, we need to recall a couple of facts from linear algebra. Let us first state the definition of the **kernel** of a linear operator.

Definition 6.3 (Kernel of linear operator) If A is a linear operator, we define its **kernel** to be the set $\ker A = \{x : Ax = 0\}$.

Chapter 6. Transfer functions and block diagrams

For example, $\ker D = \{v(t) = c; c \in \mathbb{R}\}$, that is, the kernel of D is the set of *constant* functions. Likewise, $\ker D^2 = \{v(t) = c_0 + c_1 t; c_0, c_1 t\}$, that is, the kernel of D^2 is the set of affine functions. Later we will see how we can determine the set $\ker p(D)$ of an LDO $p(D)$. We can now state a well-known result from linear algebra.

> **Theorem 6.4 (Solution of $Ax = b$)** Let A be a linear operator. Let $\tilde{x}$ be such that $A\tilde{x} = b$, that is, $\tilde{x}$ is any solution of $Ax = b$. Then, all solutions of $Ax = b$ have the form $\tilde{x} + v$, where $v \in \ker A$.

Proof. Let $x = \tilde{x} + v$ with $v \in \ker A$ and $A\tilde{x} = b$. Then $Ax = A(\tilde{x} + v) = A\tilde{x} + Av = b$, so x is a solution of $Ax = b$. Conversely, if x is a solution, define $v = x - \tilde{x}$. Then $Av = A(x - \tilde{x}) = 0$, so $v \in \ker A$. In other words, any solution x can be written as $\tilde{x} + v$, with $v \in \ker A$. □

Overall, this means that in order to determine all solutions of $p(D)x = u$ we need to be able to (i) determine any particular solution $\tilde{x}$, and (ii) determine the kernel of $p(D)$. We will do so in the following two sections.

The reason we need to be able to determine *all* solutions of $p(D)x = u$ is that when solving an IVP we will be looking for a solution that satisfies certain initial conditions.

Determination of particular solutions of $p(D)x = u$

The following result allows us to determine a solution of ODEs of the form $\dot{x} - cx = u$, that is, when $p(D) = D - c$.

> **Proposition 6.5 (A solution of $(D - c)x = u$)** Let $p(D) = D - c$, for some $c \in \mathbb{C}$. A solution of $p(D)x = u$ is
>
> $$\tilde{x}(t) = e^{ct} \int e^{-ct} u(t) dt, \qquad (6.95)$$
>
> where $\int e^{-ct} u(t) dt$ is any antiderivative of the function $e^{-ct} u(t)$.

Proof. It suffices to check that $x(t)$ is indeed a solution, i.e., that it satisfies $p(D)x = u$. This is pretty straightforward and is left to the reader as an exercise. □

Example 6.15 (Solution of ODE) We will determine a solution of the ODE
$$\dot{x} + 2x = e^{-t} \sin t. \qquad (6.96)$$
We have $p(D) = D + 2$ (so, $c = -2$) and $u(t) = e^{-t} \sin t$. From

Proposition 6.5 we have the solution

$$\tilde{x}(t) = e^{-2t} \int e^{2t} e^{-t} \sin t \, dt$$
$$= e^{-2t} \int e^t \sin \tau \, dt$$
$$= e^{-2t} \left(-\tfrac{1}{2} e^t (\cos t - \sin t)\right)$$
$$= \tfrac{1}{2} e^{-t} (\sin t - \cos t). \tag{6.97}$$

This is a particular solution of Equation (6.96), but there are more (see Example 6.17).

In case we have a higher-order polynomial, $p(D) = a_0 + a_1 D + \ldots + a_n D^n$, we know that we can write it in the form $p(D) = a_n(D - c_1)(D - c_2) \ldots (D - c_n)$, where, in general, the coefficients can be complex numbers and not necessary different from one another. Then we work as in Example 6.15. Here is an example.

Example 6.16 (Solution of second-order ODE) We need to determine a solution of the ODE

$$\ddot{x} - 3\dot{x} + 2x = e^{-t}. \tag{6.98}$$

Here we have $p(D) = D^2 - 3D + 2$ which we can factorise as $p(D) = (D-1)(D-2)$. Define $u(t) = e^{-t}$. Then we can write the given ODE as

$$(D-1)(D-2)x = u. \tag{6.99}$$

By defining $z = (D-2)x$ we have $(D-1)z = u$, which we can solve to determine a solution $\tilde{z}$. Then, we can solve $(D-2)x = \tilde{z}$ to determine a solution $\tilde{x}$.

Firstly, let us solve the ODE $(D-1)z = u$ using Proposition 6.5. We have

$$\tilde{z} = e^t \int e^{-t} e^{-t} dt = \ldots = -\tfrac{1}{2} e^{-t}. \tag{6.100}$$

Then, we solve $(D-2)x = \tilde{z}$. Again using Proposition 6.5 we have

$$\tilde{x}(t) = e^{2t} \int -e^{-2t} \tfrac{1}{2} e^{-t} dt = \ldots = \tfrac{1}{6} e^{-t}. \tag{6.101}$$

The kernel of an LDO

We start by focusing on the first-order LOD $p(D) = D - c$, for $c \in \mathbb{C}$.

Chapter 6. Transfer functions and block diagrams

Proposition 6.6 (Kernel of $D - c$) Let $c \in \mathbb{C}$. Then,
$$\ker(D - c) = \{\lambda e^{ct}, \lambda \in \mathbb{C}\}. \tag{6.102}$$

Proof. It is easy to see that every function of the form $v(t) = \lambda e^{ct}$ is in $\ker(D - c)$, i.e., $(D - c)v = 0$. Conversely, if v is such that $(D - c)v = 0$, then equivalently

$$Dv - cv = 0 \Leftrightarrow \dot{v}e^{-ct} - cve^{-ct} = 0 \quad \text{(Multiply by } e^{-ct} \neq 0\text{)}$$
$$\Leftrightarrow \dot{v}e^{-ct} + v \tfrac{d}{dt}(e^{-ct}) = 0$$
$$\Leftrightarrow \tfrac{d}{dt}(ve^{-ct}) = 0 \quad \text{(Product rule)}$$
$$\Leftrightarrow ve^{-ct} = \lambda \Leftrightarrow v = \lambda e^{ct}, \tag{6.103}$$

for some $\lambda \in \mathbb{C}$. $\square$

Proposition 6.6 enables us to detemine *all* solutions of ODEs of the form $(D - c)x = u$. Let us give an example.

Example 6.17 (Determination of all solutions) We shall determine all solutions of Equation (6.96). We know that a particular solution is given in Equation (6.97), and the kernel of $p(D) = D + 2$ is $\{\lambda e^{-2t}, \lambda \in \mathbb{C}\}$, therefore, any solution of Equation (6.96) has the form
$$x(t) = \tfrac{1}{2}e^{-t}(\sin t - \cos t) + \lambda e^{-2t}, \tag{6.104}$$
for some $\lambda \in \mathbb{C}$.

Exercise 6.5 (Solution of IVP) Solve the IVP
$$\dot{x} + 2x = e^{-t}\sin t, \tag{6.105a}$$
$$x(0) = 1. \tag{6.105b}$$

✂ ....

Hint. The set of all solutions of the given ODE is given in Equation (6.104); you just need to determine λ.

Next, we shall determine the kernel of $(D - c)^n$ for $n \in \mathbb{N}$.

Proposition 6.7 (Kernel of $(D - c)^n$) Let $c \in \mathbb{C}$. Then[a],
$$\ker((D - c)^n) = \left\{ \begin{array}{c} \lambda_1 e^{ct} + \lambda_2 t e^{ct} + \ldots + \lambda_n t^{n-1} e^{ct}, \\ \lambda_1, \ldots, \lambda_n \in \mathbb{C} \end{array} \right\}. \tag{6.106}$$

---
[a] We say that $\ker((D - c)^n)$ is the *span* of the set $\{e^{c_1 t}, t e^{c_1 t}, \ldots, t^{n-1} e^{c_1 t}\}$.

Proof. We will prove this by mathematical induction: we have shown (in Proposition 6.6) that the result holds for $n = 1$. Suppose that it holds for $n = k$. We will show that it holds for $k + 1$. We have

$$\begin{aligned}
x \in \ker(\mathrm{D} - c)^{k+1} &\Leftrightarrow (\mathrm{D} - c)^{k+1} x = 0 \\
&\Leftrightarrow (\mathrm{D} - c)^k (\mathrm{D} - c) x = 0 \\
&\Leftrightarrow (\mathrm{D} - c) x \in \ker(\mathrm{D} - c)^k \\
&\Leftrightarrow (\mathrm{D} - c) x = \lambda_1 e^{ct} + \lambda_2 t e^{ct} + \ldots + \lambda_n t^{n-1} e^{ct},
\end{aligned} \quad (6.107)$$

for some $\lambda_1, \ldots, \lambda_n \in \mathbb{C}$. Define $u(t) = \lambda_1 e^{ct} + \lambda_2 t e^{ct} + \ldots + \lambda_n t^{n-1} e^{ct}$. By Proposition 6.5, a particular solution of Equation (6.107) is

$$\begin{aligned}
\tilde{x}(t) &= e^{ct} \int e^{-ct} u(t) \mathrm{d}t \\
&= e^{ct} \int e^{-ct} (\lambda_1 e^{ct} + \lambda_2 t e^{ct} + \ldots + \lambda_n t^{n-1} e^{ct}) \mathrm{d}t \\
&= e^{ct} \int (\lambda_1 + \lambda_2 t + \ldots + \lambda_n t^{n-1}) \mathrm{d}t \\
&= e^{ct} \left(\lambda_1 t + \lambda_2 \frac{t^2}{2} + \ldots + \lambda_n \frac{t^n}{n} \right),
\end{aligned} \quad (6.108)$$

from which we see that $\tilde{x}$ is a linear combination of the functions $\{t e^{ct}, t^2 e^{ct}, \ldots, t^n e^{ct}\}$. We also know (from Proposition 6.6) that $\ker(\mathrm{D}-c) = \{\lambda e^{ct}, \lambda \in \mathbb{C}\}$. As a result, the solutions of Equation (6.107) have the form $\tilde{x} + v$ with $v = \lambda e^{ct}$, i.e., they are linear combinations of $\{e^{ct}, t e^{ct}, t^2 e^{ct}, \ldots, t^n e^{ct}\}$. This completes the proof. $\square$

Let us now use Propositions 6.5 and 6.7 to solve an IVP of the form $(\mathrm{D} - c)^n x = u$.

Example 6.18 (IVP of the form $(\mathrm{D} - c)^2 x = u$) Consider the IVP

$$\ddot{x} - 6\dot{x} + 9x = \sin t, \quad (6.109\mathrm{a})$$
$$x(0) = 2, \dot{x}(0) = -1. \quad (6.109\mathrm{b})$$

We have $p(\mathrm{D}) = \mathrm{D}^2 - 6\mathrm{D} + 9 = (\mathrm{D} - 3)^2$ and $u(t) = \sin t$. A particular solution can be determined as in Example 6.16. This is

$$\tilde{x}(t) = \tfrac{3}{50} \cos t + \tfrac{2}{25} \sin t. \quad (6.110)$$

The reader can derive this herself. Next, using Proposition 6.7 we have that

$$\ker(\mathrm{D} - 3)^2 = \{\lambda_1 e^{3t} + \lambda_2 t e^{3t}, \lambda_1, \lambda_2 \in \mathbb{C}\}, \quad (6.111)$$

so the solutions of $(\mathrm{D} - 3)^2 x = \sin t$ have the general form

$$x(t) = \tfrac{3}{50} \cos t + \tfrac{2}{25} \sin t + \lambda_1 e^{3t} + \lambda_2 t e^{3t}, \quad (6.112)$$

with $\lambda_1, \lambda_2 \in \mathbb{C}$. We must have $x(0) = 2$, so

$$\tfrac{3}{50} + \lambda_1 = 2 \Leftrightarrow \lambda_1 = \tfrac{97}{50} = 1.94. \tag{6.113}$$

We also need to have $\dot{x}(0) = -1$. By differentiating x we have $\dot{x}(t) = \tfrac{2}{25}\cos t - \tfrac{3}{50}\sin t + 3\lambda_1 e^{3t} + \lambda_2 e^{3t} + 3\lambda_2 t e^{3t}$, so

$$\dot{x}(0) = -1$$
$$\Leftrightarrow \tfrac{2}{25} + 3\lambda_1 + \lambda_2 = -1$$
$$\Leftrightarrow \lambda_2 = -1 - \tfrac{2}{25} - 3\tfrac{97}{50} = -6.9. \tag{6.114}$$

Overall, the solution of the given IVP is

$$x(t) = \tfrac{3}{50}\cos t + \tfrac{2}{25}\sin t + 1.94 e^{3t} - 6.9 t e^{3t}. \tag{6.115}$$

An LDO can be factorised as $p(\mathrm{D}) = a_n(\mathrm{D} - c_1)^{m_1}(\mathrm{D} - c_2)^{m_2} \ldots (\mathrm{D} - c_k)^{m_k}$ where $c_1, c_2, \ldots, c_k \in \mathbb{C}$ are different from one another. Then we can determine $\ker p(\mathrm{D})$ by means of the following result.

> **Proposition 6.8 (Kernel of product of LDOs)** Let $c_1, c_2 \in \mathbb{C}$ and $c_1 \neq c_2$. Let also $m_1, m_2 \in \mathbb{N}$. Then,
>
> $$\ker((\mathrm{D} - c_1)^{m_1}(\mathrm{D} - c_2)^{m_2})$$
> $$= \left\{ \begin{array}{l} \lambda_{1,1} e^{c_1 t} + \lambda_{1,2} t e^{c_1 t} + \ldots + \lambda_{1,m_1} t^{m_1 - 1} e^{c_1 t}, \\ + \lambda_{2,1} e^{c_2 t} + \lambda_{2,2} t e^{c_2 t} + \ldots + \lambda_{2,m_2} t^{m_2 - 1} e^{c_1 t}, \\ \lambda_{1,1}, \ldots, \lambda_{1,m_1}, \lambda_{2,1}, \ldots, \lambda_{2,m_2} \in \mathbb{C} \end{array} \right\}. \tag{6.116}$$

Proof. By definition, $x \in \ker((\mathrm{D} - c_1)^{m_1}(\mathrm{D} - c_2)^{m_2})$ means that

$$((\mathrm{D} - c_1)^{m_1}(\mathrm{D} - c_2)^{m_2})x = 0$$
$$\Leftrightarrow (\mathrm{D} - c_2)^{m_2} x \in \ker(\mathrm{D} - c_1)^{m_1}$$
$$\Leftrightarrow (\mathrm{D} - c_2)^{m_2} x = \lambda_1 e^{c_1 t} + \lambda_2 t e^{c_1 t} + \ldots + \lambda_{m_1} t^{m_1 - 1} e^{c_1 t}, \tag{6.117}$$

for some $\lambda_1, \ldots, \lambda_{m_1} \in \mathbb{C}$. It now suffices to determine all solutions of Equation (6.117); this is left to the reader as an exercise. $\square$

Example 6.19 (Solution of second-order IVP) Let us solve the IVP

$$\ddot{x} - 3\dot{x} + 2x = e^{-t}, \tag{6.118a}$$
$$x(0) = 0, \dot{x} = 1. \tag{6.118b}$$

Here we have $p(\mathrm{D}) = \mathrm{D}^2 - 3\mathrm{D} + 2 = (\mathrm{D} - 1)(\mathrm{D} - 2)$ and $u(t) = e^{-t}$. In Example 6.16 we found that

$$\tilde{x}(t) = \tfrac{1}{6} e^{-t}, \tag{6.119}$$

is a particular solution of the ODE. By Proposition 6.8, the kernel of $p(D)$ is
$$\ker p(D) = \{\lambda_1 e^t + \lambda_2 e^{2t}, \lambda_1, \lambda_2 \in \mathbb{C}\}, \tag{6.120}$$
so the solutions of the ODE in Equation (6.118a) have the form
$$x(t) = \tfrac{1}{6} e^{-t} + \lambda_1 e^t + \lambda_2 e^{2t}. \tag{6.121}$$
From the given initial conditions, we have
$$x(0) = 0 \Leftrightarrow \tfrac{1}{6} + \lambda_1 + \lambda_2 = 0, \tag{6.122a}$$
$$\dot{x}(0) = 1 \Leftrightarrow \lambda_1 + 2\lambda_2 - \tfrac{1}{6} = 1. \tag{6.122b}$$
By solving the above system we obtain $\lambda_1 = -3/2$ and $\lambda_2 = 4/3$, and the solution of the IVP is
$$x(t) = \tfrac{1}{6} e^{-t} - \tfrac{3}{2} e^t + \tfrac{4}{3} e^{2t}. \tag{6.123}$$

Exercise 6.6 (Third-order IVP) Solve the third-order IVP
$$x^{(3)} - 4\ddot{x} + 5\dot{x} - 2x = e^{-t}, \tag{6.124a}$$
$$x(0) = 1, \dot{x}(0) = 0, \ddot{x}(0) = 1. \tag{6.124b}$$

❊ ....

Answer. Define $p(D) = D^3 - 4D^2 + 5D - 2$. We can factorise this as $p(D) = (D-1)^2(D-2)$. Then, we can find that a particular solution is $\tilde{x}(t) = -1/12\, e^{-t}$, + $\lambda_1 e^t + \lambda_2 e^{2t} + \lambda_3 t e^t$, with $\lambda_1, \lambda_2, \lambda_3 \in \mathbb{C}$. Using the given initial conditions we find $\lambda_1 = 2$, $\lambda_2 = -11/12$, and $\lambda_3 = 3$.

Given an LDO with real coefficients, its factorisation may involve pairwise conjugate complex terms, for example $p(D) = D^2 - 2D + 2$ is factorised as $p(D) = (D - (1+j))(D - (1-j))$. However, the procedure we described above remains the same.

Exercise 6.7 (LDO involving complex terms) Solve the IVP
$$\ddot{x} - 2\dot{x} + 2x = e^{-2t}, \tag{6.125a}$$
$$x(0) = 0, \dot{x}(0) = 0. \tag{6.125b}$$

❊ ....

Answer. We can determine the particular solution $\tilde{x} = \tfrac{1}{10} e^{-2t}$, so the general form of a solution is $x(t) = \tfrac{1}{10} e^{-2t} + \lambda_1 e^{(1+j)t} + \lambda_2 e^{(1-j)t}$. We can find that $\lambda_1 = -\tfrac{1}{20} - \tfrac{j}{20}$, and $\lambda_1 = -\tfrac{1}{20} + \tfrac{j}{20}$. Note that $x(t)$ is a real-valued function.

6.5.3 Transfer functions to state representation

We now know how to recover a differential equation from a transfer function. Earlier, in Section 2.1.2, we addressed the problem of writing general differential equations that involve derivatives of the input variables in a standard state space representation. We were able to do so for simple cases that involve only first derivatives of the input, but we admitted that a generalisation was not easy. It turns out that we can address such problems in the s-domain as we will show in the following example.

Example 6.20 (Transfer function to state space representation) Consider the following transfer function

$$G(s) = \frac{X(s)}{U(s)} = \frac{5s^2 + 2s + 1}{s^3 + 3s^2 + 7s + 1}. \tag{6.126}$$

Following the procedure of Section 6.5.1 we can find that the corresponding ODE is

$$x^{(3)} + 3\ddot{x} + 7\dot{x} = 5\ddot{u} + 2\dot{u} + u. \tag{6.127}$$

Instead of trying to determine a state space representation using Equation (6.127) we will work with the transfer function in Equation (6.126) according to which there is a variable $Z(s)$ such that

$$X(s) = (5s^2 + 2s + 1)Z(s), \tag{6.128a}$$
$$U(s) = (s^3 + 3s^2 + 7s + 1)Z(s). \tag{6.128b}$$

We can then derive the ODE that corresponds to Equation (6.128b):

$$u = z^{(3)} + 3\ddot{z} + 7\dot{z} + z, \tag{6.129}$$

whose state space representation can be obtained by introducing the variables $z_1 = z$, $z_2 = \dot{z}_1$ and $z_3 = \dot{z}_2$

$$\dot{z}_1 = z_2, \tag{6.130a}$$
$$\dot{z}_2 = z_3, \tag{6.130b}$$
$$\dot{z}_3 = u - 3z_3 - 7z_2 - z_1. \tag{6.130c}$$

This is a (linear) dynamical system with state $z = (z_1, z_2, z_3)$ and input u. Note that $x(t)$ can be recovered from $z(t)$ since from Equation (6.128a) we have

$$x(t) = 5\ddot{z} + 2\dot{z} + z = z_3 + 5z_2 + 2z_1. \tag{6.131}$$

It can also be noted that the methodology presented in Example 6.20 can only be applied to *strictly proper* transfer functions, that is, rational functions $G(s) = P(s)/Q(s)$ with $\deg P < \deg Q$. The case of non-strictly proper transfer functions ($\deg P = \deg Q$) is slightly more involved — the reader can refer to Section 5.2.7.

Exercise 6.8 (Dynamical systems involving input derivatives) Construct a state space representation for the system that is described by the transfer function

$$\frac{X(s)}{U(s)} = \frac{s^3 + 2s^2 - s - 1}{s^4 + s^2 + 1}. \tag{6.132}$$

✻ ....

Answer. In terms of the state variables (z_1, z_2, z_3, z_4), the system dynamics is described by $\dot{z}_1 = z_2$, $\dot{z}_2 = z_3$, $\dot{z}_3 = z_4$ and $\dot{z}_4 = u - z_3 - z_1$. Then $x = z_4 + 2z_3 - z_2 - z_1$.

Exercise 6.9 (ODE to state space representation) Write the following dynamical system in its state space representation

$$x^{(3)} + 8\dot{x} + x = u - 3\dot{u} + 4\ddot{u}. \tag{6.133}$$

✻ ....

Answer. We have a dynamical system with state $z = (z_1, z_2, z_3)$ and input u, which is described by the ODEs $\dot{z}_1 = z_2$, $\dot{z}_2 = z_3$, $\dot{z}_3 = u - z_1 - 8z_2$. Then, x can be retrived from z via $x = z_1 - 3z_2 + 4z_3$.

6.5.4 Transfer functions of MIMO systems

Suppose that a system has m input variables and n output variables. We then need to describe the dynamic correspondence between each of the output variables, $Y_1, \ldots, Y_n$, and the input variables $U_1, \ldots, U_m$. In the Laplace domain, for each $i = 1, \ldots, n$, we have

$$Y_i(s) = G_{i,1}(s)U_1(s) + \ldots + G_{i,m}(s)U_m(s), \tag{6.134}$$

where each of the transfer functions $G_{i,j}(s)$, for $j = 1, \ldots, m$, describes the effect that input U_j has on output Y_i.

Such a multiple-input multiple-output dynamic correspondence can be described in a more concise way by introducing the input vector $U(s) = [U_1(s)\ U_2(s)\ \cdots\ U_m(s)]$ and the output vector $Y(s) = [Y_1(s)\ Y_2(s)\ \cdots\ Y_n(s)]$. Then, equations (6.134) can be concisely written as

$$Y(s) = G(s)U(s), \tag{6.135}$$

where $G(s)$ is the matrix

$$G(s) = \begin{bmatrix} G_{1,1}(s) & G_{1,2}(s) & \cdots & G_{1,m}(s) \\ G_{2,1}(s) & G_{2,2}(s) & \cdots & G_{2,m}(s) \\ \vdots & \vdots & \ddots & \vdots \\ G_{n,1}(s) & G_{n,2}(s) & \cdots & G_{n,m}(s) \end{bmatrix}. \qquad (6.136)$$

Note that $G_{i,j}(s)$ corresponds to the transfer function of a dynamical system with input $U_j(s)$ and output $Y_i(s)$ provided that all other inputs, $U_l(s)$ with $l \neq j$, are equal to zero.

Exercise 6.10 (Two inputs, one output) A system with two inputs, U_1 and U_2, and one output, Y, is described by a transfer function $G(s) = [G_1(s) \; G_2(s)]$. Sketch a block diagram for this transfer function.

Example 6.21 (Feedback control system) A feedback control system is itself a dynamical system with two inputs: the set point, $X^{\text{SP}}(s)$ and the disturbance $D(s)$. A typical feedback control system may involve the transfer functions of (i) the controlled system, G_s, (ii) the controller, G_c, (iii) the actuator, G_a, (iv) the sensor, G_m, (v) the disturbance, G_d, which are connected as shown in the following figure

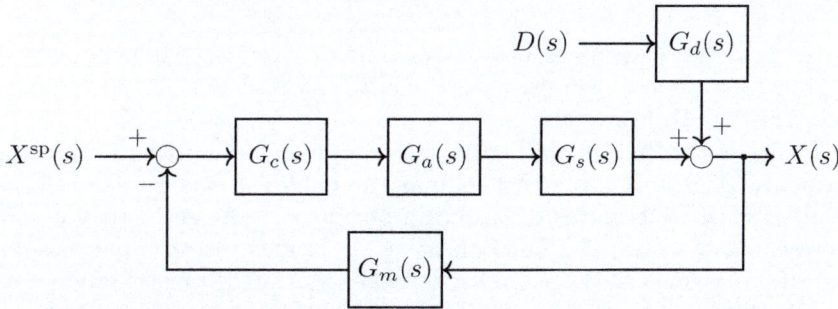

Figure 6.19: *Structure of a feedback control system.*

This is a system with two input signals ($X^{\text{SP}}(s)$ and $D(s)$) and one output ($X(s)$). In order to derive its overall transfer function, we first neglect $D(s)$ (suppose $D(s) = 0$) and we see that the transfer function from $X^{\text{SP}}(s)$ to $X(s)$ is (see Example 6.12)

$$G_{\text{SP}} = \frac{G_c G_a G_s}{1 + G_c G_a G_s G_m} \qquad (6.137)$$

Likewise, suppose that $X^{\text{SP}}(s) = 0$. Then, the transfer function from D to X is

$$G_{\text{D}} = \frac{G_d}{1 + G_c G_a G_s G_m}. \qquad (6.138)$$

The dynamic correspondence between the input signals $X^{\text{sp}}(s)$ and $D(s)$ and the output, $X(s)$, is described by

$$X(s) = G_{\text{SP}}(s) X^{\text{sp}}(s) + G_D(s) D(s), \tag{6.139}$$

that is, the transfer function of the system is

$$G(s) = \begin{bmatrix} G_{\text{SP}}(s) & G_D(s) \end{bmatrix}. \tag{6.140}$$

6.6 Impedance

In Section 2.3 we presented the basic principles of electric circuits and used them to derive models for electric circuits. Judging by the complexity of, for example, Exercise 2.22, we see that RLC circuits defy the simple rules of circuits involving only resistors. We know that the total resistance of two resistors, R_1 and R_2, connected in series is $R_1 + R_2$ and the resistance of two resistors connected in parallel is

$$R_{1\|2} = \frac{1}{\frac{1}{R_1} + \frac{1}{R_2}}.$$

With these two rules we can analyse complex circuits. After all, a large network of resistors behaves as one equivalent resistor. Do such simple rules exist for RLC circuits?

Resistors, capacitors and inductors are dynamic components: the voltage across them and the current running through them satisfies an algebraic or differential equation as discussed in Section 2.3. As such, they can be represented by a transfer function using the current as an input variable and the voltage as an output variable. The associated transfer function is, by definition,

$$Z(s) = \frac{V(s)}{I(s)}. \tag{6.141}$$

For example, for a resistor with resistance R, it is $V(t) = I(t)R$, so

$$Z_R(s) = \frac{V(s)}{I(s)} = R. \tag{6.142}$$

Similarly, for a capacitor with capacitance C and zero initial charge we know that $C\dot{V}(t) = I(t)$, so $CsV(s) = I(s)$, therefore, its current-to-voltage transfer function (its impedance) is

$$Z_C(s) = \frac{V(s)}{I(s)} = \frac{1}{Cs}, \tag{6.143}$$

Chapter 6. Transfer functions and block diagrams

Circuit	Total impedance
R (resistor)	R
L (inductor)	Ls
C (capacitor)	$\dfrac{1}{Cs}$
R, L series	$R + Ls$
R, L, C series	$R + Ls + \dfrac{1}{Cs}$
R, L parallel	$\dfrac{1}{\frac{1}{R} + \frac{1}{Ls}}$

Table 6.1: Table of impedances of a few simple RLC circuits.

and for an inductor of inductance L we have $V(t) = L\dot{I}(t)$, so, assuming zero initial conditions, $V(s) = LsI(s)$, therefore,

$$Z_L(s) = \frac{V(s)}{I(s)} = Ls. \tag{6.144}$$

This transfer function is called **impedance** and is the counterpart of resistance for general RLC circuits. The impedance is a **transfer function**, that is, it is a **complex function**. Some simple circuits and their impedances are shown in Table 6.1.

The concept of impedance allows us to analyse circuits directly in the s-domain. Most importantly, we can work with impedances the same way as we do with simple resistances. If two impedances, Z_1 and Z_2, are connected in series, the total impedance is $Z_{1,2} = Z_1 + Z_2$. If Z_1 and Z_2 are connected in parallel, then the total impedance is $Z_{1\|2} = \frac{1}{1/Z_1 + 1/Z_2}$.

Example 6.22 (Series RC circuit) Consider a resistance R and a capacitance C connected in series. Their impedances are $Z_1 = R$ and $Z_2 = \frac{1}{Cs}$ respectively. The total impedance is $Z_{1,2} = Z_1 + Z_2 =$

$R + \frac{1}{Cs}$. Let us denote by V the voltage at the two ends of the circuit and I the current that runs through it. Then,

$$Z(s) = \frac{V(s)}{I(s)} = R + \frac{1}{Cs}. \quad (6.145)$$

Equivalently,
$$CsV(s) = RCsI(s) + I(s). \quad (6.146)$$

We see that the associated differential equation is
$$C\dot{V}(t) = RC\dot{I}(t) + I(t). \quad (6.147)$$

Example 6.23 (Series RLC circuit) Let us revisit Example 2.16 using impedances.

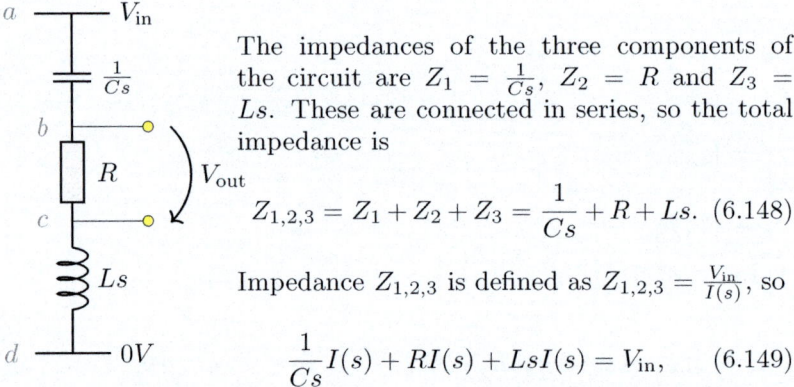

The impedances of the three components of the circuit are $Z_1 = \frac{1}{Cs}$, $Z_2 = R$ and $Z_3 = Ls$. These are connected in series, so the total impedance is

$$Z_{1,2,3} = Z_1 + Z_2 + Z_3 = \frac{1}{Cs} + R + Ls. \quad (6.148)$$

Impedance $Z_{1,2,3}$ is defined as $Z_{1,2,3} = \frac{V_{\text{in}}}{I(s)}$, so

$$\frac{1}{Cs}I(s) + RI(s) + LsI(s) = V_{\text{in}}, \quad (6.149)$$

which, in the time domain, corresponds to the differential equation

$$\frac{1}{C}\int_0^t I(\tau)d\tau + RI(t) + L\dot{I}(t) = V_{\text{in}}(t), \quad (6.150)$$

and given that $V_{\text{out}}(t) = RI(t)$, Equation (6.150) is identical to (2.86).

Moreover, from Equation (6.149) we see that the transfer function from V_{in} to I is

$$\frac{I(s)}{V_{\text{in}}(s)} = \frac{Cs}{LCs^2 + RCs + 1}, \quad (6.151)$$

so the transfer function from V_{in} to $V_{\text{out}}(s) = I(s)R$ is

$$\frac{V_{\text{out}}(s)}{V_{\text{in}}(s)} = \frac{RCs}{LCs^2 + RCs + 1}. \quad (6.152)$$

Chapter 6. Transfer functions and block diagrams

We see that the concept of impedance offers a much more systematic way of deriving mathematical models of circuits. See, for example, how easily we can derive a model for a more complex RLC circuit without even using Kirchhoff's laws.

Example 6.24 (**More complex RC circuit**) Consider the circuit of Exercise 2.22.

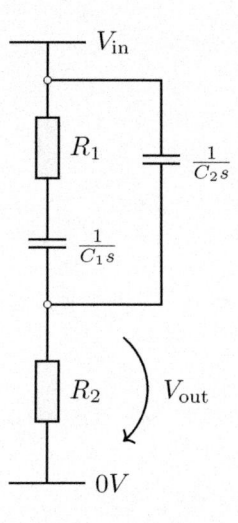

The resistance R_1 and the capacitance $Z_1 = \frac{1}{C_1 s}$ are connected in series, so the total impedance is

$$Z_{1,2} = Z_1 + Z_2 = R_1 + \frac{1}{C_1 s}. \qquad (6.153)$$

This, in turn, is connected parallel to $Z_3 = \frac{1}{C_2 s}$, so the total impedance is

$$Z_{1,2,3} = \frac{1}{\frac{1}{Z_{1,2}} + \frac{1}{C_2 s}} = \frac{1}{\frac{1}{R_1 + \frac{1}{C_1 s}} + \frac{1}{C_2 s}}. \qquad (6.154)$$

Impedance $Z_{1,2,3}$ is connected in series with $Z_4 = R_2$, so the overall impedance is

$$Z_{\text{tot}} = R_2 + \frac{1}{\frac{1}{R_1 + \frac{1}{C_1 s}} + \frac{1}{C_2 s}}. \qquad (6.155)$$

By the definition of impedance (see Equation (6.141)), this means that $Z_{\text{tot}}(s) = V_{\text{in}}(s)/I(s)$; what is the same,

$$\frac{V_{\text{in}}}{I} = R_2 + \frac{1}{\frac{1}{R_1 + \frac{1}{C_1 s}} + \frac{1}{C_2 s}}. \qquad (6.156)$$

Given that the output voltage is $V_{\text{out}} = IR_2$, we may divide both sides of the above equation by R_2 to obtain

$$\frac{V_{\text{in}}(s)}{V_{\text{out}}(s)} = 1 + \frac{1}{R_2 \frac{1}{R_1 + \frac{1}{C_1 s}} + R_2 \frac{1}{C_2 s}}. \qquad (6.157)$$

The transfer function from V_{in} (input variable) to V_{out} (ouput variable) is the reciprocal of the above, that is

$$G(s) = \frac{V_{\text{out}}(s)}{V_{\text{in}}(s)} = \frac{1}{1 + \frac{1}{R_2 \frac{1}{R_1 + \frac{1}{C_1 s}} + R_2 \frac{1}{C_2 s}}}. \qquad (6.158)$$

Exercise 6.11 (ODE and state space representation) Using the methodology of Section 6.5.1, derive the ODE that describes the dynamical system in Example 6.24 and write it in a state space representation.

6.7 Software for control systems

In this section we will present a MATLAB toolbox and a Python module, which are commonly used to define and compose transfer functions and perform simulations.

6.7.1 MATLAB's control systems toolbox

With MATLAB's Control Systems Toolbox we can define transfer functions and compute the response of systems to step, impulse, sinusoidal or even arbitrary input signals. We can also compose transfer functions as we discussed in Section 6.4.

Definition of transfer functions

There are two ways to define transfer functions in MATLAB. The first is to use `tf` to define rational transfer functions of the form

$$G(s) = \frac{a_m s^m + a_{m-1} s^{m-1} + \ldots + a_1 s + a_0}{b_n s^n + b_{n-1} s^{n-1} + \ldots + b_1 s + b_0}. \tag{6.159}$$

We may identify function G by the vectors $a = (a_n, a_{n-1}, \ldots, a_1, a_0)$ and $b = (b_n, b_{n-1}, \ldots, b_1, b_0)$ where the coefficients of the numerator and denominator polynomials are stored in *descending order*. Then, we may define G in MATLAB with `G = tf(a, b);`. For example, the transfer function $G(s) = \frac{2s+1}{3s^2+8}$ is written as

```
num_coeffs = [2 1];                          % numerator coeffs
denom_coeffs = [3 0 8];                      % denominator coeffs
G = tf(num_coeffs, denom_coeffs);            % transfer function
```

One can then perform a number of operations with transfer functions. For example, given two transfer functions G_1 and G_2 one may compute their sum, difference, product and quotient. Consider the transfer functions $G_1 = \frac{1}{s+2}$ and $G_2 = \frac{2s}{5s+3}$. Then, the transfer functions $P = G_1 G_2$, $Q = G_1/G_2$ and $F = \frac{G_1}{1+G_1 G_2}$ are

```
G1 = tf(1, [1 2]);
G2 = tf([2 0], [5 3]);
P  = G1*G2;
Q  = G1/G2;
F  = G1/(1 + P);
```

For instance, MATLAB finds that transfer function F is

```
     5 s^2 + 13 s + 6
-----------------------------
5 s^3 + 25 s^2 + 36 s + 12
```

Transfer functions with delays can be constructed as in the following example

```
td = 0.1;                              % delay
num_coeffs = [2 1];                    % numerator coefficients
denom_coeffs = [3 0 8];                % denominator coefficients
G = tf(num_coeffs, denom_coeffs, 'IODelay', td);
```

Often we need to write transfer functions in the following form, which is known as the zero-pole-gain representation

$$G(s) = K \frac{(s - z_1)(s - z_2) \ldots (s - z_m)}{(s - p_1)(s - p_2) \ldots (s - p_n)}, \qquad (6.160)$$

where $z_1, \ldots, z_m \in \mathbb{C}$ are the *zeroes* of G, $p_1, \ldots, p_n \in \mathbb{C}$ are its *poles* and $K \in \mathbb{R}$ is its *gain*. Then, we may invoke MATLAB's zpk. For example, consider the transfer function

$$G(s) = 450 \frac{(s-1)(s+0.5)}{s(s-(-1+2j))(s-(-1-2j))} \qquad (6.161)$$

This function has the poles $p_1 = 0$, $p_2 = -1 + 2j$ and $p_3 = -1 - 2j$. It has the zeroes $z_1 = 1$ and $z_2 = -0.5$ and gain $K = 450$. We can then construct the corresponding transfer function object in MATLAB as follows

```
gain_g = 450;
zeroes_g = [1 -0.5];
poles_g = [0 -1+2i -1-2i];
% Transfer function in zero-pole-gain representation
G = zpk(zeroes_g, poles_g, gain_g);
```

Then, G is the following function

```
 450 (s-1) (s+0.5)
-------------------
  s (s^2 + 2s + 5)
```

In order to obtain a representation of the form (6.159), if suffices to cast G as a tf-type object as follows:

```
G = tf(G);
```

Then, G becomes

```
  450 s^2 - 225 s - 225
-------------------------
    s^3 + 2 s^2 + 5 s
```

A third and particularly convenient way to define a transfer function is to define a symbol s akin to the symbols we introduced in Section 3.7.2. This can be done by defining s = zpk('s') or s = tf('s'). We may then define transfer functions as in the following example:

```
s = zpk('s');              % Define ZPK symbol
G = (s^2 + 3*s + 2) / ...
    (s^3 + 7*s^2 + 10*s + 1)^2;   % ZPK function
G = tf(G);                 % Transfer function
```

The resulting transfer function is

$$\frac{s^2 + 3s + 2}{s^6 + 14 s^5 + 69 s^4 + 142 s^3 + 114 s^2 + 20 s + 1}$$

Given a transfer function we can obtain a list of its poles using `pole`. This function will be particularly useful for the stability analysis of linear dynamical systems as we shall discuss in Chapter 9.

Lastly, when constructing a transfer function using either `tf` or `zpk`, it is useful to provide some additional metadata such as the names of input and output variables. For instance, for the transfer function of Example 6.23, which has the input variable V_in and output variable V_out, we can specify the input/output variable names as follows

```
R = 1; L = 1; C = 1;       % specify system parameters
num = [R*C, 0];            % numerator
den = [L*C, R*C, 1];       % denominator
G = tf(num, den, ...
       'Name', 'Series RLC circuit', ...
       'InputName', 'Vin', 'OutputName', 'Vout');
```

or, equivalently,

```
G = tf(num, den);
G.Name = 'Series RLC circuit';
G.InputName = 'Vin';
G.OutputName = 'Vout';
```

Composition of transfer functions

Transfer functions can be easily composed. For example, if two systems with transfer functions G_1 and G_2 are connected in series, the resulting transfer function is $G_{1,2} = G_1 G_2$. In MATLAB, this can be obtained by either `G1*G2` or `series(G1, G2)`.

For two transfer functions G_1 and G_2, the system obtained by connecting them in parallel, as in Figure 6.13, can be obtained using either `G1+G2` or `parallel(G1, G2)`.

If two systems with transfer functions G_1 and G_2 are connected in a negative feedback loop as shown in Figure 6.14, the resulting transfer function will be $G_{[1,2]}(s) = \frac{G_1}{1+G_1G_2}$. This can be obtained in MATLAB by `feedback(G1, G2)`.

For more complex networks of interconnected blocks, we may connect blocks of transfer functions with sum blocks using `connect`. To that end, it is necessary that all systems have a unique pair of input/output variable names. Sum blocks can be defined with `sumblk`. A complete example is available in MATLAB's documentation at https://uk.mathworks.com/help/control/ug/build-a-multi-loop-control-system.html.

Exercise 6.12 (Compose transfer functions in MATLAB) Consider the system of Example 6.11 with

$$G_1(s) = \frac{1}{s+1}, \tag{6.162a}$$

$$G_2(s) = \frac{5}{(s+1)(s+3)^4}, \tag{6.162b}$$

$$G_3(s) = 1 + s + \frac{0.1}{s}, \tag{6.162c}$$

$$G_4(s) = 1 + \frac{s}{2}, \tag{6.162d}$$

$$G_5(s) = s + 2. \tag{6.162e}$$

Use MATLAB to derive the overall transfer function.

✂ ....

Hint. Use series, parallel and feedback in MATLAB. You should find that the overall transfer function is rational with an eighth-order numeration and tenth-order denominator.

System response

MATLAB allows to simulate the response of linear dynamical systems to arbitrary input signals. Standard responses, such as the step and impulse response can be obtained by using `step` and `impulse` respectively. The reader can try the following example:

```
G = tf(2, [1 1 1]);
step(G);              % plots the step response of G
step(G, 50)           % ... for t in [0, 50]
[y, t] = step(G);     % returns step response data
plot(t, y);           % ... which we can plot
```

We may obtain the response of a system with given transfer function, G, to an arbitrary input signal u over $t \in [0,T]$ using `lsim`. Let us, for instance, simulate the response of a system with transfer function $G = \frac{1}{s^2+3s+1}$ to the input signal $u(t) = \sin^2(2t)$ over $t \in [0,5]$.

```
t_final = 5;                              % final time
t_step = 0.05;                            % time step
time_span = 0:t_step:t_final;             % time instants
input_signal = sin(2*time_span).^2;       % input signal
G = tf(1, [1 3 1]);                       % transfer function
lsim(G, input_signal, t);                 % simulates and plots
y = lsim(G, input_signal, t);             % response (no plot)
```

Exercise 6.13 (Response for given input) Write a MATLAB script and use lsim to compute the response of a system with transfer function

$$G(s) = \frac{3s}{(s+1)(s+3)}, \qquad (6.163)$$

to a sinusoidal input with period $T = 0.5\,\mathrm{s}$ and amplitude equal to 1, over the time period $[0, 5]$. Compute the response symbolically using ilaplace and compare with MATLAB's result.

Pole-zero maps

A **pole-zero map** is simply a plot that shows the locations of the poles and zeros of a transfer function on the complex plane. The zeros are shown with an × mark, and the zeros with a ○. For example, suppose we want to determine the pole-zero map of

$$G(s) = \frac{s^2 - 0.5}{s^3 + 2s^2 + 2s + 1}. \qquad (6.164)$$

We can use pzmap in MATLAB as follows

```
s = tf('s')
G = (s^2 - 0.5) / (s^3 + 2*s^2 + 2*s + 1);
pzmap(G);
```

This will produce the plot shown in Figure 6.20.

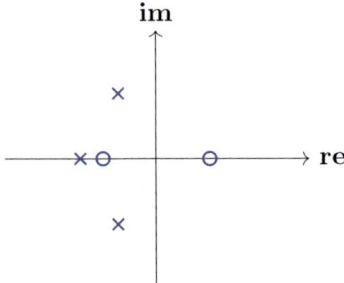

Figure 6.20: Pole-zero map of G in Equation (6.164).

Chapter 6. Transfer functions and block diagrams

6.7.2 Python's control systems module

The structure of Python's control systems module is quite similar to that of MATLAB's Control Systems Toolbox. Let us start by installing the module:

```
pip install control
```

We can then import the module

```
import control as ctrl
```

Definition of transfer functions

We can define a transfer function using the `TransferFunction` class and specifying its numerator and denominator. For example, the transfer function $G(s) = \frac{2s+1}{3s^2+8}$ is written as

```
num = [2, 1]
den = [3, 0, 8]
G = ctrl.TransferFunction(num, den)
```

Another way to define this transfer function is

```
s = ctrl.TransferFunction.s
G = (2*s+1)/(3*s**2 + 8)
```

Composition of transfer functions

We can compose transfer functions using `ctrl.series`, `ctrl.parallel` and `ctrl.feedback`. A feedback connection can have either a positive or negative feedback:

```
G = ctrl.feedback(G1, G2, sign=-1)
```

We use `sign=-1` for negative feedback loops and `sign=1` for positive ones.

> **Exercise 6.14** (Compose transfer functions in Python) Repeat Exercise 6.12 in Python.

System response

We can simulate the response of linear dynamical systems to arbitrary input signals. Standard responses, such as the step and impulse response can be obtained by using `step_response` and `impulse_response` respectively. The reader can try the following example:

```
import matplotlib.pyplot as plt

tf = ctrl.TransferFunction(1, [1, 1, 1])
step = ctrl.step_response(tf)
imp = ctrl.impulse_response(tf)
plt.plot(step.time, step.outputs)
plt.plot(imp.time, imp.outputs)
plt.show()
```

Note that in the above code snippet, the object step is an instance of the class control.timeresp.TimeResponseData and imp is an instance of the class control.timeresp.TimeResponseData. Note that we plot the step response using the time data step.time and the output data step.outputs.

We can also determine the response of a dynamical system to *arbitrary* inputs. As an example, let us determine the response of a system with transfer function

$$G = \frac{1}{0.5s^2 + 0.1s + 1}, \tag{6.165}$$

to the input signal $u(t) = \sin(5\sqrt{t})$ over $t \in [0, 20]$. To do so, we can use forced_response.

```
import control as ctrl
import matplotlib.pyplot as plt
import numpy as np

# Firstly, we define the input signal
n_points = 500
t_final = 20
t_span = np.linspace(0, t_final, n_points)
input_signal = np.sin(5*np.sqrt(t_span))

# Then we define the transfer function
tf =   ctrl.TransferFunction(1, [0.5, 0.1, 1])

result = ctrl.forced_response(tf,t_span,input_signal)
plt.plot(result.time, result.outputs)
plt.show()
```

Pole-zero maps

The control module can produce a pole-zero map using the pzmap function. For example, this is how we can construct the pole-zero map of the transfer function of Equation (6.164)

```
s = ctrl.TransferFunction.s
G = (s**2 - 0.5) / (s**3 + 2*s**2 + 2*s + 1)
poles, zeros = ctrl.pzmap(G)
```

Chapter 6. Transfer functions and block diagrams

6.8 One-minute round-up

In this chapter

1. We defined the concept of the transfer function of a linear SISO dynamical system; a complex function that describes the system dynamics

2. Typically, transfer functions are rational functions

3. The impulse, step and frequency responses of a system are the responses when the input is a Dirac pulse, a Heaviside function and a sinusoidal function respectively; these can be computed given the transfer function of the system using the inverse Laplace transform

4. A block diagram is a graphical illustration of interconnected dynamical systems such as feedback loops

5. Blocks can be combined to compute the overall transfer function

6. We gave a concise presentation of the theory of linear differential operators

7. The transfer function of a MIMO system is a matrix-valued complex function

8. The impedance of a circuit is defined as the transfer function of the circuit using as input variable the current that runs through it and the voltage across it as the output variable, that is, $Z(s) = V(s)/I(s)$

9. Impedances can be combined in exactly the same way as resistances to determine the overall impedance of a circuit; this facilitates the derivation of the circuit's transfer function

6.9 Exercises

Exercise 6.15 (True of false) True of false?

	True	False
The transfer function of a nonlinear dynamical system with input $U(s)$ and output $X(s)$ is defined as $G(s) = X(s)/U(s)$	☐	☐
If two linear dynamical systems with transfer functions G_1 and G_2 are connected in series, then the overall transfer function is $G(s) = G_1(s)G_2(s)$	☐	☐
If two linear dynamical systems with transfer functions G_1 and G_2 are connected in parallel, then the overall transfer function is $G(s) = \dfrac{1}{\frac{1}{G_1(s)} + \frac{1}{G_2(s)}}$	☐	☐
A complex number s^* is a pole of a transfer function G if $G(s^*) = 0$	☐	☐
The impulse response of a dynamical system with transfer function G is given by $\mathscr{L}^{-1}\{G\}$	☐	☐
The step response of a dynamical system is given, in the s-domain, by $sG(s)$	☐	☐
The step response of a dynamical system is given, in the s-domain, by $s^{-1}G(s)$	☐	☐
The transfer function of a system with 2 inputs and 1 output has the form $G(s) = [G_1(s)\ G_2(s)]$	☐	☐
The ODE that corresponds to the linear dynamical system with transfer function $G(s) = \dfrac{X(s)}{U(s)} = \dfrac{2}{3s+1}$ is $3\dot{x} + x = 2u$	☐	☐
The impedance of an inductor of inductance L is Ls	☐	☐
The impedance of a capacitor of capacitance C is $\dfrac{1}{Cs}$	☐	☐
Two linear circuit elements with impedance values Z_1 and Z_2 are connected in parallel. The total impedance is $\dfrac{Z_1 Z_2}{Z_1 + Z_2}$.	☐	☐

❋ ....

Answer. F (nonlinear systems do not define a transfer function), T, F (it is $G_1(s) \mp G_2(s)$ as discussed in Section 6.4.1), F, T, F, T, T, T, T, T.

Exercise 6.16 (Transfer function of inverted pendulum linearisation) Consider the dynamical system of Example 2.11 with input F and output θ. In Exercise 3.12 we derived the linearisation of this system around its upright position. Derive the transfer function of the linearised system. Can you derive the transfer function of the original nonlinear system?

Exercise 6.17 (Transfer function of multiple-input single-output system) A system with output $x(t)$ and two inputs, $u(t)$ and $d(t)$ is described by the following linear differential equation

$$\ddot{x} + 5\dot{x} - x = u + \tfrac{1}{3}\dot{d} + \tfrac{1}{10}d. \tag{6.166}$$

Derive its transfer function and sketch the corresponding block diagram.

✂··

Hint. As we defined in Section 6.5.4, the transfer function will have the form $X(s) = G_1(s)U(s) + G_2(s)D(s)$. Apply the Laplace transform to both sides of Equation 6.166 assuming zero initial conditions.

Exercise 6.18 (Impulse response of second-order transfer function with a zero) Let us consider again the circuit of Example 6.23

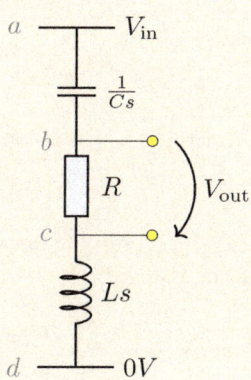

Its transfer function, from V_in to V_out is

$$\frac{V_\text{out}(s)}{V_\text{in}(s)} = \frac{RCs}{LCs^2 + RCs + 1}. \tag{6.167}$$

What is the step and impulse of the system? How do R, L and C affect the step and impulse response of the system?

✂··

Hint. The impulse response of the system is the inverse Laplace transform of its transfer function (see Section 6.2). Work as in Chapter 5.

Exercise 6.19 (Transfer function to ODE) Find the ODEs or IDEs associated with each of the following transfer functions

1. $G(s) = \frac{1}{s}$
2. $G(s) = \frac{1}{s+1}$
3. $G(s) = \frac{1}{s^n+1}$, for $n \in \mathbb{N}$
4. $G(s) = \frac{5s+2}{(s-1)(s+2)(s+3)}$
5. $G(s) = \frac{1}{s^2+3s+2}$
6. $G(s) = \frac{s+0.1}{2s+1}$
7. $G(s) = \frac{s^2+2s+0.5}{2s^3+5s^2-s+1}$
8. $G(s) = \frac{s+2}{s^3+2s^2+3s+1}$
9. $G(s) = \frac{1}{5s+1}e^{-s/4}$

✱ ....

Answer. 1. $\dot{x} = u$, or, if you prefer, $x(t) = x(0) + \int_0^t u(\tau) d\tau$. 2. $\dot{x} + x = u$, 3. $x^{(n)} + x = u$, with $n \in \mathbb{N}$. 4. Expand $G(s) = \frac{5s+2}{s^3+4s^2+s-6}$, so The corresponding ODE is $x^{(3)} + 4\ddot{x} + \dot{x} - 6x = 5\dot{u} + 2u$, questions 5–8 are similar to 1–4, 9. According to Theorem 5.6; the corresponding differential equation is $5\dot{x} + x = H_{1/4}(t) u(t - 1/4)$.

✂ ....

Exercise 6.20 (ODE to transfer function) A system with input u and output x is described by the following ODE

$$\ddot{x} + 5\dot{x} + 2x = u. \tag{6.168}$$

What is the transfer function of the system?

✱ ....

Answer. The transfer function is $\frac{X(s)}{U(s)} = \frac{1}{s^2+5s+2}$.

Exercise 6.21 (System of ODEs to transfer function) A dynamical system with input u is described by the system of equations

$$\dot{z} + 3z = u, \tag{6.169a}$$
$$\ddot{x} + 4x = \dot{z} + z. \tag{6.169b}$$

What is the transfer function $G(s) = X(s)/U(s)$?

✂ ....

Hint. Apply the Laplace transform on both equations assuming zero initial conditions. This way you will obtain a pair of algebraic equations with variables $X(s)$, $Z(s)$ and $U(s)$. Eliminate $Z(s)$ and solve for $X(s)/U(s)$ to determine the transfer function.

Exercise 6.22 (What is $G(0^+)$?) A dynamical system is described by a rational transfer function. What information does $G(0^+)$ offer about the dynamical system?

✂ ....

Hint. This is the "DC gain" of the system: indeed, by the final value theorem, provided that its conditions are satisfied (see Theorem 4.29), we have that the final value of the step response is exactly $G(0^+)$ (where the limit is taken over the real numbers).

Exercise 6.23 (Understanding impedance) Consider the following circuit

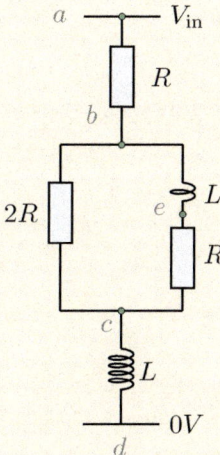

(i) Compute the impedance of the circuit between points a and d, (ii) Determine the transfer function between the input voltage V_{in} and the current that runs through the circuit, (iii) Determine the transfer function between the input voltage V_{in} and voltage V_{bc}, between points b and c, (iv) suppose that at time $t = 0$, no current is flowing through the circuit and all voltages are equal to zero. The system is excited with the input

$$V_{\text{in}}(t) = H_0(t).$$

Compute the resulting current, $I(t)$, that will run through the circuit and the resulting voltage $V_{bc}(t)$.

Exercise 6.24 (Transfer function from step response) The step response of a linear dynamical system with zero initial conditions is given by

$$x^{\text{step}}(t) = 2(1 - e^{-\frac{t}{3}}). \tag{6.170}$$

What is the transfer function of this system, and what is its frequency response?

✂ ....

Hint. The step response of a system is $x^{\text{step}}(t) = \mathscr{L}^{-1}\{G(s)/s\}(t)$; by taking the Laplace transform, $\mathscr{L}\{x^{\text{step}}(t)\}(s) = G(s)/s$, from which we can solve for $G(s)$.

Exercise 6.25 (Interesting exercise) Consider a dynamical system with transfer function
$$A(s) = \frac{K}{(\tau s + 1)^2}. \quad (6.171)$$

1. Suppose that all initial conditions are zero and $K, \tau > 0$. Upon excitation with a unit step pulse, that is, an input signal $u(t) = 1$, for all $t \geq 0$, the response of the system, $x(t)$ is such that
$$\lim_{t \to 0^+} \ddot{x}(t) = 0.5, \quad (6.172)$$
and
$$\lim_{t \to \infty} x(t) = 4. \quad (6.173)$$
Determine the values of K and τ.

2. Determine the impulse response of the system.

Exercise 6.26 (Block diagram No. 1) Consider the following system with input U and output X with $\tau_1, \tau_2 > 0$:

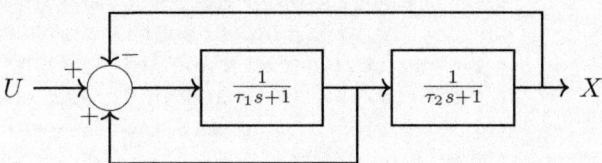

Show that the transfer function of this system has the form $G(s) = \frac{1}{\tau^2 s^2 + 2\zeta \tau s + 1}$ — known as a second-order system (we will study these systems in Chapter 8 — with $\zeta = \frac{1}{2}\sqrt{\tau_1/\tau_2}$ and $\tau = \sqrt{\tau_1 \tau_2}$.

Exercise 6.27 (Block diagram No. 2) Determine the transfer function $G(s) = X(s)/U(s)$ for the system shown in the following block diagram:

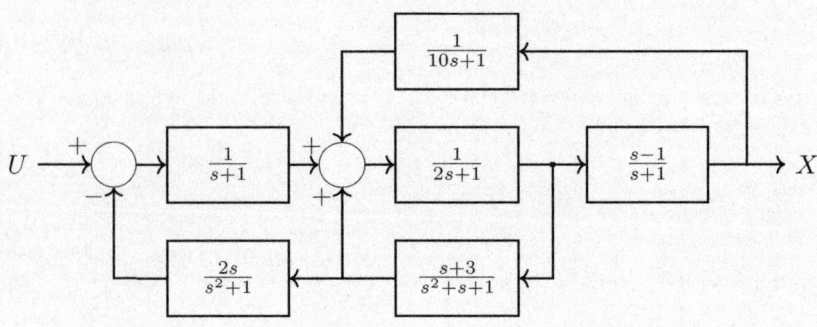

Exercise 6.28 (State space representation of proper transfer function)

A linear dynamical system has the following transfer function

$$G(s) = \frac{X(s)}{U(s)} = \frac{7s^3 + 5s^2 - 6s + 5}{s^3 + 3s^2 + s + 1}. \qquad (6.174)$$

Write it in a state space form.

✂ ....

Hint. Start by following the procedure of Example 6.20: introduce an auxiliary variable $Z(s)$ such that $X(s) = (7s^3 + 5s^2 - 6s + 5)Z(s)$ and $U(s) = \cdots$; write the corresponding ODEs and introduce three new variables: $z_1 = z$, $z_2 = \dot{z}$, $z_3 = \ddot{z}$, and write down a dynamical system in terms of the state variable (z_1, z_2, z_3). Lastly, express $x(t)$ in terms of z_1, z_2, z_3 and u.

Exercise 6.29 (Gang of four)

Consider the block diagram shown below, where G_c is the transfer function of a controller and $G_s(s)$ is the transfer function of the controlled system.

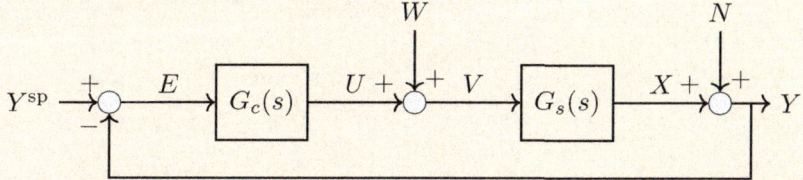

The signals W and V are additive disturbance signals that affect the control action and the output of the system, while Y^{sp} is the set point or *reference* signal. Determine the transfer function of the overall system with inputs Y^{sp}, W and N and outputs Y, X, V, U and E; fill in the following matrix:

$$\begin{bmatrix} Y \\ X \\ V \\ U \\ E \end{bmatrix} = \begin{bmatrix} & & \\ & & \\ & & \\ & & \\ & & \end{bmatrix} \begin{bmatrix} Y^{\text{sp}} \\ W \\ V \end{bmatrix}. \qquad (6.175)$$

Note that although the above matrix has 15 entries, there are just four unique transfer functions one needs to know. These are known as the **gang of four**.

Exercise 6.30 (SSR and transfer function) A dynamical system with input u is described by the following ODEs equations

$$2\ddot{x} + \ddot{y} = y\dot{y}, \quad (6.176a)$$
$$y\ddot{x} + \ddot{y} = u. \quad (6.176b)$$

1. Write the system in its state space representation and determine an equilibrium point that corresponds to $y = 1$

2. Determine the transfer function using u as the input and y as the output for the linearisation of this dynamical system at the equilibrium point (you obtained this in the previous question).

※·· ✂··························

Hint. To answer the first question we need to introduce the variables $x_1 = x$, $x_2 = \dot{x}$, $x_3 = y$, $x_4 = \dot{x}_3$. Then we will end up with a linear system in (x_2, x_4), which we need to solve.

Answer. 1. $\dot{x}_1 = x_2$, $\dot{x}_2 = \frac{2-x_3}{2x_1x_4-u}$, $\dot{x}_3 = x_4$, $\dot{x}_4 = \frac{2u-x_3^2 x_4}{2-x_3^e}$: an equilibrium point that corresponds to $y = 1$ is a tuple $(x_1^e, x_2^e, x_3^e, x_4^e, u^e)$ that satisfies $x_3^e = 1$ and

$$x_2^e = 0,$$
$$\frac{x_3^e x_4^e - u^e}{2 - x_3^e} = 0,$$
$$x_4^e = 0,$$
$$\frac{2u^e - (x_3^e)^2 x_4^e}{2 - x_3^e} = 0.$$

For example $x_1^e = 0$, $x_2^e = 0$, $x_3^e = 1$, $x_4^e = 0$, $u^e = 0$.

2. The transfer function is

$$\frac{X_3(s)}{\bar{U}(s)} = \frac{2}{s(s+1)},$$

where $X_3 = \mathscr{L}\{x_3\}(s)$ and $\bar{U}(s) = \mathscr{L}\{\bar{u}\}(s)$, where $\bar{x}_3 = x_3 - x_3^e$ and $\bar{u} = u - u^e$.

Exercise 6.31 (System response using software) Write a MATLAB or Python script to plot the impulse, step and frequency responses (with $u(t) = \sin(t)$) for the following systems (consult Section 6.7)

1. A first-order system with static gain $K = 1$ and time constant τ; perform simulations with $\tau = 1$ and $\tau = 0.1$

2. A second-order system with $K = 1$, $\tau = 1$ and $\zeta = 0.5$

3. A system with transfer function

$$G(s) = \frac{1}{(s-a)(s+1)}, \quad (6.177)$$

for $a = 0$, $a = -2$ and $a = -0.5$

4. A system with transfer function

$$G(s) = \frac{1}{(s-a)(s^2 + 0.5s + 1)}, \qquad (6.178)$$

for $a = 0$, $a = -2$ and $a = -0.5$

5. A system with transfer function

$$G(s) = \frac{s-a}{s^2 + 0.5s + 1}, \qquad (6.179)$$

for $a = 0$, $a = -0.5$, $a = -1$ and $a = 1$.

Discuss how the introduction of a pole in questions 3 and 4 and the introduction of a zero in question 6 affects the response.

Exercise 6.32 (Solution of IVP from TF) Prove that the solution of the IVP of Section 6.3 is given by Equation (6.64).

Exercise 6.33 Determine a particular solution of

$$(D-c)^\nu x = u, \qquad (6.180)$$

where $\nu \in \mathbb{N}$.

❋ ....

Answer. Apply Proposition 6.5 recursively and apply Cauchy's formula for repeated integration to obtain

$$x(t) = e^{ct} \int_{t_0}^{t} \frac{e^{-c\tau}(t-\tau)^{\nu-1}}{(\nu-1)!} u(\tau)\, d\tau, \qquad (6.181)$$

for any $a \in \mathbb{R}$.

Exercise 6.34 Complete the proof of Proposition 6.8.

7

First-order systems

In this chapter, we focus on first-order systems, that is, systems with a transfer function of the form $F(s) = K/\tau s+1$ for some constants $K, \tau > 0$. We compute their impulse, step and frequency responses and focus primarily on their qualitative characteristics. Alongside, we give examples of physical systems that exhibit first-order dynamics. First-order systems are very common in engineering applications, so this is why we dedicate a separate chapter to study them. Throughout the chapter we use known methodological tools from the previous chapters.

7.1 Introduction

A first-order system is a very simple and very common type of dynamical system. In this chapter we focus on linear dynamical systems that can be described by the first-order differential equation

$$a_0 x + a_1 \dot{x} = bu, \tag{7.1}$$

with $a_1 \neq 0$. We consider the following two cases:

Case I: $a_0 \neq 0$. We divide both sides of (7.1) by a_0 (given that $a_0 \neq 0$), to obtain

$$x + \frac{a_1}{a_0}\dot{x} = \frac{b}{a_0}u. \tag{7.2}$$

Next, we assume that (7.2) has zero initial conditions[1], that is $x(0) = 0$, and we define $K = b/a_0$, and $\tau = a_1/a_0$. Throughout this chapter we will focus on the common case where $\tau > 0$. Constant K is known as the **static**

---
[1] If not, we introduce deviation variables

gain of the system and τ is called the **time constant** of the system. This way, Equation (7.2) becomes

$$x + \tau \dot{x} = Ku. \tag{7.3}$$

We now apply the Laplace transform on both sides of (7.3) assuming zero initial conditions, which yields

$$(1 + \tau s)X(s) = KU(s), \tag{7.4}$$

where $X(s) = \mathscr{L}\{x(t)\}(s)$ and $U(s) = \mathscr{L}\{u(t)\}(s)$, therefore, the transfer function of the system is

$$G(s) = \frac{X(s)}{U(s)} = \frac{K}{\tau s + 1}. \tag{7.5}$$

Case II: $a_0 = 0$ and $a_1 \neq 0$. If $a_0 = 0$, Equation (7.1) reduces to

$$a_1 \dot{x} = bu, \tag{7.6}$$

and dividing by a_1 (since $a_1 \neq 0$) we obtain

$$\dot{x} = \frac{b}{a_1} u. \tag{7.7}$$

By defining $K' = b/a_1$, Equation (7.6) becomes

$$\dot{x} = K'u. \tag{7.8}$$

Assuming again zero initial conditions and applying the Laplace transform

$$sX(s) = K'U(s), \tag{7.9}$$

therefore, the transfer function of the system is

$$G(s) = \frac{X(s)}{U(s)} = \frac{K'}{s}. \tag{7.10}$$

An interesting observation is that if we apply the inverse Laplace transform on $X(s)$, we have

$$X(s) = G(s)U(s) \Rightarrow \mathscr{L}^{-1}\{X(s)\}(t) = \mathscr{L}^{-1}\{G(s)U(s)\}(t)$$

$$\Rightarrow x(t) = \mathscr{L}^{-1}\left\{\frac{K'U(s)}{s}\right\}(t)$$

$$\Rightarrow x(t) = K'\mathscr{L}^{-1}\left\{\frac{U(s)}{s}\right\}(t)$$

$$\Rightarrow x(t) = K' \int_0^t u(\tau)\mathrm{d}\tau. \tag{7.11}$$

So, the system output is given by the **integral** of its input. This is why such a system is called an **integrator**. The scalar K' is called the **static gain** of the integrator.

Chapter 7. First-order systems

> **Definition 7.1 (First-order systems and integrators)** A **first-order** system is one with a single input, $u(t)$, and a single output variable, $x(t)$, whose dynamics is described by Equation (7.1) with $a_0 \neq 0$. **First-order systems** are described by a transfer function of the form
> $$G(s) = \frac{K}{\tau s + 1}, \qquad (7.12)$$
> for some $\tau > 0$, $K = 0$. Systems with a transfer function
> $$G(s) = \frac{K}{s}, \qquad (7.13)$$
> with $K \neq 0$ are called (simple) **integrators**.

Hereafter, we will focus on first-order systems. Several physical and engineering systems are first-order systems. Next, we will give some examples and study their dynamic characteristics.

Example 7.1 (RL circuit) A voltage $V(t)$ is applied at the ends of an RL circuit as shown below.

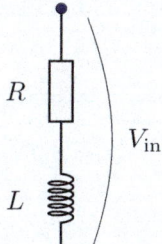

By Kirchhoff's voltage law, we have
$$V_{\text{in}} = IR + L\dot{I}. \qquad (7.14)$$

This is a first-order system with state I and input V_{in}. We assume that the initial conditions are zero, that is, $V(0) = 0$, and we apply the Laplace transform to obtain

$$V_{\text{in}}(s) = I(s)(R + Ls), \qquad (7.15)$$

therefore,
$$\frac{I(s)}{V_{\text{in}}(s)} = \frac{1}{R + Ls}. \qquad (7.16)$$

This is the transfer function of the system. In particular,
$$G(s) = \frac{1}{R + Ls} = \frac{\frac{1}{R}}{1 + \frac{L}{R}s}, \qquad (7.17)$$

so, the static gain of the system is $K = \frac{1}{R}$, and the time constant is $\tau = \frac{L}{R}$.

Note that if $R = 0$, then the RL circuit of Example 7.1 acts as an integrator. Indeed, in that case we would have

$$V(t) = L\dot{I}. \tag{7.18}$$

By applying the Laplace transform we obtain

$$V(s) = LsI(s) \Rightarrow \frac{I(s)}{V(s)} = \frac{1}{Ls} = \frac{1/L}{s}, \tag{7.19}$$

that is, the system with $R = 0$ is an integrator with static gain $K' = 1/L$.

In what follows we will study the dynamic characteristics of first-order systems with $K, \tau > 0$ and integrators with $K' > 0$.

7.2 Impulse response

The impulse response of a first-order system is its response to a Dirac pulse, $u(t) = \delta(t)$, whose Laplace transform is $U(s) = 1$. The impulse response will be

$$x(t) = \mathscr{L}^{-1}\left\{\frac{K}{\tau s + 1}\right\} = \mathscr{L}^{-1}\left\{\frac{K/\tau}{s + 1/\tau}\right\} = \frac{K}{\tau}e^{-\frac{t}{\tau}}, \tag{7.20}$$

for $t > 0$. The impulse response of a system with $K = 1$ and different values of τ are shown in Figure 7.1.

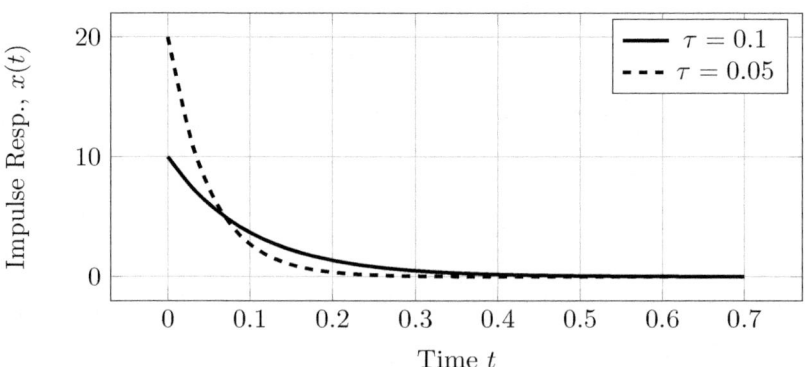

Figure 7.1: *Impulse response of a first-order system. Note that $x(t) \to 0$ as $t \to \infty$.*

The impulse response of an integrator is given by

$$x(t) = \mathscr{L}^{-1}\left\{\frac{K'}{s}\right\} = K'H_0(t). \tag{7.21}$$

Remark: *Previously we assumed that $x(0) = 0$. It is known that the solutions of IVPs that involve ODEs are continuous, so we should expect* that $x(0^+) = x(0) = 0$. However, we see from Equation (7.20) that $x(0^+) = K/\tau \neq 0$. In fact, we could have determined this limit using the initial value theorem. We see that the impulse response of a first-order system is not continuous. This is one of the exotic properties of Dirac's delta. Note that $x(t)$ in Equation 7.20 is defined over $t > 0$ and not $t \geq 0$ as $x(0) = 0$. Recall that as we discussed in Section 5.1, there are multiple time domain functions that have the same Laplace transform (but they are almost equal to one another) (see Theorem 5.2). •

7.3 Step response

Recall that the step response of a dynamical system, is the system output when the input is the Heaviside function, that is

$$u(t) = H_0(t) = \begin{cases} 0, & \text{for } t = 0 \\ 1, & \text{for } t > 0 \end{cases} \tag{7.22}$$

The corresponding Laplace transform is

$$U(s) = \frac{1}{s}. \tag{7.23}$$

If a first-order system is excited with a Heaviside pulse, then its output is

$$x(t) = \mathscr{L}^{-1}\left\{G(s)U(s)\right\} = \mathscr{L}^{-1}\underbrace{\left\{\frac{K}{(\tau s + 1)s}\right\}}_{X(s)}. \tag{7.24}$$

This function has two distinct real poles, namely, $s_1 = 0$ and $s_2 = -1/\tau$. We follow the methodology we discussed in Section 5.2.1. We expand the given function in partial fractions as follows

$$\frac{K}{(\tau s + 1)s} = \frac{A_1}{s} + \frac{A_2}{\tau s + 1}, \tag{7.25}$$

by multiplying by $s(\tau s + 1)$ we obtain

$$K = A_1(\tau s + 1) + A_2 s = (A_1 \tau + A_2)s + A_1, \tag{7.26}$$

and comparing the corresponding polynomial coefficients, we have that

$$A_1 = K, \tag{7.27a}$$
$$A_1 \tau + A_2 = 0, \tag{7.27b}$$

so $A_1 = K$ and $A_2 = -K\tau$. Therefore,

$$\frac{K}{(\tau s + 1)s} = \frac{K}{s} - \frac{K\tau}{\tau s + 1}, \tag{7.28}$$

so the inverse Laplace transform is

$$x(t) = \mathscr{L}^{-1}\left\{\frac{K}{(\tau s + 1)s}\right\} = \mathscr{L}^{-1}\left\{\frac{K}{s} - \frac{K\tau}{\tau s + 1}\right\}$$

$$= \mathscr{L}^{-1}\left\{\frac{K}{s}\right\} - \mathscr{L}^{-1}\left\{\frac{K\tau}{\tau s + 1}\right\}$$

$$= \mathscr{L}^{-1}\left\{\frac{K}{s}\right\} - \mathscr{L}^{-1}\left\{\frac{K}{s + 1/\tau}\right\}$$

$$= K - Ke^{-t/\tau} = K(1 - e^{-t/\tau}). \tag{7.29}$$

This is the step response of a first-order system. The effects of K and τ are shown in Figure 7.2.

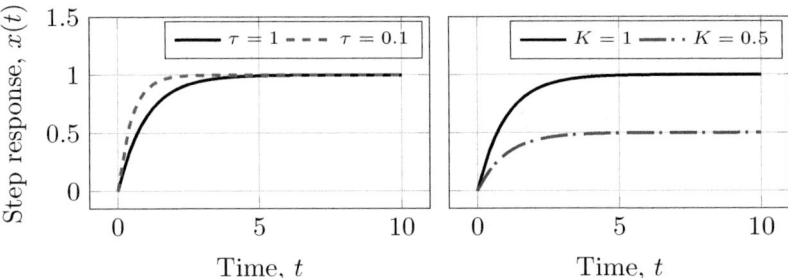

Figure 7.2: *Step response of a first-order system. (Left). Effect of τ for a constant static gain $K = 1$, (Right) Effect of the static gain for a constant $\tau = 1$.*

In order to gain some understanding of the step response of first-order systems, we first observe that

1. The step response, $x(t)$, is an increasing function.

2. The limit of $x(t)$ as t approaches infinity is

$$\lim_{t \to \infty} x(t) = \lim_{t \to \infty} K(1 + e^{-t/\tau}) = K. \tag{7.30}$$

3. We may rewrite the step response as

$$x(t) = K(1 + e^{-t/\tau}) \Leftrightarrow \frac{x(t)}{K} = 1 + e^{-t/\tau}. \tag{7.31}$$

We now define the quantities

$$x^*(t) = \frac{x(t)}{K}, \tag{7.32a}$$

$$t^* = \frac{t}{\tau}. \tag{7.32b}$$

Then the system dynamics becomes

$$x^*(t) = 1 - e^{-t^*}. \tag{7.33}$$

Chapter 7. First-order systems

This form of the system dynamics is *independent* of the system parameters K and τ.

4. By differentiating Equation (7.33), we obtain

$$\frac{dx^*(t^*)}{dt^*} = e^{-t^*}. \tag{7.34}$$

For $t^* = 0$, the derivative is

$$\frac{dx^*(0)}{dt^*} = 1. \tag{7.35}$$

This means that the slope of the plot of $x^*(t^*)$ at 0 is equal to 1.

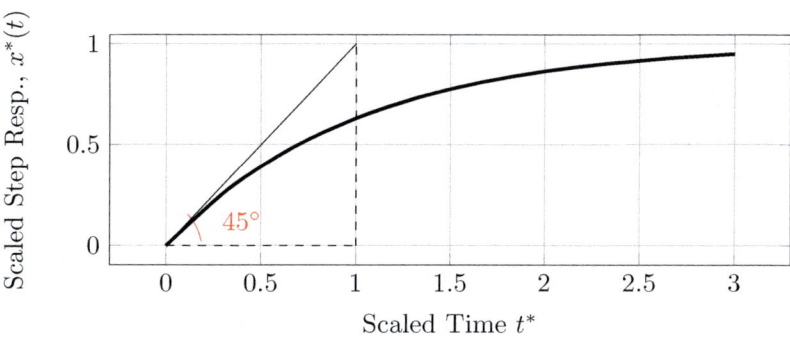

Figure 7.3: *Step response of a first-order system: plot of $x^*(t^*)$ vs t^*.*

Exercise 7.1 (Step response of integrator) What is the step response of an integrator with static gain K'?

✳ ....

Answer. It is $x_{step}(t) = K'$, defined for $t \geq 0$.

7.4 Frequency response

7.4.1 Frequency response and its asymptotic behaviour

We excite a first-order system with a sinusoidal signal of the form

$$u(t) = A\sin(\omega t), \tag{7.36}$$

which has the Laplace transform

$$U(s) = \frac{A\omega}{s^2 + \omega^2}. \tag{7.37}$$

The frequency response of the system is

$$x(t) = \mathscr{L}^{-1}\{G(s)U(s)\} = \mathscr{L}^{-1}\underbrace{\left\{\frac{K}{\tau s+1}\frac{A\omega}{s^2+\omega^2}\right\}}_{X(s)}. \quad (7.38)$$

In order to determine the above inverse Laplace transform, we expand $X(s)$ in partial fractions as follows

$$X(s) = \frac{AK\omega}{(\tau s+1)(s^2+\omega^2)} = \frac{A_1}{\tau s+1} + \frac{B_1 s + B_2}{s^2+\omega^2}. \quad (7.39)$$

We multiply both sides by $(\tau s+1)(s^2+\omega^2)$ to obtain

$$AK\omega = A_1(s^2+\omega^2) + (B_1 s + B_2)(\tau s + 1)$$
$$\Rightarrow AK\omega = A_1 s^2 + A_1\omega^2 + B_1\tau s^2 + B_1 s + B_2\tau s + B_2$$
$$\Rightarrow AK\omega = (A_1 + B_1\tau)s^2 + (B_1 + B_2\tau)s + (B_2 + A_1\omega^2). \quad (7.40)$$

By comparing the corresponding polynomial coefficients, we produce the following linear system of equations

$$A_1 + B_1\tau = 0, \quad (7.41\text{a})$$
$$B_1 + B_2\tau = 0, \quad (7.41\text{b})$$
$$A_1\omega^2 + B_2 = AK\omega. \quad (7.41\text{c})$$

The solution of this system is

$$A_1 = \frac{AK\omega\tau^2}{\tau^2\omega^2+1}, \quad (7.42\text{a})$$
$$B_1 = -\frac{AK\omega\tau}{\tau^2\omega^2+1}, \quad (7.42\text{b})$$
$$B_2 = \frac{AK\omega}{\tau^2\omega^2+1}, \quad (7.42\text{c})$$

therefore,

$$X(s) = \frac{AK\omega}{(\tau s+1)(s^2+\omega^2)}$$
$$= \frac{\frac{AK\omega\tau^2}{\tau^2\omega^2+1}}{\tau s+1} + \frac{-\frac{AK\omega\tau}{\tau^2\omega^2+1}s + \frac{AK\omega}{\tau^2\omega^2+1}}{s^2+\omega^2}$$
$$= \frac{KA\omega}{\tau^2\omega^2+1}\left[\tau^2\frac{1}{\tau s+1} + \frac{-\tau s+1}{s^2+\omega^2}\right]$$
$$= \frac{KA\omega}{\tau^2\omega^2+\omega\frac{1}{\omega}}\left[\tau^2\frac{1/\tau}{s+1/\tau} + \frac{-\tau s+1}{s^2+\omega^2}\right]. \quad (7.43)$$

The inverse Laplace transform of $X(s)$ is

$$x(t) = \frac{AK\omega}{\tau^2\omega^2+1}\left[\tau e^{-t/\tau} + \sqrt{(-\tau)^2 + \frac{1}{\omega^2}}\sin(\omega t + \phi_0)\right]$$

$$= \frac{AK\omega}{\tau^2\omega^2 + 1}\left[\tau e^{-t/\tau} + \frac{1}{\omega}\sqrt{\tau^2\omega^2 + 1}\sin(\omega t + \phi_0)\right], \quad (7.44)$$

with $\phi_0 = \arctan(-\tau\omega)$. We have proven the following result:

> **Theorem 7.2 (Frequency response of first-order systems)** When a first-order system with static gain $K > 0$ and time constant $\tau > 0$ is excited with a sinusoidal input signal,
>
> $$u(t) = A\sin(\omega t), \quad (7.45)$$
>
> the response is
>
> $$x(t) = \frac{AK\omega}{\tau^2\omega^2 + 1}\left[\tau e^{-t/\tau} + \frac{1}{\omega}\sqrt{\tau^2\omega^2 + 1}\sin(\omega t + \phi_0)\right], \quad (7.46a)$$
>
> with
>
> $$\phi_0 = \arctan(-\tau\omega). \quad (7.46b)$$

We may observe that for large times ($t \gg 1$), the frequency response shown in Figure 7.4 is approximately sinusoidal.

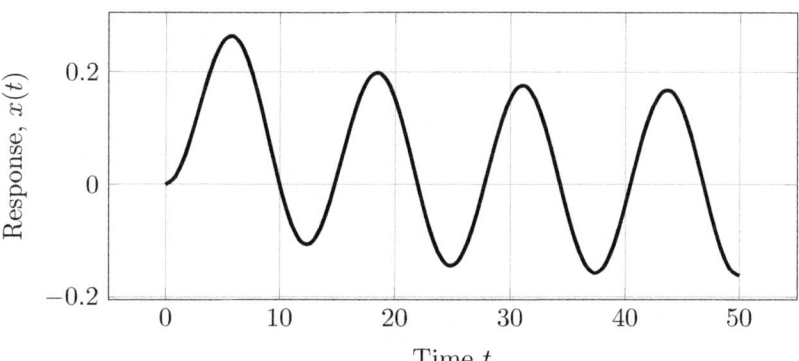

Figure 7.4: *Frequency response of first-order system ($A = 1$, $K = 1$, $\omega = 0.5$, and $\tau = 15$).*

Indeed, for large t, the exponential part of (7.44) is approximately zero, that is

$$\begin{aligned}
x(t) &\approx \frac{AK\omega}{\tau^2\omega^2 + 1}\left[\cancel{\tau e^{-t/\tau}} + \frac{1}{\omega}\sqrt{\tau^2\omega^2 + 1}\sin(\omega t + \phi_0),\right] \\
&= \frac{AK\cancel{\omega}}{\tau^2\omega^2 + 1}\frac{1}{\cancel{\omega}}\sqrt{\tau^2\omega^2 + 1}\sin(\omega t + \phi_0) \\
&= \frac{AK}{\tau^2\omega^2 + 1}\sqrt{\tau^2\omega^2 + 1}\sin(\omega t + \phi_0)
\end{aligned}$$

$$= \underbrace{\frac{AK}{\sqrt{\tau^2\omega^2+1}}\sin(\omega t + \phi_0)}_{x_\infty(t)}. \tag{7.47}$$

We denote this eventual response of the system by $x_\infty(t)$. We shall refer to this as the **steady state frequency response** of the system. This is illustrated in the following figure.

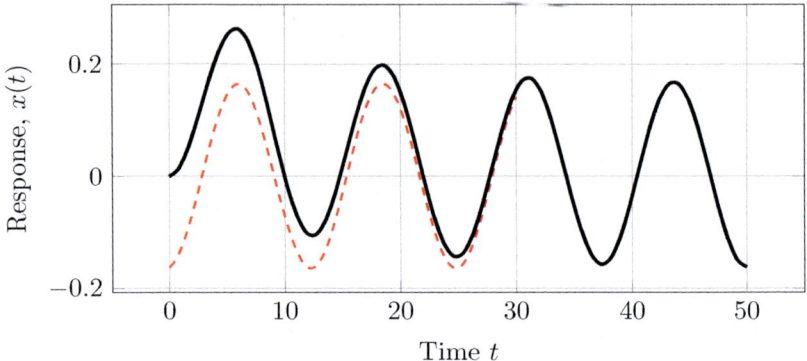

Figure 7.5: *Frequency response of first-order system ($A = 1$, $K = 1$, $\omega = 0.5$, $\tau = 15$).*

Exercise 7.2 (Integrator) An integrator with static gain K' is excited with a sinusoidal signal $u(t) = A\sin(\omega t)$. Determine its output. The initial conditions of the system are zero.

※ ....

Answer. Working as above we obtain $x(t) = AK'\omega^{-1}(1 - \cos(\omega t))$, for $t \geq 0$.

7.4.2 Amplitude ratio and its asymptotes

When a first-order system is excited with a sinusoidal input $u(t) = A\sin(\omega t)$ with amplitude A, the response after a large time period has passed (at $t \gg 1$) will have amplitude

$$A_\infty = \frac{AK}{\sqrt{\tau^2\omega^2+1}}.$$

The output will also have the same frequency as the input signal as shown in Equation (7.47) and a **phase lag** ϕ_0 given in Equation (7.46b).

The **amplitude ratio** between the input and output of the system is defined as

$$R_\infty = \frac{A_\infty}{A} = \frac{K}{\sqrt{\tau^2\omega^2+1}}. \tag{7.48}$$

The amplitude ratio depends only on the parameters of the system (the static gain and time constant) and the frequency of the excitation signal. Note that for very low frequency signals (with $\omega\tau \ll 1$), the amplitude ratio is approximately equal to K. On the other hand, for large frequencies, (i.e., for $\omega\tau \gg 1$) the amplitude ratio is approximately zero. In other words, first-order systems cut off high frequency signals, i.e., they serve as **low-pass filters**. Let us illustrate this with an example.

Example 7.2 (RL circuit as low-pass filter) Consider the series RL circuit of Example 7.1 with $R = 10\,\Omega$ and $L = 50mH$. The system is of first order with static gain

$$K = \frac{1}{R} = 0.1\,\Omega^{-1}, \tag{7.49}$$

and time constant

$$\tau = \frac{L}{R} = 5\,\text{ms} (= 0.005\,\text{s}). \tag{7.50}$$

Suppose that the system is actuated with a sinusoidal signal with amplitude $A = 1\,\text{V}$, that is, $V_{\text{in}} = A\sin(\omega t)$.

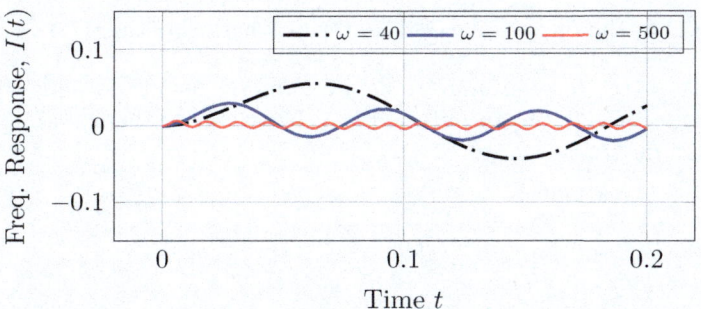

Figure 7.6: *Frequency response of series RL circuit for signals of three different frequencies.*

ω	Freq. (Hz)	R_∞
$\omega \to 0$	0	0.1
10	1.592	0.0998
100	15.92	0.0894
1000	159.2	0.0196
5000	795.8	0.0040
10000	1592	0.0020
50000	7957	0.0004

We observe that for high values of ω, the amplitude ratio decreases and approaches zero.

Next, we state a very important result: the large-time asymptotic behaviour of first-order systems can be derived directly from their transfer function without the need to perform the above inverse Laplace transform.

The value of this result is that it can be applied to systems of higher order as well.

> **Theorem 7.3** (Steady state frequency response of first-order systems)
> The amplitude ratio and the phase lag of a first-order system are given by
> $$R_\infty = |G(j\omega)|, \qquad (7.51\text{a})$$
> $$\phi_0 = \arg G(j\omega). \qquad (7.51\text{b})$$

Proof. It is
$$|G(j\omega)| = \left|\frac{K}{j\omega\tau + 1}\right| = \frac{K}{|j\omega\tau + 1|} = \frac{K}{\sqrt{\tau^2\omega^2 + 1}}, \qquad (7.52)$$

which is equal to the amplitude ratio (see Equation (7.48)). The argument of $G(j\omega)$ is

$$\arg G(j\omega) = \arg \frac{K}{j\omega\tau + 1} = -\arg(j\omega\tau + 1) = \arctan(-\omega\tau), \qquad (7.53)$$

which is equal to the phase lag (see Theorem 7.2 and Equation (7.47)). □

We will now try to gain more understanding on how the frequency of the input signal affects the amplitude ratio. Firstly, note that by applying the logarithm on R_∞ we have

$$R_\infty = \frac{K}{\sqrt{\tau^2\omega^2 + 1}}$$
$$\Rightarrow \frac{R_\infty}{K} = \frac{1}{\sqrt{(\tau\omega)^2 + 1}}$$
$$\Rightarrow \log \frac{R_\infty}{K} = -\tfrac{1}{2}\log((\tau\omega)^2 + 1). \qquad (7.54)$$

We may now plot $\frac{R_\infty}{K}$ versus $\omega\tau$ in logarithmically scaled axes. In order to do so, we will first study the behaviour of $\frac{R_\infty}{K}$ at very low and very high frequencies.

At very low frequencies, when $\tau\omega \ll 1$,

$$\log \frac{R_\infty}{K} \approx 0, \text{ for } \tau\omega \ll 1, \qquad (7.55)$$

which means that the low-frequency asymptote is a constant. At high frequencies, that is, when $\tau\omega \gg 1$,

$$\log \frac{R_\infty}{K} = -\log(\tau\omega), \text{ for } \tau\omega \gg 1. \qquad (7.56)$$

The high-frequency asymptote is a line with slope -1.

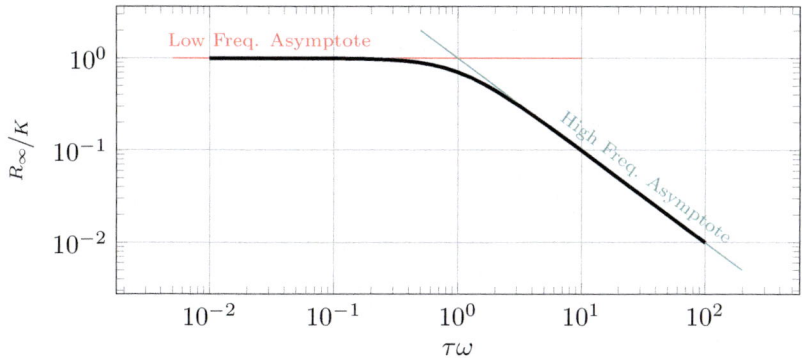

Figure 7.7: Plot of R_∞/K vs $\tau\omega$ in logarithmically scaled axes. The low and high frequency asymptotes are shown. The amplitude ratio is well approximated by its low-frequency asymptote for $\tau\omega \ll 1$, that is, for $\omega \ll 1/\tau$. Its high-frequency asymptote provides a good approximation for $\omega \gg 1/\tau$.

The above plot describes the dependence of the amplitude ratio on the time constant, τ. We may plot R_∞ against ω directly.

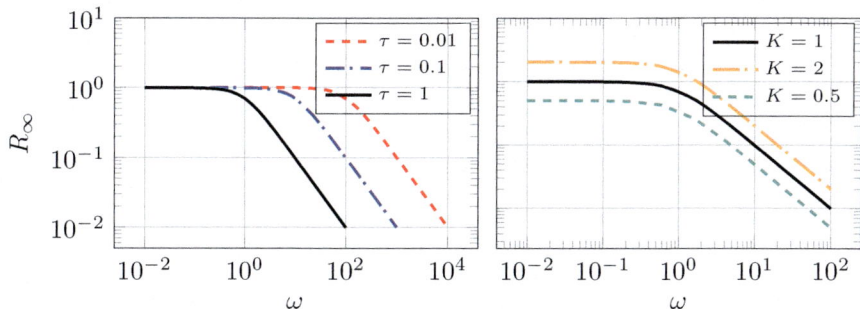

Figure 7.8: Plot of R_∞ vs ω in logarithmically scaled axes: (Left) Plots for different values of τ and fixed $K = 1$, (Right) Plots for different values of K and fixed $\tau = 1$.

We will study the dependence of the phase lag on ω in Chapter 11, where we will focus on the frequency response of more general systems.

Example 7.3 (RC low-pass filter) The aim of this example is to design an RC circuit which serves as a **low-pass filter**, that is, it attenuates high-frequency signals and allows low-frequency ones to pass through. Such circuits are very useful in sound amplifiers, power supply devices, communication systems and more.

The requirement of our design is that the amplitude ratio at the frequency of 1 kHz should be no larger than 0.001.

The correspondence between V_{in} and V_{out} is cast by the following ODE

$$\dot{V}_{\text{out}} = \frac{1}{CR}(V_{\text{in}} - V_{\text{out}}). \tag{7.57}$$

Applying the Laplace transform (and assuming zero initial conditions), we have

$$sV_{\text{out}} = \frac{1}{CR}(V_{\text{in}} - V_{\text{out}})$$
$$\Rightarrow sCRV_{\text{out}} = V_{\text{in}} - V_{\text{out}}$$
$$\Rightarrow (1 + sCR)V_{\text{out}} = V_{\text{in}}$$
$$\Rightarrow \frac{V_{\text{out}}}{V_{\text{in}}} = \frac{1}{1 + sCR}. \tag{7.58}$$

So, the transfer function of this system is

$$G(s) = \frac{1}{1 + CRs}, \tag{7.59}$$

which is a first-order system with time constant $\tau = RC$ and static gain $K = 1$.

By Equation (7.48), the amplitude ratio at angular frequency ω is given by

$$R_\infty = \frac{K}{\sqrt{\tau^2 \omega^2 + 1}} = \frac{1}{\sqrt{R^2 C^2 \omega^2 + 1}} \tag{7.60}$$

Recall that the angular frequency, ω, is related to the frequency f via $f = \omega/2\pi$, so the frequency of 1 kHz corresponds to $\omega = 2\pi f = 6283.18$ rad/s. The requirement is that

$$R_\infty \leq 0.001 \Leftrightarrow \frac{1}{\sqrt{R^2 C^2 \cdot 6283.18^2 + 1}} \leq 0.001$$
$$\Leftrightarrow RC \geq 0.15 \tag{7.61}$$

So, if we have a capacitor with $C = 100\,\mu\text{F}$ we need to choose a resistor with $R \geq 1500\,\Omega$.

Remark: *In the example above we saw that the voltage across the capacitor in an RC circuit filters out the high-frequency components of the input voltage V_{in}. We say that the RC circuit acts as a **low-pass filter** (from V_{in} to V_{out}).*

The same circuit can act as a **high-pass filter** if we take the output voltage to be the voltage across the resistor. You can see this yourselves by following the steps of the following exercise. •

Exercise 7.3 (High-pass RC filter) Consider the RC circuit shown in the following schematic, where the input voltage, V_{in} is the same as before and the output voltage, V_R, is the voltage across the resistor.

Assume zero initial conditions.

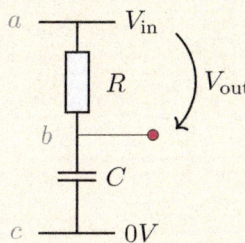

i. Show that the transfer function of the system (between V_{in} and V_R is)

$$G(s) = \frac{RCs}{1 + RCs}.$$

ii. Determine the frequency response of the system, i.e., the response $V_R(t)$ of the system when $V_{in}(t) = A\sin(\omega t)$. Hint: work as in Section 7.4.1.

iii. Show that the amplitude ratio at $t \gg 1$ is given by

$$R_\infty = \frac{RC\omega}{\sqrt{1 + R^2C^2\omega^2}}.$$

Hint: work as in Section 7.4.2.

iv. Determine the low-frequency and high-frequency asymptotes of R_∞ and plot R_∞/RC against $RC\omega$.

7.5 One-minute round-up

In this chapter

1. We studied first-order systems which have a transfer function of the form $G(s) = K/{\tau s+1}$ with $K, \tau > 0$; K is called the static gain and τ is the time constant of the system

2. We derived the impulse, step and frequency responses using the inverse Laplace transform

3. When a first-order system is excited with a sinusoidal input of the form $u(t) = \sin(\omega t)$ then after a long time, the system output will

have amplitude $R_\infty = |G(j\omega)|$ and phase lag $\phi_0 = \arg G(j\omega)$ approximately, that is, it will be approximately equal to $x_\infty(t) = R_\infty \sin(\omega t + \phi_0)$

7.6 Exercises

Exercise 7.4 (Application of FVT) Let $x(t)$ be the impulse response of a first-order system. Use the final value theorem (Theorem 4.29) to show that $\lim_{t \to \infty} x(t) = 0$. Assume that $x(0) = 0$.

Exercise 7.5 (Asymptotics of step response of first-order system) Let $x_{\text{step}}(t)$ denote the unit step response of a first-order system with static gain K and time constant τ. Use the IVT and FVT to show that

$$\lim_{t \to \infty} x_{\text{step}}(t) = K, \tag{7.62a}$$

$$\lim_{t \to 0^+} \dot{x}_{\text{step}}(t) = \frac{K}{\tau}. \tag{7.62b}$$

Exercise 7.6 (Another application of FVT) A first order system is excited with an input signal defined by

$$u(t) = \begin{cases} 1, & \text{for } 0 \leq t \leq 5 \\ 0, & \text{for } t > 5 \end{cases} \tag{7.63}$$

Use the final value theorem to find the value of $\lim_{t \to \infty} x(t)$. Assume that $x(0) = 0$.

Exercise 7.7 (Identification of first-order system) A first-order system with unknown static gain K and time constant τ is excited with a step response of amplitude $A = 1$, that is, $u(t) = H_0(t)$. The system response is shown below.

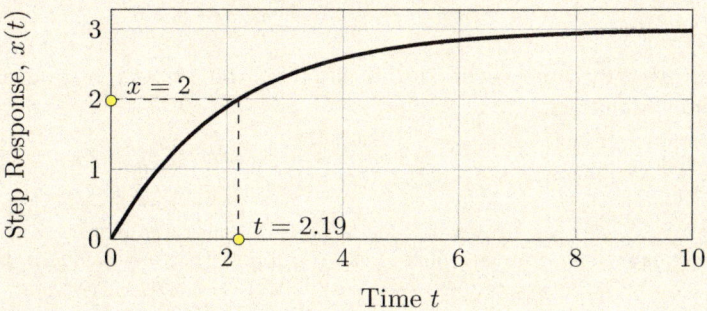

Figure 7.9: *Step response of first-order system with unknown parameters.*

We find that at time $t = 2.19$, the system output is $x = 2$. Determine K and τ.

✂ ....

Hint. Note that the value $\lim_{t\to\infty} x(t)$ can be inferred from Figure 7.9. This can be used to determine the value of K. Be careful when applying the final value theorem: you always need to make sure that the necessary conditions are satisfied. From the figure we also see that $x(2.19) = 2$; this can be used to determine τ.

Exercise 7.8 (Industrial motor) The thrust, x, of an industrial motor is controlled by the voltage, u applied to it. Once a unit step change $(u(t) = H_0(t))$ is applied to the input voltage, it takes 0.8 s for the thrust to reach its new equilibrium value. Determine the time constant of the motor.

Exercise 7.9 (Identification of first-order system) A first-order system with unknown static gain K and time constant τ is excited with a step response of amplitude $A = 1$, that is, $u(t) = H_0(t)$. The system response is shown below.

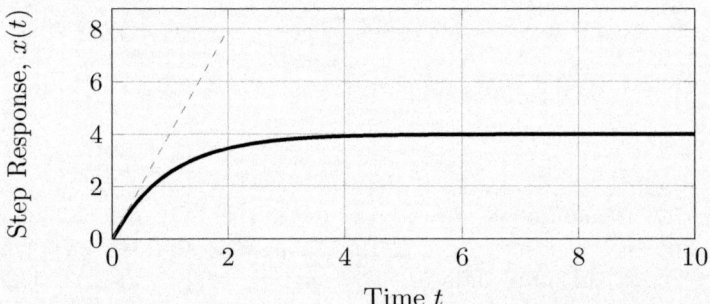

Figure 7.10: Step response of first-order system with unknown parameters.

The dashed grey line is the initial slope of the system response. Determine K and τ.

✂ ....

Hint. Solve Exercise 7.5 first. We can see in Figure 7.10 that the slope of the system response at time $t = 0^+$ is equal to 4, that is, $\dot{x}(0^+) = 8/2 = 4$. We also see that $\lim_{t\to\infty} x(t) = 4$.

Exercise 7.10 (Initial value theorem) Consider a first-order system with transfer function

$$G(s) = \frac{K}{\tau s + 1}, \qquad (7.64)$$

with $K, \tau > 0$.

1. Show that it is not possible to apply the initial value theorem to determine the initial value of the derivative of the impulse response of the system, that is $\dot{x}(0^+)$ for $u(t) = \delta(t)$

2. Determine the Laplace transform of the derivative of the impulse response of this system and

3. Expand the above Laplace transform into partial fractions

4. Show the following auxiliary result: let $y(t) = y_1(t) + y_2(t)$ and suppose that y_1 has the Laplace transforms $Y_1(s)$ and Y_2 satisfies the conditions of the initial value theorem (see Theorem 4.28). Then,

$$\lim_{t \to 0^+} y(t) = \lim_{t \to 0^+} y_1(t) + \lim_{s \to \infty} sY_2(s). \qquad (7.65)$$

5. Determine the initial value of the derivative of the impulse response of this system by combining the above results.

Exercise 7.11 (Settling time of first-order system) Let $x_{\text{step}}(t)$ denote the step response of a first-order system with static gain K and time constant τ. The 5%-*settling time*, $t_{s,0.05}$, is defined as the time when the step response becomes 95% of its final value. Determine $t_{s,0.05}$ as a function of K and τ.

Answer. It is $t_{s,0.05} = -\tau \ln 0.05$.

8
Second-order systems

Spring-mass-damper systems, RLC circuits and pendula are typical examples of second-order dynamical systems. Second-order systems are omnipresent in engineering practice. Their transfer function has the general form $G(s) = \frac{K}{\tau^2 s^2 + 2\zeta\tau s + 1}$, where K, τ, ζ are positive constants. Depending on the value of ζ, the system exhibits a qualitatively different behaviour. In this chapter, we employ the Laplace transform and its inverse to compute the step, impulse and frequency response of second-order systems, and we study their characteristics. We give particular emphasis on the step response, especially for low values of ζ when it is oscillatory.

8.1 Introduction

In this chapter we study the dynamic characteristics of single-input single-output systems which are governed by differential equations of the form

$$a_0 x(t) + a_1 \dot{x}(t) + a_2 \ddot{x}(t) = bu(t), \tag{8.1}$$

with $a_2 \neq 0$. If $a_0 \neq 0$, the system dynamics can be written as

$$x(t) + \frac{a_1}{a_0}\dot{x}(t) + \frac{a_2}{a_0}\ddot{x}(t) = \frac{b}{a_0}u(t). \tag{8.2}$$

We have already encountered examples of systems described by such dynamics such as spring-mass-damper systems, RLC circuits and systems of two connected tanks.

Hereafter we will focus exclusively on the common case where $a_0, a_1, a_2 > 0$. We then define the following characteristic quantities

$$\tau^2 = \frac{a_2}{a_0}, \tag{8.3a}$$

$$2\zeta\tau = \frac{a_1}{a_0}, \tag{8.3b}$$

$$K = \frac{b}{a_0}, \tag{8.3c}$$

where τ and ζ are positive constants and $K \neq 0$.

These three parameters determine the qualitative behaviour of a second-order system. This is better understood through the study of the step response of second-order systems. Roughly speaking, ζ plays the role of a damping component; the larger the value of ζ, the more "damped" the system is and the less likely to exhibit oscillatory behaviour. The time constant, τ, determines how fast the system response is; the higher the value of τ, the slower the system response. Lastly, the static gain, K, serves as an amplification or attenuation factor. The effect of K can be understood through the following exercise.

Exercise 8.1 (Final value of step response of SOS) A second-order system is excited with a Heaviside function, $u(t) = H_0(t)$. Let $x(t)$ be the system response. Use the final value theorem to find the final value of its response, that is

$$\lim_{t \to \infty} x(t). \tag{8.4}$$

Hereafter we assume that the initial conditions of the system are zero[1], and we apply the Laplace transform to obtain the following algebraic equation

$$X + 2\zeta\tau s X(s) + \tau^2 s^2 X(s) = K U(s), \tag{8.5}$$

where $X(s) = \mathscr{L}\{x(t)\}(s)$ and $U(s) = \mathscr{L}\{u(t)\}(s)$, so, the transfer function is

$$G(s) = \frac{X(s)}{U(s)} = \frac{K}{\tau^2 s^2 + 2\zeta\tau s + 1}. \tag{8.6}$$

This leads to the definition of the class of second-order systems

Definition 8.1 (Second-order systems) A dynamical system with input u and output x which can be written in the form

$$x + 2\zeta\tau \dot{x} + \tau^2 \ddot{x} = Ku, \tag{8.7}$$

for some $\zeta, \tau > 0$ and $K \neq 0$ is called a **second-order system**. Its transfer function is

$$G(s) = \frac{K}{\tau^2 s^2 + 2\zeta\tau s + 1}. \tag{8.8}$$

We call τ the **time constant** of the system, ζ the **damping factor** and K the **static gain**.

---

[1] If not, we introduce deviation variables.

Chapter 8. Second-order systems

The poles of G play a very important role in the study of the dynamic behaviour of the system. The poles of G are the roots of the quadratic equation

$$\tau^2 s^2 + 2\zeta\tau s + 1 = 0. \tag{8.9}$$

The roots of this equation are

$$s_1 = -\frac{\zeta}{\tau} + \frac{\sqrt{\zeta^2 - 1}}{\tau}, \tag{8.10a}$$

$$s_2 = -\frac{\zeta}{\tau} - \frac{\sqrt{\zeta^2 - 1}}{\tau}. \tag{8.10b}$$

We distinguish three cases:

> **Definition 8.2 (Overdamped, underdamped, critically damped)** A second-order system is called (i) **overdamped**, if $\zeta > 1$, (ii) **underdamped**, if $\zeta < 1$ and (iii) **critically damped**, if $\zeta = 1$.

If G is overdamped, it has two distinct real poles ($s_1 \neq s_2$). If it is underdamped it has a pair of complex conjugate (non-real) poles and if it is critically damped, it has a double real pole. We will study these three cases separately as they are linked with different qualitative properties of the system dynamics.

8.2 Impulse response

Before we determine the impulse response of a second-order system we can use the final value theorem (Theorem 4.29) to determine the limit $\lim_{t\to\infty} x(t)$. We can verify that the conditions of the theorem are satisfied[2], so

$$\lim_{t\to\infty} x(t) = \lim_{s\to 0^+} sX(s) = \lim_{s\to 0^+} sG(s) = \lim_{s\to 0^+} \frac{Ks}{\tau^2 s^2 + 2\zeta\tau s + 1} = 0. \tag{8.11}$$

The impulse response of a second-order system with $\zeta \neq 1$ (which has simple real or complex poles) can be derived following the methodology described in Section 5.2.4. We have

$$G(s) = \frac{K}{\tau^2 s^2 + 2\zeta\tau s + 1} = \frac{A_1}{s - s_1} + \frac{A_2}{s - s_2}. \tag{8.12}$$

The reader can determine A_1 and A_2 and verify that the inverse Laplace transform of G — the impulse response of the system — is

$$x(t) = \frac{1}{\tau\sqrt{\zeta^2 - 1}} e^{-\frac{\zeta}{\tau}t} \sinh\left(\frac{\sqrt{\zeta^2 - 1}}{\tau} t\right). \tag{8.13}$$

---

[2] The read can verify that the poles of G have strictly negative real part.

In the particular case where $\zeta < 1$, $\sqrt{\zeta^2 - 1}$ is an imaginary number. We may then use the property $\sinh(jx) = j\sin(x)$ to simplify the impulse response.

> **Exercise 8.2 (Impsulse response of SOS)** Find the impulse response of a second-order system with $0 < \zeta < 1$.

If $\zeta = 1$, the system has a double real pole (a pole of multiplicity 2). Then, following the methodology presented in Section 5.2.2, the impulse response can be found to be

$$x(t) = \frac{K}{\tau^2} t e^{-t/\tau}, \qquad (8.14)$$

for all $t \geq 0$.

8.3 Step response

The value of ζ determines the qualitative behaviour of a second-order system.

8.3.1 Overdamped systems ($\zeta > 1$)

A second-order system with $\zeta > 1$ is called **overdamped**. If excited with a Heaviside pulse of amplitude A, that is $u(t) = AH_0(t)$, it produces the output

$$X(s) = G(s)U(s) \stackrel{U=A/s}{=} \frac{AK}{s(\tau^2 s^2 + 2\zeta\tau s + 1)}, \qquad (8.15)$$

and given that the poles s_1 and s_2 are real and distinct we may use the following partial fraction expansion

$$X(s) = \frac{A_1}{s} + \frac{A_2}{s - s_1} + \frac{A_3}{s - s_2}. \qquad (8.16)$$

We can then determine the coefficients A_1, A_2 and A_3 and determine the inverse Laplace transform of $X(s)$, which is

$$x(t) = AK\left[1 - e^{-\zeta\frac{t}{\tau}}\left(\cosh\left(\frac{\sqrt{\zeta^2 - 1}}{\tau}t\right) + \frac{\zeta}{\sqrt{\zeta^2 - 1}}\sinh\left(\frac{\sqrt{\zeta^2 - 1}}{\tau}t\right)\right)\right], \qquad (8.17)$$

for $t \geq 0$.

Chapter 8. Second-order systems

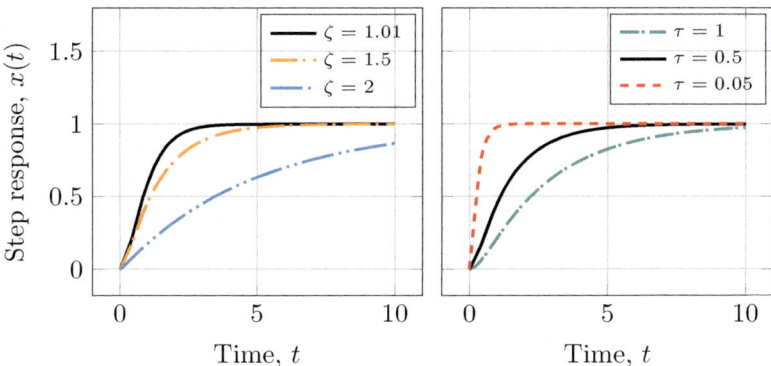

Figure 8.1: *Step response of an overdamped system for $A = 1$ and $K = 1$; (Left) Step response for $\tau = 1$ and different values of ζ, (Right) Step response profiles for $\zeta = 1.5$ and different values of τ.*

As shown in Figure 8.1, high values of the damping factor, ζ, leads to a more sluggish response.

The step response of an overdamped second-order system resembles that of a first-order system; yet, there is a distinctive difference. For a first order system, the value of $\dot{x}(0^+)$ is[3]

$$\begin{aligned}
\lim_{t \to 0^+} \dot{x}(t) &= \lim_{s \to \infty} s\mathscr{L}\{\dot{x}\}(s) \\
&= \lim_{s \to \infty} s^2 X(s) \\
&= \lim_{s \to \infty} s^2 \frac{G(s)}{s} \\
&= \lim_{s \to \infty} \frac{Ks}{\tau s + 1} \\
&= \lim_{s \to \infty} \frac{K}{\tau + \frac{1}{s}} = \frac{K}{\tau}.
\end{aligned} \qquad (8.18)$$

On the other hand, for a second-order system, the derivative of the output as $t \to 0^+$ is

$$\lim_{t \to 0^+} \dot{x}(t) = \lim_{s \to \infty} \frac{Ks}{\tau^2 s^2 + 2\zeta\tau s + 1} = \lim_{s \to \infty} \frac{K}{\tau^2 s + 2\zeta\tau + \frac{1}{s}} = 0. \qquad (8.19)$$

---

[3] We assume zero initial conditions and apply the initial value theorem.

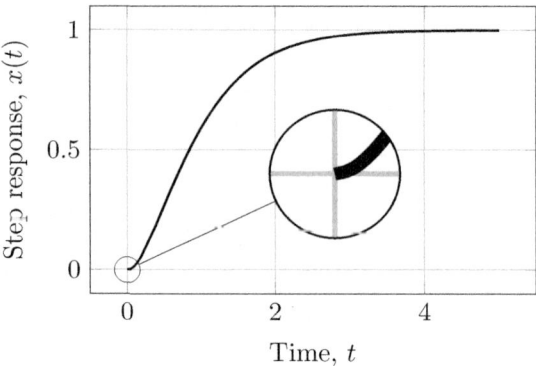

Figure 8.2: Detail of step response of overdamped system ($\tau = 0.5$, $\zeta = 1.01$ and $K = 1$). As shown in (8.19) we can observe that $\dot{x}(0^+) = 0$.

8.3.2 Critically damped systems ($\zeta = 1$)

In this case, the system has a double real pole at $s_{1,2} = -\zeta/\tau$. Indeed, for $\zeta = 1$, we see that the denominator of the transfer function can be written as $\tau^2 s^2 + 2\zeta\tau s + 1 = \tau^2 s^2 + 2\tau s + 1 = (\tau s + 1)^2$ Its step response is given by the inverse Laplace transform

$$\begin{aligned}
x(t) &= \mathscr{L}^{-1}\left\{\frac{K}{s(\tau s + 1)^2}\right\} \\
&= \mathscr{L}^{-1}\left\{\frac{K}{s\tau^2(s + \frac{1}{\tau})^2}\right\} \\
&= \mathscr{L}^{-1}\left\{\frac{\frac{K}{\tau^2}}{s(s + \frac{1}{\tau})^2}\right\}.
\end{aligned} \quad (8.20)$$

Next, we expand the function is partial fractions as discussed in Section 5.2.2 and obtain

$$\begin{aligned}
x(t) &= K\mathscr{L}^{-1}\left\{\frac{1}{s} - \frac{1}{s + \frac{1}{\tau}} - \frac{1}{\tau(s + \frac{1}{\tau})^2}\right\} \\
&= K\left[1 - \left(1 + \frac{t}{\tau}\right)e^{-\frac{t}{\tau}}\right],
\end{aligned} \quad (8.21)$$

for $t \geq 0$.

Chapter 8. Second-order systems

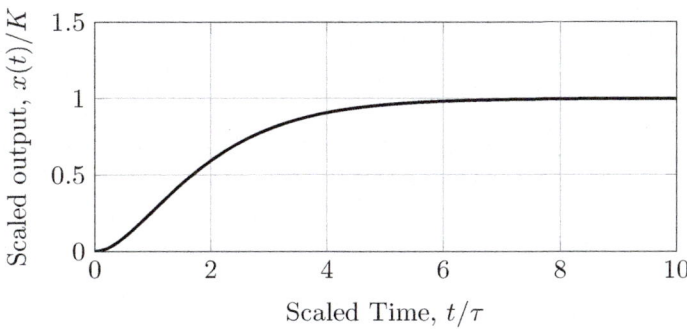

Figure 8.3: *Step response of critically damped system. Observe that the final value of $x(t)$ is $x(\infty) = K$ and the initial value of the derivative is $\dot{x}(0^+) = K/\tau$.*

8.3.3 Underdamped systems ($\zeta < 1$)

This is the most interesting case because the step response exhibits an oscillatory behaviour. In particular, it is

$$x(t) = AK \left[1 - \frac{1}{\sqrt{1-\zeta^2}} \cdot \underbrace{e^{-\frac{\zeta}{\tau}t}}_{\text{damping}} \cdot \underbrace{\sin(\omega_d t + \phi_0)}_{\text{oscillatory}} \right], \qquad (8.22)$$

where

$$\omega_d = \frac{\sqrt{1-\zeta^2}}{\tau}, \qquad (8.23a)$$

$$\phi_0 = \arctan \frac{\sqrt{1-\zeta^2}}{\zeta}. \qquad (8.23b)$$

We observe that the step response of an underdamped system involves an oscillatory and a damping component. The damping component drives the system to its final value, $\lim_{t \to \infty} x(t) = AK$. The oscillatory component has an angular frequency ω_d, which is called the **damped frequency of the oscillation** of the system.

Smaller values of ζ lead to a weaker decay rate and oscillations of higher frequency. Indeed, as we see in Figure 8.4, lower values of ζ lead to oscillations of higher amplitude and frequency.

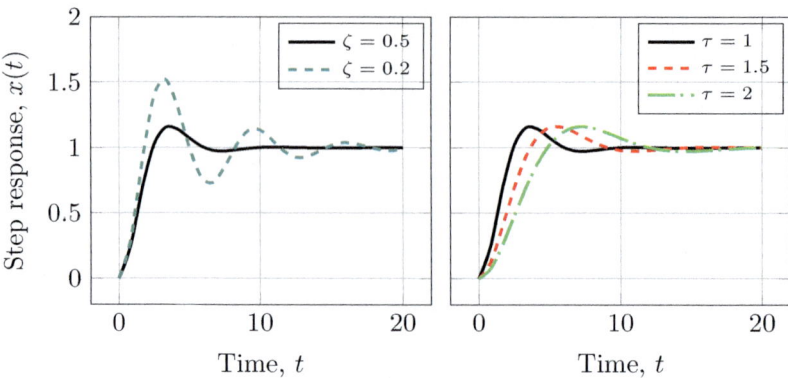

Figure 8.4: *Step response of an underdamped system for $A = 1$ and $K = 1$; (Left) Step response for $\tau = 1$ and different values of ζ, (Right) Step response profiles for $\zeta = 0.5$ and different values of τ.*

8.4 Characteristics of underdamped systems

For the step response of an underdamped second-order system we define the following characteristic quantities

i. The **peak time**, which is the time at which the first peak occurs

ii. The **overshoot**, which quantifies how far above its final value, AK, the response can get; the overshoot is a measure of the intensity of the oscillations of the response.

iii. The **rise time**, which is the time needed for the response to reach its equilibrium value for the first time

iv. The **decay ratio** which is a measure of the energy dissipation of the system

v. The **settling time** which measures how fast the system will be approximately at equilibrium

8.4.1 Peak time

The peak time of an underdamped step response is the time required for $x(t)$ to reach its first maximum. This is given in the following proposition.

> **Proposition 8.3 (Peak time)** The peak time of an underdamped step response is
> $$t_{\text{peak}} = \frac{\pi}{\omega_d}.$$

Chapter 8. Second-order systems

Proof. In order to determine the peak time, t_{peak}, we first compute the time-derivative of $x(t)$. By Equation (8.22) we have that

$$\dot{x}(t) = \frac{d}{dt}AK\left[1 - \frac{1}{\sqrt{1-\zeta^2}}e^{-\frac{\zeta}{\tau}t}\sin(\omega_d t + \phi_0)\right]$$

$$= -AK\frac{1}{\sqrt{1-\zeta^2}}\frac{d}{dt}\left(e^{-\frac{\zeta}{\tau}t}\sin(\omega_d t + \phi_0)\right)$$

$$= -AK\frac{1}{\sqrt{1-\zeta^2}}\left(-\frac{\zeta}{\tau}e^{-\frac{\zeta}{\tau}t}\sin(\omega_d t + \phi_0) + e^{-\frac{\zeta}{\tau}t}\omega_d\cos(\omega_d t + \phi_0)\right)$$

$$= -AK\frac{1}{\sqrt{1-\zeta^2}}e^{-\frac{\zeta}{\tau}t}\left(-\frac{\zeta}{\tau}\sin(\omega_d t + \phi_0) + \omega_d\cos(\omega_d t + \phi_0)\right) \quad (8.24)$$

Since t_{peak} is a local maximum of $x(t)$ and $x(t)$ is differentiable at that point, by Fermat's Theorem, $\dot{x}(t_{\text{peak}}) = 0$, that is

$$\dot{x}(t_{\text{peak}}) = 0 \Rightarrow -AK\frac{1}{\sqrt{1-\zeta^2}}e^{-\frac{\zeta}{\tau}t}\left(-\frac{\zeta}{\tau}\sin(\omega_d t_{\text{peak}} + \phi_0)\right. \quad (8.25)$$

$$\left. + \omega_d\cos(\omega_d t_{\text{peak}} + \phi_0)\right) = 0$$

$$\Rightarrow -\frac{\zeta}{\tau}\sin(\omega_d t_{\text{peak}} + \phi_0) + \omega_d\cos(\omega_d t_{\text{peak}} + \phi_0) = 0$$

$$\Rightarrow \frac{\zeta}{\tau}\sin(\omega_d t_{\text{peak}} + \phi_0) = \omega_d\cos(\omega_d t_{\text{peak}} + \phi_0), \quad (8.26)$$

and since t_{peak} cannot be such that $\cos(\omega_d t_{\text{peak}} + \phi_0) = 0$, we can divide both sides by $\cos(\omega_d t_{\text{peak}} + \phi_0)$ to obtain

$$\tan(\omega_d t_{\text{peak}} + \phi_0) = \omega_d \frac{\tau}{\zeta}, \quad (8.27)$$

and it can be seen that $\omega_d \frac{\tau}{\zeta} = \frac{\sqrt{1-\zeta^2}}{\zeta} = \tan\phi_0$, so

$$\tan(\omega_d t_{\text{peak}} + \phi_0) = \tan\phi_0, \quad (8.28)$$

which means that $\omega_d t_{\text{peak}} = k\pi$, for $k \in \mathbb{N}$. We are interested in the case $k = 1$ since we are looking for the first peak, so $t_{\text{peak}} = \frac{\pi}{\omega_d}$, which completes the proof. $\square$

8.4.2 Overshoot

We call **overshoot** the ratio of the first and highest peak of the response over the final value of $x(t)$. In other words, overshoot is defined as the ratio $\frac{B}{AK}$ in the following figure:

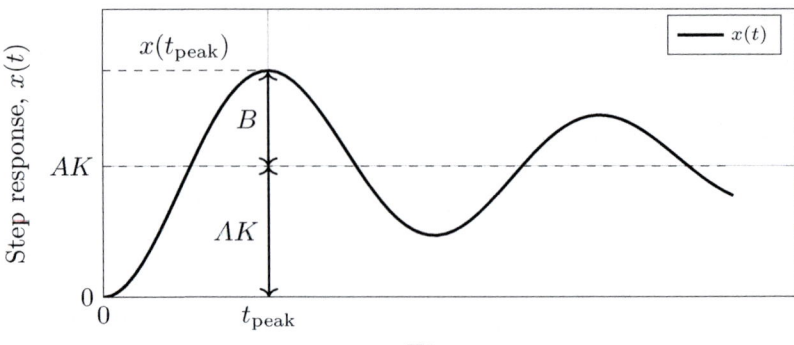

Figure 8.5: *Definition of overshoot: it is the ratio B/AK.*

Since we know that the first peak is observed at time $t = t_{\text{peak}}$, the corresponding value of the output will be $x(t_{\text{peak}})$ as shown in Figure 8.5. This means that $B + AK = x(t_{\text{peak}})$. This observation allows us to compute the overshoot.

> **Proposition 8.4 (Overshoot)** The overshoot of an underdamped step response is
> $$\text{Overshoot} = \frac{B}{AK} = e^{-\frac{\pi \zeta}{\sqrt{1-\zeta^2}}}. \tag{8.29}$$

Proof. By definition

$$\text{Overshoot} = \frac{B}{AK} = \frac{x(t_{\text{peak}}) - AK}{AK} = \frac{x(t_{\text{peak}})}{AK} - 1. \tag{8.30}$$

We substitute $t_{\text{peak}} = \pi/\omega_d$ (Proposition 8.3) into $x(t)$ (see Equation (8.22)) and we compute

$$\frac{x(t_{\text{peak}})}{AK} = 1 - \frac{1}{\sqrt{1-\zeta^2}} e^{-\frac{\zeta}{\tau} t_{\text{peak}}} \sin(\omega_d t_{\text{peak}} + \phi_0). \tag{8.31}$$

The result follows by carrying out the operations and using the identity $\sin(\arctan x) = \frac{x}{\sqrt{1+x^2}}$, which holds for $x \in (-\frac{\pi}{2}, \frac{\pi}{2})$. □

Remark: *Often, the overshoot is expressed in percent (%). An overshoot value of 10% means that the first and largest peak of the step response will be 10% larger than the final value of $x(t)$, that is, it will reach $1.1 \cdot AK$. The overshoot is a measure of the intensity of the oscillations of the response. The overshoot is defined only for underdamped systems.* •

Chapter 8. Second-order systems

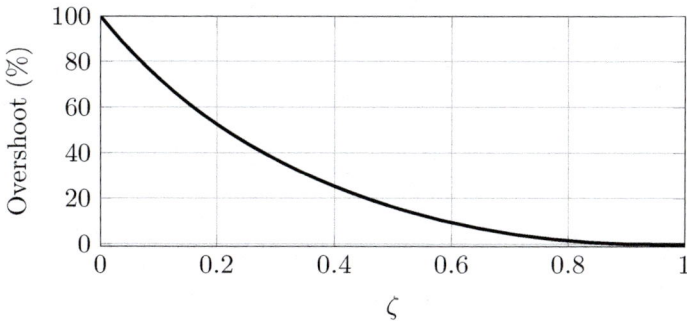

Figure 8.6: *Dependence of the overshoot on ζ.*

8.4.3 Rise time

The **rise time** is defined as the time required for an underdamped step response, $x(t)$, to reach its final value for the first time. By definition, the rise time, t_{rise}, is the smallest time that satisfies the equation $x(t_{\text{rise}}) = AK$.

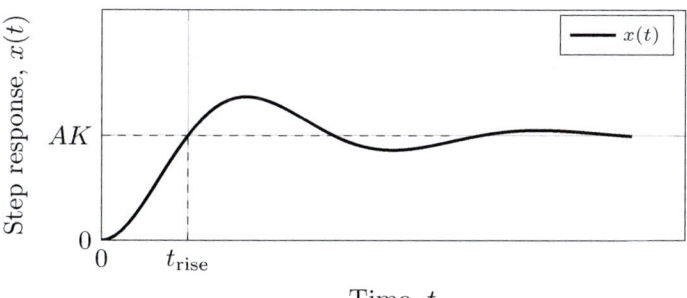

Figure 8.7: *Definition of rise time.*

> **Proposition 8.5 (Rise time)** The rise time of an underdamped step response is given by
> $$t_{\text{rise}} = \frac{\pi - \phi_0}{\omega_d}. \qquad (8.32)$$

Proof. We know that t_{rise} is the smallest time that satisfies the equation $x(t_{\text{rise}}) = AK$. By virtue of Equation (8.22),

$$x(t_{\text{rise}}) = AK \Rightarrow \cancel{AK}^{1}\left[1 - \frac{1}{\sqrt{1-\zeta^2}}e^{-\frac{\zeta}{\tau}t_{\text{rise}}}\sin(\omega_d t_{\text{rise}} + \phi_0)\right] = \cancel{AK}^{1}$$

$$\Rightarrow \cancel{1} - \frac{1}{\sqrt{1-\zeta^2}}e^{-\frac{\zeta}{\tau}t_{\text{rise}}}\sin(\omega_d t_{\text{rise}} + \phi_0) = \cancel{1}$$

$$\Rightarrow \sin(\omega_d t_{\text{rise}} + \phi_0) = 0, \qquad (8.33)$$

therefore, $\omega_d t_{\text{rise}} + \phi_0 = k\pi$ for $k \in \mathbb{N}$, but since we are looking for the first such time instant, we set $k = 1$, therefore,

$$\omega_d t_{\text{rise}} + \phi_0 = \pi, \tag{8.34}$$

and the assertion follows. $\square$

8.4.4 Decay ratio

The decay ratio of the step response of an underdamped second-order system is a measure of how fast the system loses energy. It is defined as the ratio C/B as show in Figure 8.8.

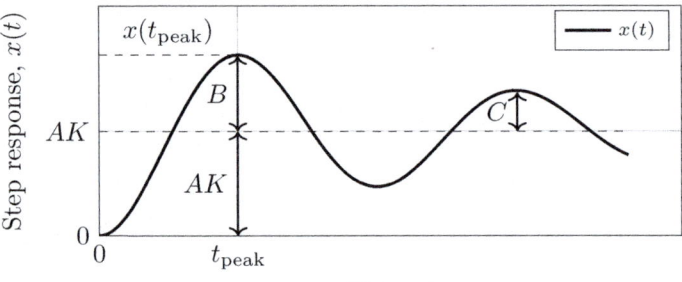

Figure 8.8: *Definition of the decay ratio: it is the ratio C/B.*

Proposition 8.6 (Decay ratio) The decay ratio is given by

$$\text{DR} = \frac{C}{B} = \text{Overshoot}^2. \tag{8.35}$$

Proof. An easy way to show this is to use the fact that the second peak occurs at time equal to $t_2 = \frac{3T_d}{2}$, where T_d is the natural period of oscillation which corresponds to ω_d, that is, $T_d = 2\pi/\omega_d$. Then,

$$x(3T_d/2) = AK + C. \tag{8.36}$$

The rest of the proof is easy and is left to the reader as an exercise. $\square$

8.4.5 Upper and lower envelopes

The upper and lower envelopes of the step response of an underdamped second-order system exponential functions which are tight upper and lower bounds of the response. In other words the lower envelope $S^-(t)$ and the upper envelope $S^+(t)$ are exponential functions with

$$S^-(t) \leq x(t) \leq S^+(t). \tag{8.37}$$

Chapter 8. Second-order systems

These are given by[4]

$$S^-(t) = AK\left(1 - \frac{e^{-\frac{\zeta}{\tau}t}}{\sqrt{1-\zeta^2}}\right), \quad (8.38a)$$

$$S^+(t) = AK\left(1 + \frac{e^{-\frac{\zeta}{\tau}t}}{\sqrt{1-\zeta^2}}\right). \quad (8.38b)$$

These are tight bounds as illustrated in the following figure

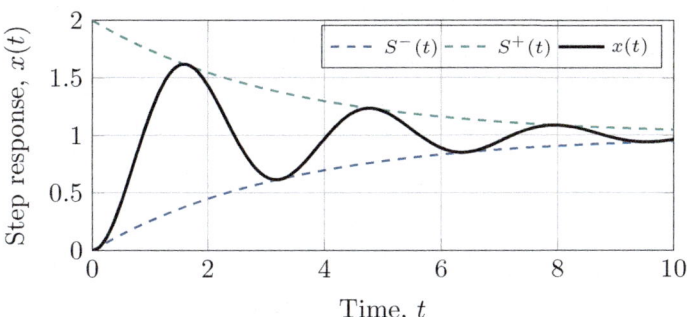

Figure 8.9: *Upper and lower envelopes for the step response of an underdamped second-order system with $\zeta = 0.5$, $\tau = 0.2$ and $A = K = 1$.*

8.4.6 Settling time

The **settling time** of $x(t)$ is a time after which the signal will remain within $\pm 5\%$ of its limit value. The 5%-settling time is defined in terms of the envelopes of the response (see Section 8.4.5) as the time $t_{s,0.05}$ at which

$$S^+(t_{s,0.05}) - S^-(t_{s,0.05}) = 10\% \cdot AK. \quad (8.39)$$

The settling time is given in the following proposition.

> **Proposition 8.7 (Settling time)** The settling time, $t_{s,0.05}$, is given by
>
> $$t_{s,0.05} = -\frac{\tau}{\zeta}\ln(0.05\sqrt{1-\zeta^2}). \quad (8.40)$$

Proof. Because of symmetry, this is equivalent to

$$S^+(t_{s,0.05}) - AK = 5\% \cdot AK, \quad (8.41)$$

---

[4] The fact that $S^-(t) \leq x(t) \leq S^+(t)$ follows from the inequality $-1 \leq \sin(x) \leq 1$. Note also that the upper envelope, $S^+(t)$, touches the response, $x(t)$, at times t with $\omega_d t + \phi_0 = 2k\pi + \pi/2$ for all $k \in \mathbb{N}$. Likewise, the lower envelope, $S^-(t)$, touches the response at times $\omega_d t + \phi_0 = 2k\pi - \pi/2$. The points where the envelopes touch the response are not the local maximum/minimum points of $x(t)$.

and by substituting Equation (8.38b) we have

$$\cancel{AK}^1\left(1 + \frac{e^{-\frac{\zeta}{\tau}t_{s,0.05}}}{\sqrt{1-\zeta^2}}\right) - \cancel{AK}^1 = 0.05 \cdot \cancel{AK}^1$$

$$\Rightarrow \cancel{1} + \frac{e^{-\frac{\zeta}{\tau}t_{s,0.05}}}{\sqrt{1-\zeta^2}} - \cancel{1} = 0.05$$

$$\Rightarrow e^{-\frac{\zeta}{\tau}t_{s,0.05}} = 0.05\sqrt{1-\zeta^2}$$

$$\Rightarrow -\frac{\zeta}{\tau}t_{s,0.05} = \ln(0.05\sqrt{1-\zeta^2})$$

$$\Rightarrow t_{s,0.05} = -\frac{\tau}{\zeta}\ln(0.05\sqrt{1-\zeta^2}). \qquad (8.42)$$

This completes the proof. $\square$

Example 8.1 (Characteristics of second-order underdamped systems) In this example we consider the transfer function

$$G(s) = \frac{1}{s^2 + 0.4s + 1}. \qquad (8.43)$$

We will verify that this is an underdamped second-order system and determine the characteristics of its step response (with $A = 1$). Firstly, $G(s)$ can be written as

$$G(s) = \frac{1}{1^2 \cdot s^2 + 2 \cdot 0.2 \cdot s + 1}, \qquad (8.44)$$

therefore, the static gain is $K = 1$, the time constant is $\tau = 1$ and the damping factor is $\zeta = 0.2$. Since $0 < \zeta < 1$, this is an underdamped second-order system.

The damped frequency of the oscillation is $\omega_d = 0.980$, the corresponding period is $T_d = 6.413$ and $\phi_0 = 1.369\,\text{rad} = 78.46°$. The overshoot is

$$\text{Overshoot} = e^{-\frac{\pi\zeta}{\sqrt{1-\zeta^2}}} = e^{-\frac{0.2\pi}{\sqrt{1-0.2^2}}} = 0.527 = 52.5\%. \qquad (8.45)$$

The decay ratio is

$$\text{DR} = \text{Overshoot} = 0.527^2 = 0.277. \qquad (8.46)$$

The peak time is $t_{\text{peak}} = \frac{\pi}{\omega_d} = 3.206$, the rise time is $t_{\text{rise}} = \frac{\pi - \phi_0}{\omega_d} = 1.809$ and the settling time is $t_{s,0.05} = 15.08$. Lastly, the response is given by

$$x(t) = K\left[1 - \frac{1}{\sqrt{1-\zeta^2}}e^{-\frac{\zeta}{\tau}t}\sin(\omega_d t + \phi_0)\right]$$

$$= 1 - 1.02e^{-0.2t}\sin(0.98t + 1.369). \tag{8.47}$$

The system response is shown in the figure below.

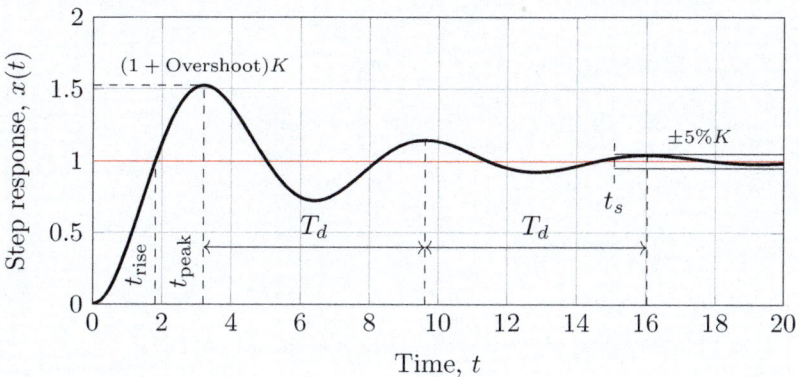

Figure 8.10: *Characteristics of step response of underdamped second-order system*

Example 8.2 (Robotic arm) The elbow angle, $x(t)$, of a robotic arm is manipulated with an electric signal of voltage $u(t)$.

The system dynamics is described by an underdamped second-order model with unknown parameters.

We know that an increase of 10 V to the input voltage leads to a change of the angle by 0.12 rad after infinite time. We also know that the overshoot is 30% and the peak time is 0.43 s. We need to determine the differential equation that describes the system dynamics.

Firstly, we will determine the parameters of the system, that is K, ζ and τ.

We know that $\lim_{t\to\infty} x(t) = AK$, so

$$K = \frac{\lim_{t\to\infty} x(t)}{A} = \frac{0.12}{10} = 0.012. \tag{8.48}$$

By Proposition 8.4

$$\text{Overshoot} = e^{-\frac{\pi\zeta}{\sqrt{1-\zeta^2}}}$$

$$\Rightarrow \ln(\text{Overshoot}) = -\frac{\pi\zeta}{\sqrt{1-\zeta^2}}$$

$$\Rightarrow \ln(\text{Overshoot})\sqrt{1-\zeta^2} = -\pi\zeta.$$

Using the fact that $-\ln(x) = \ln(x^{-1})$ for all $x > 0$,

$$\Rightarrow \ln(\text{Overshoot}^{-1})\sqrt{1-\zeta^2} = \pi\zeta,$$

and squaring both sides,

$$\Rightarrow \ln(\text{Overshoot}^{-1})^2 - \ln(\text{Overshoot}^{-1})^2\zeta^2 = \pi^2\zeta^2$$

$$\Rightarrow \ln(\text{Overshoot}^{-1})^2 = \left[\pi^2 + \ln(\text{Overshoot}^{-1})^2\right]\zeta^2,$$

so, solving for ζ we have

$$\Rightarrow \zeta = \frac{\ln(\text{Overshoot}^{-1})}{\sqrt{\pi^2 + \ln(\text{Overshoot}^{-1})^2}} = \frac{\ln(1/0.3)}{\sqrt{\pi^2 + \ln(1/0.3)^2}} \approx 0.362. \tag{8.49}$$

By Proposition 8.3 and Equation (8.23a),

$$t_{\text{peak}} = \frac{\pi}{\omega_d} = \frac{\pi}{\frac{\sqrt{1-\zeta^2}}{\tau}}$$

$$\Rightarrow \tau = \frac{t_{\text{peak}}\sqrt{1-\zeta^2}}{\pi} = \frac{0.43\sqrt{1-0.362^2}}{\pi} = 0.127\,\text{s}. \tag{8.50}$$

The transfer function of the system is

$$G(s) = \frac{X(s)}{U(s)} = \frac{K}{\tau^2 s^2 + 2\zeta\tau s + 1} = \frac{0.012}{0.016 s^2 + 0.0919 s + 1}. \tag{8.51}$$

In order to find the differential equation of the system we replace s with the derivative operator D in the transfer function and change X and U to x and u respectively, that is

$$\frac{x(t)}{u(t)} = \frac{0.012}{0.016 D^2 + 0.0919 D + 1}$$

$$\Rightarrow 0.016 \cdot D^2 x(t) + 0.0919 \cdot D x(t) + x(t) = 0.012 u(t)$$

$$\Rightarrow 0.016 \cdot \ddot{x}(t) + 0.0919 \cdot \dot{x}(t) + x(t) = 0.012 u(t). \tag{8.52}$$

Example 8.3 (First-order systems in series) In this example we will show that when two first-order systems are connected in series, then the resulting system is of second order, but cannot be underdamped.

Suppose that the two first-order systems have transfer functions

$$G_1(s) = \frac{K_1}{\tau_1 s + 1}, \tag{8.53a}$$

$$G_2(s) = \frac{K_2}{\tau_2 s + 1}. \tag{8.53b}$$

Since the two systems are connected in series, the overall transfer function is

$$G(s) = G_1(s)G_2(s) = \frac{K_1 K_2}{(\tau_1 s + 1)(\tau_2 s + 1)}. \tag{8.54}$$

Clearly, the poles of G are $-1/\tau_1$ and $-1/\tau_2$, both real numbers, therefore, G cannot be underdamped. We will now write G in the standard form of a second order system

$$G(s) = \frac{K_1 K_2}{(\tau_1 s + 1)(\tau_2 s + 1)} = \frac{K_1 K_s}{\tau_1 \tau_2 s^2 + (\tau_1 + \tau_2)s + 1}, \tag{8.55}$$

therefore, the static gain is $K = K_1 K_2$, the time constant is

$$\tau^2 = \tau_1 \tau_2 \Rightarrow \tau = \sqrt{\tau_1 \tau_2}, \tag{8.56}$$

and the damping factor is

$$2\zeta\tau = \tau_1 + \tau_2 \Rightarrow \zeta = \frac{\tau_1 + \tau_2}{2\sqrt{\tau_1 \tau_2}}. \tag{8.57}$$

According to the arithmetic-geometric mean inequality (AGM), $\frac{\tau_1+\tau_2}{2} \geq \sqrt{\tau_1 \tau_2}$, so, $\zeta \geq 1$. The value $\zeta = 1$ corresponds to the case $\tau_1 = \tau_2$.

8.5 Frequency response

In order to determine the frequency response of a second-order system we work as in Section 7.4. Our objective is to determine the response of a second-order system to input signals of the form $u(t) = A\sin(\omega t)$. The corresponding Laplace transform is

$$U(s) = \frac{A\omega}{s^2 + \omega^2}. \tag{8.58}$$

The system output, in the s-domain, will be

$$X(s) = \frac{AK\omega}{(s^2+\omega^2)(\tau^2 s^2 + 2\zeta\tau s + 1)}. \tag{8.59}$$

We may then use a partial fraction expansion and apply the inverse Laplace transform to determine the frequency response of the system. In the underdamped and the overdamped case[5]

$$X(s) = \frac{A_1}{s-s_1} + \frac{A_2}{s-s_2} + \frac{B}{s-j\omega} + \frac{\bar{B}}{s+j\omega}, \tag{8.60}$$

where s_1 and s_2 are the roots of the quadratic equation $\tau^2 s^2 + 2\zeta\tau s + 1 = 0$ and $\pm j\omega$ are the roots of $s^2 + \omega^2 = 0$. Without determining the coefficients A_1, A_2 and B we can tell that the inverse Laplace transforms of the first two terms will be

$$A_1 e^{s_1 t} + A_2 e^{s_2 t}. \tag{8.61}$$

Since the real parts of s_1 and s_2 are negative (see Section 8.1), the two exponential terms will converge to zero (exponentially fast). Therefore, asymptotically as $t \to \infty$ (i.e., at $t \gg 1$), the contribution of $A_1 e^{s_1 t} + A_2 e^{s_2 t}$ will be negligible, and the dominant part will be the inverse Laplace for the third and fourth terms. In other words, the frequency response can be approximated by

$$x_\infty(t) = Be^{j\omega t} + \bar{B}e^{-j\omega t}, \tag{8.62}$$

asymptotically as $t \to \infty$. As in the case of first-order systems, this is a sinusoidal function, and it can be shown that it is described by $x_\infty(t) = AR_\infty \sin(\omega t + \phi_0)$, where $R_\infty = |G(j\omega)|$ and $\phi_0 = \arg G(j\omega)$, that is, Theorem 7.3 holds for second-order systems too.

We state the following result without a proof because we will show in Chapter 11 that this property holds for more general and higher order systems.

> **Theorem 8.8 (Steady state frequency response)** The amplitude ratio and the phase lag of a second-order system are given by
>
> $$R_\infty = |G(j\omega)|, \tag{8.63a}$$
> $$\phi_0 = \arg G(j\omega), \tag{8.63b}$$
>
> and, at $t \gg 1$, the system output can be approximated by[a]
>
> $$x_\infty = AR_\infty \sin(\omega t + \phi_0). \tag{8.64}$$
>
> ---
> [a]in the sense that $|x(t) - x_\infty(t)| \to 0$ as $t \to \infty$.

---

[5]For the sake of simplicity we exclude the case of critically damped systems where $s_1 = s_2$. As discussed in Section 5.2.4, the cases of single real and complex poles can be treated the same way.

Chapter 8. Second-order systems

The amplitude ratio and the phase lag of a second-order system can be computed by expanding $G(j\omega)$ in Theorem 8.8.

Corollary 8.9 (Amplitude ratio and phase lag of second-order systems)
The amplitude ratio and phase lag of a second-order system are given by

$$R_\infty = \frac{K}{\sqrt{(1-\tau^2\omega^2)^2 + 4\zeta^2\tau^2\omega^2}}, \qquad (8.65a)$$

$$\phi_0 = \operatorname{atan2}\left(-2\zeta\tau\omega, 1 - \tau^2\omega^2\right). \qquad (8.65b)$$

Exercise 8.3 (Prove Corollary 8.9) Prove Corollary 8.9 using Theorem 8.8.

✂ ....

Hint. Use the following facts: (i) the modulus and argument of the complex number $z = a + jb$ are $|z| = \sqrt{a^2 + b^2}$ and $\arg z = \operatorname{atan2}(b, a)$, respectively; (ii) for $z_1, z_2 \in \mathbb{C}$ with $z_2 \neq 0$, $\left|\frac{z_1}{z_2}\right| = \frac{|z_1|}{|z_2|}$ and (iii) $\arg \frac{z_1}{z_2} = \arg z_1 - \arg z_2$.

In Chapter 11 we will provide a more detailed analysis of the frequency response of second-order (as well as higher-order) systems. For how, we can observe that at low frequencies ($\tau^2\omega^2 \ll 1$) the amplitude ratio is approximately

$$R_\infty \approx K, \text{ for } \tau^2\omega^2 \ll 1. \qquad (8.66)$$

At very high frequencies, that is, for $\tau^2\omega^2 \gg 1$, we have that $1 - \tau^2\omega^2 \approx \tau^2\omega^2$, therefore,

$$R_\infty = \frac{K}{\sqrt{(1-\tau^2\omega^2)^2 + 4\zeta^2\tau^2\omega^2}}$$

$$\approx \frac{K}{\sqrt{(\tau^2\omega^2)^2 + 4\zeta^2\tau^2\omega^2}}$$

since $\tau^2\omega^2 \gg 1$ and $0 < \zeta < 1$, it is $(\tau^2\omega^2)^2 \gg 4\zeta^2\tau^2\omega^2$, so

$$\approx \frac{K}{\sqrt{(\tau^2\omega^2)^2}} = \frac{K}{\tau^2\omega^2}, \qquad (8.67)$$

Then, by rearranging the terms and applying the logarithm on both sides we obtain

$$\log \frac{R_\infty}{K} = \log \frac{1}{(\omega\tau)^2} = -2\log(\tau\omega). \qquad (8.68)$$

This means that the high-frequency asymptote of a second order system drops faster than that of a first-order system (compare to Equation (7.56)). The following figure shows the dependence of R_∞ on ω.

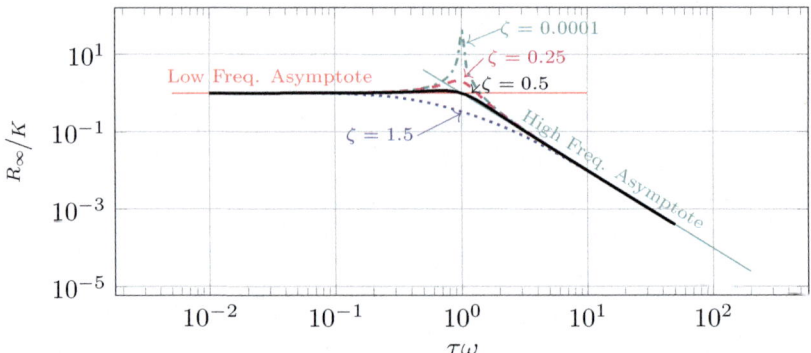

Figure 8.11: *Amplitude ratio of second-order system for different values of ζ.*

8.6 One-minute round-up

In this chapter

1. We studied the dynamic properties of second-order systems which have a transfer function of the form $G(s) = \frac{K}{\tau^2 s^2 + 2\zeta\tau s + 1}$, with $K, \zeta, \tau > 0$; K is called the *static gain*, ζ is the *damping factor* and τ is the *time constant* of the system

2. We considered three cases: (i) underdamped systems ($\zeta \in (0,1)$), (ii) overdamped systems ($\zeta > 1$) and (iii) critically damped systems ($\zeta = 1$)

3. Underdamped systems have two complex conjugate (non-real) poles with a negative real part; overdamped systems have two distinct negative real poles; critically damped systems have a double negative real pole

4. We derived the impulse, step and frequency responses of second-order systems. Same as for first-order systems, the amplitude ratio and the phase lag are given by $R_\infty = |G(j\omega)|$ and $\phi_0 = \arg G(j\omega)$

5. Our main focus was on the step response of underdamped systems, which exhibit oscillatory behaviour. We defined five characteristic quantities to describe such oscillatory behaviour: (i) the peak time, (ii) overshoot, (iii) rise time, (iv) decay ratio and (v) settling time.

Chapter 8. Second-order systems

8.7 Exercises

Exercise 8.4 (Multiple choice) Two second-order systems have the same time constant, τ. The first system has a damping factor $\zeta_1 = 0.5$ and the second one has $\zeta_2 = 0.25$. Both systems are underdamped.

1. In regard to the overshoot:

    (a) The overshoot of the first system will be half of the overshoot of the second one

    (b) The overshoot of the second system will be larger than that of the first one

    (c) We cannot draw a conclusion because we do not know the static gains of the two systems

2. In regard to the decay ratio:

    (a) The decay ratio of the first system is larger than that of the second one

    (b) The decay ratio of the second system is larger than that of the first one

    (c) We cannot know that because we have no information about the static gains of the two systems

3. Which system has a larger rise time?

    (a) The first one

    (b) The second one

    (c) They are equal

Choose one correct answer in each question.

❋ ......

Answer: 1-b, 2-b, 3-a.

Exercise 8.5 (Choose the correct answer) Which one is correct?

1. The overshoot of an underdamped second-order system can take any positive value

2. If the step response of a dynamical system exhibits an oscillatory behaviour, then this is an underdamped second-order system

3. In the step response of an underdamped second-order system, the time from one peak to the next one is equal to $2\pi/\omega_d$

4. The upper and lower envelopes, S^+ and S^-, touch the response of an underdamped second-order system at its peaks

Exercise 8.6 (True or false?) Choose the right answer

	True	False
The system with $G(s) = \frac{1}{(s+1)^2+1}$ is underdamped [Note: you do not need to expand the denominator]	☐	☐
We call overshoot the ratio between the peak value of the step response of an underdamped second-order system over its equilibrium value	☐	☐
A first-order system has a single negative real pole	☐	☐
A second-order system in the sense of Definition 8.1 has two poles with a negative real part	☐	☐
An underdamped second-order system has a pair of complex conjugate (not real) poles	☐	☐
The system with transfer function $G(s) = \frac{1}{(s+1)^2}$ is a critically damped second-order system	☐	☐
A spring-mass-damper system is an overdamped second-order system	☐	☐
A system is of second order in the sense of Definition 8.1 if and only if G has two poles	☐	☐
The transfer function of a second order system is the product of the transfer functions of two first-order systems	☐	☐
The step response of an underdamped second-order system meets its upper envelope at t_peak (see Figures 8.5 and 8.9)	☐	☐

✳ ....

Answer: T, T, T, T, T, T, F (it is a second-order system, but not necessarily overdamped), F (for example $G(s) = \frac{1-s}{1}$ is not a second-order system in the sense of Definition 8.1), F (this is not the case for underdamped systems), F (the upper envelope meets the step response at a time "close" to t_peak, but not exactly there).

Chapter 8. Second-order systems

Exercise 8.7 (Plots) Write a Python or MATLAB script to plot the overshoot and the decay ration against $\zeta \in (0,1)$. Likewise, plot the rise time against ζ for different values of τ.

Exercise 8.8 (Select the correct answers) Given a second-order system, suppose that we keep K and $\zeta \in (0,1)$ constant and double the value of τ. Then:

- ☐ i. The overshoot will increase
- ☐ ii. The overshoot will decrease
- ☐ iii. The peak time will double
- ☐ iv. The decay ratio will not be affected
- ☐ v. The rise time will double
- ☐ vi. The settling time will decrease
- ☐ vii. The system will remain underdamped

Tick all correct answers.

※ ....

Answer. The correct ones are: iii, iv, v, and vii.

Exercise 8.9 (Understanding of second-order systems) A second-order system has static gain $K = 1$ and two complex conjugate poles $a \pm jb$, with $a \leq 0$.

1. Under what conditions is the system underdamped, critically damped and overdamped?

2. Assuming that a and b are such that the system is underdamped, what is its damped frequency of oscillation? What is the damping factor and the time constant of this system?

3. Two first-order systems have transfer functions $G_1(s) = \frac{1}{(s+1)^2+2}$ and $G_2(s) = \frac{1}{(s+1)^2+3}$; which system has a larger overshoot?

4. What is the damped frequency of oscillation of a second-order system with transfer function $G(s) = \frac{1}{(s+2)^2+100}$?

Exercise 8.10 (System of interconnected tanks) In Example 3.13 we derived a linearised dynamical model for a system of interconnected tanks. Determine the transfer function $G(s)$ that associates the input variable F_{in} with the output variable h_2. Is this a second-order system? What are the values of τ, ζ and K? Under what conditions is the system underdamped/critically damped/overdamped?

✂ ....

Hint. Apply the Laplace transform to both sides of Equation (3.74) (assuming zero initial conditions) and eliminate $H_1(s) = \mathscr{L}\{h_1\}(s)$ to determine the transfer function of the system, $G(s) = H_2(s)/F_{\text{in}}(s)$.

Exercise 8.11 (Third-order systems) Show that any higher-order system with a transfer function of the form

$$\frac{K}{Q(s)}, \qquad (8.69)$$

where K is a positive scalar and Q is a polynomial of degree $m \geq 3$, can be written as the product of first and second-order transfer functions. Verify that this is true for the third-order system with transfer function

$$G(s) = \frac{1}{s^3 + 2s^2 + 2s + 1}. \qquad (8.70)$$

Then determine the impulse response and the step response of G.

✂ ....

Hint. Employ the fundamental theorem of algebra (Theorem A.7 and the complex conjugate root theorem (Theorem A.8). For the system of Equation (8.70) we can observe that -1 is a pole of G, so using Horner's method we can factorise the denominator as $(s+1)(s^2 + s + 1)$.

Exercise 8.12 (Programming exercise) In Section 8.3.1 we gave the step response of an overdamped system ($\zeta > 1$). Let us define the **settling time** for an overdamped system as the unique time $t_{s,0.05}$ so that

$$x(t_{s,0.05}) = 0.95AK. \qquad (8.71)$$

Write a Python of MATLAB function that computes the settling time given the parameters ζ, τ. Plot $t_{s,0.05}$ against ζ for $\tau = 1$. Plot $t_{s,0.05}$ against τ for $\zeta = 2$.

Chapter 8. Second-order systems

Exercise 8.13 (Identification of parameters) An underdamped second-order linear dynamical system has the step response shown in Figure 8.12. Determine the static gain, time constant and damping factor of the system.

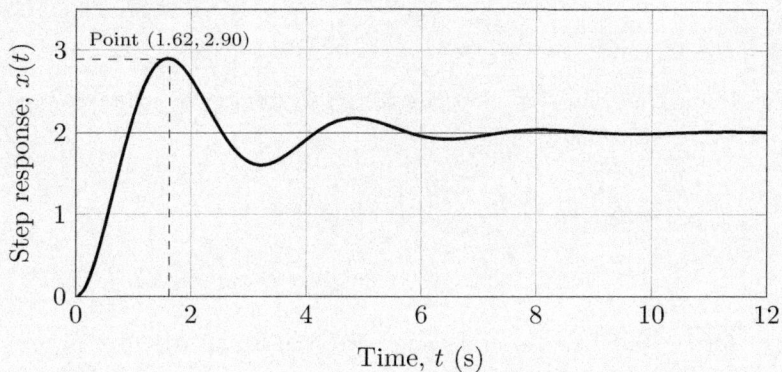

Figure 8.12: Step response of second-order system with unknown parameters

✳ ....

Answer. The values of the parameters are $K = 2$, $\tau = 0.5$, $\zeta = 0.25$.

Exercise 8.14 (Impulse response of second-order system with a zero) Consider a dynamical system with transfer function

$$G(s) = \frac{K(1+\alpha s)}{\tau^2 s^2 + 2\zeta\tau s + 1}, \qquad (8.72)$$

where K, α, τ are positive constants and $\zeta \in (0,1)$. Determine the impulse response of this system by using the results of Section 8.2; note that you do not need to determine any inverse Laplace transforms.

✳ ....

Answer. Let ω_d be as in Equation (8.23a) and let $\phi = \operatorname{atan2}(\alpha\omega_d, 1 - \alpha\frac{1}{\tau})$. Then, the impulse response is

$$x(t) = \frac{\sqrt{\left(1 - \frac{\alpha}{\tau}\right)^2 + \alpha^2\omega_d^2}}{\tau\sqrt{1-\zeta^2}} e^{-\frac{t}{\tau}} \sin(\omega_d t + \phi), \qquad (8.73)$$

for $t \geq 0$.

✂ ....

Hint. Use Theorem 4.19 and Equation (A.37).

Exercise 8.15 (Step response of second-order system with a zero) Consider a dynamical system with transfer function

$$G(s) = \frac{K(1+\alpha s)}{\tau^2 s^2 + 2\zeta\tau s + 1}, \qquad (8.74)$$

where K, α, τ are positive constants and $\zeta \in (0,1)$.

1. Show that the step response of this system (when the input is $u(t) = H_0(t)$) is

$$x(t) = K\left[1 - \frac{1}{\sqrt{1-\zeta^2}} e^{-\frac{\zeta}{\tau}t}\left(\sigma \sin(\omega_d t + \phi_0)\right.\right.$$
$$\left.\left. + \alpha\omega_d \cos(\omega_d t + \phi_0)\right)\right], \quad (8.75)$$

for $t \geq 0$, where ω_d and ϕ_0 are given in Equations (8.23a) and (8.23b) respectively and $\sigma = 1 - \frac{\alpha\zeta}{\tau}$.

2. Use (A.37) to determine constants c and ϕ_* such that the step response given in Equation (8.75) is written is the following form

$$x(t) = K\left[1 - c e^{-\frac{\zeta}{\tau}t} \sin(\omega_d t + \phi_*)\right], \qquad (8.76)$$

※ ┈┈┈┈┈┈┈┈┈┈┈┈┈┈┈┈┈┈┈┈┈┈┈┈┈┈┈┈┈┈┈┈

Answer: 2. it is $c = \sqrt{A^2 + B^2}$, where $A = \frac{\sigma}{\sqrt{1-\zeta^2}}$ and $B = \frac{\alpha\omega_d}{\sqrt{1-\zeta^2}}$, $\phi_* = \phi_0 + \theta_0$, where $\theta_0 = \mathrm{atan2}(B, A)$.

Exercise 8.16 (Characteristics of the step response of G in Exercise 8.14) Consider a dynamical system with transfer function

$$G(s) = \frac{K(1+\alpha s)}{\tau^2 s^2 + 2\zeta\tau s + 1}, \qquad (8.77)$$

where K, α, τ are positive constants and $\zeta \in (0,1)$. According to Equation (8.76), the step response of this system is of the same form as that of a standard second-order system. Determine

1. the peak time,

2. the overshoot, and

3. the rise time

Lastly, write a short program Python or MATLAB to compute the above quantities and plot the step response of this system.

Chapter 8. Second-order systems

Answer. The peak time is $t_{\text{peak}} = \frac{\pi}{\omega - \theta_0}$. The overshoot is

$$\text{Overshoot} = c\sqrt{1-\zeta^2} \exp\left(-\frac{\zeta(\pi-\theta_0)}{\sqrt{1-\zeta^2}}\right). \tag{8.78}$$

The rise time is given by $t_{\text{rise}} = \frac{\pi - \phi^*}{\omega_d}$.

※

9

BIBO stability

*In this chapter we ask what is the long-term effect that an input can have on the system output and introduce the concept of bounded-input bounded-output (BIBO) **stability**. BIBO stability is a concept of great importance in control theory and a typical minimum requirement is that a control is BIBO-stable. We first give the definition of BIBO stability, which is straightforward. Next, we give some useful criteria that allow us to check whether a system is BIBO-stable by using its transfer function.*

9.1 The concept of BIBO stability

Before we can state the definition of bounded-input bounded-output (BIBO) stability, we need to introduce the concept of a bounded signal. Roughly speaking, a signal is bounded if it remains within certain bounds at all times and its magnitude never grows to infinity.

> **Definition 9.1 (Bounded signal)** Let $x : [0, \infty) \to \mathbb{R}$ be a signal (a function in the time domain). We say that x is **bounded** if there is an $M \geq 0$ such that
> $$|x(t)| \leq M, \qquad (9.1)$$
> for all $t \geq 0$. A signal which is not bounded is called **unbounded**.

For example, the signals $x_1(t) = t$, $x_2(t) = e^t$ and $x_3(t) = t\sin(t)$ are not bounded — they are unbounded. The signals $x_4(t) = 100$, $x_5(t) = e^{-t}$ and $x_6(t) = \sin(t)$ are bounded.

Next, we give the definition of an absolutely integrable function.

> **Definition 9.2 (Absolutely integrable)** A signal $x : [0, \infty) \to \mathbb{R}$ is called **absolutely integrable** if
> $$\int_0^\infty |x(\tau)|\, d\tau < \infty. \qquad (9.2)$$

In this chapter we will be dealing with signals $x(t)$ which do not have *vertical asymptotes*, that is, we assume that all signals x are such that there is no $t_0 \geq 0$ such that $\lim_{t \to t_0^-} |x(t)| = \infty$ or $\lim_{t \to t_0^+} |x(t)| = \infty$. Note that the integral in the definition of absolute integrability is an improper integral of the first kind (see Section A.6.3).

The signals $x_1(t) = 0$, $x_2(t) = e^{-x}$, $x_3(t) = (t+1)^{-2}$ are absolutely integrable, whereas, $x_4(t) = 1$ are $x_5(t) = \sin t$ are not absolutely integrable. Boundedness is a necessary condition for absolute integrability; if a signal is unbounded, it will not be absolutely integrable. Moreover, for a signal to be absolutely bounded, it is necessary to converge to zero. However, convergence to zero is not sufficient for absolute integrability. The signal $x(t) = \frac{1}{t+1}$, for $t \geq 0$, is bounded and converges to zero, but it is not absolutely integrable. Roughly speaking, an absolutely integrable signal converges to zero *adequately fast*.

We are now ready to introduce the very important notion of bounded-input bounded-output (BIBO) stability. We say that a dynamical system is BIBO-stable if a bounded input signal will lead to bounded output. This means that the only way for the output of the system to grow unbounded is to put enough effort to drive the input to infinity. Otherwise, if the input is kept within certain bounds (it is a *bounded signal*), then the output will also remain bounded.

> **Definition 9.3 (BIBO stability)** A dynamical system with input u and output x is called **BIBO-stable** if its output is bounded whenever its input is bounded[a].
>
> ---
> [a]We implicitly consider only those input signals which are "sufficiently regular," in the sense that a response, $x(t)$, exists.

For linear dynamical systems it suffices to show that the response is bounded for all input signals, u, with $|u(t)| \leq 1$ for all $t \geq 0$, and assuming zero initial conditions (why?). Let us give the definition of an *unstable* dynamical system.

> **Definition 9.4 (BIBO instability)** A system that is not BIBO-stable is called **unstable** in the BIBO sense or **BIBO-unstable** for short[a].
>
> ---
> [a]For linear systems, we may state the definition more rigorously as follows: a linear system with zero initial conditions is BIBO-unstable if for all $M_x \geq 0$ there is an input signal with $|u(t)| \leq 1$, for $t \geq 0$ such that there is a time $t_1 \geq 0$

Chapter 9. BIBO stability

such that $|x(t_1)| \geq M_x$. Note that u is allowed to depend on M_x.

Note that the definition of BIBO stability postulates that the output remains bounded *for every* bounded input. Therefore, if we have observed that one particular output, $x(t)$, for a particular input, $u(t)$, is bounded, this **does not mean** that the system is necessarily BIBO-stable. The said property should hold for every bounded input. In other words, *every* bounded input should produce a bounded output.

A BIBO **unstable** system is one which produces an unbounded output for a bounded input. Here is such an example.

Example 9.1 (How not to check whether a system is stable) Consider a system with transfer function

$$G(s) = \frac{1}{s^2 + 1}. \qquad (9.3)$$

The Heaviside function, $u(t) = H_0(t)$, is a bounded function. Indeed, $|u(t)| = |H_0(t)| \leq 1$ for all $t \geq 1$. The Laplace transform of u is $U(s) = 1/s$ and the corresponding output is

$$X(s) = \frac{1}{s(s^2 + 1)} \Rightarrow x(t) = 1 - \cos(t), \qquad (9.4)$$

which is a bounded function. Indeed,

$$|x(t)| = |1 - \cos(t)| \leq 2,$$

for all $t \geq 0$. However, we should not jump into the conclusion that this system is BIBO-stable. If we try a different bounded input, namely $u(t) = 2\sin t$, with $U(s) = \frac{2}{s^2+1}$, the output becomes

$$X(s) = \frac{2}{(s^2 + 1)^2} \Rightarrow x(t) = \sin t - t\cos t,$$

which is not bounded because $\lim_{t \to \infty} |x(t)| = \infty$. Therefore, the system is not BIBO-stable.

Naturally, there arises a question: How can we tell whether a system is BIBO-stable? We cannot test against all possible bounded input functions (there are infinitely many!). In order to answer this question we need a convenient representation of the system output in terms of its input.

Given the transfer function, G, of a system, its output is given by

$$X(s) = G(s)U(s), \qquad (9.5)$$

where U is the Laplace transform of its input. In order to obtain the system output in the time domain, we apply the inverse Laplace transform

$$x(t) = \mathscr{L}^{-1}\{G(s)U(s)\}(t). \tag{9.6}$$

We will now leverage on the convolution property of the Laplace transform. According to Theorem 5.9,

$$x(t) = \mathscr{L}^{-1}\{G(s)\} * \mathscr{L}^{-1}\{U(s)\} = \mathscr{L}^{-1}\{G(s)\} * u(t). \tag{9.7}$$

Now recall that $\mathscr{L}^{-1}\{G(s)\}$ is the *impulse response* of the system (see Section 6.2). Let us denote this by

$$g(t) := \mathscr{L}^{-1}\{G(s)\}(t). \tag{9.8}$$

Therefore,

$$x(t) = g(t) * u(t), \tag{9.9}$$

where $g(t) = \mathscr{L}^{-1}\{G(s)\}(t)$. The following theorem allows us to determine whether a system is BIBO stable by using only its impulse response.

> **Theorem 9.5 (BIBO stability #1)** A linear dynamical system with transfer function G is BIBO-stable if and only if its impulse response, g, is absolutely integrable.

Proof. ($\Leftarrow$). We will first show that if g is absolutely integrable, then the system is BIBO-stable. Suppose that u is some bounded input with $|u(t)| \leq M_u$ for all $t \geq 0$ for some constant $M_u \geq 0$. By virtue of Proposition 6.2, the corresponding output is $x(t) = g(t) * u(t)$ and

$$\begin{aligned}|x(t)| = |(g*u)(t)| &= \left|\int_0^\infty g(\tau)u(t-\tau)\mathrm{d}\tau\right| \\ &\leq \int_0^\infty |g(\tau)u(t-\tau)|\,\mathrm{d}\tau \\ &\leq \int_0^\infty |g(\tau)| \cdot |u(t-\tau)|\mathrm{d}\tau \\ &\leq M_u \int_0^\infty |g(\tau)|\mathrm{d}\tau,\end{aligned}$$

and since g is absolutely integrable, there is an M_g so that $\int_0^\infty |g(\tau)|\mathrm{d}\tau \leq M_g$

$$\leq M_u M_g, \tag{9.10}$$

so x is bounded.

($\Rightarrow$). This part is a bit more technical. We will show that if g is not absolutely integrable, then x is unbounded for some bounded input. If g

Chapter 9. BIBO stability

is not absolutely integrable, for every $M > 0$ there is a $T = T(M) > 0$ so that

$$\int_0^T |g(\tau)|d\tau > M. \tag{9.11}$$

We now choose the input u defined as

$$u(T-t) = \text{sign}[g(t)] = \begin{cases} 1, & \text{if } g(t) \geq 0, \\ -1, & \text{if } g(t) < 0 \end{cases} \tag{9.12}$$

defined for $0 \leq t \leq T$. This is a bounded input signal with $|u(t)| \leq 1$. The output at time $t = T(M)$ is

$$\begin{aligned} x(T) &= \int_0^T g(\tau)u(T-\tau)d\tau \\ &= \int_0^T g(\tau)\,\text{sign}[g(\tau)]d\tau \\ &= \int_0^T |g(\tau)|d\tau > M, \end{aligned} \tag{9.13}$$

implying the system is unstable in the BIBO sense according to Definition 9.4. □

Exercise 9.1 (BIBO stability via absolute integrability) Use Theorem 9.5 to show that

1. $G_1(s) = 1/s$ is BIBO-unstable
2. $G_2(s) = 1/{s+1}$ is BIBO-stable
3. $G_3(s) = 1/{s-1}$ is BIBO-unstable
4. $G_4(s) = 1/{s^2+1}$ is BIBO-unstable

✳ ....

Answer: (i) the impulse response of G_1 is $x_{\text{imp}}(t) = 1$, which is not absolutely integrable, therefore, G_1 is BIBO-unstable, (ii) $\mathscr{L}^{-1}\{G_2(s)\}(t) = e^{-t}$, $t \geq 0$ and $\int_0^\infty |e^{-t}|d\tau = 1$, therefore, G_2 is BIBO-stable, (iii) $\mathscr{L}^{-1}\{G_3(s)\}(t) = e^t$, $t \geq 0$, which is clearly not integrable, $\int_0^\infty |e^t|d\tau \to \infty$ as $t \to \infty$ (iv) $\mathscr{L}^{-1}\{G_4(s)\}(t) = \sin t$, $t \geq 0$ and $\int_0^\infty |\sin t|d\tau$ diverges, therefore G_4 is BIBO-unstable.

We have shown that a system is BIBO-stable if and only if the integral $\int_0^\infty |g(\tau)|d\tau$ converges. But, if it converges, **what does its value tell us about the system?** It turns out that the value $M := \int_0^\infty |g(\tau)|d\tau$ quantifies the effect that the magnitude of u can have on the system output.

Proposition 9.6 (Bound on system response) Suppose that g is the (absolutely integrable) impulse response of a BIBO-stable linear dynamical system. Define $M := \int_0^\infty |g(\tau)| d\tau$ and[a]

$$\bar{u} := \sup_{t \geq 0} |u(t)|, \qquad (9.14)$$

Let $x(t)$ be the system response when the system input is $u(t)$. Then, for all $t \geq 0$,

$$|x(t)| \leq M\bar{u}. \qquad (9.15)$$

---
[a] We denote the *supremum* by sup. See Section A.9 in Appendix A for a detailed discussion.

Proof. This is similar to the proof of Theorem 9.5. It is

$$\begin{aligned}|x(t)| &= \left| \int_0^t g(\tau) u(t-\tau) d\tau \right| \\ &\leq \int_0^t |g(\tau)| |u(t-\tau)| d\tau \\ &\leq \bar{u} \int_0^t |g(\tau)| d\tau \\ &\leq \bar{u} \int_0^\infty |g(\tau)| d\tau \\ &\leq M\bar{u}, \end{aligned} \qquad (9.16)$$

which completes the proof. □

Exercise 9.2 (Application of Proposition 9.6) The impulse of a linear dynamical system is

$$g(t) = e^{-(t-1)} \sin(10(t-1)) H_1(t), \qquad (9.17)$$

for $t \geq 0$. The system is actuated with some piecewise continuous input $u(t)$ with $|u(t)| \leq 1$ for all $t \geq 0$. Determine an upper bound for $|x(t)|$ over $t \geq 0$.

✂ ....

Hint. Firstly, note that the question is well posed since the Laplace transform of u exists (why?). Next, you need to determine the value of M in Proposition 9.6.

Example 9.2 (First-order systems are stable) First-order systems are BIBO-stable. Indeed, the impulse response of a first-order system is

(see Section 7.2)

$$g(t) = \frac{K}{\tau} e^{-\frac{t}{\tau}}. \tag{9.18}$$

This is an absolutely integrable function. Indeed,[a]

$$\begin{aligned}
\int_0^\infty |g(\xi)| d\xi &= \int_0^\infty \left| \frac{K}{\tau} e^{-\frac{\xi}{\tau}} \right| d\xi \\
&= \frac{K}{\tau} \int_0^\infty e^{-\frac{\xi}{\tau}} d\xi \\
&= -K e^{-\frac{\xi}{\tau}} \Big|_0^\infty \\
&= K < \infty.
\end{aligned} \tag{9.19}$$

The impulse response of a first-order system is integrable, therefore, a bounded input signal will produce a bounded output signal. Moreover, according to Theorem 9.6, if the system is excited with an input signal u with $\bar{u} := \sup_{t \geq 0} |u(t)|$, then the output x will satisfy $|x(t)| \leq K\bar{u}$.

---

[a] We need to be careful with the notation here: τ is the time constant of the system, not the integration variable. Instead, we denote the integration variable by ξ.

Example 9.3 (Second-order systems are stable) A second-order system is BIBO-stable. We will show this for overdamped systems. The impulse response is

$$g(t) = \frac{1}{\tau\sqrt{\zeta^2 - 1}} e^{-\frac{\zeta}{\tau} t} \sinh\left(\frac{\sqrt{\zeta^2 - 1}}{\tau} t \right). \tag{9.20}$$

This is absolutely integrable[a]:

$$\begin{aligned}
\int_0^\infty |g(\xi)| d\xi &= \int_0^\infty \left| \frac{1}{\tau\sqrt{\zeta^2 - 1}} e^{-\frac{\zeta}{\tau} \xi} \sinh\left(\frac{\sqrt{\zeta^2 - 1}}{\tau} \xi \right) \right| d\xi \\
&= \frac{1}{\tau\sqrt{\zeta^2 - 1}} \int_0^\infty e^{-\frac{\zeta}{\tau} \xi} \sinh\left(\frac{\sqrt{\zeta^2 - 1}}{\tau} \xi \right) d\xi \\
&= \frac{1}{\tau\sqrt{\zeta^2 - 1}} \int_0^\infty \tfrac{1}{2} e^{-\frac{\zeta}{\tau} \xi} \left(e^{\frac{\sqrt{\zeta^2 - 1}}{\tau} \xi} - e^{-\frac{\sqrt{\zeta^2 - 1}}{\tau} \xi} \right) d\xi \\
&= \frac{1}{2\tau\sqrt{\zeta^2 - 1}} \int_0^\infty e^{\frac{\sqrt{\zeta^2 - 1} - \zeta}{\tau} \xi} - e^{-\frac{\sqrt{\zeta^2 - 1} + \zeta}{\tau} \xi} d\xi.
\end{aligned} \tag{9.21}$$

We do not need to carry out the full computation of the integral. We are only interested in determining whether it converges. Both exponentials have negative exponents (e.g., in the first exponential, the exponent $\sqrt{\zeta^2 - 1} - \zeta$ is negative for all $\zeta > 1^b$). Therefore, the integral converges and an overdamped second-order system is BIBO-stable.

---
[a] Again, we denote the integration variable by ξ. We cannot use τ because this denotes the time constant of the system.
[b] This is because $\sqrt{\zeta^2 - 1} < \sqrt{\zeta^2}$

Example 9.4 (Systems with imaginary poles are unstable) In this example we apply Theorem 9.5 to show that a system with transfer function

$$G(s) = \frac{1}{s^2 + 1}, \qquad (9.22)$$

is BIBO-unstable. Its impulse response is

$$g(t) = \sin t, \qquad (9.23)$$

and we see that $|\sin t|$ does not converge to zero, therefore the integral $\int_0^\infty |\sin \xi| \mathrm{d}\xi$ necessarily diverges[a], so the system is BIBO-unstable.

---
[a] The converse is not true. If a function $|f(t)|$ converges to zero, this does not imply that its integral, $\int_0^\infty |f(\xi)| \mathrm{d}\xi$ converges.

We have now a tool to check whether a dynamical system is BIBO-stable. Nevertheless, this comes with two drawbacks. Firstly, we need to compute the impulse response of the system, which might (will) be difficult for systems with high-order transfer functions. Secondly, we have to compute the integral of the impulse response, which can be rather tedious a task.

9.2 BIBO stability criterion based on the poles of G

Let us start by giving the following definition

> **Definition 9.7 (Characteristic polynomial/equation & poles)** Consider a rational transfer function of the form
>
> $$G(s) = \frac{P(s)}{Q(s)}. \qquad (9.24)$$
>
> The polynomial Q is called the **characteristic polynomial** of the transfer function and the equation
>
> $$Q(s) = 0, \qquad (9.25)$$
>
> is called the **characteristic equation**. The roots of the characteristic equation are called the **poles** of G.

Chapter 9. BIBO stability

> **Definition 9.8 ((Strictly) proper transfer function)** If the degree of the numerator of the transfer function of a system is less than or equal to the degree of its denominator ($\deg P \leq \deg Q$) we say that the transfer function is **proper**. If $\deg P < \deg Q$ we say that the transfer function is **strictly proper**.

Let us have a look at some simple transfer functions. Let us start with functions with a **simple real pole**,

$$G(s) = \frac{K}{s-a}. \qquad (9.26)$$

The impulse response is $g(t) = Ke^{at}$. This is absolutely integrable if and only if $a < 0$.

If G has a **double real pole**, that is

$$G(s) = \frac{K}{(s-a)^2}, \qquad (9.27)$$

then, its impulse response is $g(t) = Kte^{at}$. Likewise, this is absolutely integrable if and only if $a < 0$.

If G has a **pair of imaginary poles**, that is,

$$G(s) = \frac{K}{s^2 + b^2}, \qquad (9.28)$$

then, the inverse Laplace transform is $g(t) = K/b \sin(bt)$, which is not absolutely integrable, so it is BIBO-unstable for all $b \in \mathbb{R}$.

If G has a **pair of complex conjugate poles**,

$$G(s) = \frac{K}{(s-a)^2 + b^2}, \qquad (9.29)$$

then, its impulse response is $g(t) = K/b \, e^{at} \sin(bt)$, which is absolutely integrable for $a < 0$.

If G has **double complex conjugate poles**,

$$G(s) = \frac{K}{[(s-a)^2 + b^2]^2}, \qquad (9.30)$$

then, its impulse response is $g(t) = K/2b^3 (\sin(bt) - bt \cos(bt))e^{at}$ (see Example 5.4). We can verify that this is absolutely integrable for $a < 0$.

In all these examples, the system was found to be BIBO-stable if the poles of the transfer function have a negative real part. This is in fact true for systems with general rational transfer functions as stated in the following theorem.

9.2. BIBO stability criterion based on the poles of G

Theorem 9.9 (BIBO stability #2) Suppose a dynamical system is described by a proper rational transfer function $G(s)$. The system is BIBO-stable if and only if all poles of G have a negative real part.

Proof. ($\Rightarrow$). Suppose that the system is BIBO-stable (see Definition 9.3). We will show that if $s \in \mathbb{C}$ has a positive real part, it cannot be a pole of G. By Theorem 9.5, the impulse response of G, function $g(t) = \mathcal{L}^{-1}\{G(s)\}$, is absolutely integrable, therefore there is an $M > 0$ such that

$$M > \int_0^\infty |g(\tau)| d\tau,$$

so for $s \in \mathbb{C}$ with $\mathbf{re}(s) > 0$,

$$> \int_0^\infty |g(\tau)| e^{-\mathbf{re}(s)\tau} d\tau,$$

since $0 < e^{-\mathbf{re}(s)\tau} < 1$ for all $\tau > 0$,

$$> \int_0^\infty |g(\tau)||e^{-s\tau}| d\tau > \left|\int_0^\infty g(\tau)e^{-s\tau} d\tau\right| = |G(s)|. \tag{9.31}$$

This means that $G(s)$ is uniformly bounded over $\{s \in \mathbb{C} : \mathbf{re}(s) > 0\}$. This makes it impossible for s to be a pole of G.

($\Leftarrow$). Suppose that all poles of G have a negative real part; we will use Theorem 9.5 to show that g is absolutely integrable. In Section 5.2.6 we saw that the inverse Laplace transform of G has the general form

$$g(t) = A_0 \delta(t) + \sum_{i=1}^n \sum_{k=1}^{\kappa_i} A_{i,k} t^{k-1} e^{s_i t}, \tag{9.32}$$

where $A_{i,k} \in \mathbb{C}$ are constant complex coefficients, $A_0 \in \mathbb{R}$, and $s_i \in \mathbb{C}$ are the poles of G (the roots of Q). Let $s_i = \sigma_i + jb_i$ (σ_i is the real part of s_i). If $\sigma_i < 0$ for all i, $g(t)$ is absolutely integrable because

$$|g(t)| = \left|A_0 \delta(t) + \sum_{i=1}^n \sum_{k=1}^{\kappa_i} A_{i,k} t^{k-1} e^{s_i t}\right|$$

$$\leq |A_0|\delta(t) + \sum_{i=1}^n \sum_{k=1}^{\kappa_i} |A_{i,k}||t^{k-1}||e^{s_i t}|$$

$$= |A_0|\delta(t) + \sum_{i=1}^n \sum_{k=1}^{\kappa_i} |A_{i,k}||t^{k-1}|e^{\sigma_i t}, \tag{9.33}$$

which is absolutely integrable (recall that $\int_0^\infty |A_0|\delta(\tau) d\tau = |A_0|$). $\square$

Chapter 9. BIBO stability

Example 9.5 (BIBO stability based on poles of transfer function) By Theorem 9.9,

- $G(s) = \frac{s-2}{(s+1)(s+2)(s+3)}$ is BIBO-stable because the poles of G are negative

- $G(s) = \frac{5}{s(s+1)(s+5)}$ is BIBO-unstable because it has a pole at the origin ($s=0$)

- $G(s) = \frac{1}{s^2+2s+1}$ is BIBO-stable because this is a second-order system with $K=1$, $\tau=1$, $\zeta=1$; it has a double negative real pole because $s^2+2s+1 = (s+1)^2$, so it is BIBO-stable

- $G(s) = \frac{1}{(s+1)(s-2)}$ is BIBO-unstable because it has a positive pole (at $s=2$)

- $G(s) = \frac{s^2+5s-12}{(s+1)(s+7)(s^2+5)}$ is BIBO-unstable because it has a pair of conjugate imaginary poles ($\pm j\sqrt{5}$), which have a zero real part; note that the zeros of the numerator play no role

- $G(s) = \frac{1}{(s+1)^2+3}$ is BIBO-stable because it has two complex conjugate poles, $s_1 = -1+j\sqrt{3}$ and $s_2 = -1-j\sqrt{3}$, which have a negative real part

- $G(s) = \frac{1}{[(s+1)^2+3]^4}$ is BIBO-stable because it has a pair of complex conjugate poles of multiplicity 4 and negative real part

- $G(s) = \frac{1}{s^5+3s^4+2s^3+7s^2+10s}$ is unstable; although it is difficult to determine all poles of G, clearly $s=0$ is a pole.

It seems that to use Theorem 9.9 we need to be able to determine all poles of the transfer function. This is possible for polynomials of up to second order and, in certain cases, for higher-order polynomials by using Horner's scheme. In the general case, however, the task of determining all poles of G is not easy. Nevertheless, the exact values of the poles are not important. We only need to know whether the real part of all poles is negative.

It can be verified that if (i) not all coefficients Q have the same sign or (ii) one or more coefficients of Q are zero, then the system is BIBO-unstable. We state this in the following theorem.

> **Theorem 9.10 (BIBO instability)** Let $Q(s) = a_0 + a_1 s + \ldots + a_n s^n$ be a polynomial of degree n (that is, $a_n \neq 0$). If at least one coefficient is equal to zero and/or not all coefficients share the same sign, then Q has at least one root with nonnegative real part.

Proof. Without loss of generality, we shall assume that $a_n = 1$, that is $Q(s) = a_0 + a_1 s + \ldots + a_{n-1} s^{n-1} + s^n$. Such polynomials are called *monic*.

We will now show that if a monic polynomial has roots with negative real part, then its coefficients must be positive. Suppose that Q has $2k$ complex conjugate roots $s_1, \bar{s}_1, s_2, \bar{s}_2, \ldots, s_k, \bar{s}_k$ and $n - 2k$ real (negative) roots $-s_{2k+1}, \ldots, -s_n$. Suppose that $s_1 = -a_1 + jb_1$, $s_2 = -a_2 + jb_2$, $\ldots$, $s_k = -a_k + jb_k$ with $a_1, \ldots, a_k < 0$. We can then factorise Q as follows

$$Q(s) = \prod_{j=1}^{k}[(s+a_j)^2 + b_j^2] \cdot \prod_{l=2k+1}^{n}(s+s_l)$$
$$= \prod_{j=1}^{k}[s^2 + 2a_j s + a_j^2 + b_j^2] \cdot \prod_{l=2k+1}^{n}(s+s_l). \qquad (9.34)$$

Since Q is the product of polynomials with positive coefficients, it must have positive coefficients. □

Exercise 9.3 (Stable or unstable?) Are the following transfer functions stable or unstable?

	Stable	Unstable
$G(s) = \dfrac{1}{s-1}$	☐	☐
$G(s) = \dfrac{1}{s+1}$	☐	☐
$G(s) = \dfrac{2s+3}{s^2+1}$	☐	☐
$G(s) = \dfrac{2020}{s^2-1}$	☐	☐
$G(s) = \dfrac{s^2+3s+5}{(s+1)(s+2)^2(s+3)^3}$	☐	☐
$G(s) = \dfrac{s-10}{(s+1)^2+1}$	☐	☐

※ ....

Answer: U, S, U, U, S, S

Example 9.6 (Quick instability checks) Without determining the poles of $G(s) = P(s)/Q(s)$ we can tell that

- If $Q(s) = x^6 + 3x^5 + 2x^4 + x^2 + 1$, the system is unstable because the coefficient of x^3 vanishes

- If $Q(s) = x^4 + 2x^3 - 8x^2 + x + 5$, the system is unstable because not all coefficients have the same sign.

The instability conditions provided in Theorem 9.10 are sufficient, but not necessary. A polynomial Q with positive coefficients may have roots with nonnegative real part. Let us give an example.

Example 9.7 (Counterexample for the converse of Theorem 9.10) Consider the following polynomial

$$Q(s) = (s+1)^3((s-1)^2 + 3), \qquad (9.35)$$

which has a triple stable root at -1 and the pair of complex conjugate roots $1 \pm j\sqrt{3}$. If we expand Q we obtain

$$Q(s) = s^5 + s^4 + s^3 + 7s^2 + 10s + 4, \qquad (9.36)$$

where all coefficients have the same sign and no coefficients are equal to zero.

If a delay is added to our transfer function, its stability properties are not altered, i.e., if $G(s)$ is BIBO-stable, then and only then $G(s)e^{-t_d s}$ is also stable. Indeed, for any input $U(s)$ the output of the system without delay will be

$$x_{\text{no delay}}(t) = \mathscr{L}^{-1}\left\{U(s)G(s)\right\}(t), \qquad (9.37)$$

and the output of the system with delay will be

$$x_{\text{delay}}(t) = \mathscr{L}^{-1}\left\{U(s)G(s)e^{-t_d s}\right\}(t) = H_{t_d}(t)x_{\text{no delay}}(t - t_d), \qquad (9.38)$$

because of Proposition 5.6. Clearly $x_{\text{delay}}(t)$ is bounded if and only if $x_{\text{no delay}}(t)$ is bounded. We can state this as a corollary of the above stability criterion (Theorem 9.9).

Corollary 9.11 (BIBO stability of systems with delays) A system with transfer function

$$G_2(s) = G_1(s)e^{-t_d s}, \qquad (9.39)$$

for some $t_d \geq 0$, is BIBO-stable if and only if G_1 is BIBO stable.

Proof. Follows from Theorem 9.9. □

9.3 Routh's method

9.3.1 Routh's stability criterion

Routh's tabulation is a method that can serve as a test for BIBO stability. It enables us to tell whether a given polynomial has any roots with nonnegative real part and, if so, gives us the number of such unstable roots.

Consider a polynomial of degree n of the form

$$Q(s) = a_0 + a_1 s + a_2 s^2 + \ldots + a_n s^n. \tag{9.40}$$

In order to construct Routh's tabulation, we organise the coefficients of Q in the first two rows of a table as shown below:

s^n	a_n	a_{n-2}	a_{n-4}	$\ldots$
s^{n-1}	a_{n-1}	a_{n-3}	a_{n-5}	$\ldots$
s^{n-2}	$a_{3,1}$	$a_{3,2}$	$a_{3,3}$	$\ldots$
s^{n-3}	$a_{4,1}$	$a_{4,2}$	$a_{4,3}$	$\ldots$
$\vdots$	$\vdots$	$\vdots$	$\vdots$	$\vdots$
s^0	$a_{n+1,1}$			

If any of the coefficients are missing we simply write 0. We then proceed by computing the coefficients of the third row of the table. These are

$$a_{3,1} = \frac{a_{n-1} a_{n-2} - a_n a_{n-3}}{a_{n-1}}, \tag{9.41a}$$

$$a_{3,2} = \frac{a_{n-1} a_{n-4} - a_n a_{n-5}}{a_{n-1}}, \tag{9.41b}$$

and similarly we determine $a_{3,3}$, $a_{3,4}$ and so on. Next, we compute the entries of the fourth column in a similar fashion

$$a_{4,1} = \frac{a_{3,1} a_{n-3} - a_{n-1} a_{3,2}}{a_{3,1}}, \tag{9.42a}$$

$$a_{4,2} = \frac{a_{3,1} a_{n-5} - a_{n-1} a_{3,3}}{a_{3,1}}, \tag{9.42b}$$

and we continue till we determine $a_{n+1,1}$. Note that in order to compute the coefficients of Routh's tabulation we must assume that the entries of the first column, from a_{n-1} to $a_{n,1}$ are nonzero. We will cover the case of zero coefficients in the next section. We give Routh's stability criterion below without a proof.

Chapter 9. BIBO stability

> **Theorem 9.12 (Routh's stability criterion)** Let Q be a polynomial of degree n given by (9.40). Suppose there are no zeroes in the first column of Routh's tabulation. The number of roots with positive real part is equal to the number of sign changes in the first column.

Proof. The proof can be found in [C24] and [C23]. □

IMPORTANT NOTE: In Routh's stability criterion we assume that there are no zeros in the first column. This version of Routh's stability criterion covers cases where there are no roots on the imaginary axis (meaning there are no purely imaginary roots and no zero roots). If the given polynomial has a pair of imaginary roots, then (not *only* then) there will be a zero in the first column. Such cases are covered in the following sections (see Sections 9.3.2 and 9.3.3) where we introduce a variant of this criterion.

Example 9.8 (Routh's tabulation) We will use Routh's criterion to determine whether the polynomial

$$Q(s) = 1 + 2x + 5x^2 + 3x^3 + 4x^4, \tag{9.43}$$

has all its roots in the open left half-plane (i.e., whether they all have a negative real part). We have $a_0 = 1$, $a_1 = 2$, $a_2 = 5$, $a_3 = 3$ and $a_4 = 4$. We introduce the coefficients of Q into Routh's tabulation as shown below

s^4	4	5	1	
s^3	3	2	0	← *write* 0 *in lieu of* a_{-1}
s^2	$a_{3,1}$	$a_{3,2}$		
s^1	$a_{4,1}$	$a_{4,2}$		
s^0	$a_{5,1}$			

We then find that $a_{3,1} = \frac{3 \cdot 5 - 4 \cdot 2}{3} = 7/3$ and $a_{3,2} = \frac{3 \cdot 1 - 4 \cdot 0}{3} = 1$, so Routh's tabulation becomes

s^4	4	5	1
s^3	3	2	0
s^2	7/3	1	0
s^1	$a_{4,1}$	$a_{4,2}$	
s^0	$a_{5,1}$		

and $a_{4,1} = \frac{7/3 \cdot 2 - 3 \cdot 1}{7/3} = 5/7$ and $a_{4,2} = 0$, so the table becomes

s^4	4	5	1
s^3	3	2	0
s^2	7/3	1	0
s^1	5/7	0	
s^0	$a_{5,1}$		

Lastly, $a_{5,1} = \frac{5/7 \cdot 1 - 7/3 \cdot 0}{5/7} = 1$, so the complete table is

s^4	4	5	1
s^3	3	2	0
s^2	2.33	1	0
s^1	0.71	0	
s^0	1		

Since all entries of the first column are positive, all roots of Q have a negative real part. Indeed, we can use MATLAB's roots or Python's numpy.roots to find that the roots of Q are approximately equal to $-0.8609 \pm 1.6040j$ and $-0.1391 \pm 1.0898j$.

Example 9.9 (**Third-order polynomial**) Under what conditions on $a, b, c \in \mathbb{R}$ do all roots of the following polynomial have a negative real part?
$$Q(s) = s^3 + as^2 + bs + c. \tag{9.44}$$

We will use Routh's criterion. Routh's tabulation is

s^3	1	b
s^2	a	c
s^1	$b - c/a$	
s^0	c	

According to Routh's criterion, we require that $a > 0$, $b - c/a > 0$ and $c > 0$, that is
$$a > 0 \text{ and } 0 < c < ab. \tag{9.45}$$

Example 9.10 (**Range of K for BIBO stability**) A dynamical system with input $U(s)$ and output $X(s)$ is described by the following block diagram

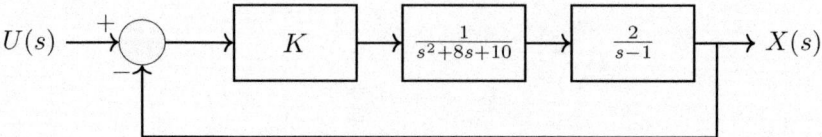

where $K \in \mathbb{R}$ is a parameter. We need to find the range of values of K so that the system is BIBO-stable. Firstly, we will construct the transfer function of the system. Following the methodology outlined in Section 6.4.1, this is

$$G(s) = \frac{K \frac{1}{s^2+8s+10} \frac{2}{s-1}}{1 + K \frac{1}{s^2+8s+10} \frac{2}{s-1}} = \frac{2K}{(s^2+8s+10)(s-1)+2K}$$

$$= \frac{2K}{s^3 + 7s^2 + 2s - 10 + 2K} \qquad (9.46)$$

We now construct Routh's tabulation for the denominator of G, that is for $Q(s) = s^3 + 7s^2 + 2s - 10 + 2K$

s^3	1	2
s^2	7	$2K - 10$
s^1	$\frac{14-(2K-10)}{7}$	
s^0	$2K - 10$	

In order for all entries in the first column of Routh's tabulation to be positive we require that

$$\frac{14 - (2K - 10)}{7} > 0 \Leftrightarrow K < 12, \qquad (9.47)$$

and

$$2K - 10 > 0 \Leftrightarrow K > 5, \qquad (9.48)$$

so K needs to be in the interval $(5, 12)$ so that Q has all its roots in the left half plane and thus the system is BIBO-stable.

Exercise 9.4 (**Unstable system for all** K) A dynamical system with input $U(s)$ and output $X(s)$ is described by the following block diagram

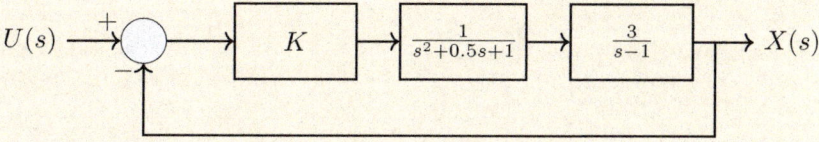

where $K \in \mathbb{R}$ is a parameter. Show that this system will be unstable for all values of K.

Answer. The denominator of the transfer function $G(s) = X(s)/U(s)$, that is, the "characteristic polynomial", is $Q(s) = s^3 - 0.5s^2 + 0.5s + 3K - 1$. The coefficients of s^3 and s^2 have opposite signs; therefore, G is always unstable; this can be easily confirmed with Routh's method too.

Exercise 9.5 (Routh's tabulation) For the polynomial

$$Q(s) = 2s^5 - 0.6s^4 - 35.2s^3 - 54s^2 - 24.4s - 3, \qquad (9.49)$$

fill in Routh's tabulation

s^5	2	-35.2	-24.4
s^4	-0.6	-54	-3
s^3	___	___	___
s^2	___	___	___
s^1	___	___	___
s^0	___	___	___

How many of the roots of Q have a positive real part?

Answer. Q has one root with a positive real part. Confirm that $Q(5) = 0$.

Exercise 9.6 (Routh's tabulation) Use Routh's criterion to determine the number of roots of the following polynomials which are in the right closed half plane (i.e., they have a positive real part):

i. $Q_1(s) = 5s^4 - 31s^3 + 61s^2 - 41s + 6$

ii. $Q_2(s) = 6s^4 + 29s^3 + 39s^2 + 19s + 3$

iii. $Q_3(s) = s^5 + s^4 + 2s^3 + s^2 + 3s + 1$

Answer. (i) All the roots of Q_1 have a positive real part, (ii) all roots of Q_2 are in the open left half-plane, (iii) Two of the roots of Q_3 have a positive real part.

9.3.2 A zero in the first column

If an entry in the first column of Routh's tabulation that corresponds to the rows s^n to s^1 is zero, we cannot proceed with the computation of the subsequent entries. Such a situation occurs, for example, for

$$Q(s) = s^4 + s^3 + s^2 + s + 1, \qquad (9.50)$$

where Routh's tabulation is

s^4	1	1	1
s^3	1	1	0
s^2	0	1	
s^1	?	?	
s^0	?		

In case an entry in the first column is zero, but there exist nonzero entries in the same row, we can replace 0 by a (small) positive parameter ϵ. Then, we proceed with filling in the subsequent values in Routh's tabulation. This way, the above tabulation becomes

s^4	1	1	1
s^3	1	1	0
s^2	ϵ	1	
s^1	$1 - 1/\epsilon$	0	
s^0	1		

In the first column we have two changes of sign, namely from $\epsilon > 0$ to $1 - 1/\epsilon < 0$ and again from $1 - 1/\epsilon < 0$ to 1, therefore Q has two roots with positive nonnegative part.

When such a situation occurs, that is, when a zero shows up in the first column of Routh's tabulation, there will be at least one root of Q in the closed right half plane. If we are constructing Routh's tabulation in order to check whether a given system is BIBO-stable, we can stop and conclude that it is unstable since, typically, we do not need to determine the number of unstable poles of the transfer function.

Example 9.11 (ϵ-relexation) We will apply Routh's tabulation to

$$Q(s) = s^4 + 2s^3 + 2s^2 + 4s + 1. \qquad (9.51)$$

s^4	1	2	1
s^3	2	4	0
s^2	0	1	← zero entry (not all zeroes)
s^1	?	?	
s^0	?		

We replace the zero entry with a small positive parameter $\epsilon > 0$; this allows us to proceed

s^4	1	2	1
s^3	2	4	0
s^2	ϵ	1	0
s^1	$4 - \frac{2}{\epsilon}$	0	
s^0	1		

And we observe that there are two changes of sign (because $\epsilon > 0$, $4 - \frac{2}{\epsilon} < 0$ and $1 > 0$), meaning that Q has two roots with nonnegative real part. Indeed, we can use MATLAB's roots or Python's numpy.roots to find that the roots of Q are approximately equal to -1.9070, $0.0933 \pm 1.3660j$ and -0.2797.

9.3.3 A zero row

In some special cases, a whole row of Routh's tabulation will have zero entries. This situation arises only when the polynomial has roots which are radially symmetric on the complex plane; that is, if the polynomial

i. has a pair of pure imaginary roots ($\pm jb$),

ii. has two real roots with opposite signs ($\pm b$),

iii. has two pairs of complex roots of the form $a \pm jb$ and $-a \pm jb$.

When it comes to testing whether a transfer function is stable, we see that a zero row implies instability of the associated system.

When a row of Routh's matrix is zero we cannot apply the ϵ-relaxation of the previous section. Instead, we need to use the row above the zero row which stores the coefficients of a polynomial we call the **auxiliary polynomial**, $A(s)$. The auxiliary polynomial is an *even polynomial*[1] whose degree is the degree of the corresponding row; for example, if it is in row s^2, its degree is equal to 2. The zero row is the replaced with the coefficients of the derivative of $A(s)$. The roots of the auxiliary polynomial are also roots of the original one.

This procedure is better demonstrated in the following example.

Example 9.12 (Case i) We want to determine the number of roots with nonnegative real part of the following polynomial

$$Q(s) = s^3 + 3s^2 + s + 3. \qquad (9.52)$$

We formulate Routh's tabulation:

---

[1] A polynomial is called even if its variable, s, appears only in even powers. For example, $A(s) = s^4 + 5s^2 + 1$ is an even polynomial.

Chapter 9. BIBO stability

s^3	1	1	
s^2	3	3	← *auxiliary polynomial*
s^1	0	0	← *zero row*
s^0	?		

Here we have a zero row and the ϵ-relaxation trick cannot be applied. The row above the zero contains the coefficients of the zero polynomial, $A(s)$. Since the auxiliary polynomial is in the row of s^2, its degree is equal to 2, and given that A is an even polynomial, it is

$$A(s) = 3s^2 + 3. \tag{9.53}$$

The roots of A are the pure imaginary numbers $\pm j$. These are **also roots of** Q.

The derivative of A is

$$A'(s) = 6s + 0. \tag{9.54}$$

We now replace the zero row with the coefficients of A'.

s^3	1	1	
s^2	3	3	
s^1	6	0	← *coefficients of A' (previously, zero row)*
s^0	3		

Although we already know that Q has a pair of imaginary roots, the fact that the above Routh tabulation has no sign flips in the first column means that there are no roots in the right half plane, therefore, it has roots on the imaginary axis. Indeed, we can use MATLAB's `roots` or Python's `numpy.roots` to find that the roots of Q are -3, j and $-j$.

Example 9.13 (Case ii) Consider the polynomial

$$Q(s) = s^5 + s^4 - 7s^3 - s^2 + 6s. \tag{9.55}$$

We observe that Q has a zero constant term, therefore, we can write

$$Q(s) = s(s^4 + s^3 - 7s^2 - s + 6), \tag{9.56}$$

so zero is a root of Q ($Q(0) = 0$).

We define the polynomial

$$R(s) = s^4 + s^3 - 7s^2 - s + 6, \tag{9.57}$$

and we formulate its Routh tabulation

s^4	1	-7	6	
s^3	1	-1	0	
s^2	-6	6	0	← auxiliary polynomial
s^1	0	0	0	← zero row
s^0	cannot proceed...			

The auxiliary polynomial is $A(s) = -6s^2 + 6$. The roots of A are $s_1 = -1$ and $s_2 = 1$. These are also roots of R and, of course, roots of Q.

Moreover, the derivative of A is $A'(s) = -12s + 0$. We replace the zero row with the coefficients of A':

s^4	1	-7	6	
s^3	1	-1	0	
s^2	-6	6	0	
s^1	-12	0	0	← coefficients of A'
s^0	6			

We see that we have two sign flips, so R has two roots on the right half plane.
Overall, Q has

1. A zero root, $s_1 = 0$,

2. Two *real* roots on the right half plane ($s_2, s_3 > 0$),

3. A real root, $s_4 = -s_2$, which is the reason for the zero row in Routh's tabulation.

Indeed, we can use MATLAB's `roots` or Python's `numpy.roots` to find that the roots of Q are $0, -3, -1, 1,$ and 2.

Example 9.14 (Case iii) Consider the polynomial

$$Q(s) = s^5 + 2s^4 + 4s + 8. \tag{9.58}$$

Routh's tabulation of Q is

s^5	1	0	4	
s^4	2	0	8	← auxiliary polynomial
s^3	0	0	0	← zero row

The degree of the auxiliary polynomial is 4 and is given by

$$A(s) = 2s^4 + 0s^2 + 8 = 2(s^4 + 4). \tag{9.59}$$

Recall that the roots of A are also roots of Q. In order to solve the equation $2(s^4 + 4) = 0$ we introduce the variable $y = s^2$, so the equation becomes

$$y^2 + 4 = 0, \tag{9.60}$$

from which we have the roots $y_1 = 2j$ and $y_2 = -2j$. These correspond to the roots $s_{1,2} = \sqrt{2j} = \pm(1+j)$ and $s_{3,4} = \pm(1-j)$. Clearly, two of these roots have a positive real part.

Let us now continue with Routh's tabulation to determine the sign of the remaining root of Q (apparently, Q has two pairs of complex conjugate roots and one real root).

The derivative of the auxiliary polynomial is

$$A'(s) = 8s^3. \tag{9.61}$$

We replace the zero row above with the coefficients of A'

s^5	1	0	4	
s^4	2	0	8	
s^3	8	0	0	← coefficients of A'
s^2	1	8	0	
s^1	−64	0		
s^0	1			

We observe that there are two sign flips in the first column, so Q has two roots with positive real part (which we knew already). It follows that Q has

1. the four roots $s_{1,2} = \sqrt{2j} = \pm(1+j)$ and $s_{3,4} = \pm(1-j)$ we mentioned above (the roots of A),

2. another (real) negative root (the reader can verify that -2 is a root of Q).

Exercise 9.7 (Confirm Routh's method) Use Routh's method to show that $Q(s) = s^4 + 5s^3 + 5s^2 - 5s - 6$ has one root with positive real part. Then confirm this by finding all roots of Q.

※ . . . .

> **Answer.** The roots of Q are $+1, -1, -2, -3$. The s^1-row of Routh's tabulation is a zero row and the corresponding auxiliary polynomial is $A(s) = 6s^2 - 6$. The roots of A, i.e., ∓ 1, are also roots of Q. The remaining roots of Q can be determined using Horner's method.

9.3.4 Poles with sufficiently negative real part

If the poles of a transfer function have a negative real part, then the system is MIMO stable. If, on the other hand, there is a pole on the imaginary axis or a pole in the right half plane, then the system is BIBO-unstable. Roughly speaking, if a pole is very close to the imaginary axis, but on its left-hand side the system will be BIBO-stable, but not "sufficiently stable". For example, if some parameters of the system have been misestimated, some of the actual poles may be unstable. Therefore, we are often interested in determining whether all poles of G are "sufficiently stable", e.g., $\mathbf{re}(s_i) < -c < 0$ for some positive constant c as shown below.

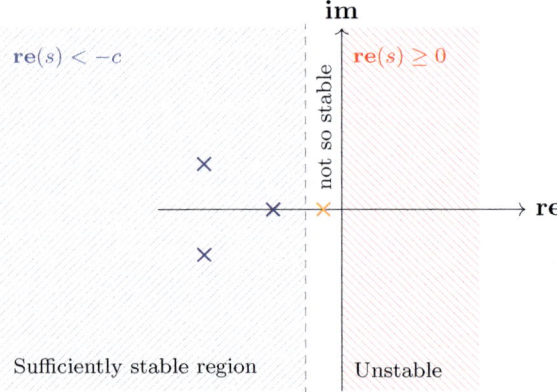

Figure 9.1: *Poles too close to the imaginary axis are not "sufficiently stable". The grey shaded are is a sufficiently stable region defined by $\{s \in \mathbb{C} : \mathbf{re}(s) < -c\}$.*

In such cases we can use Routh's criterion applied to $Q(s-c)$. This is stated in the following proposition. The proof is simple is based on the simple fact that s is in the sufficiently stable region shown in Figure 9.1 if and only if $s + c$ is in the open left half-plane of the complex plane.

> **Proposition 9.13 (Sufficiently negative real parts)** Let Q be a polynomial and $c > 0$. All roots of Q have a real part strictly less than $-c$ if and only if the roots of $\tilde{Q}(s) = Q(s-c)$ have a negative real part.

Proof. Let s^* be a root of Q. Define $\zeta = s^* + c$ and note that $\mathbf{re}(s) < -c$

if and only if $\mathbf{re}(\zeta) < 0$. Moreover, $0 = Q(s^*) = Q(s^* + c - c) = Q(\zeta - c) = \tilde{Q}(\zeta)$, so ζ is a root of $\tilde{Q}$. This completes the proof. □

Example 9.15 (**Application of Proposition 9.13**) We need to check whether all roots of $Q(s) = 5s^3 + 16s^2 + 13s + 2$ have a real part which is less than -0.1. According to Proposition 9.13, this holds if $\tilde{Q}(s) = Q(s - 0.1)$ has all its roots in the open left half-plane. We compute $\tilde{Q}(s)$:

$$\tilde{Q}(s) = 5(s - 0.1)^3 + 16(s - 0.1)^2 + 13(s - 0.1) + 2$$
$$= 5s^3 + 14.5s^2 + 9.95s + 0.855. \tag{9.62}$$

Routh's tabulation for $\tilde{Q}$ is

s^3	5	9.95
s^2	14.5	0.855
s^1	9.66	0
s^0	0.855	

Since all entries in the first column of Routh's tabulation are positive, all roots of $\tilde{Q}$ have a negative real part, so all roots of Q have a real part strictly less than -0.1. To verify this result we can use MATLAB's roots or Python's numpy.roots to find that the roots of Q are equal to -3, -1, and -0.2.

Example 9.16 (**Choice of K to have sufficiently stable poles**) We will determine the range of values of K so that the following polynomial

$$Q(s) = s^2 + 2Ks + 4 + K, \tag{9.63}$$

has roots with real part less than -1. We define $\tilde{Q}(s) = Q(s - 1)$. This is

$$\tilde{Q}(s) = Q(s - 1) = (s - 1)^2 + 2K(s - 1) + 4 + K$$
$$= s^2 + 2(K - 1)s - K + 5. \tag{9.64}$$

By Proposition 9.10, it is necessary, but not sufficient, that all coefficients of Q be positive, that is, it must be $1 < K < 5$. By using Routh's stability criterion, we have

s^2	1	$5 - K$
s^1	$2(K - 1)$	0
s^0	$5 - K$	

For $\tilde{Q}$ to have all its roots in the open left half-plane of the complex plane — what is the same, for Q to have all its roots with a real part no larger than -1 — all entries in the first column of the above tabulation must be positive, that is, $2(K-1) > 0$ and $5 - K > 0$. Equivalently, $1 < K < 5$.

9.3.5 Poles in desirable region

Even if all roots of Q are in the sufficiently stable region shown in Figure 9.1 does not imply that the system has desirable dynamic characteristics. For example, if the real part of the poles is sufficiently negative, but there are poles with a large imaginary part, the step response of the system will be oscillatory with a very large overshoot. Let us give a motivating example.

Example 9.17 (**Sufficiently stable, but bad dynamics**) Consider a dynamical system with transfer function

$$G(s) = \frac{10^6}{((s+0.7)^2 + 20^2)((s+0.6)^2 + 10^2)}, \quad (9.65)$$

which has the poles $-0.7 \pm 20j$ and $-0.6 \pm 10j$. We can determine the transfer function of the system in Python as follows

```
import control as ctrl
import matplotlib.pyplot as plt

# Convenient way to define TFs:
s = ctrl.TransferFunction.s
G = 1e6/((s+0.7)**2 + 400)/((s+0.6)**2 + 100)
t_step, y_step = ctrl.step_response(G)
plt.plot(t_step, y_step)
plt.show()
```

The step response of the system is shown below.

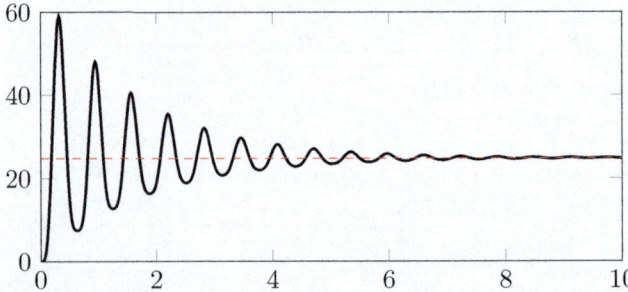

Figure 9.2: *Overly oscillatory response; the overshoot is* 137%.

We see that the response is wildly oscillatory.

In such cases we may want to allow for large imaginary parts of the poles of the transfer function, but provided that the corresponding real parts are appropriately negative. The reader can compare the step response that we obtained in Example 9.17 with that of

$$G(s) = G(s) = \frac{10^6}{((s+7)^2 + 20^2)((s+6)^2 + 10^2)}, \tag{9.66}$$

which has an overshoot of about 26%. One may want to impose (or check) the requirement that the poles of G are in the region shown below.

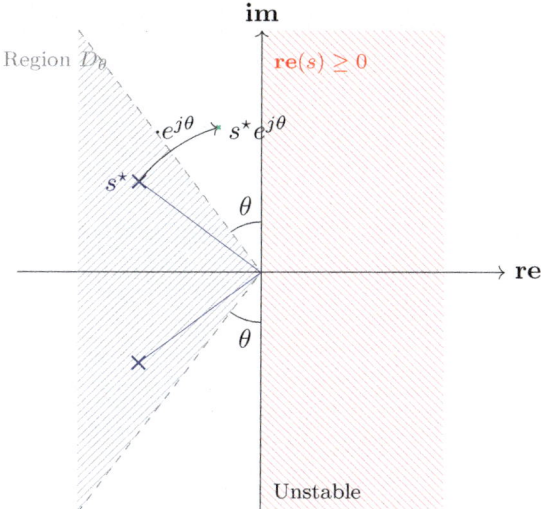

Figure 9.3: The desirable region is defined to be the grey shaded cone in the open left half-plane. This region is identified by the angle θ it forms with the imaginary axis. A point $s^\star$ is in the desirable region, D_θ, if $s^\star e^{j\theta}$ and $s^\star e^{-j\theta}$ are in the open left half-plane.

We define the desirable region, D_θ, as follows.

Definition 9.14 (Region D_θ) For $0 < \theta < \frac{\pi}{2}$, a point s is said to be in region D_θ if the real part of $se^{j\theta}$ and $se^{-j\theta}$ is negative.

We can then check whether all roots of a polynomial Q are in D_θ using the following proposition.

Proposition 9.15 (Conditions for the roots of Q to be in D_θ) Given a polynomial Q and $0 < \theta < \frac{\pi}{2}$ we define the polynomial

$$\widetilde{Q}(s) = Q(se^{j\theta})Q(se^{-j\theta}). \tag{9.67}$$

All roots of Q are in D_θ if and only if all roots of $\widetilde{Q}$ have a negative real part.

Proof. The proof is left to the reader as an exercise (Exercise 9.32). □

Note that both $Q(se^{j\theta})$ and $Q(se^{-j\theta})$ are polynomials with complex coefficients, but $\widetilde{Q}$ is a polynomial with real coefficients (why?), so we can use Routh's tabulation to check whether its roots have negative real parts. Moreover, note that the degree of $\widetilde{Q}$ is twice that of Q. Let us give an example.

Example 9.18 (Application of Proposition 9.15) Consider the polynomial $Q(s) = s^2 + s + 1$. We can easily determine that the two roots of Q are $-\frac{1}{2} \pm \frac{\sqrt{3}}{2}j$ which are shown in Figure 9.4.

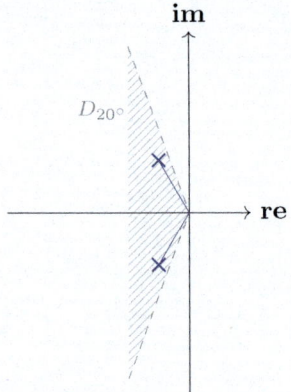

Figure 9.4: *Position of roots of Q and $D_{20°}$.*

Let us now use Proposition 9.15 to check whether the roots of Q are inside $D_{20°}$ (it is $20° = 0.3491\,\text{rad}$). We can use the fact that $e^{j0.3491} = \cos(0.3491) + j\sin(0.3491) = 0.9397 + j0.3420$ to determine

$$\widetilde{Q}(s) = s^4 + 1.732068s^3 + 2.00006s^2 + 1.732068s + 1. \qquad (9.68)$$

Routh's tabulation for $\widetilde{Q}$ is

s^4	1	2.00006	1
s^3	1.732068	1.732068	0
s^2	1.00006	1	
s^1	$1.04713 \cdot 10^{-4}$		
s^0	1		

Since all entries of the first column are positive, all roots of $\widetilde{Q}$ have a negative real part, therefore, all roots of Q are in $D_{20°}$.

Example 9.19 (Tuning) Suppose that the characteristic polynomial of a closed-loop system with a P-controller ($G_c(s) = K$) is

$$Q(s) = s^2 + s + K. \tag{9.69}$$

We ask under what conditions (on K) all roots of Q are in $D_{30°}$. We shall follow the exact same procedure as in the previous example: we will determine $\widetilde{Q}$ and construct its Routh's tabulation. To that end, we can use the following MATLAB script

```
syms s K
theta = pi/6;     % 30 degrees
Q = s^2 + s + K;
Q1 = subs(Q, s, s*exp(-theta*1i));
Q2 = subs(Q, s, s*exp(theta*1i));
F = simplify(Q1 * Q2)
R = simplify(myRouth(coeffs(F, s)))
```

Function myRouth is available at https://gist.github.com/alphaville/107a04c9f44af76a2170584208b93620 (developed by Ismail Ilker Delice — available on MATLAB Central File Exchange — minor modifications by Pantelis Sopasakis; follow the link for detailed information). This MATLAB script produces the following tabulation (we are showing only the first column):

s^4	K^2
s^3	$\sqrt{3}K$
s^2	1
s^1	$\sqrt{3}(1-K)$
s^0	1

Clearly, the required conditions are $0 < K < 1$.

Exercise 9.8 (Location of roots of Q) Write a MATLAB or Python script to plot the maximum angle θ so that the roots of Q in Equation (9.69) are in D_θ.

9.4 Stability of systems with time delays

A system with a transfer function which involves terms of the form $e^{-t_d s}$ corresponds to a system with time delays (see Section 5.2.8). In such cases, we may not be able to use Routh's stability criterion to test whether the system is BIBO-stable. One approach is to approximate the terms of the form $e^{-t_d s}$ by a rational function.

Similar to the concept of linearisation, a Padé approximation is the *best approximation* of a function by a rational function with a numerator and denominator of a certain order. The simplest Padé approximation is that of the form

$$e^{-t_d s} \approx \frac{1 - \frac{t_d}{2} s}{1 + \frac{t_d}{2} s}. \tag{9.70}$$

This is the so-called [1/1] Padé approximant. The error of the [1/1]-Padé approximant is shown in Figure 9.5. Note that the Padé approximant is quite accurate when $t_d s \in \mathbb{C}$ is close to 0.

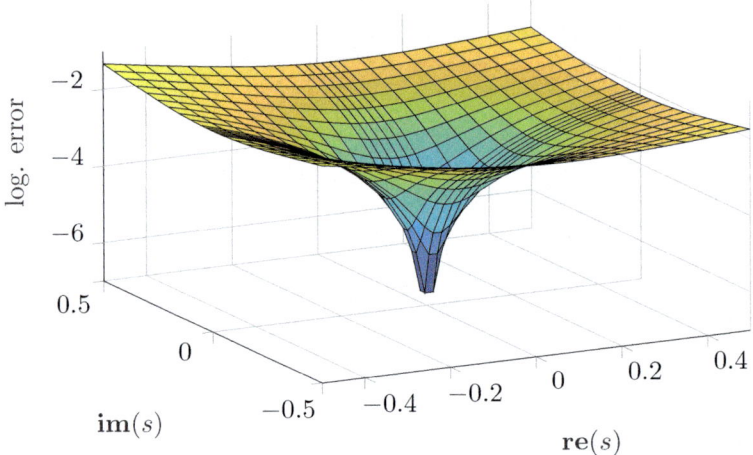

Figure 9.5: *Plot of* $\log_{10} \left| e^{-s} - \frac{1-s/2}{1+s/2} \right|$.

Example 9.20 (Padé approximation) The following block diagram shows a feedback control system which involves a controller with transfer function $G_c(s) = K$, a system with $G_p(s) = \frac{1}{s-1}$ and a sensor which returns a measurement after a delay of $t_d = 0.1$.

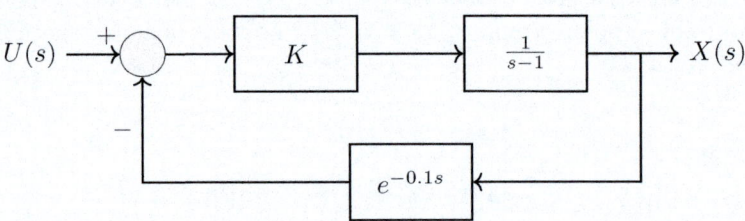

The transfer function of the system, from U to X, is given by

$$G(s) = \frac{K \frac{1}{s-1}}{1 + K \frac{1}{s-1} e^{-0.1s}} = \frac{K}{s - 1 + K e^{-0.1s}}$$

and by using the Padé approximation of order [1/1],

$$\approx \frac{K}{s - 1 + K\frac{1-0.05s}{1+0.05s}} = \frac{K(1 + 0.05s)}{(s - 1)(1 + 0.05s) + K(1 - 0.05s)}$$

$$= \frac{K(1 + 0.05s)}{0.05s^2 + (0.95 - 0.05K)s + K - 1}. \tag{9.71}$$

The denominator of the (approximate) transfer function is the polynomial $Q(s) = 0.05s^2 + (0.95 - 0.05K)s + K - 1$. The associated Routh's tabulation is

s^2	0.05	$K - 1$
s^1	$0.95 - 0.05K$	0
s^0	$K - 1$	

Therefore, in order for Q to have all its roots in the open left half-plane, the conditions $K - 1 > 0$ and $0.95 - 0.05K > 0$ must be satisfied, that is, $1 < K < 19$.

Remark: In Example 9.20 we used a [1/1]-Padé approximation of approximate $e^{-0.1s}$ by a rational function. This allowed us to approximate the closed-loop transfer function of the system by a rational function. **This was a bad idea!** The reason is that we determined conditions under which the approximate transfer function is stable, but the actual transfer function can be unstable (and vice versa). In Example 11.15 we give an example where the above approximation can lead to erroneous results. •

Higher-order Padé approximants, such as the [2/2]-Padé approximant can lead to a lower approximation error. The [2/2]-approximant is given by

$$e^{-t_d s} \approx \frac{(3 - t_d s)^2 + 3}{(3 + t_d s)^2 + 3}. \tag{9.72}$$

Exercise 9.9 (Approximation error of Padé approximants) Use Python or MATLAB to plot the logarithm of the modulus of the approximation error of the [2/2]-Padé approximation and compare your result to that the [1/1]-Padé approximation in Figure 9.5.

9.5 One-minute round-up

In this chapter

1. We introduced the notion of bounded-input bounded-output (BIBO) stability

2. We established an equivalence between BIBO stability and the absolute integrability of the impulse response of a system

3. We established a bound on the system response of a BIBO-stable system by using the value of $\int_0^\infty |g(\tau)|\mathrm{d}\tau$, where g is the impulse response of the system

4. The most important result in this chapter was perhaps the realisation that a system is BIBO-stable if and only if all of its poles have a negative real part

5. We stated and proved a quick test for instability: if any of the coefficients of the characteristic polynomial of a transfer change sign or vanish, then the system is BIBO-unstable

6. We presented Routh's tabulation and stability criterion which gives the number of poles with a nonnegative real part

7. We gave a trick that can be used to check whether all poles of a transfer function are "adequately" negative

8. We presented the Padé approximation of the exponential function, which can be used to enable us to apply Routh's stability criterion to functions that involve the exponential function and do not fall in the special case of Corollary 9.11.

Later we will present another two stability criteria, namely the Bode and Nyquist criteria, which do not need to use the Padé, or any other approximations, for exponentials and allow the designer to quantify how robust a BIBO-stable system is.

Chapter 9. BIBO stability

9.6 Exercises

Exercise 9.10 (Basic comprehension) True of false?

	True	False
A system is BIBO-stable if and only if its impulse response is absolutely integrable	☐	☐
The signal $x(t) = \sin(3t)$ is bounded	☐	☐
The signal $x(t) = t^2 - t$ is bounded	☐	☐
The signal $x(t) = t$ is unbounded	☐	☐
The function $f(t) = e^{-t}$, defined for $t \geq 0$ is absolutely integrable	☐	☐
A system is excited with the input $u(t) = \sin(t)$ and its output is $y(t) = e^t - \cos t - \sin t$; the system is BIBO-stable.	☐	☐
If the (rational) transfer function of a system has a pole at zero, the system is (BIBO) unstable	☐	☐
If the (rational) transfer function of a system has a pair of imaginary poles, the system is unstable in the BIBO sense	☐	☐
If the (rational) transfer function of a system has a double real pole, the system is BIBO-stable	☐	☐
A system with a rational transfer function is BIBO-stable if and only if all its poles have a negative real part	☐	☐
A system with non-real poles cannot be BIBO-stable	☐	☐
The transfer function of a system is $G(s) = P(s)/Q(s)$ and $Q(s) = s^4 + s + 1$; the system is unstable	☐	☐
The transfer function of a system is $G(s) = P(s)/Q(s)$ and $Q(s) = s^3 + 3s^2 - 5s + 10$; the system is unstable	☐	☐
All roots of a polynomial Q have a real part less than $-c$ if and only if the roots of $Q(s-c)$ have a negative real part	☐	☐

	True	False
The linear dynamical system $2\dot{x} + x = u$ is BIBO-stable	☐	☐
The linear dynamical system $\ddot{x}+\dot{x}+x = u$ is BIBO-stable	☐	☐
If G is a BIBO-stable transfer function, the $\frac{G}{1+G}$ is also stable	☐	☐
If G and H are stable transfer functions, then GH is also stable	☐	☐

✻ ....

Answer. T, T, F, T, F, T, T, T, F, T, F, F, T, T, T, F, F, T

Exercise 9.11 (BIBO stability) Determine whether the following system, with input U and output X, is BIBO-stable

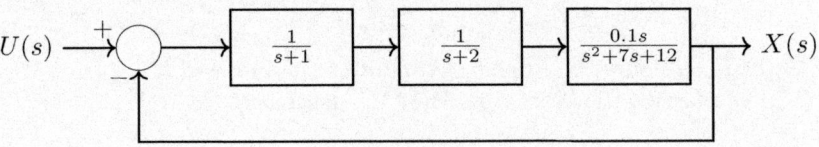

✂ ....

Hint. Determine the closed-loop transfer function of the system and apply Routh's stability criterion.

✻ ....

Answer. The closed-loop transfer function is

$$G_{cl}(s) = \frac{10s^4 + 100s^3 + 350s^2 + 501s + 240}{s}, \qquad (9.73)$$

and the elements of the first column of its Routh tabulation are (10, 100, 299.9, 420.97, 240), so G_{cl} is BIBO-stable.

Exercise 9.12 (Stability of ODE) A dynamical system is described by the ODE

$$x^{(3)} + 2\ddot{x} = u - 10\dot{x} - x. \qquad (9.74)$$

Is this system BIBO-stable?

✻ ....

Answer: Yes, this system is BIBO-stable.

Exercise 9.13 (Routh's tabulation) Complete Routh's tabulation for the polynomial:

$$Q(s) = 15s^5 + 38s^4 + 32s^3 + 40s^2 + 17s + 2. \qquad (9.75)$$

s^5 ___ ___ ___
s^4 ___ ___ ___
s^3 ___ ___ ___
s^2 ___ ___ ___
s^1 ___ ___ ___
s^0 ___ ___ ___

How many roots does it have with positive real part? How many imaginary roots does it have?

※ ....

Answer: Q has zero roots with positive real part and has one pair of imaginary roots. The reason is that the s^1-row turns out to be a zero row and the s^2-row gives the auxiliary polynomial $A(s) = 2s^2 + 2$, the roots of which (namely $\pm j$) are also roots of Q.

Exercise 9.14 (Routh's tabulation) Consider the following polynomial

$$Q(s) = s^5 + s^4 + 4s + 4. \qquad (9.76)$$

Complete the following tabulation:

s^5 ___ ___ ___
s^4 ___ ___ ___
s^3 ___ ___ ___
s^2 ___ ___ ___
s^1 ___ ___ ___
s^0 ___ ___ ___

What can you tell about the roots of Q? Can you determine its roots?

✳ ....

Answer. The s^3-row of this tabulation is zero and the auxiliary polynomial is $A(s) = s^4 + 4$. The four roots of A are also roots of Q; the roots of A are $1 \mp j$ and $-1 \mp j$ (to determine the roots of A let $s = \rho e^{j\theta}$). The elements of the first column of the tabulation are $(1, 1, 4, \epsilon, -16/\epsilon, 4)$. We see that there are two sign changes, so Q has two roots with a positive real part (we already know that these are $1 \mp j$). The remaining root of Q can be found by dividing Q by $s^4 + 4$; the result is $Q(s) = (s^4 + 4)(s + 1)$, so -1 is a root of Q.

Exercise 9.15 (Routh's tabulation) Complete the following tabulation:

s^7	1	-3	2	2
s^6	1	-3	2	2
s^5	___	___	___	
s^4	___	___	___	
s^3	___	___	___	
s^2	___	___		
s^1	___			
s^0	___			

What is the associated polynomial? How many roots does it have with positive real part? How many imaginary roots does it have?

✳ ....

Answer. The associated polynomial is $p(s) = s^7 + s^6 - 3s^5 - 3s^4 + 2s^3 + 2s^2 + 2$; it has a pair of imaginary roots and two roots with a positive real part. Note that the s^5-row turns out to be a zero row; the s^6-row defines the even polynomial $A(s) = s^6 - 3s^4 + 2s^2 + 2$ with $A'(s) = 6s^5 - 12s^3 + 4s$, so the entries of the s^5-row become 6, -12 and 4.

Exercise 9.16 (Conditions for second and third order polynomials) Under what conditions does the second-order polynomial $p(s) = s^2 + a_1 s + a_0$, with $a_0, a_1 \in \mathbb{R}$, have all of its roots on the left of the imaginary axis? What about the third-order polynomial $q(s) = s^3 + a_2 s^2 + a_1 s + a_0$, with $a_0, a_1, a_2 \in \mathbb{R}$?

✳ ....

Chapter 9. BIBO stability

Answer: A necessary and sufficient condition for p is $a_0, a_1 > 0$. The necessary and sufficient conditions for q are $a_0 > 0$, $a_2 > 0$ and $a_1 a_2 > a_0$.

Exercise 9.17 (BIBO stability) Show that the system with transfer function

$$G(s) = \frac{s^3 + s^2 - 5s - 1}{8s^5 + 38s^4 + 65s^3 + 50s^2 + 17s + 2}, \qquad (9.77)$$

is BIBO-stable. Additionally, show that the poles of G have a real part which is less that -0.2.

✳ ....

Answer. The characteristic polynomial of the transfer function is $Q(s) = 8s^5 + 38s^4 + 65s^3 + 50s^2 + 17s + 2$. The first column of its Routh tabulation is $(8, 38, 54.47, 38.43, 13.74, 2)$, so G is stable. Let $\tilde{Q}(s) = Q(s-0.2) = 8s^5 + 30s^4 + 37.8s^3 + 19.48s^2 + 3.648s + 0.13824$; the first column of its Routh tabulation is $(8, 30, 32.6, 16.16, 3.33, 0.138)$, so the poles of G have a real part which is less that -0.2.

Exercise 9.18 (Largest real part) Show that all roots of

$$Q(s) = s^3 + \tfrac{7}{2}s^2 + \tfrac{7}{2}s + 1, \qquad (9.78)$$

have a real part that is no larger than -0.5.

✂ ....

Hint: Use Proposition 9.13. You do not have to determine all roots of Q, but if you want to, you can use Horner's method.

Exercise 9.19 (Unstable circuit) Can you construct a circuit that involves only resistors, capacitors and inductors such that its transfer function — with the input being the voltage across the circuit and the output being the current that runs through it — is unstable in the BIBO sense?

✳ ....

Answer. Derive the transfer function of a series LC circuit. In practice, the inevitable presence of nonzero resistances makes it impossible to have unstable circuits.

Exercise 9.20 (BIBO stability) Find the range of values of K so that the following system is BIBO-stable

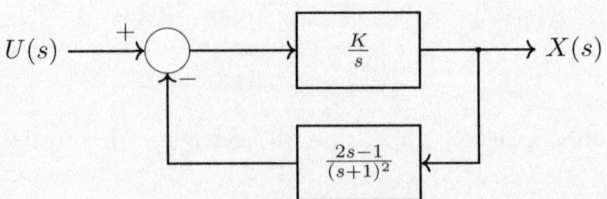

Answer. The closed-loop transfer function is $G(s) = K \cdot \frac{s^3+2s^2+1}{s^3+2s^2+(2K+1)s-K}$. By applying Routh's method, we can find the range of values of K so G is BIBO-stable is $-0.4 < K < 0$.

✂ ··

Exercise 9.21 (**Application of Theorem 9.5**) A dynamical system is described by the transfer function

$$G(s) = \frac{s}{s^3 + 3s^2 + 4s + 2}, \tag{9.79}$$

Apply Theorem 9.5 to tell whether the system is BIBO-stable. Verify your answer by using Routh's tabulation.

✂ ··

Hint. Start by determining the poles of G, compute the impulse response of G (you can use Python or MATLAB for that) and compute its integral from 0 to $+\infty$.

Exercise 9.22 (**Programming exercise**) Write a function in Python or MATLAB that constructs the Routh tabulation for a given polynomial and returns the number of poles in the right half plane.

Exercise 9.23 (**BIBO stability condition for rational transfer functions**) Suppose that G is a strictly proper rational function. Show that a necessary and sufficient condition for BIBO stability is that

$$\lim_{t \to \infty} \mathscr{L}^{-1}\{G(s)\}(t) = 0. \tag{9.80}$$

✂ ··

Hint. To show this you can combine Theorem 9.5 with the discussion in Section 5.2.6. Show that Equation (9.80) implies the absolute integrability condition of Theorem 9.5.

Chapter 9. BIBO stability

Exercise 9.24 (Observation on Theorem 9.5) We know that a dynamical system with transfer function G is BIBO-stable if and only if its impulse response, g, is absolutely integrable (see Theorem 9.5) Find a transfer function, G, such that its impulse response converges to 0 as $t \to \infty$, but it is not BIBO-stable.

✂ ....

Hint. According to Exercise 9.23, you should not be looking for a rational transfer function. Think of a function $x(t)$ in the time domain that converges to zero, but is not absolutely integrable. That would be a function that converges to zero slower than e^{-ct}, $c > 0$. Then take its Laplace transform to find the desired system.

Exercise 9.25 (Feedback can have a destabilising effect) Consider a feedback system such as the one shown below:

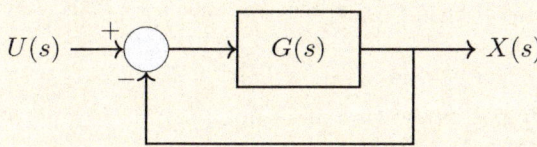

Give an example of a BIBO-stable transfer function, G, so that the above system with input U and output X is BIBO-unstable.

✳ ....

Answer. A very simple example is to take $G(s) = -\frac{s+1}{2}$, which is a stable system; the transfer function of the above closed-loop system is $H(s) = \frac{X(s)}{U(s)} = \frac{G(s)}{1+G(s)} = -\frac{s-1}{2}$, which is unstable.

Exercise 9.26 (Stability of differentiator) A system has transfer function $G(s) = s$. Is it stable?

✳ ....

Answer. We cannot use Theorem 9.9 because G is not a proper rational function. Instead, we will use the definition: G is not stable because the input $u(t) = \sin t^2$, which is bounded, leads to the output $x(t) = \dot{u}(t) = 2t \cos t^2$, which is unbounded.

Exercise 9.27 (**That's impossible!**) Someone claims that the impulse response of a linear dynamical system is given by $x^{\text{imp}}(t) = \sin(3t)e^{-2t} + e^{-5t}$, $t \geq 0$ and the response of system to a square wave of amplitude 10 has the property $x^{\text{square}}(1) = 10$. Show that this is impossible.

※ ....

Answer. The impulse response of the system is $x^{\text{imp}}(t) = \sin(3t)e^{-2t} + e^{-5t}$, therefore

$$M = \int_0^\infty \left| \sin(3t)e^{-2t} + e^{-5t} \right| d\tau$$

$$\leq \int_0^\infty \left(|\sin(3\tau)e^{-2\tau}| + |e^{-5\tau}| \right) d\tau$$

$$\leq \int_0^\infty \left(e^{-2\tau} + e^{-5\tau} \right) d\tau = 0.7. \tag{9.81}$$

The input has an amplitude of 10, that is, $|u(t)| \leq 10$ for all $t \geq 0$. By Theorem 9.6, the output, x^{square}, when the input is must satisfy $|x^{\text{square}}(t)| \leq 10 \cdot 0.7 = 7$, but $x^{\text{square}}(1) = 10$, which cannot be the case.

Exercise 9.28 (**Roots in D_θ**) Show that it is not possible to find a $K \in \mathbb{R}$ such that all roots of

$$Q(s) = s^3 + s^2 + s + K, \tag{9.82}$$

are in $D_{45°}$. What is the largest θ so that all roots of Q are in D_θ? You may use MATLAB or Python to perform the necessary computations (symbolically).

※ ....

Answer. The supremum of all θ such that all roots are in D_θ is $\theta = 30°$, but this is not attained for any K. For $K = 0.1$, all roots are in $D_{27.91°}$. For $K = 0.01$, all roots are in $D_{29.831°}$. For $K = 0.0001$, all roots are in $D_{29.998°}$. However, for $K = 0$, there is a root at the origin.

Exercise 9.29 (**Roots in D_θ implies bounded overshoot**) Consider a second-order system with static gain $K > 0$, time constant $\tau > 0$ and damping factor $\zeta > 0$. (i) Suppose that the poles of the transfer function, G, are in $G_{30°}$. What is the maximum overshoot that the system may have? (ii) Determine a $0 < \theta < \pi/2$ such that if all the poles of G are in G_θ, the overshoot is less than 10%.

※ ....

Answer. (i) the maximum overshoot is approximately 16.3%, (ii) it is $\theta = 36.239°$.

Chapter 9. BIBO stability

✂ ....

Hint. Since G is a second-order transfer function, it has two poles, $a \pm bj$. Define $\kappa = |b/a|$. Show that the overshoot is given by $\exp(-\pi/\kappa)$.

Exercise 9.30 **(Roots in $D_{20°}$)** Show that all the roots of

$$Q(s) = s^5 + 3.1s^4 + 7.5s^3 + 7.2s^2 + 5.7s + 0.5, \qquad (9.83)$$

are in $D_{20°}$ and have real parts less than -0.05.

✂ ....

Hint. Use Propositions 9.13 and 9.15.

Advanced Exercise 9.31 **(Roots in horizontal strip)** Describe a procedure to determine whether all the roots of a given polynomial with real coefficients are in the shaded area shown below

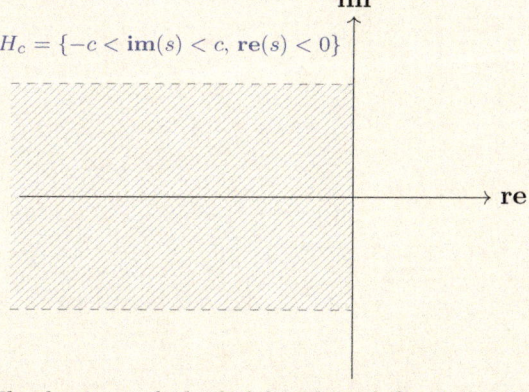

$H_c = \{-c < \mathrm{im}(s) < c, \ \mathrm{re}(s) < 0\}$

The horizontal shaded band is defined by $H_c = \{s \in \mathbb{C} : \ -c < \mathrm{im}(s) < c, \ \mathrm{re}(s) < 0\}$, where c is a given positive constant.

✂ ....

Hint. According to the Complex Conjugate Root Theorem (see Theorem A.8), the roots of a polynomial are in H_c if and only if they satisfy $\mathrm{re}(s) < 0$ and $\mathrm{im}(s) < -c$. What is the relationship between the region $H_c^+ = \{s \in \mathbb{C} : -c < \mathrm{im}(s), \mathrm{re}(s) < 0\}$ and the "vertical" space $V_c^+ = \{s \in \mathbb{C} : \mathrm{re}(s) < -c\}$? Can you think of a mapping from H_c^+ to V_c^+?

Exercise 9.32 **(Proof of Proposition 9.15)** Prove Proposition 9.15.

10

PID controllers

*The PID is one of the simplest controllers. Perhaps for exactly this reason, it remains to this day the most widely used controller in industrial practice. In this chapter we will first give some intuition using an example from every day life: a vehicle that needs to drive in the middle of its lane. Next, we will state the two main problems in control engineering: **regulation** and **disturbance rejection** and use Routh's stability criterion and the final value theorem to design appropriate controllers.*

10.1 Intuition

10.1.1 Conceptual construction of the PID controller

In order to understand how a PID controller works, we will study a familiar dynamical system: the steering of a car. In fact one of the first applications of PID control was Elmer Sperry's steering controller for ships[1].

Suppose we need to keep a vehicle exactly at the middle of the lane (set point) by manipulating the steering angle (control action). We will assume that the linear velocity of the car is constant. Let y denote the (signed) distance from the middle of the lane. It makes sense that the farther away we are from our target, the more we need to steer towards it, while when we are exactly at the set point, the steering angle should be equal to zero. This leads naturally to the adoption of the following controller

$$u(t) = K_p e(t), \qquad (10.1)$$

where $u(t)$ is the steering angle (with the convention that steering to the left is positive) and $e(t)$ is the control error defined as $e(t) = -y(t)$. This

---

[1] Elmer Sperry came up with a PID-type controller (1911) that later led to what is currently known as the PID controller; the theoretical foundations of PID control started to be laid in 1922 by Nicolas Minorsky (1922) [C28].

is called a proportional controller (for short, **P controller**), and K_p is the **controller gain**.

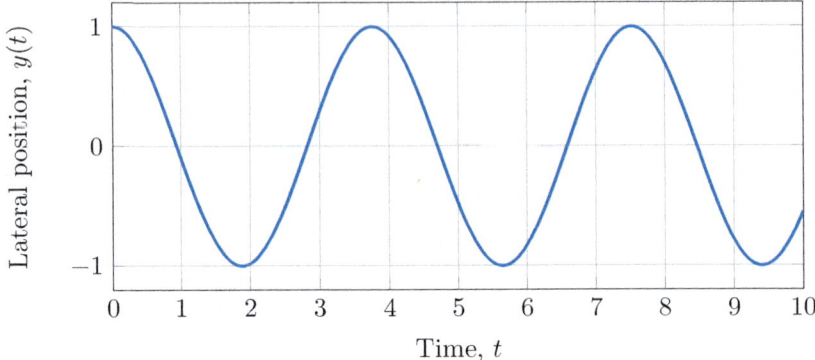

Figure 10.1: Control of lateral position of the vehicle with a P controller. The P controller cannot make the car converge to the set point.

As we see in Figure 10.1, this leads to a heavily oscillatory behaviour and the vehicle's later position does not converge to the set point. Observe that the vehicle plunges towards the target position, but instead of slowing down in order to converge to it, it moves past it. As soon as it moves past the set point, the controller will start to steer the wheels to the opposite direction, but it will only be after some distance that the vehicle will actually turn towards the set point again. Reducing or increasing the controller gain will only make the vehicle oscillate at a lower or higher frequency, but will not make it converge to the middle of the lane. It is worth noting that the control action does not depend on the lateral speed of the car and is not affected by whether the car is moving towards or away from the set point. The controller will decide the steering angle based on the distance from the set point; whether the car moves too fast or too slow towards the target will make no difference to a P controller.

In order to make the trajectory of the vehicle converge to the set point, we need to make it slow down as it approaches the set point. In other words, we need to penalise large lateral velocities. To that end, we introduce the following controller

$$u(t) = K_p e(t) + K_d \dot{e}(t), \qquad (10.2)$$

where $\dot{e}$ is the time-derivative of the control error and K_d is a coefficient known as the *derivative gain*. This is known as a **PD controller** (proportional derivative controller) and the behaviour of the closed-loop system is shown in Figure 10.2.

The PD controller has two tuning knobs: the designer needs to choose the values of K_p and K_d to achieve a desired closed-loop behaviour. The BIBO stability of the closed-loop system is a *conditio sine qua non*[2], that

---

[2] meaning an indispensable and essential condition.

Chapter 10. PID controllers

is, K_p and K_d should be chosen so that bounded changes of the set-point cause a bounded error response.

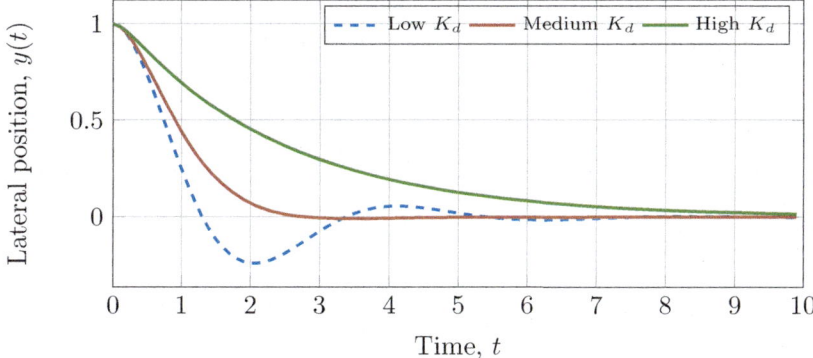

Figure 10.2: *Control of lateral position of the vehicle with a PD controller with a fixed K_p and different values of K_d. Low K_d values will lead to a more oscillatory behaviour, whereas if K_d becomes too high, the convergence becomes sluggish.*

The selection of controller parameters is known as **controller tuning**, and it is both an art and a science. The designer needs to take into account multiple requirements such as (i) stability (always), (ii) how oscillatory the system response should be, (iii) how fast the response converges to the set point, (iv) safety and technical constraints (e.g., the steering angle should not be allowed to be larger that 30°), (v) the fact that system model typically captures only part of the system dynamics and comes with assumptions that may not be fulfilled in practice (all models are wrong[3]).

It now seems that we have come up with a controller that can keep a vehicle in the middle of the road nicely and smoothly. There is one more step before we can test it on an actual car: we need to study what happens if there is some discrepancy between the model and the actual system. There are several reasons why this may happen. For example, the actuator might be imperfect. The fact that the controller commands the wheel to steer at $+5°$ does not mean that the steering system has the capacity to achieve this exact steering angle; for all we know the actual steering angle can be $+3°$ or $+8°$. This inaccuracy can be mitigated with proper calibration and high quality hardware, but can never be eliminated. Secondly, the sensors are imperfect. In this example, we assume we measure the distance of the car from the middle of the road. This can be done by some computer vision system from which we should expect a standard error of up to a few centimetres. Lastly, the model may be (and typically is) too simple to capture the underlying complex system dynamics. In our case for example we have not accounted for cases of slip, side winds, uneven roads and so on.

---

[3]"...but some are useful," (George Box, 1976) is a common maxim in statistics.

Let us do a little experiment. Suppose that the controller commands a steering angle u according to Equation (10.2), but the actuator has a constant bias of merely 1°. The closed-loop simulation of the PD-controlled system is shown in Figure 10.3.

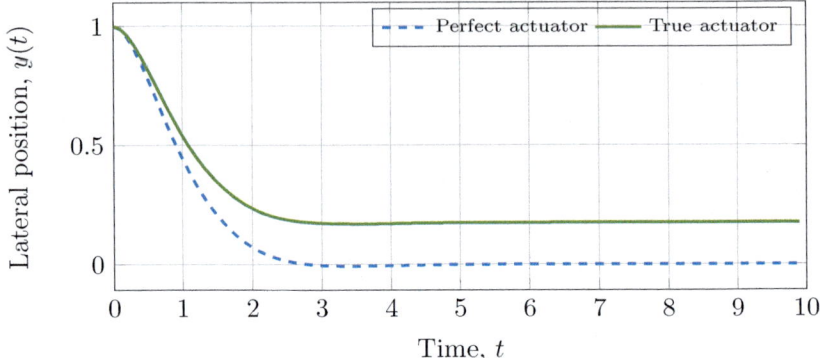

Figure 10.3: *Closed-loop system with the actual actuator (with a bias of 1°) compared to the ideal actuator. Simulations using a PD controller.*

We see that because of the bias of the actuator, the car does not converge to the set point, but remains 17.5 cm away from it. The difference between the set point and the limit of the system output is called **offset**. We give the following definition

> **Definition 10.1 (Offset)** The offset of a closed-loop system relative to a set point signal $x^{\text{sp}}(t)$ and a disturbance signal $d(t)$ is given by the following limit, if it exists
>
> $$\text{offset} = \lim_{t \to \infty} x^{\text{sp}}(t) - x(t). \qquad (10.3)$$

The PD controller cannot do much about nonzero offsets. Eventually, it will command the car to steer 1° towards the set point, but because of the actuation bias the car will not follow. Neither the P nor the D element of the PD controller can do anything to correct the offset.

We need to find a way to tell the controller that the output of the system has been staying away from the set point despite its commands. In other words, we need to construct a term that keeps growing as long as the output of the system is away from the set point. This will be the time-integral of the error,

$$I(t) = \int_0^t e(\tau) \mathrm{d}\tau. \qquad (10.4)$$

We can then plug the integral into the controller to define the PID controller

Chapter 10. PID controllers

(proportional integral derivative)

$$u(t) = K_p e(t) + K_d \dot{e}(t) + K_i I(t). \tag{10.5}$$

The PID controller achieves zero offset as illustrated in Figure 10.4.

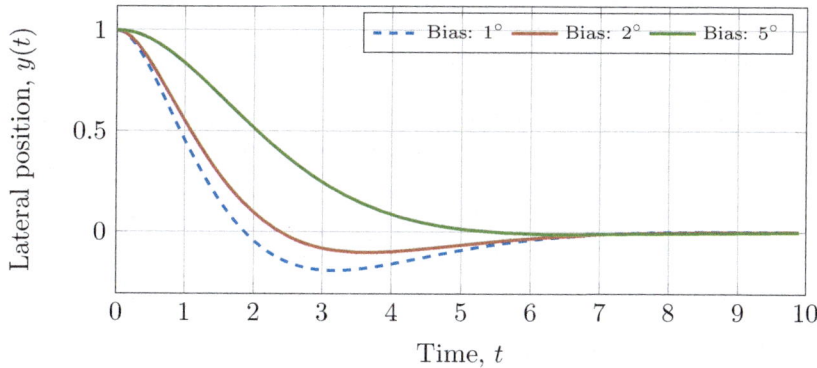

Figure 10.4: *Elimination of offset with a PID controller; closed-loop behaviour in presence of different levels of bias, namely, $1°$, $2°$ and $5°$.*

It remains to see how the three parameters of the PID controller affect the closed-loop behaviour.

10.1.2 Additional material

The reader can access animations at https://alphaville.github.io/qub/pid-101 (you can scan the QR code of Figure 10.5).

Figure 10.5: *Animations: vehicle control of vehicle with PID controller.*

10.1.3 Effect of PID parameters

Although different dynamical systems have significantly different behaviours, there are certain qualitative commonalities. We will try to give a qualitative understanding of how K_p, K_d and K_i may affect the closed-loop behaviour, how fast/slow the output converges to the set point and how oscillatory the system response is.

In Figure 10.6 we demonstrate the effect of K_p on the closed-loop behaviour. In general, higher values of K_p lead to a more aggressive behaviour, while low values lead to a sluggish convergence speed. However, too high values of K_p may lead to oscillations.

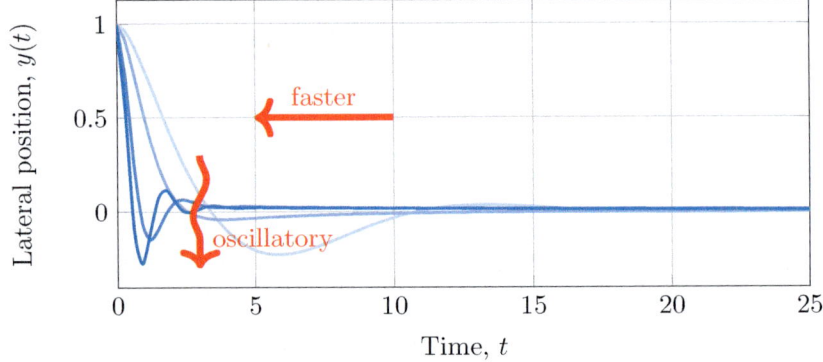

Figure 10.6: *Closed-loop behaviour with different values of K_p. Increasing K_p makes the system react faster and steer more aggressively towards the set point. If K_p increases a lot, the response becomes overly oscillatory.*

As we discussed above, the derivative of the error (the "D" component of PID) is introduced to control the oscillations. By increasing K_d we can reduce the oscillations, but we should not overdo as it will make the closed-loop system slow. The effect of K_d on the closed-loop behaviour is illustrated in Figure 10.7.

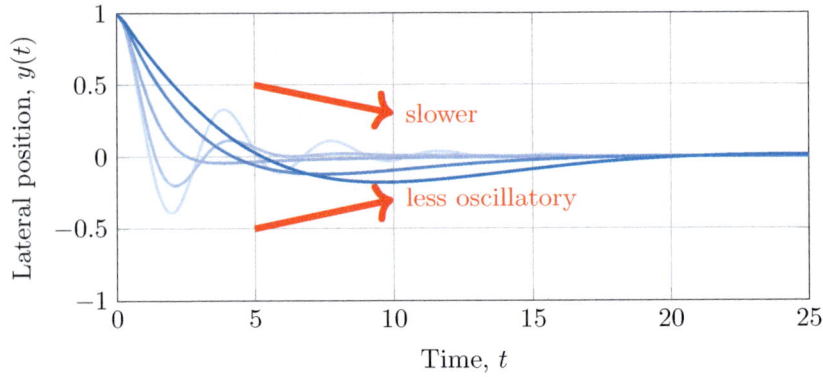

Chapter 10. PID controllers

Figure 10.7: *Closed-loop behaviour with different values of K_d. We can suppress the oscillations by increasing K_d, however, too high values of K_d can lead to a very sluggish response. Increasing K_d tends to make the closed-loop response less oscillatory, but slower.*

The integral, $I(t)$ (the "I" component of PID) was introduced to attenuate the effect of modelling inaccuracies or other external disturbances. Again, we need to select an appropriate value for the respective gain (in this case, K_i). A very low gain may[4] lead to the slow convergence towards the set point, while too large a K_i may destabilise the system.

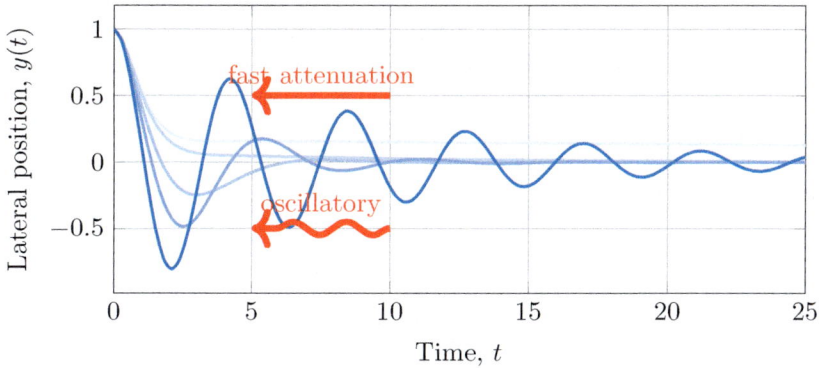

Figure 10.8: *Closed-loop behaviour with different values of K_i. Very low values of K_i may lead to a very slow attenuation of offset. Too high values may lead to strong oscillations.*

The PID controller is very often written in the following form, which is equivalent to Equation (10.2)[5]

$$u(t) = K_c \left(e(t) + \tau_D \dot{e}(t) + \frac{1}{\tau_I} I(t) \right). \tag{10.6}$$

This is the form we will be using in what follows. We will call parameter K_c the **controller gain**, τ_D is the **derivative time** constant and τ_I is the **integral time** constant.

> **Exercise 10.1 (Transfer function of PID)** Apply the Laplace transform to both sides of Equation (10.6). Given that a controller is a linear dynamical system with input $e(t)$ and output $u(t)$, what is its transfer function?

---

[4]The reason we insist in using the modal verb "may" is that different systems behave in different ways; the discussion in this session is to offer an intuition about the effect that the parameters of a PID controller may have on the closed-loop system, and not a rigorous framework to analyse the closed-loop dynamics.

[5]Apparently, $K_c = K_p$, $K_d = K_c \tau_D$ and $K_i = K_c/\tau_I$.

Answer. We assume zero initial conditions; the transfer function is $\frac{U(s)}{E(s)} = K_c(1 + \tau_D s + \frac{1}{\tau_I s})$.

※

10.1.4 Outlook

Through the lane keeping example in the previous sections, we introduced the PID controller and discussed the effect that the controller parameters may have on the closed-loop behaviour. The designer needs to make certain choices such as whether to use a P, a PD, a PI or a PID controller, and then she needs to choose appropriate values for the controller parameters to achieve certain dynamic characteristics for the closed-loop system. This procedure is known as **tuning** (we introduced this concept in Chapter 1).

As designers, we will need to impose two sets of requirements. First, assuming that the disturbance acting on the system is constant, the requirements in regard to the set point are

A-1 **BIBO stability**: in the sense that bounded changes in the set point will cause bounded changes in the output; in the example of lane keeping, this means that if we command the car to move 0.1 m to the right, it should not follow an unbounded trajectory (e.g., one that escapes to infinity)! Secondly,

A-2 **Zero offset**: after a step change in the set point, the output of the system should eventually converge to the set point. For example, if we command the car to move 0.1 m to the right, it must eventually converge to that position.

Second, in regard to the disturbance and assuming that the set point remains constant, we will need to address the following requirements

B-1 **BIBO stability**: the dynamical system with input $D(s)$ and output $X(s)$, needs to be BIBO stable, that is, any bounded change in the disturbance signal should cause a bounded change in the output and

B-2 **Zero offset**: Step changes in the disturbance should be eventually eliminated leading to zero offset.

10.2 Regulation and disturbance rejection

10.2.1 Structure of feedback control systems

We first introduced the structure of a feedback control system in Chapter 1. Here we will revisit this topic and describe the structure of feedback control loops using transfer functions. Recall that a basic control loop involves (i) the controlled system with input $U(s)$ and output $X(s)$, which is described by a transfer function G_s, (ii) the sensor with transfer function G_m, (iii) the controller, with transfer function G_c, and the actuator, G_a. The system may be subject to an external disturbance $D(s)$, which may capture either an actual external variable that affects the system dynamics (e.g., ambient temperature, electromagnetic interference, etc) or any other unmodelled component or modelling error. We assume that D acts on the system output as shown in Figure 10.9.

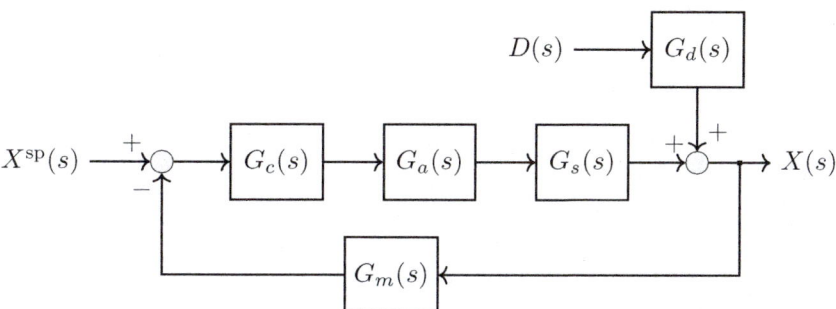

Figure 10.9: *Structure of a feedback control system with a disturbance $D(s)$ acting on the system output.*

The correspondence between the system inputs $X^{\text{sp}}(s)$ and $D(s)$ and the output $X(s)$ is described by

$$X(s) = G_{\text{cl}}(s)X^{\text{sp}}(s) + G_L(s)D(s), \tag{10.7}$$

where $G_{\text{cl}}(s)$ and $G_L(s)$ are the **closed-loop transfer function** and the **load transfer function** respectively, which we define below.

Definition 10.2 (Closed-loop and load transfer functions) The closed-loop transfer function of the closed-loop system in Figure 10.9 is given by

$$G_{\text{cl}}(s) = \frac{G_c(s)G_a(s)G_s(s)}{1 + G_c(s)G_a(s)G_s(s)G_m(s)}, \tag{10.8}$$

and the load transfer function is given by

$$G_L(s) = \frac{G_d(s)}{1 + G_c(s)G_a(s)G_s(s)G_m(s)}. \tag{10.9}$$

Note that $D(s)$ acts on the system output through a transfer function $G_d(s)$. In the example with the car we discussed in the beginning of this chapter, D can be the speed of a strong side wind or the lateral inclination of the road which have not been accounted for in the system model. Nevertheless, there exist other sources of disturbance such as actuator imperfections — those imperfections that motivated the use of the integral in Section 10.1.1. In such a case, the disturbance $U^{\text{bias}}(s)$ acts on the system as shown in Figure 10.10. This, however, is equivalent to the block diagram shown in Figure 10.9 with $D(s) = U^{\text{bias}}(s)$ and $G_d(s) = G_s(s)$. Hereafter, we will be referring to the standard block diagram of Figure 10.9.

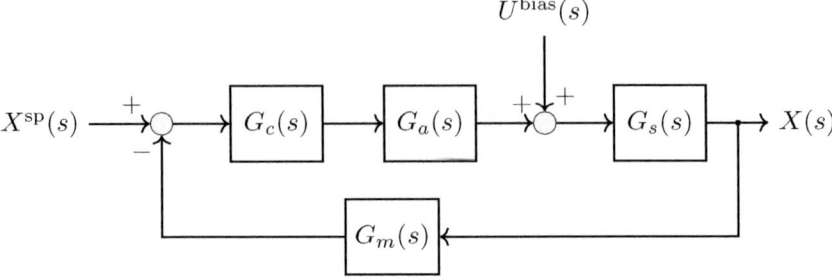

Figure 10.10: *Structure of a feedback control system with a disturbance $U^{\text{bias}}(s)$ acting on the control action. This is equivalent to the block diagram in Figure 10.9 with $D(s) = U^{\text{bias}}(s)$ and $G_d(s) = G_s(s)$.*

10.2.2 Regulation

The **regulation problem** consists in assuming that no disturbance is acting on the system, that is $D(s)$ and $X(s) = G_{\text{cl}}(s) X^{\text{sp}}(s)$, and the objective is to choose an appropriate controller (e.g., P, PI, PD or PID) and appropriate controller parameters (K_c, τ_I and/or τ_D) so that conditions A-1 and A-2 be satisfied. Our tools will be

1. Routh's stability criterion (Theorem 9.12) to tell whether the closed-loop transfer function, G_{cl}, is BIBO stable,

2. the final value theorem (Theorem 4.29) that will allow us to compute the offset without taking any inverse Laplace transforms.

Let us give a few examples.

Example 10.1 (P Controller) Consider a controlled system with first-order dynamics with static gain $K = 10$ and time constant $\tau = 1$. The output of the system is measured using a sensor with transfer function
$$G_m(s) = \frac{1}{0.1s + 1}. \tag{10.10}$$
The system is controlled with a P controller with gain K_c. The actuator's transfer function is $G_a(s) = 1$. Our objective is to determine the range of values of K_c so that conditions A-1 and A-2 be satisfied. Throughout, we assume that $D(s) = 0$.

The transfer function of the controlled system is
$$G_s(s) = \frac{10}{s+1}. \tag{10.11}$$
The transfer function of the controller is $G_c(s) = K_c$. The closed-

loop transfer function is

$$G_{cl}(s) = \frac{G_c(s)G_a(s)G_s(s)}{1+G_c(s)G_a(s)G_s(s)G_m(s)}$$

$$= \frac{K_c \frac{10}{s+1}}{1 + K_c \frac{10}{s+1}\frac{1}{0.1s+1}}$$

$$= \frac{K_c(0.1s+1)}{(s+1)(0.1s+1)+10K_c}$$

$$= \frac{K_c(0.1s+1)}{0.1s^2 + 1.1s + 1 + 10K_c}. \quad (10.12)$$

Routh's tabulation for the characteristic polynomial of G_{cl}, $Q(s) = 0.1s^2 + 1.1s + 1 + 10K_c$, is

s^2	0.1	$1 + 10K_c$
s^1	1.1	
s^0	$1 + 10K_c$	

Function G_{cl} is BIBO stable (condition A-1) if all elements in the first column of Routh's tabulation are positive. The reader can easily verify that this is always the case since here $K_c > 0$.

We shall now compute the offset after a unit step change in the set point, i.e., when $x^{\text{sp}}(t) = 1$ or $X^{\text{sp}}(s) = 1/s$. The offset is

$$\lim_{t \to \infty}(x^{\text{sp}}(t) - x(t)) = \lim_{t \to \infty}(1 - x(t)) = 1 - \lim_{t \to \infty} x(t), \quad (10.13)$$

where $x(t)$ is the inverse Laplace transform of

$$X(s) = \frac{1}{s}G_{cl}(s). \quad (10.14)$$

All conditions of the final value theorem are satisfied, so

$$\lim_{t \to \infty} x(t) = \lim_{s \to 0^+} sX(s) = \lim_{s \to 0^+} s\frac{1}{s}G_{cl}(s)$$

$$= \lim_{s \to 0^+} \frac{K_c(0.1s+1)}{0.1s^2 + 1.1s + 1 + 10K_c}$$

$$= \frac{K_c}{1 + 10K_c}, \quad (10.15)$$

so according to Equation (10.13), the offset is

$$\text{offset} = 1 - \frac{K_c}{1 + 10K_c}. \quad (10.16)$$

Note that not only is it impossible to find a controller gain that leads to zero offset, but the offset cannot drop below 0.1, so it is not possible to satisfy condition A-2 for this system using a P controller.

Exercise 10.2 (Propose appropriate controller) In Example 10.1 we saw the P controller cannot eliminate the offset. What type of controller would you propose for this system and why? What are the values of the parameters of your controller so that the closed-loop system satisfies conditions A-1 and A-2.

Example 10.2 (P controller works) The transfer function of a mechanical system is given by

$$G_s(s) = \frac{1}{s(\tau s + 1)}. \qquad (10.17)$$

The system is controlled with an actuator with transfer function $G_a(s) = 1$ and the output is measured with a sensor with transfer function $G_m(s) = 1$. The system is controlled with a P controller with gain $K_c > 0$. Our first objective is to determine the range of values of K_c so that conditions A-1 and A-2 be satisfied.

The closed-loop transfer function is

$$G_{\text{cl}}(s) = \frac{K_c \frac{1}{s(\tau s+1)}}{1 + K_c \frac{1}{s(\tau s+1)}} = \frac{K_c}{s(\tau s+1) + K_c} = \frac{K_c}{s^2 + s + K_c}. \qquad (10.18)$$

Routh's tabulation for the characteristic polynomial of the closed-loop transfer function, $Q(s) = s^2 + s + K_c$ is

s^2	1	K_c
s^1	1	
s^0	K_c	

Clearly, the closed-loop system is stable for all $K_c > 0$.

We shall now compute the offset after a unit step change in the set point, i.e., when $x^{\text{sp}}(t) = 1$ or $X^{\text{sp}}(s) = 1/s$. The offset is

$$\lim_{t \to \infty} (x^{\text{sp}}(t) - x(t)) = \lim_{t \to \infty} (1 - x(t)) = 1 - \lim_{t \to \infty} x(t), \qquad (10.19)$$

where $x(t)$ is the inverse Laplace transform of

$$X(s) = \frac{1}{s} G_{\text{cl}}(s). \qquad (10.20)$$

All conditions of the final value theorem are satisfied, so

$$\lim_{t \to \infty} x(t) = \lim_{s \to 0^+} sX(s) = \lim_{s \to 0^+} s \frac{1}{s} G_{\text{cl}}(s)$$

$$= \lim_{s \to 0^+} \frac{K_c}{s^2 + s + K_c} = 1, \qquad (10.21)$$

so according to Equation (10.19), the offset is

$$\text{offset} = 1 - 1 = 0. \qquad (10.22)$$

Exercise 10.3 (**Example 10.1 vs Example 10.2**) In Example 10.1 the P controller was unable to lead to eliminate the offset, but it guaranteed a zero-offset operation for the system of Example 10.2. Why do you think that is?

Suppose that $G_m(s) = 1$ and $G_a(s) = 1$. Put a tick next to those systems for which the P controller lead to the satisfaction of conditions A-1 and A-2.

☐ $G(s) = \dfrac{2}{3s+1}$

☐ $G(s) = \dfrac{1}{s^2 + 5s}$

☐ $G(s) = \dfrac{1}{(s+1)^2 + 1}$

☐ $G(s) = \dfrac{s + 31}{11s^2 + 5s}$

Confirm your hypotheses.

Exercise 10.4 (**Effect of K_c**) Use Python or MATLAB to produce a plot of the step response of the closed-loop system of Exercise 10.2. How does the value of K_c affect the closed-loop behaviour?

Next, we will see how the PI controller can eliminate the offset in cases where the P controller fails to do so. We start by revisiting Example 10.1.

Example 10.3 (**PI controller to the rescue**) Consider the closed-loop system of Example 10.1 with a PI controller with transfer function

$$G_c(s) = K_c \left(1 + \frac{1}{\tau_I s}\right). \qquad (10.23)$$

The closed-loop transfer function is

$$G_{cl}(s) = \frac{G_c(s) G_a(s) G_s(s)}{1 + G_c(s) G_a(s) G_s(s) G_m(s)}$$

$$= \frac{K_c\left(1+\frac{1}{\tau_I s}\right)\frac{10}{s+1}}{1+K_c\left(1+\frac{1}{\tau_I s}\right)\frac{10}{s+1}\frac{1}{0.1s+1}}$$

multiply the numerator and denominator by $\tau_I s(s+1)(0.1s+1)$

$$= \frac{K_c(\tau_I s+1)10(0.1s+1)}{\tau_I s(s+1)(0.1s+1)+10K_c(\tau_I s+1)}$$

$$= \frac{10K_c(\tau_I s+1)(0.1s+1)}{\tau_I s(0.1s^2+1.1s+1)+10K_c\tau_I s+10K_c}$$

$$= \frac{10K_c(\tau_I s+1)(0.1s+1)}{0.1\tau_I s^3+1.1\tau_I s^2+\tau_I s+10K_c\tau_I s+10K_c}$$

$$= \frac{10K_c(\tau_I s+1)(0.1s+1)}{0.1\tau_I s^3+1.1\tau_I s^2+\tau_I(1+10K_c)s+10K_c}. \tag{10.24}$$

First, we need to derive conditions under which G_{cl} is BIBO stable (condition A-1). Routh's tabulation for the characteristic polynomial of G_{cl}, that is $Q(s)=0.1\tau_I s^3+1.1\tau_I s^2+\tau_I(1+10K_c)s+10K_c$, is as follows

s^3	$0.1\tau_I$	$\tau_I(1+10K_c)$
s^2	$1.1\tau_I$	$10K_c$
s^1	$a_{3,1}$	
s^0	$a_{4,1}$	

where

$$a_{3,1} = \frac{1.1\tau_I^2(1+10K_c)-\tau_I K_c}{1.1\tau_I}, \tag{10.25}$$

and $a_{4,1}=10K_c$. Since $K_c>0$ and assuming $\tau_I>0$, function G_{cl} is BIBO stable if and only if $a_{3,1}>0$, or equivalently, if and only if

$$1.1\tau_I(1+10K_c)-K_c>0. \tag{10.26}$$

Next, following the same procedure as in the previous examples in this section, we may find that the offset relative to the set point signal $x^{\text{sp}}(t)=H_0(t)$ (equivalently, $X^{\text{sp}}(s)=1/s$) is

$$\text{offset} = 1 - \lim_{s\to 0^+} G_{\text{cl}}(s)$$

$$= 1 - \lim_{s\to 0^+}\frac{10K_c(\tau_I s+1)(0.1s+1)}{0.1\tau_I s^3+1.1\tau_I s^2+\tau_I(1+10K_c)s+10K_c} = 0. \tag{10.27}$$

Therefore, conditions A-1 and A-2 are satisfied if $1.1\tau_I(1+10K_c)-K_c>0$.

Example 10.4 (PI controller: another example) Consider a controlled system with transfer function

$$G_s(s) = \frac{1}{(s+1)^2 + 1}. \tag{10.28}$$

Suppose that the actuator has the transfer function $G_a(s) = K_a$ and the sensor's transfer function is $G_m(s) = 1$. The system is controlled with a PI controller with controller gain K_c and integral time τ_I. Our objective is to derive conditions on K_c and τ_I (assuming that $K_c, \tau_I > 0$) so that conditions A-1 and A-2 hold.

Firstly, let us compute the closed-loop transfer function

$$\begin{aligned} G_{\text{cl}}(s) &= \frac{G_c(s)G_a(s)G_s(s)}{1 + G_c(s)G_a(s)G_s(s)G_m(s)} \\ &= \frac{K_c\left(1 + \frac{1}{\tau_I s}\right) K_a \frac{1}{(s+1)^2+1}}{1 + K_c\left(1 + \frac{1}{\tau_I s}\right) K_a \frac{1}{(s+1)^2+1}}, \end{aligned}$$

we multiply the numerator and the denominator by $\tau_I s[(s+1)^2 + 1]$

$$\begin{aligned} &= \frac{K_c K_a (\tau_I s + 1)}{\tau_I s[(s+1)^2 + 1] + K_c K_a (\tau_I s + 1)} \\ &= \frac{K_c K_a (\tau_I s + 1)}{\tau_I s(s^2 + 2s + 2) + K_c K_a (\tau_I s + 1)} \\ &= \frac{K_c K_a (\tau_I s + 1)}{\tau_I s^3 + 2\tau_I s^2 + 2\tau_I s + K_c K_a \tau_I s + K_c K_a} \\ &= \frac{K_c K_a (\tau_I s + 1)}{\tau_I s^3 + 2\tau_I s^2 + \tau_I (2 + K_c K_a)s + K_c K_a}. \tag{10.29} \end{aligned}$$

The characteristic polynomial of $G_{\text{cl}}(s)$ is $Q(s) = \tau_I s^3 + 2\tau_I s^2 + \tau_I(2 + K_c K_a)s + K_c K_a$. Let us construct Routh's tabulation

s^3	τ_I	$\tau_I(2 + K_c K_a)$
s^2	$2\tau_I$	$K_c K_a$
s^1	$a_{3,1}$	
s^0	$a_{4,1}$	

where

$$a_{3,1} = \frac{2\tau_I^2(1 + K_c K_a) - K_c K_a}{2\tau_I} = \frac{1}{2}\left(2\tau_I(2 + K_c K_a) - K_c K_a\right). \tag{10.30}$$

and $a_{4,1} = K_c K_a$. By virtue of Routh's stability criterion, we conclude that a necessary and sufficient condition for condition A-1 to hold is
$$2\tau_I(2 + K_c K_a) - K_c K_a > 0. \tag{10.31}$$
Let us now compute the offset when the set point signal is $x^{\text{sp}}(t) = H_0(t)$ or $X^{\text{sp}}(s) = 1/s$. It is

$$\text{offset} = 1 - \lim_{s \to 0^+} G_{\text{cl}}(s)$$
$$= 1 - \lim_{s \to 0^+} \frac{K_c K_a (\tau_I s + 1)}{\tau_I s^3 + 2\tau_I s^2 + \tau_I(2 + K_c K_a)s + K_c K_a}$$
$$= 1 - 1 = 0, \tag{10.32}$$

therefore, the offset is always zero.

Example 10.5 (PID controller design) A system with transfer function
$$G_s(s) = \frac{1}{s^2 - 1}, \tag{10.33}$$
is controlled with a PID controller. Suppose that $G_a(s) = 1$ and $G_m(s) = 1$. Our objective is to derive conditions on the PID parameters, K_c, τ_D and τ_I, so that the closed-loop system satisfies conditions A-1 and A-2. Hereafter we shall assume that $K_c, \tau_I, \tau_D > 0$ for simplicity.

Firstly, let us derive the closed-loop transfer function of the system. It is,
$$G_{\text{cl}}(s) = \frac{G_c(s)G_a(s)G_s(s)}{1 + G_c(s)G_a(s)G_s(s)G_m(s)}$$
$$= \frac{K_c \left(1 + \tau_D s + \frac{1}{\tau_I s}\right) \frac{1}{s^2 - 1}}{1 + K_c \left(1 + \tau_D s + \frac{1}{\tau_I s}\right) \frac{1}{s^2 - 1}},$$
and we now multiply the numerator and denominator by $(s^2 - 1)\tau_I s$
$$= \frac{K_c \left(\tau_I s + \tau_I \tau_D s^2 + 1\right)}{(s^2 - 1)\tau_I s + K_c \left(\tau_I s + \tau_I \tau_D s^2 + 1\right)}$$
$$= \frac{K_c \left(\tau_I s + \tau_I \tau_D s^2 + 1\right)}{\tau_I s^3 - \tau_I s + K_c \tau_I s + K_c \tau_I \tau_D s^2 + K_c}$$
$$= \frac{K_c \left(\tau_I s + \tau_I \tau_D s^2 + 1\right)}{\tau_I s^3 + K_c \tau_I \tau_D s^2 + \tau_I(K_c - 1)s + K_c}. \tag{10.34}$$

The characteristic polynomial of G_{cl} is $Q(s) = \tau_I s^3 + K_c \tau_I \tau_D s^2 + \tau_I(K_c - 1)s + K_c$. Before we proceed with the application of Routh's stability criterion, we can make a quick observation: recall that G_{cl} is unstable if any of the coefficients of Q is nonpositive (Theorem 9.10). Therefore, we must require that $\tau_I(K_c-1) > 0$ or equivalently, given that $\tau_I > 0$,

$$K_c > 1. \tag{10.35}$$

Next, we apply Routh's stability criterion. Routh's tabulation is

s^3	τ_I	$\tau_I(K_c - 1)$
s^2	$K_c\tau_I\tau_D$	K_c
s^1	$a_{3,1}$	
s^0	$a_{4,1}$	

where

$$a_{3,1} = \frac{K_c\tau_I^2\tau_D(K_c - 1) - K_c\tau_I}{K_c\tau_I\tau_D} = \frac{\tau_I\tau_D(K_c - 1) - 1}{\tau_D}, \tag{10.36}$$

and $a_{4,1} = K_c$. We conclude that the necessary and sufficient conditions for stability are

$$K_c > 1, \tag{10.37a}$$
$$\tau_I\tau_D(K_c - 1) > 1, \tag{10.37b}$$

where the first condition is redundant (the second condition implies the first one).

Let us now compute the offset when the set point signal is $x^{\text{sp}}(t) = H_0(t)$ or $X^{\text{sp}}(s) = 1/s$. As before, using the final value theorem,

$$\begin{aligned}\text{offset} &= 1 - \lim_{s \to 0^+} G_{\text{cl}}(s) \\ &= 1 - \lim_{s \to 0^+} \frac{K_c\left(\tau_I s + \tau_I\tau_D s^2 + 1\right)}{\tau_I s^3 + K_c\tau_I\tau_D s^2 + \tau_I(K_c - 1)s + K_c} = 0,\end{aligned} \tag{10.38}$$

so the offset is always zero.

Next, we will give an example of a system for which neither a P nor a PI controller can solve the regulation problem, i.e., lead to the satisfaction of conditions A-1 and A-2.

Example 10.6 (P and PI controllers do not work; the PID controller does) Consider the following system

$$G_s(s) = \frac{1}{(s - a)^2 + b^2}, \tag{10.39}$$

where $a \geq 0$ and $b > 0$ are constants. Suppose that $G_a(s) = 1$ and $G_m(s) = 1$.

I. P controller. We will first attempt to control this system with a P controller with gain K_c. The closed-loop function is

$$G_{\text{cl}}(s) = \frac{G_c(s)G_a(s)G_s(s)}{1 + G_c(s)G_a(s)G_s(s)G_m(s)} = \frac{K_c \frac{1}{(s-a)^2 + b^2}}{1 + K_c \frac{1}{(s-a)^2 + b^2}},$$

we multiply the numerator and denominator by $(s-a)^2 + b^2$

$$= \frac{K_c}{(s-a)^2 + b^2 + K_c} = \frac{K_c}{s^2 - 2as + a^2 + b^2 + K_c}. \quad (10.40)$$

Regardless of the sign of $a^2 + b^2 + Kc$, the closed-loop system is unstable in the BIBO sense because of Theorem 9.10.

II. PI controller. Let us now use a PI controller with controller gain K_c and integral time constant τ_I. The closed-loop transfer function becomes

$$G_{\text{cl}}(s) = \frac{K_c \left(1 + \frac{1}{\tau_I s}\right) \frac{1}{(s-a)^2 + b^2}}{1 + K_c \left(1 + \frac{1}{\tau_I s}\right) \frac{1}{(s-a)^2 + b^2}},$$

and we multiply the numerator and denominator by $\tau_I s[(s-a)^2 + b^2]$

$$= \frac{K_c(\tau_I s + 1)}{\tau_I s[(s-a)^2 + b^2] + K_c(\tau_I s + 1)}$$
$$= \frac{K_c(\tau_I s + 1)}{\tau_I s(s^2 - 2as + a^2 + b^2) + K_c \tau_I s + K_c}$$
$$= \frac{K_c(\tau_I s + 1)}{\tau_I s^3 - 2a\tau_I s^2 + (a^2 + b^2)\tau_I s + K_c \tau_I s + K_c}$$
$$= \frac{K_c(\tau_I s + 1)}{\tau_I s^3 - 2a\tau_I s^2 + (a^2 + b^2 + K_c)\tau_I s + K_c}, \quad (10.41)$$

but here we see that the coefficients of s^3 and s^2, namely τ_I and $-2a\tau_I$, have always opposite signs, so by Theorem 9.10, the closed-loop system is always BIBO unstable.

III. PID controller. Let us now use a PID controller with controller gain K_c, integral time constant τ_I and derivative time constant τ_D. The closed-loop transfer function is

$$G_{\text{cl}}(s) = \frac{K_c \left(1 + \tau_D s + \frac{1}{\tau_I s}\right) \frac{1}{(s-a)^2 + b^2}}{1 + K_c \left(1 + \tau_D s + \frac{1}{\tau_I s}\right) \frac{1}{(s-a)^2 + b^2}}$$

we multiply the numerator and denominator by $\tau_I s[(s-a)^2 + b^2]$

$$= \frac{K_c\left(\tau_I s + \tau_I \tau_D s^2 + 1\right)}{\tau_I s[(s-a)^2 + b^2] + K_c\left(\tau_I s + \tau_I \tau_D s^2 + 1\right)}$$

$$= \frac{K_c\left(\tau_I s + \tau_I \tau_D s^2 + 1\right)}{\tau_I s[s^2 - 2as + a^2 + b^2] + K_c \tau_I s + K_c \tau_I \tau_D s^2 + K_c}$$

$$= \frac{K_c\left(\tau_I s + \tau_I \tau_D s^2 + 1\right)}{\tau_I s^3 - 2a\tau_I s^2 + (a^2+b^2)\tau_I s + K_c \tau_I s + K_c \tau_I \tau_D s^2 + K_c}$$

$$= \frac{K_c\left(\tau_I s + \tau_I \tau_D s^2 + 1\right)}{\tau_I s^3 + (K_c \tau_D - 2a)\tau_I s^2 + (a^2 + b^2 + K_c)\tau_I s + K_c}. \quad (10.42)$$

Before we proceed with Routh's tabulation, note that for G_{cl} to be BIBO stable, the coefficients of its characteristic polynomial must have the same sign (see Theorem 9.10). We will focus on the case where $K_c > 0$ (so all coefficients must be positive): we must require that $K_c \tau_D - 2a > 0$. Routh's tabulation for the characteristic polynomial of G_{cl} is

s^3	τ_I	$\tau_I(a^2 + b^2 + K_c)$
s^2	$\tau_I(K_c \tau_D - 2a)$	K_c
s^1	$a_{3,1}$	
s^0	$a_{4,1}$	

where $a_{3,1}$ is

$$a_{3,1} = \frac{\tau_I^2 (K_c \tau_D - 2a)(a^2 + b^2 + K_c) - \tau_I K_c}{\tau_I (K_c \tau_D - 2a)}, \quad (10.43)$$

and $a_{4,1} = K_c$. According to Routh's stability criterion, G_{cl} is BIBO stable if and only if all entries of the first column of the above tabulation are positive, that is

$$K_c > 0, \quad (10.44\text{a})$$
$$\tau_I > 0, \quad (10.44\text{b})$$
$$K_c \tau_D - 2a > 0, \quad (10.44\text{c})$$
$$\tau_I (K_c \tau_D - 2a)(a^2 + b^2 + K_c) - K_c > 0. \quad (10.44\text{d})$$

The reader can examine the case $K_c < 0$ as an exercise.

It remains to check whether condition A-2 is satisfied. We will compute the offset when the set point signal is $x^{\text{sp}}(t) = H_0(t)$ or $X^{\text{sp}}(s) = 1/s$. Using the final value theorem,

$$\text{offset} = 1 - \lim_{s \to 0^+} G_{\text{cl}}(s)$$

$$= 1 - \lim_{s \to 0^+} \frac{K_c \left(\tau_I s + \tau_I \tau_D s^2 + 1\right)}{\tau_I s^3 + (K_c \tau_D - 2a)\tau_I s^2 + (a^2 + b^2 + K_c)\tau_I s + K_c}$$
$$= 1 - 1 = 0. \tag{10.45}$$

Therefore, the PID controller can solve the regulation problem for this system.

We may now summarise a few conclusions about the PID controller: (i) The I component can eliminate the offset, (ii) The D component suppresses the oscillations of the output variable; we observed that in the example in Section 10.1.1, (iii) The PID controller can solve the regulation problem for systems where both P and PI prove to be inadequate, such as the system of Example 10.6.

It is natural to ask under what conditions should we opt for a P, PI, PD or PID controller. We can of course repeat the above procedure for each controller and check whether they can solve the regulation problem, but we have not been able to come up with a rule that will guide us. Unfortunately, we cannot do this now — the reader will need to wait until the end of Chapter 11.

10.2.3 Disturbance rejection

We have so far been studying systems without disturbances. However, as we demonstrated in Section 10.1.1, unless we take into account the indispensable and ubiquitous presence of disturbances, we are in for big surprises. In this section we address the disturbance rejection problem: assuming that the set point does not change, $X^{\text{sp}}(s) = 0$, we require that conditions B-1 and B-2 be satisfied. Namely, the load transfer function needs to be stable and the system output needs to return to its equilibrium value after any bounded change in the disturbance.

We saw in Example 10.1 that the P controller cannot solve the regulation problem, but as shown in Example 10.3, the PI controller can lead to a BIBO stable closed-loop system with zero offset. We will study whether and under what conditions, a PI controller can solve the disturbance rejection problem.

Example 10.7 (PI controller for disturbance rejection) Consider a first-order system which is described by

$$X(s) = \underbrace{\frac{K}{\tau s + 1}}_{G_s(s)} U(s) + \underbrace{\frac{K}{\tau s + 1}}_{G_d(s)} D(s), \tag{10.46}$$

for some constants $K, \tau > 0$. Suppose that the set point signal remains equal to its equilibrium value, that is $X^{\text{sp}}(s) = 0$ and let $G_a(s) = 1$, $G_m(s) = 1$. The system is controlled with a PI controller with controller gain K_c and integral time τ_I. The block diagram of the closed-loop system is shown below.

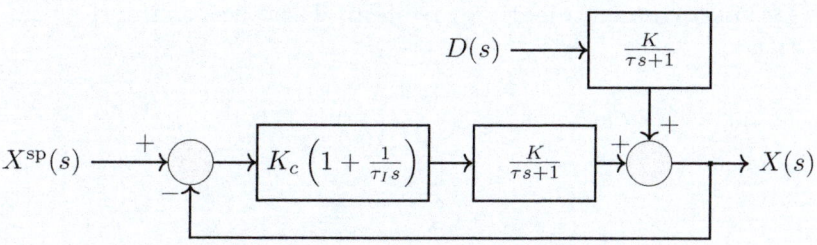

Figure 10.11: *Block diagram of closed-loop system.*

The dynamics of the closed-loop system is described by $X(s) = G_{cl}(s)X^{sp}(s) + G_L(s)D(s)$.

I. Regulation problem: It is left to the reader as an exercise to check for what values of K_c and τ_I, conditions A-1 and A-2 are satisfied.

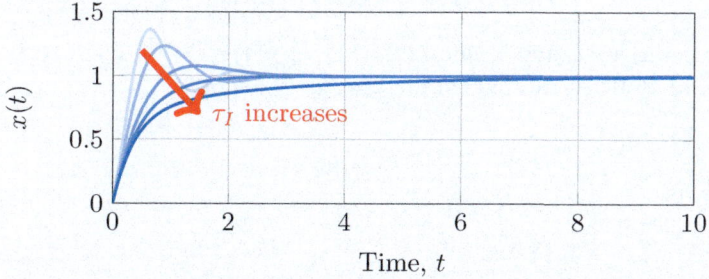

Figure 10.12: *Response of $x(t)$ for $x^{sp}(t) = H_0(t)$, with $D(s) = 0$, for different values of τ_I and fixed K_c (here assumed positive); $x(t)$, converges to the set point. Small values of τ_I lead to a more oscillatory behaviour, whereas high values of τ_I yield a sluggish convergence.*

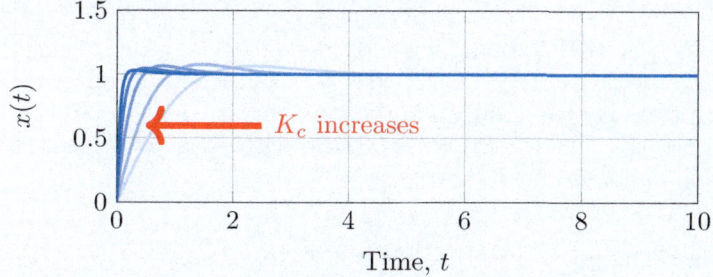

Figure 10.13: *Response of $x(t)$ for $x^{sp}(t) = H_0(t)$, with $D(s) = 0$, for different values of K_c and fixed τ_I (both assumed positive); $x(t)$,*

converges to the set point, while K_c controls how fast the output converges.

II. Disturbance rejection problem: Then the load transfer function is

$$G_L(s) = \frac{G_d(s)}{1 + G_c(s)G_a(s)G_m(s)G_s(s)} \qquad (10.47)$$

$$= \frac{\frac{K}{\tau s + 1}}{1 + K_c\left(1 + \frac{1}{\tau_I s}\right)\frac{K}{\tau s + 1}},$$

by multiplying the numerator and denominator by $\tau_I s(\tau s + 1)$ we have

$$= \frac{K\tau_I s}{\tau_I s(\tau s + 1) + KK_c(\tau_I s + 1)}$$

$$= \frac{K\tau_I s}{\tau_I \tau s^2 + \tau_I s + KK_c \tau_I s + KK_c}$$

$$= \frac{K\tau_I s}{\tau_I \tau s^2 + \tau_I(1 + KK_c)s + KK_c}. \qquad (10.48)$$

The characteristic polynomial of G_L is $Q(s) = \tau_I \tau s^2 + \tau_I(1 + KK_c)s + KK_c$ and Routh's tabulation for Q is given below

s^2	$\tau\tau_I$	KK_c
s^1	$\tau_I(1 + KK_c)$	
s^0	KK_c	

We can see that if $K_c, \tau_I > 0$, all entries in the first column are positive and G_L is BIBO stable (i.e., condition B-1 is satisfied). Condition B-1 is also satisfied if all entries in the first column are negative. Since $K, \tau_I > 0$, this holds true if and only if $\tau_I < 0$ and $-1/K < K_c < 0$.

Suppose G_L is BIBO-stable. Then, the offset is again defined as $\lim_{t\to\infty} x^{\mathrm{sp}}(t) - x(t)$, but $x^{\mathrm{sp}}(t) = 0$, therefore, offset $= -\lim_{t\to\infty} x(t)$. By the final value theorem this is equal to offset $= -\lim_{t\to\infty} sX(s)$, where $X(s) = G_L(s)D(s)$ with $D(s) = 1/s$ (see the description of condition B-2)

$$\text{offset} = -\lim_{t\to\infty} sX(s) = -\lim_{t\to\infty} \not{s}\frac{G_L(s)}{\not{s}}$$

$$= -\lim_{t\to\infty} \frac{K\tau_I s}{\tau_I \tau s^2 + \tau_I(1 + KK_c)s + KK_c} = 0. \qquad (10.49)$$

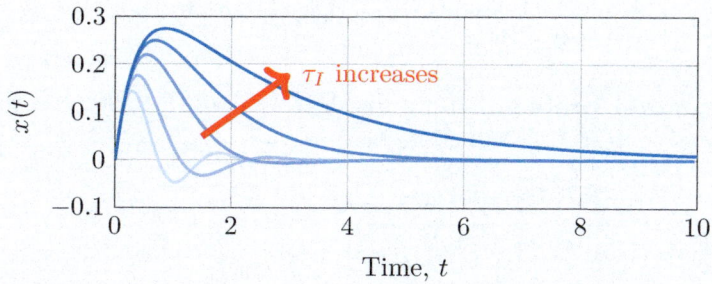

Figure 10.14: *Response of $x(t)$ for $d(t) = H_0(t)$ for different values of τ_I and a fixed K_c. The output returns (converges) to the set point after the action of a step change in the disturbance.*

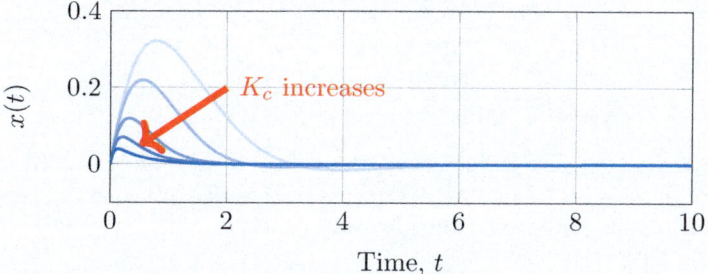

Figure 10.15: *Response of $x(t)$ for $d(t) = H_0(t)$ for different values of K_c and a fixed τ_I.*

Example 10.8 **(Regulation and Disturbance Rejection: second-order system)** Consider the system shown in the following block diagram

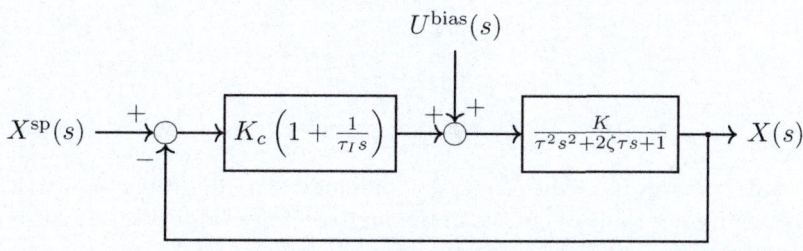

Figure 10.16: *Block diagram of closed-loop system.*

Hereafter we assume that $K, K_c, \tau_I > 0$ for simplicity. The closed-loop system involves the controlled system, with transfer function $G_s(s) = \frac{K}{\tau^2 s^2 + 2\zeta\tau s + 1}$ — a second-order system — and a PI controller. An external disturbance, denoted by U^{bias}, acts on the system as shown in Figure 10.16. Our objective is to derive the

I. Regulation problem. Here we shall assume that $U^{\text{bias}} = 0$. The closed-loop transfer function is

$$G_{\text{cl}}(s) = \frac{K_c\left(1 + \frac{1}{\tau_I s}\right) \frac{K}{\tau^2 s^2 + 2\zeta\tau s + 1}}{1 + K_c\left(1 + \frac{1}{\tau_I s}\right) \frac{K}{\tau^2 s^2 + 2\zeta\tau s + 1}}$$

we multiply the numerator and denominator by $\tau_I s(\tau^2 s^2 + 2\zeta\tau s + 1)$

$$= \frac{KK_c(\tau_I s + 1)}{\tau_I s(\tau^2 s^2 + 2\zeta\tau s + 1) + KK_c(\tau_I s + 1)}$$

$$= \frac{KK_c(\tau_I s + 1)}{\tau_I \tau^2 s^3 + 2\zeta\tau\tau_I s^2 + \tau_I s + KK_c\tau_I s + KK_c}$$

$$= \frac{KK_c(\tau_I s + 1)}{\tau_I \tau^2 s^3 + 2\zeta\tau\tau_I s^2 + \tau_I(1 + KK_c)s + KK_c} \tag{10.50}$$

At a first glance, all coefficients of the characteristic polynomial of G_{cl}, that is $Q(s) = \tau_I \tau^2 s^3 + 2\zeta\tau\tau_I s^2 + \tau_I(1 + KK_c)s + KK_c$, are positive. Let us construct Routh's tabulation

s^3	$\tau_I \tau^2$	$\tau_I(1 + KK_c)$
s^2	$2\zeta\tau\tau_I$	KK_c
s^1	$a_{3,1}$	
s^0	$a_{4,1}$	

where $a_{3,1}$ is given by

$$a_{3,1} = \frac{2\zeta\tau\tau_I^2(1 + KK_c) - KK_c\tau_I\tau^2}{2\zeta\tau\tau_I}$$

$$= \frac{2\zeta\tau_I(1 + KK_c) - KK_c\tau}{2\zeta} \tag{10.51}$$

and $a_{4,1} = KK_c$. The necessary and sufficient conditions for BIBO stability are that all entries in the first column of Routh's tabulation are positive. Therefore, we only need to require that

$$2\zeta\tau_I(1 + KK_c) - KK_c\tau > 0. \tag{10.52}$$

Let us now compute the offset when the set point is $x^{\text{sp}}(t) = H_0(t)$, or, what is the same, $X^{\text{sp}}(s) = 1/s$. Working as in all previous examples we have

$$\text{offset} = 1 - \lim_{s \to 0^+} G_{\text{cl}}(s)$$

$$= 1 - \lim_{s \to 0^+} \frac{KK_c(\tau_I s + 1)}{\tau_I \tau^2 s^3 + 2\zeta\tau\tau_I s^2 + \tau_I(1 + KK_c)s + KK_c} = 0. \tag{10.53}$$

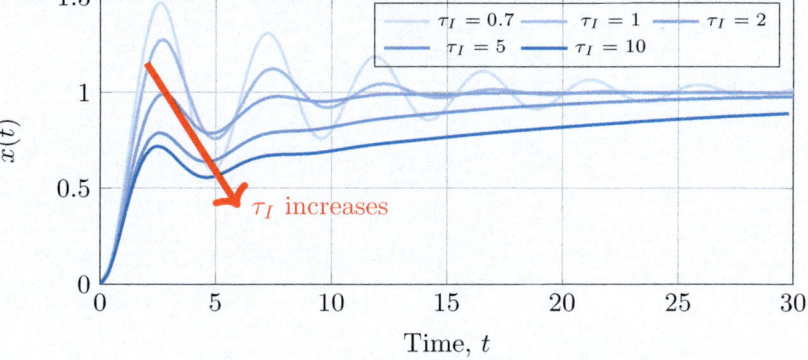

Figure 10.17: *Closed-loop simulations for a system with $K = 1$, $\tau = 1$ and $\zeta = 0.5$ using a PI controller with $K_c = 1$ and different τ_I.*

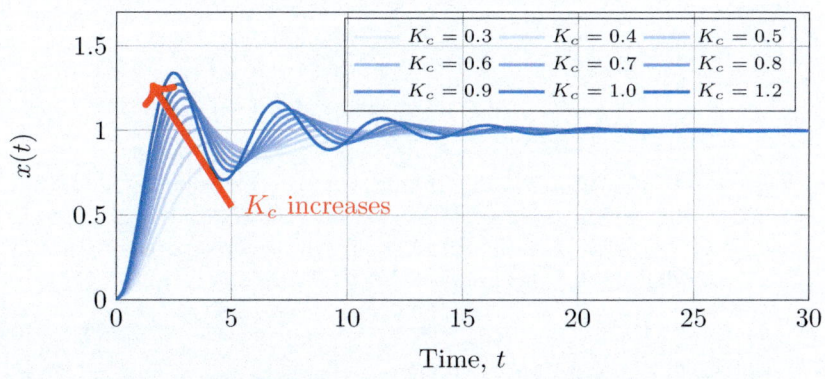

Figure 10.18: *Closed-loop simulations for a system with $K = 1$, $\tau = 1$ and $\zeta = 0.5$ using a PI controller with $\tau_I = 1$ and different values of K_c.*

II. Disturbance rejection problem: Now suppose that $X^{\text{sp}}(s) = 0$. The load transfer function is given by

$$G_L(s) = \frac{\frac{K}{\tau^2 s^2 + 2\zeta\tau s + 1}}{1 + K_c\left(1 + \frac{1}{\tau_I s}\right)\frac{K}{\tau^2 s^2 + 2\zeta\tau s + 1}},$$

we multiply the numerator and denominator by $\tau_I s(\tau^2 s^2 + 2\zeta\tau s + 1)$

$$= \frac{K\tau_I s}{\tau_I s(\tau^2 s^2 + 2\zeta\tau s + 1) + KK_c(\tau_I s + 1)}$$

$$= \frac{K\tau_I s}{\tau_I \tau^2 s^3 + 2\zeta\tau\tau_I s^2 + \tau_I(1 + KK_c)s + KK_c}. \tag{10.54}$$

The characteristic polynomial of G_L is identical to that of G_{cl}, therefore, G_L is BIBO stable if and only if G_{cl} is BIBO stable. The offset, when $d(t) = H_0(t)$ — equivalently, $D(s) = 1/s$ — is given by

$$\text{offset} = -\lim_{s \to 0^+} G_L(s)$$

$$= -\lim_{s \to 0^+} \frac{K\tau_I s}{\tau_I \tau^2 s^3 + 2\zeta\tau\tau_I s^2 + \tau_I(1 + KK_c)s + KK_c} = 0. \tag{10.55}$$

The response of the system is shown in Figures 10.19 and 10.20.

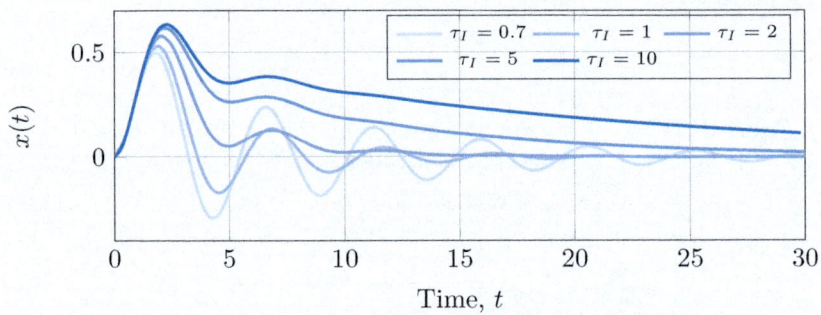

Figure 10.19: Closed-loop simulations for a system with $K = 1$, $\tau = 1$ and $\zeta = 0.5$ using a PI controller with $K_c = 1$ and different values of τ_I. The plot shows the response of the system to a step change in the disturbance, $d(t) = H_0(t)$, while $x^{\text{sp}}(t) = 0$.

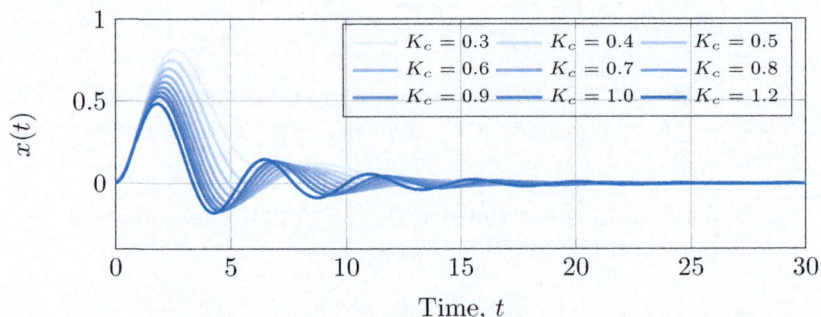

Figure 10.20: Closed-loop simulations for a system with $K = 1$, $\tau = 1$ and $\zeta = 0.5$ using a PI controller with $\tau_I = 1$ and different values of K_c. The plot shows the response of the system to a step change in the disturbance, $d(t) = H_0(t)$, while $x^{\text{sp}}(t) = 0$.

10.3 Tuning

The controller parameters need to be chosen to guarantee stability for the closed-loop system and lead to zero offset. Additionally, it is desirable that the closed-loop system behaviour is neither too aggressive nor too sluggish. If the step response of the closed-loop system is oscillatory, then the overshoot should be low and the oscillations should decay adequately fast.

The designer aims at choosing the PID controller parameters to (i) achieve a BIBO-stable closed loop, (ii) eliminate any offset, (iii) achieve a desired level of rise time, (iv) a reasonable overshoot, (v) a desired settling time, etc.

In what follows we describe three popular tuning methods: manual tuning and the two Ziegler-Nichols methods. It is important to note that tuning methods and guidelines should be taken with a grain of salt as the tuning criteria depend on several application-specific requirements that determine how responsive the closed-loop system should be, what levels of overshoot are acceptable and so on.

10.3.1 Feeling lucky?

Tuning can be performed empirically following the guidelines of Table 10.1. One typically starts by setting the integral and derivative gains to zero and increases the proportional gain, K_p, until the output oscillates. Next, we increase K_i to eliminate any offset adequately fast. The integral element should be used carefully and sparingly as high values can cause high overshoots, large settling times and instability. Lastly, we increase the derivative gain to suppress the overshoot and reduce the sensitivity of the closed-loop to disturbances.

Table 10.1: Expert controller tuning.

	Rise Time	Overshoot	Settling time	Offset
K_p	↓	↑	—	↓
K_i	↓	↑	↑	Eliminates
K_d	—	↓	↓	—

Exercise 10.5 (Manual tuning) A dynamical system with input variable $U(s)$, output variable $X(s)$ and disturbance $D(s)$ is described by the equation

$$X(s) = \frac{1}{(s+1)^4(0.1s+1)}(U(s) + D(s)). \tag{10.56}$$

Tune P, PI, PD or PID controllers for this system; plot the response to a step change in the set point and a step change in the disturbance.

10.3.2 Ziegler-Nichols first method

The *open-loop* step response of several systems resembles that of a first-order system with time delay as the one shown in Figure 10.21. The first Ziegler-Nichols method proposes a choice of parameters for P, PI and PID controllers aiming at achieving a *closed-loop* response with a decay ratio close to 0.25.

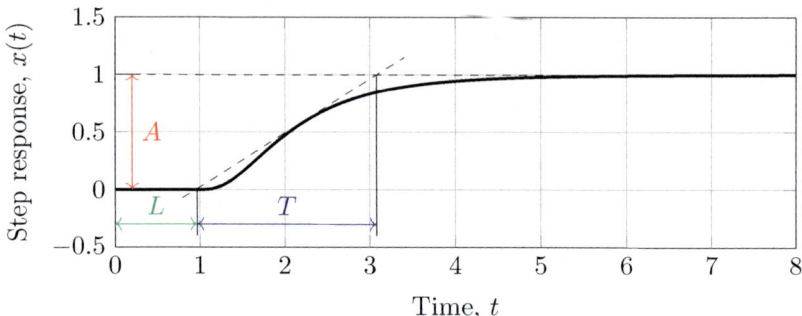

Figure 10.21: *Tangent line at the inflection point of the step response of a system.*

The design needs to obtain an S-shaped step response as the one in Figure 10.21, take the tangent line at its inflection point and measure the three parameters A, L and T as shown in the figure. The slope of the tangent line is

$$R = \frac{A}{T}. \tag{10.57}$$

The first Ziegler-Nichols method proposes the use of the parameters shown in Table 10.2.

Table 10.2: Parameter tuning with the first Ziegler-Nichols method: Recommended values of K_c, τ_I and τ_D for P, PI and PID controllers (see Figure 10.21).

Controller	K_c	τ_I	τ_D
P	$1/RL$	—	—
PI	$0.9/RL$	$L/0.3$	—
PID	$1.2/RL$	$2L$	$0.5L$

10.3.3 Ziegler-Nichols ultimate sensitivity method

The Ziegler-Nichols ultimate sensitivity method is the most popular method for tuning PID controllers. This method requires the following experiment: initially we use a P controller and increase the gain, K_p, aiming at determining the *ultimate gain*, K_u, that is, the gain value at which the output performs oscillations which neither dissipate to zero nor diverge to infinity. If we increase the gain beyond K_u, the system will become unstable. The period of oscillation of the output is called *ultimate period* and is denoted by T_u. Then, the parameters of a P, PI, PD or PID controller are determined as in Table 10.3.

Table 10.3: Parameter tuning with the Ziegler-Nichols method: P, PI, PD and PID controllers. The PID controller parameters are determined from the ultimate gain, K_u, and ultimate period, T_u.

Controller	K_p	τ_I	τ_D
P	$0.5K_u$	—	—
PI	$0.45K_u$	$T_u/1.2$	—
PD	$0.8K_u$	—	$T_u/8$
PID	$0.6K_u$	$T_u/2$	$T_u/8$

In certain cases, but not always, it is possible to derive explicit expressions for the ultimate gain and ultimate period of sustained oscillation. Let us give an example.

Example 10.9 (Controller tuning for third-order system) In this example, we will apply the Ziegler-Nichols method to design a P, PI, PD and PID controller for a third-order system with transfer function

$$G(s) = \frac{1}{(\tau_1 s + 1)(\tau_2 s + 1)(\tau_3 s + 1)}, \tag{10.58}$$

with $\tau_1, \tau_2, \tau_3 > 0$.

The main objective is to determine the ultimate gain and ultimate period of oscillation. The ultimate gain is the gain of a P controller, $G_c = K$, above which the closed-loop system becomes unstable and at which the step response is oscillatory.

The closed-loop transfer function is

$$G_{\text{cl}}(s) = \frac{KG(s)}{1 + KG(s)}$$
$$= \frac{K}{\tau_1\tau_2\tau_3 s^3 + (\tau_1\tau_2 + \tau_2\tau_3 + \tau_1\tau_3)s^2 + (\tau_1 + \tau_2 + \tau_3)s + (1 + K)}. \tag{10.59}$$

Routh's tabulation for this system is

s^3	$\tau_1\tau_2\tau_3$	$\tau_1 + \tau_2 + \tau_3$
s^2	$\tau_1\tau_2 + \tau_2\tau_3 + \tau_1\tau_3$	$1 + K$
s^1	$a_{3,1}$	
s^0	$a_{4,1}$	

where

$$a_{3,1} = \frac{(\tau_1\tau_2 + \tau_2\tau_3 + \tau_1\tau_3)(\tau_1 + \tau_2 + \tau_3) - \tau_1\tau_2\tau_3(1+K)}{\tau_1\tau_2 + \tau_2\tau_3 + \tau_1\tau_3}, \quad (10.60)$$

and

$$a_{4,1} = 1 + K. \quad (10.61)$$

The closed-loop system is, therefore, stable if and only if $a_{3,1} > 0$, or equivalently

$$\frac{(\tau_1\tau_2 + \tau_2\tau_3 + \tau_1\tau_3)(\tau_1 + \tau_2 + \tau_3) - \tau_1\tau_2\tau_3(1+K)}{\tau_1\tau_2 + \tau_2\tau_3 + \tau_1\tau_3} > 0$$

$$\Leftrightarrow (\tau_1\tau_2 + \tau_2\tau_3 + \tau_1\tau_3)(\tau_1 + \tau_2 + \tau_3) > \tau_1\tau_2\tau_3(1+K)$$

$$\Leftrightarrow \left(\frac{1}{\tau_1} + \frac{1}{\tau_2} + \frac{1}{\tau_3}\right)(\tau_1 + \tau_2 + \tau_3) > 1 + K$$

$$\Leftrightarrow K < \left(\frac{1}{\tau_1} + \frac{1}{\tau_2} + \frac{1}{\tau_3}\right)(\tau_1 + \tau_2 + \tau_3) - 1, \quad (10.62)$$

so that ultimate gain is

$$K_u = \left(\frac{1}{\tau_1} + \frac{1}{\tau_2} + \frac{1}{\tau_3}\right)(\tau_1 + \tau_2 + \tau_3) - 1. \quad (10.63)$$

In order to determine the period of the sustained oscillation we proceed as follows: We know from Chapter 5 (revise Section 5.2.6) that the inverse Laplace transform of an s-domain function exhibits sustained oscillations if and only if the function has a (single) pair of imaginary poles, $\pm j\omega$, alongside other *stable poles* (with negative real part) and at most one pole at the origin.

The step response of a closed-loop system with transfer function G_{cl} is given by $\mathcal{L}^{-1}\{G(s)/s\}$. For $\mathcal{L}^{-1}\{G(s)/s\}$ to exhibit sustained oscillations it must be

$$\frac{G(s)}{s} = \frac{P(s)}{(s^2 + \omega^2)Q(s)}, \quad (10.64)$$

where P and Q are polynomials, $\deg P < \deg Q + 2$ and Q has no poles with positive real part, it has no imaginary roots and has at most one root at the origin. Then, at large times (as t goes to ∞), $\mathscr{L}^{-1}\{G(s)/s\}$ will exhibit sustained oscillations with period $2\pi/\omega$.

In this example, the closed-loop system will exhibit sustained oscillations with period $2\pi/\omega_u$ if $j\omega_u$ is a root of the characteristic polynomial of G_{cl}, that is, if

$$\tau_1\tau_2\tau_3(j\omega_u)^3 + (\tau_1\tau_2 + \tau_2\tau_3 + \tau_1\tau_3)(j\omega_u)^2$$
$$+ (\tau_1 + \tau_2 + \tau_3)j\omega_u + (1 + K_u) = 0$$
$$\Leftrightarrow \left[-(\tau_1\tau_2 + \tau_2\tau_3 + \tau_1\tau_2)\omega_u^2 + 1 + K_u\right]$$
$$+ j\left[-\tau_1\tau_2\tau_3\omega_u^3 + (\tau_1 + \tau_2 + \tau_3)\omega_u\right] = 0. \quad (10.65)$$

By choosing K_u as in Equation (10.63) and by taking

$$\omega_u = \sqrt{\frac{\tau_1 + \tau_2 + \tau_3}{\tau_1\tau_2\tau_3}}, \quad (10.66)$$

both the real and imaginary parts in Equation (10.65) vanish. Then,

$$T_u = \frac{2\pi}{\sqrt{\frac{\tau_1+\tau_2+\tau_3}{\tau_1\tau_2\tau_3}}}. \quad (10.67)$$

We have computed K_u and T_u, therefore, we may use the guidelines of Table 10.3 to determine the parameters of a P, PI, PD or PID controller. For example, for a PID controller, we will choose:

$$K_p = 0.6K_u, \quad (10.68a)$$

$$\tau_I = \tfrac{1}{2}T_u = \frac{\pi}{\sqrt{\frac{\tau_1+\tau_2+\tau_3}{\tau_1\tau_2\tau_3}}}, \quad (10.68b)$$

$$\tau_D = \tfrac{1}{8}T_u = \frac{\pi}{4\sqrt{\frac{\tau_1+\tau_2+\tau_3}{\tau_1\tau_2\tau_3}}}. \quad (10.68c)$$

Now let $\tau_1 = 1$, $\tau_2 = 2$, and $\tau_3 = 3$. We then find that $K_u = 10$, $\omega_u = 1$, and $T_u = 2\pi \approx 6.2832$. For a PID controller, the Ziegler-Nichols method gives $K_c = 6$, $\tau_I = 3.1416$, and $\tau_D = 0.7854$.

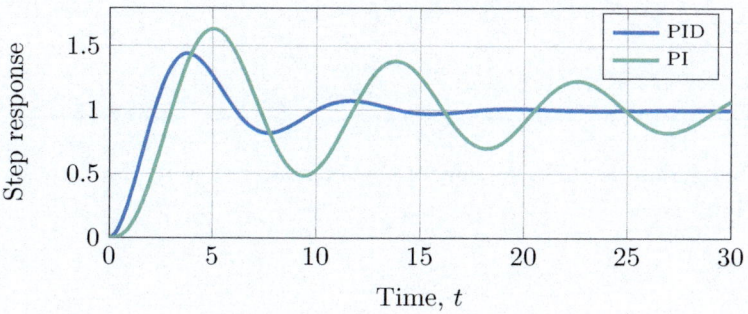

Figure 10.22: *Step response of the closed-loop system with PI and PID controllers tuned by using the Ziegle-Nichols method. The P and PD controllers would lead to a nonzero offset.*

The Ziegler-Nichols ultimate sensitivity method is not always applicable. For example, if G is a stable first or second-order transfer function, the closed-loop function, $G_{\text{cl}}(s) = \frac{KG(s)}{1+KG(s)}$, will be stable for all $K > 0$ and the step response of the closed-loop system will never exhibit sustained oscillations. The method cannot be applied to systems with transfer function $G(s) = \frac{1}{s^2 + a^2}$ as the step response of the corresponding P-controlled closed-loop system leads to sustained oscillations for all $K > 0$, that is, there is no ultimate gain. The reader can show that this method cannot be applied to systems with transfer function

$$G(s) = \frac{1}{s-a}, \tag{10.69}$$

with $a > 0$; the closed-loop system is unstable for $K \leq 1$ and stable for $K > 1$, but it never exhibits sustained oscillations.

The following theorem allows the determination of the ultimate gain, K_u and the associated period, T_u.

> **Proposition 10.3 (Determination of K_u and T_u)** A system with a rational transfer function, $G(s) = P(s)/Q(s)$ is controlled with a P controller with gain K. The closed-loop transfer function,
>
> $$G_{\text{cl}}(s) = \frac{KG(s)}{1 + KG(s)} = \frac{KP(s)}{Q(s) + KP(s)}, \tag{10.70}$$
>
> has the characteristic polynomial $Q(s) + KP(s)$. Suppose that the characteristic polynomial (i) has no roots with positive real part, (ii) has at most a single zero root, (iii) has one pair of imaginary roots. Let K and ω be such that $Q(j\omega) + KP(j\omega) = 0$. Then, the ultimate gain and ultimate period of sustained oscillation are given by
>
> $$K_u = K, \tag{10.71a}$$
>
> $$T_u = \frac{2\pi}{\omega}. \tag{10.71b}$$

Proof. See Example 10.9. We encourage the reader to write down the proof as an exercise. □

Chapter 10. PID controllers

Exercise 10.6 (Ziegler-Nichols exercise) A dynamical system has the transfer function
$$G(s) = \frac{1}{(s+1)^n}, \qquad (10.72)$$
with $n \in \mathbb{N}$.

1. Show that the second Ziegler-Nichols method can only be applied to systems with $n \geq 3$

2. Determine the ultimate gain and ultimate period of sustained oscillations, as a function of n, for systems with $n \geq 3$

3. Tune a PID controller for a system with $n = 5$

✂ ⋯⋯

Hint: (i) this is similar to Example 10.9. (ii) recall that complex numbers of the form $z = a + jb$ can be written in exponential form as $re^{j\theta}$, where $r = |z| = \sqrt{a^2 + b^2}$ and $\theta = \operatorname{atan2}(b, a)$. See Section A.1.

Exercise 10.7 (Computational exercise) A dynamical system has the transfer function
$$G(s) = \frac{1}{(s+1)(2s+1)(3s+1)}. \qquad (10.73)$$

Write a Python or MATLAB script to determine the parameters of a P, PI, PD and PID controller using the second Ziegler-Nichols method and plot the step response of the closed-loop systems.

In most cases, the ultimate gain (and ultimate frequency) are difficult to derive analytically. Then, one needs to determine the ultimate gain by a trial and error procedure and measure the ultimate period of oscillation experimentally as in the following exercise.

Exercise 10.8 (Ziegler-Nichols tuning) Tune a PID controller using the ultimate sensitivity method of Ziegler and Nichols for the following system
$$G(s) = \frac{1}{(s+1)(2s+1)} e^{-0.05s}. \qquad (10.74)$$
Use Python or MATLAB to perform closed-loop simulations.

Remark: It should be noted that tuning methods such as the Ziegler-Nichols two methods discussed above lead to feedback loops which may behave either too aggressively or too sluggishly for a particular application. What is the same, the disturbance rejection characteristics of the designed controllers may be suboptimal for the system at hand. That said, one should use the values obtained by Ziegler-Nichols only as initial values and perform additional tuning. •

10.4 Implementation of PID controllers

The vast majority of control systems nowadays are digital, and the control action is computed in a digital hardware device (a computer, microcontroller, FPGA, or other application-specific integrated circuit).

In a digital control system, the closed loop is driven by a *clock* which runs at a constant frequency. The clock ticks at times $0, T_s, 2T_s, \ldots$, and defines a **discrete** timescale, $(kT_s)_{k \in \mathbb{N}}$. The period between two successive discrete times is called **sampling time** or **sampling period**. At every tick of the clock, the controller receives an error measurement, $e(kT_s) = y^{\mathrm{sp}} - y(kT_s)$, computes a control action, $u(kT_s)$, and transmits the control command to the actuator. The controller does not have access to the error until the next sampling time, that is, the values $e(t)$ for $kT_s < t < (k+1)T_s$ are unknown to the controller. Likewise, the controller can update the control action, u, only at the discrete time instants $(kT_s)_{k \in \mathbb{N}}$. Typically, the control action is kept *constant* throughout the period between successive discrete time instants — this method is known as a **zero-order hold**.

Hereafter, we shall adopt the convenient notation $e_k := e(kT_s)$ for discrete-time signals. Similarly, we denote $\dot{e}_k := \dot{e}(kT_s)$.

Given the discrete nature of digital control systems, the derivative and integral of the error cannot be computed. Instead, we need to resort to approximations. Modern computing systems can achieve a very high sampling frequency (over 100 samples per second) allowing for a good approximation of the derivative and integral of the error, provided that the error changes much slower compared to the sampling period.

A common approximation of the derivative of the error is given by its **backward Euler difference**. The derivative of $e(t)$ at time t is defined as

$$\dot{e}(t) = \lim_{h \to 0} \frac{e(t) - e(t-h)}{h}. \tag{10.75}$$

Then, the derivative of e at $t = kT_s$ can be approximated by

$$\dot{e}_k \approx \frac{e_k - e_{k-1}}{T_s}. \tag{10.76}$$

If the sampling period, T_s, is very small, then we will have a good approximation of the derivative. According to Equation (10.76), the derivative of e at time kT_s can be approximated by the successive differences of the error divided by the sampling period.

Likewise, the integral of e over $[0, t]$ can be approximated by a numerical integration method such as the **rectangle rule**, according to which

$$\int_{(k-1)T_s}^{kT_s} e(\tau)d\tau \approx T_s e_k. \tag{10.77}$$

As a result, we have that for $k \geq 0$,

$$\int_0^{kT_s} e(\tau)d\tau \approx T_s \sum_{j=1}^{k} e_j \tag{10.78}$$

The **trapezoidal rule** can also be used to achieve a better approximation of the integral

$$\int_{(k-1)T_s}^{kT_s} e(\tau)d\tau \approx \frac{T_s}{2}(e_k + e_{k-1}), \tag{10.79}$$

leading to

$$\int_0^{kT_s} e(\tau)d\tau \approx \frac{T_s}{2}\sum_{j=1}^{k} e_j + e_{j-1} \tag{10.80}$$

Remark: We may derive the Euler approximation equation using Taylor's theorem (Theorem 3.1) according to which, if e is differentiable at $t = kT_s$, then

$$e_{k-1} = e_k + \dot{e}_k T_s + o(T_s), \tag{10.81}$$

and rearranging the terms and dividing both sides by T_s we obtain

$$\frac{e_k - e_{k-1}}{T_s} = \dot{e}_k + \frac{o(T_s)}{T_s}. \tag{10.82}$$

If e is twice differentiable at $t = kT_s$ we actually have

$$\frac{e_k - e_{k-1}}{T_s} = \dot{e}_k + \mathcal{O}(T_s). \tag{10.83}$$

•

The continuous-time PID controller can then be approximated by the following discrete-time variant

$$u_k = K_p e_k + K_d \frac{e_k - e_{k-1}}{T_s} + K_i T_s \sum_{j=1}^{k} e_j, \tag{10.84}$$

using the rectangle rule, or

$$u_k = K_p e_k + K_d \frac{e_k - e_{k-1}}{T_s} + K_i \frac{T_s}{2}\sum_{j=1}^{k} e_j - e_{j-1}, \tag{10.85}$$

using the trapezoidal rule.

A simple C implementation of a discrete-time PID controller is available at https://bit.ly/pid_c (you can scan the QR code of Figure 10.23).

Figure 10.23: *Discrete-time PID implementation in C (https://bit.ly/pid_c).*

Remark: *We should bear in mind that the discrete version of the PID controller calls for a separate stability analysis. The stability properties of the continuous-time PID controller are not necessarily inherited by its discretised counterpart — especially if the sampling time is large.* •

10.5 One-minute round-up

In this chapter we introduced the **PID controller** — hitherto, the most popular controller in industrial practice — and discussed the role of each of its components. The PID control actions are given by $u(t) = K_p e(t) + K_d \dot{e}(t) + K_i \int_0^t e(\tau) d\tau$, or, what is the same, $u(t) = K_c(e(t) + \tau_D \dot{e}(t) + \frac{1}{\tau_I} \int_0^t e(\tau) d\tau)$. We saw that

1. The **P** element is the most essential component of the PID controller. It can be used to control how aggressive the closed-loop system is.

2. The **I** element is used to eliminate **offset**.

3. The **D** element can be used to suppress oscillations. We will see in Chapter 12 that in certain cases, it can be used to guarantee stability (see also: Example 10.6). High derivative gains can, however, lead to a sluggish closed-loop behaviour.

We presented the two main problems in feedback control: the regulation and disturbance rejection problems. We used heavily the **Routh stability criterion** to derive conditions on the controller parameters so that the closed-loop is BIBO stable, and the **final value theorem** to check whether there is any offset.

Lastly, we discussed three methods for tuning P, PI, PD and PID controllers, namely, manual tuning, the open-loop (aka first) Ziegler-Nichols method and the ultimate sensitivity (aka second) Ziegler-Nichols method. The first Ziegler-Nichols method is applied to systems whose step response resembles the step response of a delayed first-order system. The second

Ziegler-Nichols method can be applied to P-controlled closed-loop systems which exhibits sustained oscillations at some gain K_u. To determine the ultimate gain, K_u, and the ultimate period of sustained oscillations, T_u we need to find an **imaginary root** of the **characteristic polynomial** of the closed-loop transfer function.

We briefly discussed how to implement a PID controller on a digital computer by approximating the derivative and the integral by using the backward Euler difference scheme and numerical integration schemes respectively.

It is still not entirely clear how we can design stabilising controllers — or even derive stability conditions — for systems with time delays. It is also not clear how the PID parameters affect the stability of the closed-loop system.

10.6 Exercises

Exercise 10.9 (Literature review) Find (online, in books, scientific articles, etc) a few examples of control systems where the PID controller is used. What is the controlled system? What are the output and input variables? What are the possible disturbances?

Exercise 10.10 (PI(D) controller on nonlinear system) Consider a nonlinear dynamical system described, in state space, by $\dot{x} = f(x, u)$. The system is controlled with a PI controller. Suppose that the set point is $x^{\text{sp}}(t) = 0$. Write the closed-loop system in a state space representation. Note: you would not be able to do the same with a PID controller; can you see why?

✂ ......

Hint: The PI controller is a dynamical system. Write it in its state space representation by introducing the state variable $z(t) = \int_0^t e(\tau)\mathrm{d}\tau$. Then, by the fundamental theorem of calculus, $\dot{z} = e$.

Exercise 10.11 (Control of a DC motor) A DC motor is controlled via the voltage across its armature, V. Let the moment of inertia of its rotor be $I = 0.014\,\text{kg m}^2$, the viscous friction coefficient of the motor is $b = 0.08\,\text{Nms}$, the electromotive force constant is $k_e = 0.011\,\text{V/rad s}$, the torque constant of the motor is $k_t = 0.012\,\text{Nm/A}$, the electric resistance is $R = 2.2\,\Omega$ and the inductance is $L = 460\,\text{mH}$. The transfer function between V and the speed of rotation, $\omega(t)$, is

$$G_s(s) = \frac{k_t}{(Ls + R)(Is + b) + k_t k_e}. \tag{10.86}$$

1. Show that a P controller can make the closed-loop system BIBO stable, but cannot eliminate the offset
2. Show that a PI controller can eliminate the offset

Exercise 10.12 (First-order system with PI controller) A first-order system with gain $K_s > 0$ and time constant $\tau_s > 0$ is controlled with a PI controller with gain $K_c > 0$ and integral time $\tau_I > 0$. Determine the closed-loop transfer function, G_{cl}, and determine the output of the closed-loop system upon a unit step change of the set point.

✳ ......

Chapter 10. PID controllers

Answer. The transfer function of the closed-loop system is

$$G_{cl}(s) = \frac{\tau_I \tau_s s^2 + \tau_I(K_c K_s + 1)s + K_c K_s}{\underbrace{\tau_I \tau_s s^2 + \tau_I(K_c K_s + 1)s + K_c K_s}_{\text{first term}}} + \underbrace{\frac{K_c K_s \tau_I s}{\tau_I \tau_s s^2 + \tau_I(K_c K_s + 1)s + K_c K_s}}_{\text{second term}}.$$
(10.87)

which is *similar*, but not identical, to the standard second-order systems we studied in Chapter 8. The step response of G_{cl} that corresponds to the "first" term" above is given in Section 8.3.3. The step response that corresponds to the second term can be determined from the first term using Theorem 4.19 (note that the second term is equal to the first term time $\tau_I s$).

Exercise 10.13 (True/False) True of false?

	True	False
A PID controller computes its control actions according to $u(t) = K_c\left(e(t) + \tau_D \dot{e}(t) + \frac{1}{\tau_I} I(t)\right)$, where $e(t)$ is the control error	☐	☐
When designing a PID controller, one is only interested in guaranteeing closed-loop stability	☐	☐
The limit of the error as time, t, goes to infinity is called offset	☐	☐
The main reason for using the integral in PID is to attenuate oscillations	☐	☐
If we decrease the proportional gain in a PID controller, *ceteris paribus*, we expect weaker oscillations	☐	☐

Answer. T, F (we also need, at least, to have zero offset), T, F, F.

Exercise 10.14 (Unstable second-order system with P controller) A system with transfer function

$$G(s) = \frac{1}{s^2 + 2s - 1}, \qquad (10.88)$$

is controlled with a P controller. Determine the range of controller gain values so that the closed-loop system is BIBO stable. What is the value of the controller gain so that the step response of the closed-loop system is oscillatory with a decay ratio of 25%.

---

should find that $G_{cl}(s) = \frac{K_c}{s^2+2s+K_c+1}$. Then use Propositions 8.4 and 8.6.

※

Answer. The closed-loop system is BIBO-stable for all $K_c > -1$. To achieve the desired decay ratio we need to use $K_c \approx 20.542288$.

Exercise 10.15 (Effect of τ_D on offset) Suppose that the transfer function G_{cl} is BIBO stable and the system is controlled with a PD controller with controller gain K_c and derivative time τ_D. Show that changing τ_D does not affect the offset.

For the sake of simplicity, you may assume that $G_a(s) = 1$, $G_m(s) = 1$ and that $G_s(s)$ is a proper rational function.

※

Answer. First, confirm that the conditions of the final value theorem are satisfied. Suppose that $G_s(s) = \frac{Q(s)}{P(s)}$, where P and Q are polynomials with $\deg P \leq \deg Q$. The closed-loop transfer function is $G_{cl}(s) = \frac{Q(s)+K_c(1+\tau_D s)P(s)}{Q(s)+K_c(1+\tau_D s)P(s)}$ where the degree of its numerator is $n = \deg P + 1$ and the degree of its denominator is $m = \max\{\deg Q, \deg P + 1\} \geq \deg P + 1$, so G_{cl} is proper. By FVT, the offset is $1 - \lim_{s\to\infty} G_{cl}(s)$. Since G_{cl} is proper, $\lim_{s\to\infty} G_{cl}(s)$ is finite and equal to $\frac{Q(0)+K_c P(0)}{K_c P(0)}$, which does not depend on τ_D.

Exercise 10.16 (Tuning methods) Discuss the advantages and limitations of the tuning methods in Section 10.3.

Exercise 10.17 (Tuning) Tune a PID controller for the following system

$$G(s) = \frac{0.5s+1}{(s+1)^3(0.1s+1)}, \qquad (10.89)$$

using the first Ziegler-Nichols method and the Ziegler-Nichols ultimate sensitivity method.

※

Answer. For the first Ziegler-Nichols method we have that $R \approx 0.26$ and $L \approx 0.4$, so $K_c = 11.53$, $\tau_I = 0.8$ and $\tau_D = 0.2$. Using the Ziegler-Nichols ultimate sensitivity method we have that $K_u \approx 23.74$, $T_u \approx 1.8$ and we find $K_c = 14.24$, $\tau_I = 0.9$ and $\tau_D = 0.225$.

Exercise 10.18 (Tuning by pole placement) A linear system with transfer function

$$G(s) = \frac{s+2}{s^2(s+5)} \quad (10.90)$$

is controlled with a P controller with gain K. Show that the closed-loop transfer function of the system, G_{cl}, is of third order (its characteristic polynomial, χ, is of degree 3) and determine the value of K so that χ has a pair of complex conjugate roots $a \pm jb$ with $a < 0$ and

$$\left|\frac{b}{a}\right| = 10. \quad (10.91)$$

✲ ....

Answer: The characteristic polynomial of the closed-loop system is $\chi(s) = s^3 + 5s^2 + Ks + 2K$. Using Routh's tabulation we can see that for every $K > 0$ the closed-loop system is BIBO-stable. It suffices to have $\chi(a + bj) = 0$ with $b = 10a$, i.e., $\chi(a + 10aj) = 0$, from which we find that

$$K = \frac{a^2(a(299 + 970j) + (495 - 100j))}{2 + a(1 + 10j)}, \quad (10.92)$$

but we also need to require that $K \in \mathbb{R}$, i.e., $\text{im}K = 0$, from which we find that the two possible values of a are $a_1 \approx -1.472358$ and $a_2 = -0.067246$. The corresponding values of K are $K_1 = 225.0$ and $K_2 = 1.111$.

Exercise 10.19 (Can this work?) Fill in the following table following the example in the first two rows. In what follows, a is a positive constant.

System	P	PI	PID
$G_s(s) = \dfrac{1}{s^2 + b^2}$	U	U	✓
$G_s(s) = \dfrac{1}{s + a}$	S + O	✓	✓
$G_s(s) = \dfrac{1}{s(s + a)}$			
$G_s(s) = \dfrac{1}{(s + a)^2}$			
$G_s(s) = \dfrac{1}{s - a}$			
$G_s(s) = \dfrac{1}{s^2 - a}$			

In particular, write "U" if the system is always unstable regardless of the controller parameters, write "S + O" if the controller parameters can be chosen so that the closed-loop system be stable, but there will always be some nonzero offset and write "✓" if the controller parameters can be such that the closed-loop system is BIBO stable and the offset is zero.

Answer. (Third row) ✓, ✓; (Fourth row) S+O, ✓, ✓; (Fifth row) S+O, ✓; (Last row) U, U, S+O, ✓.

Exercise 10.20 (**First-order system with PI controller: tuning**) A first-order system with gain 0.5 and time constant 5 is controlled with a PI controller. Choose the parameters of the PI controller so that the step response of the closed-loop transfer function exhibits an overshoot of 30%.

Hint. See Exercise 8.16. You may fix the value of K_c, e.g., take $K_c = 1$ and determine a τ_I such that the overshoot is 30%. The resulting equation is highly nonlinear and cannot be solved for τ_I, so you will need to resort to a numerical method, e.g., MATLAB's fsolve. You can also plot the overshoot against τ_I to obtain an understanding of how τ_I affects the overshoot.

Answer. Choose $K_c = 1$ and $\tau_I = 0.596356$.

Exercise 10.21 (**Stability analysis**) Consider a system with transfer function

$$G_s(s) = \frac{2-s}{(s+1)(s+4)}, \qquad (10.93)$$

whose output is measured with a sensor with transfer function $G_m(s) = 1$ and its input is manipulated with an actuator with transfer function $G_a(s) = 1$. The system is controlled with a P controller.

1. Determine the range of values of the controller gain so that the closed-loop system is stable in the BIBO sense.

2. Determine the range of values of the controller gain so that all poles of the closed-loop transfer function have a real part no larger than -0.5.

3. The system is controlled with a P controller with controller gain equal to the 80% of the maximum value you determined

Chapter 10. PID controllers

in part 1. Determine the final value of the output of the system upon a step change of the set point of amplitude 2.

4. Design a controller so that the closed-loop system is BIBO stable and the offset upon a step change of the set point is equal to zero.

❋ ....

Answer. The ranges of values are (i) $-2 < K_c < 5$ and (ii) $-0.7 < K_c < 4$. In the third question, the final value is 2/3. In the last question, we opt for a PI controller in order to be able to eliminate the offset. We can assume that $K_c\tau_I > 0$ (although it is not necessary). Any pair of (K_c, τ_I) with $K_c < 5$, and $(5 - K_c)(4\tau_I - K_c - 2K_c\tau_I) - 2K_c > 0$ lead to a stable closed loop with zero offset. Note that by the BIBO instability criterion, we must also have $4\tau_I - K_c + 2K_c\tau_I > 0$. We can then employ the final value theorem to show that the offset upon a step change of the set point is zero.

11

Frequency response and design of stable closed loops

*When a stable linear system is excited with a sinusoidal input, its output, after adequately long time, is a scaled and shifted sinusoidal signal. In this chapter, we study the large-time asymptotic behaviour of the frequency response of BIBO-stable linear systems as well as for systems with a single pole at zero and a pair of conjugate imaginary poles. We show that in the case of linear systems, the amplitude of the frequency response at large time is given by $|G(j\omega)|$ and its phase lag is given by $\arg G(j\omega)$. It turns out that the plot of $|G(j\omega)|$ and $\arg G(j\omega)$ vs ω — the so-called **Bode diagram** — can be used for the stability analysis of linear systems. Bode diagrams can be sketched easily, can be used for systems with time delays without the need for a Padé or other approximation, and provide information about "how much stable/unstable" a system is.*

11.1 Frequency response of general linear systems

11.1.1 Frequency response theorem

In Chapters 7 and 8 we saw that when a first or second order system is excited with a sinusoidal input, $u(t) = A\sin(\omega t)$, then, at large times the output of the system can be approximated by

$$x_\infty(t) = AR_\infty \sin(\omega t + \phi_0), \qquad (11.1)$$

where $R_\infty = |G(j\omega)|$ and $\phi_0 = \arg G(j\omega)$, and where G is the transfer function of the system. This approximation is meant in the sense

$$\lim_{t\to\infty} |x(t) - x_\infty(t)| = 0. \qquad (11.2)$$

This approximation is known as the **steady state frequency response** of the system. Note that the frequency of the output is equal to the frequency of the input signal. We call R_∞ the **amplitude ratio** and ϕ_0 the **phase lag** of the response. It turns out that this result holds true for all stable linear systems as shown in the following theorem.

> **Theorem 11.1 (Steady state frequency response)** Let G be the transfer function of a BIBO-stable linear dynamical system which is excited with a sinusoidal input,
>
> $$u(t) = A\sin(\omega t). \tag{11.3}$$
>
> Then, the steady state frequency response of the system is given by
>
> $$x_\infty(t) = AR_\infty \sin(\omega t + \phi_0), \tag{11.4}$$
>
> with $R_\infty = |G(j\omega)|$ and $\phi_0 = \arg G(j\omega)$.

Proof. Let $g(t)$ be the impulse response of the system. Then, the response to the input signal in Equation (11.3) will be $x(t) = \int_0^t g(\tau) A \sin(\omega(t-\tau)) H_0(t-\tau) \mathrm{d}\tau$. We have explicitly written $H_0(t-\tau)$ for clarity[1]. This integral can be written as

$$x(t) = \underbrace{\int_0^\infty g(\tau) A \sin(\omega(t-\tau))\mathrm{d}\tau}_{x_\infty} - \int_t^\infty g(\tau) A \sin(\omega(t-\tau))\mathrm{d}\tau$$

Since the system is assumed to be BIBO-stable, the second integral converges to zero as $t \to \infty$[2], so for large t it can be ignored and $x(t)$ can be approximated by the first integral, which we denote by $x_\infty(t)$.

Now recall Euler's formula that allows us to trade trigonometric functions for an exponential:

$$e^{jx} = \cos x + j \sin x,$$

so, we observe that $\mathbf{im}(e^{jx}) = \sin x$. This allows us to write

$$x_\infty(t) = \int_0^t g(\tau) A \sin(\omega(t-\tau)) \mathrm{d}\tau$$

$$= A \int_0^\infty g(\tau) \mathbf{im}[e^{j\omega(t-\tau)}] \mathrm{d}\tau$$

and given that $g(\tau)$ is a real number

$$= A \int_0^\infty \mathbf{im}[g(\tau) e^{j\omega(t-\tau)}] \mathrm{d}\tau$$

---

[1] So far, we have been omitting H_0. We have been tacitly assuming that all involved function in the time domain are defined over $[0, \infty)$ vanish for $t < 0$ (read again the discussion at the beginning of Section 4.1)

[2] Why? The reader can verify this statement as an exercise. Hint: Use Theorem 9.5.

Chapter 11. Frequency response and design of stable closed loops

$$= A \operatorname{im} \left[\int_0^\infty g(\tau) e^{j\omega(t-\tau)} d\tau \right]$$
$$= A \operatorname{im} \left[e^{j\omega t} \int_0^\infty g(\tau) e^{-j\omega\tau} d\tau \right]$$
$$= A \operatorname{im} \left[e^{j\omega t} G(j\omega) \right],$$

where $G(j\omega)$ can be written as $G(j\omega) = |G(j\omega)| e^{j \arg G(j\omega)}$, so

$$= A|G(j\omega)| \operatorname{im} \left[e^{j(\omega t + \arg G(j\omega))} \right]$$
$$= A|G(j\omega)| \sin(\omega t + \arg G(j\omega)), \tag{11.5}$$

which completes the proof. □

In order to facilitate our computations, we will use the following properties of complex numbers (see Section A.1). For complex numbers $z_1, z_2, \ldots, z_n$, the modulus and argument of their product, $z = z_1 z_2 \ldots z_n$ is given by

$$|z| = |z_1| \cdots |z_2| \cdot \ldots \cdot |z_n|, \tag{11.6a}$$
$$\arg z = \arg z_1 + \arg z_2 + \ldots + \arg z_n. \tag{11.6b}$$

These properties can be derived from the exponential representation of the complex numbers $z_1, z_2, \ldots, z_n$ as discussed in Section A.1.4. Similarly, the modulus and argument of the complex number z_1/z_2, with $z_2 \neq 0$, is

$$\left| \frac{z_1}{z_2} \right| = \frac{|z_1|}{|z_2|}, \tag{11.7a}$$
$$\arg \frac{z_1}{z_2} = \arg z_1 - \arg z_2. \tag{11.7b}$$

These properties facilitate greatly the computation of the modulus and argument of $G(j\omega)$.

We have already determined the steady state frequency response of first-order systems (see Section 7.4, Theorem 7.3). Let us now apply Theorems 9.9 and 11.1 to determine the steady state frequency response of a stable dynamical system with a triple real pole.

Example 11.1 (First example of frequency respose) Consider a system with transfer function

$$G_{\text{ol}}(s) = \frac{10}{(s+1)^3}. \tag{11.8}$$

This system has a triple pole at -1, therefore, it is BIBO-stable (Theorem 9.9), therefore, we can apply Theorem 11.1.

The amplitude ratio of the system is

$$|G(j\omega)| = \left|\frac{10}{(j\omega+1)^3}\right|$$
$$= \frac{10}{|(j\omega+1)^3|} = \frac{10}{|j\omega+1|^3} = \frac{10}{\sqrt{\omega^2+1}^3}, \quad (11.9)$$

where we used the fact that $|z^3| = |z \cdot z \cdot z| = |z| \cdot |z| \cdot |z| = |z|^3$.

The phase lag of G is

$$\arg G(j\omega) = \arg \frac{10}{(j\omega+1)^3}$$
$$= -\arg(j\omega+1)^3$$
$$= -3\arg(j\omega+1) = -3\arctan\omega. \quad (11.10)$$

According to Theorem 11.1, at large times, the system output can be approximated by

$$x_\infty(t) = A|G(j\omega)|\sin(\omega t + \arg G(j\omega)), \quad (11.11)$$

where $|G(j\omega)|$ and $\arg G(j\omega)$ are given by Equations (11.9) and (11.10) respectively. The frequency response, $x(t)$, and the steady state frequence response, $x_\infty(t)$, with $A = 1$ and $\omega = 10$ are shown in Figure 11.1.

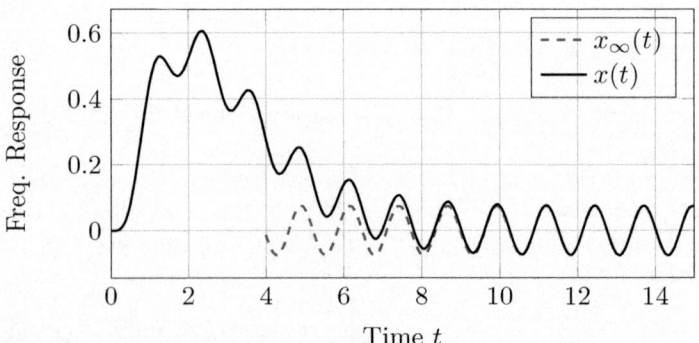

Figure 11.1: *Frequency response with $A = 1$ and $\omega = 10$. Note that $x(t) - x_\infty(t) \to 0$ as $t \to \infty$.*

Chapter 11. Frequency response and design of stable closed loops

Example 11.2 (Frequency response: two distinct real poles, one zero) We will determine the amplitude ratio and the phase lag for the following system

$$G(s) = \frac{2s - 1}{(s+1)(s+2)}. \tag{11.12}$$

The system is BIBO-stable (because of Theorem 9.9), so Theorem 11.1 applies. The amplitude ratio at frequency ω is given by $|G(j\omega)|$ and the phase lag is $\arg G(j\omega)$. Let us compute $G(j\omega)$.

$$G(j\omega) = \frac{2j\omega - 1}{(j\omega + 1)(j\omega + 2)}. \tag{11.13}$$

The modulus of $G(j\omega)$ is

$$\begin{aligned}
|G(j\omega)| &= \left| \frac{2j\omega - 1}{(j\omega + 1)(j\omega + 2)} \right| \\
&= \frac{|2j\omega - 1|}{|j\omega + 1||j\omega + 2|} \\
&= \frac{\sqrt{4\omega^2 + 1}}{\sqrt{\omega^2 + 1}\sqrt{\omega^2 + 4}}. \tag{11.14}
\end{aligned}$$

The argument of $G(j\omega)$ is

$$\begin{aligned}
\arg G(j\omega) &= \arg \frac{2j\omega - 1}{(j\omega + 1)(j\omega + 2)} \\
&= \arg(2j\omega - 1) - \arg(j\omega + 1) - \arg(j\omega + 2) \\
&= \operatorname{atan2}(2\omega, -1) - \operatorname{atan2}(\omega, 1) - \operatorname{atan2}(\omega, 2). \\
&= -\arctan(2\omega) + \pi - \arctan \omega - \arctan \tfrac{\omega}{2}. \tag{11.15}
\end{aligned}$$

According to Theorem 11.1, at large times, the system output can be approximated by

$$x_\infty(t) = A|G(j\omega)| \sin(\omega t + \arg G(j\omega)), \tag{11.16}$$

where $|G(j\omega)|$ and $\arg G(j\omega)$ are given by equations (11.14) and (11.15) respectively. We should note that we determined the frequency response characteristics of the system (for large t) without determining the exact system response using the inverse Laplace transform.

Exercise 11.1 (Frequency response) Similar to Example 11.1 (Figure 11.1), write a Python or MATLAB script to plot the frequency response and the steady frequency response of the system in Example 11.2.

Let us now move on to an example which involves a delay. We will determine the amplitude ratio and phase lag without approximating the exponential.

Example 11.3 (Frequency response: delay) We will now determine the frequency response characteristics of a dynamical system with transfer function

$$G(s) = \frac{(s-2)e^{-3s}}{(s+1)^2 + 1}. \quad (11.17)$$

This is the product of a rational function with an exponential, so we can invoke Corollary 9.11. Function G has a pair of complex conjugate poles, $-1 \pm j$, which have a negative real part, so according to Theorem 9.9 it is BIBO-stable, therefore, Theorem 11.1 applies. In order to compute $|G(j\omega)|$ and $\arg G(j\omega)$ we observe that $G(s)$ can be written as

$$G(s) = \frac{G_1(s)G_2(s)}{G_3(s)}, \quad (11.18)$$

where $G_1(s) = s - 2$, $G_2(s) = e^{-3s}$ and $G_3(s) = (s+1)^2 + 1$. Then,

$$|G(j\omega)| = \frac{|G_1(j\omega)||G_2(j\omega)|}{|G_3(j\omega)|}. \quad (11.19)$$

Let us compute the moduli $|G_1(j\omega)|$, $|G_2(j\omega)|$ and $|G_3(j\omega)|$.

$$|G_1(j\omega)| = |j\omega - 2| = \sqrt{\omega^2 + 4}, \quad (11.20)$$

similarly,

$$|G_2(j\omega)| = |e^{-3j\omega}| = 1, \quad (11.21)$$

and

$$|G_3(j\omega)| = |(j\omega+1)^2 + 1| = |-\omega^2 + 2j\omega + 1 + 1|$$
$$= |(2-\omega^2) + 2j\omega| = \sqrt{(2-\omega^2)^2 + 4\omega^2}. \quad (11.22)$$

By virtue of Equation (11.19),

$$|G(j\omega)| = \frac{\sqrt{\omega^2 - 4}}{\sqrt{(2-\omega^2)^2 + 4\omega^2}}. \quad (11.23)$$

The arguments of $G_1(j\omega)$, $G_2(j\omega)$ and $G_3(j\omega)$ are

$$\arg G_1(j\omega) = \arg(j\omega - 2) = \text{atan2}(\omega, -2)$$
$$= \pi - \arctan\frac{\omega}{2}, \quad (11.24a)$$
$$\arg G_2(j\omega) = \arg e^{-3j\omega} = -3\omega, \quad (11.24b)$$

$$\arg G_3(j\omega) = \arg((2-\omega^2) + 2j\omega)$$
$$= \text{atan2}\,(2\omega, 2-\omega^2). \qquad (11.24c)$$

The argument of $G(j\omega)$ is, therefore,

$$\arg G(j\omega) = \arg G_1(j\omega) + \arg G_3(j\omega) - \arg G_3(j\omega)$$
$$= \pi - \arctan\frac{\omega}{2} - 3\omega - \text{atan2}\,(2\omega, 2-\omega^2). \qquad (11.25)$$

The quantities $|G(j\omega)|$ and $\arg G(j\omega)$ characterise the frequency response of the system at large times. If the system is excited with an input $u(t) = A\sin(\omega t)$, then, after sufficiently long time, its response can be approximated by

$$x_\infty(t) = A|G(j\omega)|\sin(\omega t + \arg G(j\omega)). \qquad (11.26)$$

The frequency response of the system, $x(t)$, and the steady state frequency response, $x_\infty(t)$, are shown in Figure 11.2.

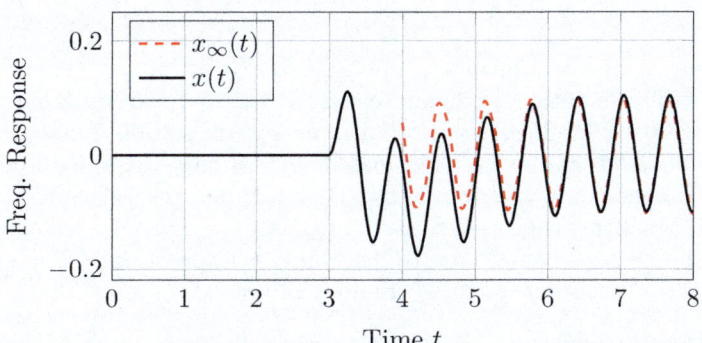

Figure 11.2: *Frequency response of the above system with $A = 1$ and $\omega = 10$.*

Exercise 11.2 (Determine $|G(j\omega)|$ and $\arg G(j\omega)$) Compute $|G(j\omega)|$ and $\arg G(j\omega)$ in each of the following cases:

1. $G(s) = \frac{e^{-s}(s+1)}{s-1}$

2. $G(s) = \frac{2s+3}{(s+1)^2(3s+1)^{10}}$

3. $G(s) = \frac{s}{(s+1)^5}$

4. $G(s) = \frac{2s+1}{(s^2+s+1)^2(s+1)(s+3)}$

What do you conclude about the frequency response of the system?

✂ ....

Hint. When a term is raised to a power, we can use the fact that for all $n \in \mathbb{N}$, $|z^n| = |z|^n$, which follows from Equation (11.6a). The argument of z^n is $\arg z^n = n \arg z$, which is because of Equation (11.6b). The answer to the last question, regarding the frequency response, is not trivial. Check the conditions of Theorem 11.1 for each case.

Exercise 11.3 (Frequency response) Determine the frequency response of a dynamical system described by the transfer function

$$G(s) = \frac{e^{-t_d s}}{(s+a)^\nu}, \qquad (11.27)$$

where $a, t_d > 0$, $\nu \in \mathbb{N}$, $\nu \geq 1$.

✳ ....

Answer. Firstly, we should note that the system is BIBO-stable. If the system is excited with an input $u = \sin(\omega t)$, the output, after a long time, can be approximated by $x_\infty(t) = (\omega^2 + a^2)^{-\nu/2} \sin\left(\omega(t - t_d) - \nu \arctan\frac{\omega}{a}\right)$.

When computing the frequency response characteristics of a dynamical system, its amplitude ratio and its phase lag, we have to make sure that the system is BIBO-stable. If it is not stable, the system response may grow unbounded, but the amplitude ratio may well be finite. In other words, if we fail to observe that a system is not BIBO-stable, the values $|G(j\omega)|$ and $\arg G(j\omega)$ will provide no relevant information about the frequency response of the system. Note that BIBO stability is a condition of Theorem 11.1. Let us have a look at a simple counterexample.

Example 11.4 (Unstable system) Consider the following unstable system

$$G(s) = \frac{1}{s-1}. \qquad (11.28)$$

This is unstable because it has a pole at $s = 1 > 0$ (Theorem 9.9).

Let us compute $|G(j\omega)|$ and $\arg G(j\omega)$. It is

$$|G(j\omega)| = \frac{1}{|j\omega - 1|} = \frac{1}{\sqrt{\omega^2 + 1}}, \qquad (11.29)$$

and

$$\begin{aligned}
\arg G(j\omega) &= \arg 1 - \arg(j\omega - 1) \\
&= -\operatorname{atan2}(\omega, -1) \\
&= -(\arctan(-\omega) + \pi) \\
&= \arctan(\omega) - \pi. \qquad (11.30)
\end{aligned}$$

Chapter 11. Frequency response and design of stable closed loops

However, we cannot conclude that if the system is excited with $u(t) = A\sin(\omega t)$ the output will be approximated by $x_\infty(t) = A|G(j\omega)|\sin(\omega t + \arg G(j\omega))$. On the contrary, we know that x will grow unbounded.

The reader can verify that

$$x(t) = \mathscr{L}^{-1}\left\{G(s)\frac{A\omega}{s^2+\omega^2}\right\}(t)$$
$$= \frac{A}{\omega^2+1}\left(\omega e^t - \sin(\omega t) - \omega\cos(\omega t)\right), \qquad (11.31)$$

for $t \geq 0$, which means that $\lim_{t\to\infty} x(t) = \infty$; the output is not sinusoidal in this case.

11.1.2 Special case: systems with a pole at zero

The methodology of Section 11.1.1 can be extended to systems with a pole at zero. This case is not covered by Theorem 11.1 since it is required that the system is BIBO-stable, that is, all its poles must have a negative real part.

Nevertheless, we know that if a simple integrator is excited with a sinusoidal input, $u(t) = A\sin(\omega t)$, the output will be oscillatory. In particular,

$$x(t) = \mathscr{L}^{-1}\left\{\frac{1}{s}\frac{A\omega}{s^2+\omega^2}\right\}(t) = \frac{A}{\omega}(1-\cos(\omega t)), \qquad (11.32)$$

for $t \geq 0$. Note that the response of the system is offset by A/ω.

As a second example, consider a dynamical system with transfer function

$$G(s) = \frac{1}{s(s+1)}, \qquad (11.33)$$

which has a stable pole at -1 and a pole at zero. The reader can confirm that the response of this system to $u(t) = A\sin(\omega t)$ is

$$x(t) = \mathscr{L}^{-1}\left\{\frac{1}{s(s+1)}\frac{A\omega}{s^2+\omega^2}\right\}(t)$$
$$= A\left[\frac{1}{\omega} - \frac{\cos(\omega t) + \omega\sin(\omega t)}{\omega(\omega^2+1)} - \frac{\omega e^{-t}}{\omega^2+1}\right], \qquad (11.34)$$

for $t \geq 0$. Again we see that the system output is oscillatory and at large t, the term e^{-t} becomes negligible, so the system response can be approximated by

$$x_\infty(t) = A\left[\frac{1}{\omega} - \frac{\cos(\omega t) + \omega\sin(\omega t)}{\omega(\omega^2+1)}\right], \qquad (11.35)$$

for $t \geq 0$. Note that, again, the steady state frequency response is offset by A/ω. Overall, we have reasons to believe that if a system has a *single* pole at zero and *all other poles* have a negative real part, then the system response is oscillatory. Let us have a closer look.

Suppose that a transfer function has the form

$$G(s) = \frac{H(s)}{s}, \qquad (11.36)$$

where H is a rational function with its poles in the open left half plane (negative real parts). If the system with transfer function G is excited with a sinusoidal input of the form $u(t) = A\sin(\omega t)$, $A, \omega > 0$, the output will be

$$X(s) = G(s)\frac{A\omega}{s^2 + \omega^2} = \frac{H(s)}{s}\frac{A\omega}{s^2 + \omega^2}. \qquad (11.37)$$

In order to determine $x(t)$ we first need to expand the above in partial fractions. As we discussed in Section 5.2, this is

$$\frac{H(s)}{s}\frac{A\omega}{s^2 + \omega^2} = \frac{C}{s} + \frac{B_1 s + B_2}{s^2 + \omega^2} + R(s), \qquad (11.38)$$

where $R(s)$ accounts for all other terms which corresponds to the stable poles of H. We multiply both sides of Equation (11.38) by $s(s^2 + \omega^2)$ — assuming $s(s^2 + \omega^2) \neq 0$ — to obtain

$$A\omega H(s) = C(s^2 + \omega^2) + (B_1 s + B_2)s + R(s)s(s^2 + \omega^2), \qquad (11.39)$$

Note that R has no poles at 0 or $\pm j\omega$. By taking $s \to 0$ we have

$$A\omega H(0) = C\omega^2 \Rightarrow C = \tfrac{A}{\omega}H(0). \qquad (11.40)$$

In order to determine B_1 and B_2, we take $s \to j\omega$ and $s \to -j\omega$, and we have

$$A\omega H(j\omega) = (B_1 j\omega + B_2)j\omega, \qquad (11.41a)$$
$$A\omega H(-j\omega) = (B_1(-j\omega) + B_2)(-j\omega). \qquad (11.41b)$$

By adding and subtracting the two equations by parts and using the fact that $H(-j\omega) = \overline{H(j\omega)}$ we have

$$A(H(j\omega) + \overline{H(j\omega)}) = -2B_1\omega, \qquad (11.42a)$$
$$A(H(j\omega) - \overline{H(j\omega)}) = 2jB_2. \qquad (11.42b)$$

By Equation (A.10), we have that $H(j\omega) + \overline{H(j\omega)} = 2\mathrm{re}[H(j\omega)]$ and $H(j\omega) - \overline{H(j\omega)} = 2j\mathrm{im}[H(j\omega)]$, so B_1 and B_2 are given by

$$B_1 = -\frac{A}{\omega}\mathrm{re}[H(j\omega)], \qquad (11.43a)$$
$$B_2 = A\mathrm{im}[H(j\omega)]. \qquad (11.43b)$$

By Proposition 5.4, the inverse Laplace transform of $X(s)$, ignoring the stable terms, i.e., $R(s)$, which will decay to zero, gives $x_\infty(t)$

$$x_\infty(t) = \tfrac{A}{\omega} H(0) + K \sin(\omega t + \phi_0), \tag{11.44}$$

where (see Proposition 5.4)

$$\begin{aligned} K &= \tfrac{1}{\omega}\sqrt{(\omega B_1)^2 + B_2^2} \\ &= \tfrac{1}{\omega}\sqrt{\left(-\omega\tfrac{A}{\omega}\mathbf{re}[H(j\omega)]\right)^2 + (A\mathbf{im}[H(j\omega)])^2} \\ &= \tfrac{A}{\omega}|H(j\omega)| = A|G(j\omega)|, \end{aligned} \tag{11.45}$$

and the phase lag is

$$\begin{aligned} \phi_0 &= \mathrm{atan2}(\omega B_1, B_2) \\ &= \mathrm{atan2}(-\mathbf{re}[H(j\omega)], \mathbf{im}[H(j\omega)]), \end{aligned}$$

and using the fact that $\mathrm{atan2}(x, y) + \mathrm{atan2}(y, x) = \pi/2$ for $x, y \in \mathbb{R}$, not both zero (see Equation (A.44)),

$$= \frac{\pi}{2} - \mathrm{atan2}(\mathbf{im}[H(j\omega)], -\mathbf{re}[H(j\omega)]),$$

and since $\mathrm{atan2}(y, -x) = \pi - \mathrm{atan2}(y, x)$

$$\begin{aligned} &= \mathrm{atan2}(\mathbf{im}[H(j\omega)], \mathbf{re}[H(j\omega)]) - \pi/2 \\ &= \arg H(j\omega) - \pi/2 = \arg G(j\omega). \end{aligned} \tag{11.46}$$

Note: To be more precise, $\mathrm{atan2}(x, y) + \mathrm{atan2}(y, x) = \pi/2$ if at least one of x, y is not negative. If $x, y < 0$, then $\mathrm{atan2}(x, y) + \mathrm{atan2}(y, x) = \pi/2 - 2\pi$, however, the angles $\pi/2$ and $\pi/2 - 2\pi$ are equivalent for the purpose of determining the frequency response (they only show up inside a trigonometric function[3]).

---

[3] Recall that $\sin(x + 2\pi) = \sin(x)$, $\sin(x - 2\pi) = \sin(x)$ and, in general, $\sin(x + 2k\pi) = \sin(x)$ for all $k \in \mathbb{Z}$.

Theorem 11.2 (Steady state frequency response with single zero pole) Suppose that a dynamical system has the transfer function

$$G(s) = \frac{H(s)}{s}, \qquad (11.47)$$

where H is a BIBO-stable transfer function. If the system is excited with a sinusoidal input of the form $u(t) = A\sin(\omega t)$, $A, \omega > 0$, then at large times, the output can be approximated by

$$x_\infty(t) = \tfrac{A}{\omega} H(0) + A R_\infty \sin(\omega t + \phi_0), \qquad (11.48)$$

with

$$R_\infty = |G(j\omega)|, \qquad (11.49\text{a})$$
$$\phi_0 = \arg G(j\omega). \qquad (11.49\text{b})$$

Proof. See above. □

Example 11.5 (Frequency response with single pole at zero) Consider the system

$$G(s) = \frac{1}{s(s+1)}. \qquad (11.50)$$

This is in the form discussed above: it has a single pole at zero and a stable pole at -1. The system can be written in the form of Equation (11.47) with $H(s) = \frac{1}{s+1}$. The value of H at zero is

$$H(0) = 1. \qquad (11.51)$$

The modulus of $H(j\omega)$ is

$$|H(j\omega)| = \left|\frac{1}{j\omega+1}\right| = \frac{1}{\sqrt{\omega^2+1}}, \qquad (11.52)$$

so the amplitude ratio is $R_\infty = \frac{1}{\omega}|H(j\omega)| = \frac{1}{\omega\sqrt{\omega^2+1}}$ and the phase lag is

$$\phi_0 = \arg G(j\omega) = \arg \frac{1}{j\omega(j\omega+1)} = -\tfrac{\pi}{2} - \arctan \omega. \qquad (11.53)$$

We have shown that if a system with transfer function $G(s) = \frac{1}{s(s+1)}$ is excited with input $u(t) = A\sin(\omega t)$, the output, at large t, can be approximated by

$$x_\infty(t) = \frac{A}{\omega} + \frac{A}{\omega\sqrt{\omega^2+1}} \sin\left(\omega t - \arctan \omega - \frac{\pi}{2}\right)$$

$$= \frac{A}{\omega} - \frac{A}{\omega\sqrt{\omega^2+1}} \cos(\omega t - \arctan \omega). \tag{11.54}$$

The reader can verify that this result is the same as the one we derived in Equation (11.35). The response of the system and its steady state frequency response are shown in Figure 11.3.

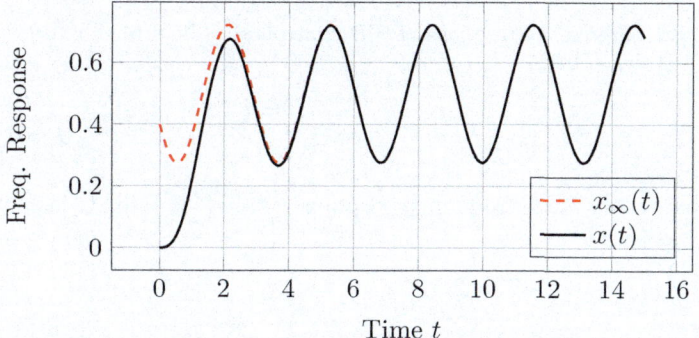

Figure 11.3: *System response, $x(t)$, and steady state frequency response, $x_\infty(t)$ with $A = 1$, and $\omega = 2$.*

Exercise 11.4 (Application of Theorem 11.2) A dynamical system is described by the transfer function

$$G(s) = \frac{s-3}{s(s+1)(2s+1)}. \tag{11.55}$$

Is this system BIBO-stable? The system is excited with a sinusoidal input of the form $u(t) = A\sin(\omega t)$. Approximate its output at large times using Theorem 11.2.

✂ ....

Hint: Start by defining $H(s) = \frac{(s+1)(2s+1)}{s-3}$ so that $G(s) = \frac{s}{H(s)}$. Determine the value $H(0)$ and apply Theorem 11.2.

Remark: *Theorem 11.2 does not hold if G has a multiple pole at zero or if it has any poles with positive real part. If G has a multiple zero pole, its frequency response will not be bounded, even if all other poles of G are in the open left complex plane. That said, Theorem 11.2 does not apply to*

$$G(s) = \frac{1}{s^2(s+1)}. \tag{11.56}$$

Using the inverse Laplace transform we can verify that if the input is $u(t) =

$A\sin(\omega t)$, with $A, \omega > 0$, then the output is given by

$$x(t) = \frac{A}{\omega}\left[t - 1 - \frac{1}{\omega(\omega^2+1)}(\sin(\omega t) - \omega\cos(\omega t))\right] + \frac{A\omega}{\omega^2+1}e^{-t}. \quad (11.57)$$

And it is easy to see that $\lim_{t\to\infty} x(t) = \infty$. •

Exercise 11.5 (Frequency response of PID controller) The transfer function of a PID controller is

$$G_c(s) = K_c\left(1 + \tau_D s + \frac{1}{\tau_I s}\right). \quad (11.58)$$

What is its steady state frequency response? Note that this transfer function has a pole at zero.

※ ....

Answer. The amplitude ratio of G_c at frequency ω is given by

$$|G(j\omega)| = K_c\sqrt{1 + \left(\tau_D\omega - \frac{\tau_I}{\omega}\right)^2}. \quad (11.59)$$

The phase lag is given by the argument of $G(j\omega)$, that is

$$\arg G(j\omega) = \arctan\left(\tau_D\omega - \frac{\tau_I}{\omega}\right). \quad (11.60)$$

The steady-state frequency response is given by Theorem 11.2.

11.1.3 Special case: simple pairs of imaginary poles

In the beginning of this section we characterised the steady state frequency response of a BIBO-stable dynamical system (Theorem 11.1). We discussed that BIBO stability is an essential requirement: if the system is not BIBO-stable its frequency *may* not be bounded.

In Section 11.1.2 we discussed an exception: if the system has a *single* pole at zero, the output will be bounded and at large times it can be approximated by a sinusoidal signal of amplitude $|G(j\omega)|$ and phase lag $\arg G(j\omega)$ plus a constant (see Theorem 11.2). There exists yet another special case where the frequency response is bounded. That is the case where the input signal is

$$u(t) = A\sin(\omega t), \quad (11.61)$$

with $A, \omega > 0$, and the transfer function of the system has N distinct pairs ($N \geq 1$) of imaginary poles $j\beta_\nu$, $\nu = 1, \ldots, N$ and $\beta_\nu > 0$, but $\omega \neq \beta_\nu$ for all ν. In this case, the system response is the sum of $N + 1$ sinusoidal functions with frequencies $\omega, \beta_1, \ldots, \beta_N$.

Chapter 11. Frequency response and design of stable closed loops

The input signal in the s domain is given by $U(s) = \frac{A\omega}{s^2+\omega^2}$. For the sake of simplicity, we will focus on the case where the transfer function has one pair of poles at $\pm j\beta$, that is, it has the form

$$G(s) = \frac{H(s)}{s^2 + \beta^2}, \tag{11.62}$$

where H is a BIBO-stable transfer function. Then, the system response, in the s-domain, is given by

$$X(s) = G(s)U(s) = \frac{H(s)}{s^2 + \beta^2} \frac{A\omega}{s^2 + \omega^2}. \tag{11.63}$$

We proceed as in Section 11.1.2; we first expand $X(s)$ in partial fractions.

$$\frac{H(s)}{s^2 + \beta^2} \frac{A\omega}{s^2 + \omega^2} = \frac{B_1 s + B_2}{s^2 + \beta^2} + \frac{C_1 s + C_2}{s^2 + \omega^2} + R(s), \tag{11.64}$$

where R corresponds to the stable poles of $H(s)$. In order to determine the values of B_1, B_2, C_1 and C_2 we first multiply both sides by $(s^2+\beta^2)(s^2+\omega^2)$

$$A\omega H(s) = (B_1 s + B_2)(s^2 + \omega^2) + (C_1 s + C_2)(s^2 + \beta^2)$$
$$+ R(s)(s^2 + \beta^2)(s^2 + \omega^2). \tag{11.65}$$

We first take $s \to \pm j\beta$ in Equation (11.65) to obtain

$$A\omega H(j\beta) = (B_1 j\beta + B_2)(\omega^2 - \beta^2), \tag{11.66a}$$
$$A\omega \overline{H(j\beta)} = (-B_1 j\beta + B_2)(\omega^2 - \beta^2). \tag{11.66b}$$

Again, we used the property $H(-j\beta) = \overline{H(j\beta)}$. We can solve for B_1 and B_2 to obtain

$$B_1 = A\frac{\omega}{\beta(\omega^2 - \beta^2)}\operatorname{im}[H(j\beta)], \tag{11.67a}$$

$$B_2 = A\frac{\omega}{(\omega^2 - \beta^2)}\operatorname{re}[H(j\beta)]. \tag{11.67b}$$

Similarly, we take $s \to \pm j\omega$ in Equation (11.65) and solve for C_1 and C_2 to obtain

$$C_1 = A\frac{1}{\beta^2 - \omega^2}\operatorname{im}[H(j\omega)], \tag{11.68a}$$

$$C_2 = A\frac{\omega}{\beta^2 - \omega^2}\operatorname{re}[H(j\omega)]. \tag{11.68b}$$

We ignore the contributions of R because it corresponds to stable poles and $\mathscr{L}^{-1}\{R\}(t) \to 0$ as $t \to \infty$, so, by Proposition 5.4, the system response can be approximated by

$$x_\infty(t) = \mathscr{L}^{-1}\left\{\frac{B_1 s + B_2}{s^2 + \beta^2}\right\}(t) + \mathscr{L}^{-1}\left\{\frac{C_1 s + C_2}{s^2 + \omega^2}\right\}(t)$$

$$= K_\omega \sin(\omega t + \vartheta_\omega) + K_\beta \sin(\beta t + \vartheta_\beta), \tag{11.69}$$

where, after carrying out some algebraic operations,

$$K_\omega = A|G(j\omega)|, \tag{11.70a}$$
$$\vartheta_\omega = \arg H(j\beta), \tag{11.70b}$$
$$K_\beta = \frac{A}{|\omega^2 - \beta^2|}|H(j\beta)|, \tag{11.70c}$$
$$\vartheta_\beta = \arg H(j\omega). \tag{11.70d}$$

We have proven the following result.

> **Theorem 11.3 (Steady state frequency response: simple pair of imaginary poles)** Suppose that a dynamical system is described by the transfer function
> $$G(s) = \frac{H(s)}{s^2 + \beta^2}, \tag{11.71}$$
> for some $\beta > 0$, where H is a BIBO-stable transfer function. The system is excited with an input signal of the form
> $$u(t) = A\sin(\omega t), \tag{11.72}$$
> with $A, \omega > 0$, where $\omega \neq \beta$. Then, at large times, the system response can be approximated by
> $$x_\infty(t) = K_\omega \sin(\omega t + \vartheta_\omega) + K_\beta \sin(\beta t + \vartheta_\beta), \tag{11.73}$$
> with $K_\omega, \vartheta_\omega, K_\beta, \vartheta_\beta$ given by Equation (11.70).

Note that as $\omega \to \beta$, $K_\omega \to \infty$ and $K_\beta \to \infty$; as a result $x_\infty \to \infty$ (why?). This effect is known as **resonance**, and we encountered it for the first time in Example 5.20.

Exercise 11.6 (Several pairs of imaginary poles) Generalise the above theorem to the case where the system is described by a transfer function of the form

$$G(s) = \frac{H(s)}{(s^2 + \beta_1^2) \cdot (s^2 + \beta_2^2) \cdot \ldots \cdot (s^2 + \beta_N^2)}, \tag{11.74}$$

and H is a BIBO-stable rational function (with all its poles in the open left half plane) and $\beta_\nu > 0$, $\nu = 1, \ldots, N$ are different from one another. The system is excited with a sinusoidal input of the form $u(t) = A\sin(\omega t)$ with $\omega \neq \beta_\nu$ for all $\nu = 1, \ldots, N$. Determine the steady state frequency response of the system.

✂ ....

Chapter 11. Frequency response and design of stable closed loops

$$G(s) = \sum_{i=1}^{N} \frac{s^2 + \beta_i^2}{B_{1,i} s + B_{2,i}} + \frac{s^2 + \omega^2}{C_1 s + C_2} + H(s), \qquad (11.75)$$

where $B_{1,i}$, $B_{2,i}$ for $i = 1, \ldots, N$ and C_1, C_2 are parameters to be determined. Now multiply both sides by $(s^2 + \omega^2) \prod_{i=1}^{N}(s^2 + \beta_i^2)$. Can you take it from there?

Hint. Follow the exact same steps as in the beginning of the section. For example, the generalisation of Equation (11.64) is

Example 11.6 (Pair of imaginary poles) Consider a dynamical system with transfer function

$$G(s) = \frac{10}{(s^2 + 5^2)(s + 0.5)}, \qquad (11.76)$$

and a sinusoidal input signal with

$$U(s) = \frac{1}{s^2 + 6^2}. \qquad (11.77)$$

Function G has a stable pole at -0.5 and a pair of imaginary poles at $\pm j5$, therefore, its response at large times can be approximated by Theorem 11.3. In particular, G is in the form of Equation (11.71) with $\beta = 5$, $H(s) = \frac{10}{s+0.5}$ and U is in the form of Equation (11.72) with $\omega = 6 \neq \beta$.

Its frequency response at large times is then given by Equation (11.72), that is,

$$x_\infty(t) = 0.1809 \sin(5t - 1.471) - 0.1258 \sin(6t - 1.488). \qquad (11.78)$$

The system response, $x(t)$, and $x_\infty(t)$ are shown in the figure below.

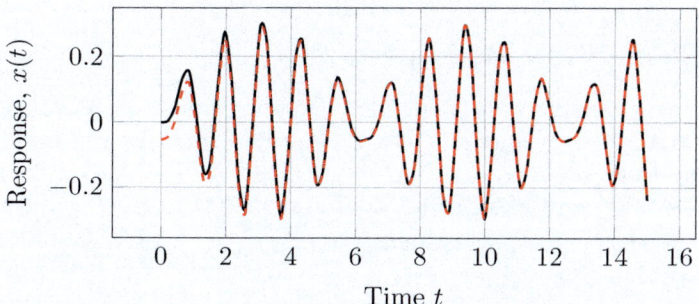

Figure 11.4: System response, $x(t)$, (black, solid line) and its approximation at large times, $x_\infty(t)$ (red, dashed line) given by Equation (11.78).

11.2 Bode plots

A Bode plot comprises two plots which show the dependence of $|G(j\omega)|$ and $\arg G(j\omega)$ on ω. These two plots are known as the Bode magnitude plot and the Bode phase plot respectively. The horizontal axis is logarithmic (we plot against $\log \omega$ instead of ω) and $|G(j\omega)|$ is typically measured in decibel (dB). In order to convert $|G(j\omega)|$ in dB we use

$$|G(j\omega)|_{\mathrm{dB}} = 20\log_{10}|G(j\omega)|. \tag{11.79}$$

You may have guessed why we scale $|G(j\omega)|$ logarithmically and why we plot against $\log \omega$. As discussed previously, in Sections 7.4 and 8.5, when plotting in logarithmic axes makes it easy to compute and draw low and high frequency asymptotes. This facilitates the drawing of Bode plots. Moreover, it offers great insight about the behaviour of $|G(j\omega)|$ at low and high frequencies.

Bode plots give information about the frequency response of **BIBO-stable systems**. Indeed, if the system is BIBO-stable, then $|G(j\omega)|$ and $\arg G(j\omega)$ give the amplitude ratio and phase lag of the system's frequency response as shown in Theorem 11.1. If the system has a single pole at zero (see Section 11.1.2) or if it has simple pairs of imaginary poles (see Section 11.1.3), then $|G(j\omega)|$ and $\arg G(j\omega)$ can provide some information about the frequency response at large times as we discussed in the previous sections. If, however, function G is **not BIBO-stable** and does not fall in the special cases of Sections 11.1.2 or 11.1.3, Bode plots **cannot provide information about its frequency response**.

Unless we know whether the underlying system is BIBO-stable, we cannot use a Bode plot to infer its frequency response characteristics. This has been demonstrated in Example 11.4. Bode plots, however, are useful for determining whether a closed-loop system is BIBO-stable using its open-loop transfer function. We will elaborate on that in Section 11.3.

11.2.1 Elementary Bode plots

In this section we construct the Bode plots for some simple systems. Next, we will see that these elementary Bode plots can serve as the building blocks for any Bode plot.

We will focus on four cases: first-order systems ($G(s) = \frac{K}{\tau s+1}$), simple integrators ($G(s) = \frac{1}{s}$), delay systems ($G(s) = e^{-t_d s}$) and second-order systems ($G(s) = \frac{K}{\tau^2 s^2 + 2\tau\zeta s + 1}$).

Example 11.7 (First-order systems) Consider a first-order system with transfer function

$$G(s) = \frac{K}{\tau s + 1}, \tag{11.80}$$

Chapter 11. Frequency response and design of stable closed loops 445

with $K, \tau > 0$. Then,

$$|G(j\omega)| = \left|\frac{K}{j\omega\tau + 1}\right| = \frac{K}{|j\omega\tau + 1|} = \frac{K}{\sqrt{\tau^2\omega^2 + 1}}, \qquad (11.81)$$

so, the common logarithm of $G(j\omega)$ is

$$\log_{10}|G(j\omega)| = \log_{10} K - \tfrac{1}{2}\log_{10}(\tau^2\omega^2 + 1). \qquad (11.82)$$

The argument of $G(j\omega)$ is given by

$$\arg G(j\omega) = \arg K - \arg(j\tau\omega + 1) = -\arctan(\tau\omega). \qquad (11.83)$$

In order to construct the plot of $\log_{10}|G(j\omega)|$ vs $\log_{10}\omega$ we follow the following steps

- **Step 1.** We construct the low-frequency asymptote (for $\omega\tau \ll 1$). Then, $\tau^2\omega^2 + 1 \approx 1$, so $\log_{10}|G(j\omega)|$ can be approximated by

$$\log_{10}|G(j\omega)| \approx \log_{10} K. \qquad (11.84)$$

  This is called the low-frequency asymptote of the Bode magnitude plot.

- **Step 2.** We construct the high-frequency asymptote (for $\omega\tau \gg 1$). At high frequencies we see that $\tau^2\omega^2 + 1 \approx \tau^2\omega^2$, so $\log_{10}|G(j\omega)|$ can be approximated by

$$\log_{10}|G(j\omega)| \approx \log_{10} K - \log_{10}(\tau\omega). \qquad (11.85)$$

  This is called the high-frequency asymptote of the Bode magnitude plot.

- **Step 3.** The two asymptotes meet at $\omega_c = \frac{1}{\tau}$. This frequency is called the **corner frequency**. For frequencies well below ω_c, the magnitude is approximated by the low-frequency asymptote and for frequencies much larger than ω_c, the high-frequency asymptote provides a good approximation of $\log_{10}|G(j\omega)|$.

Similarly, we construct asymptotes for the Bode phase plot.

- **Step 1.** For $\omega\tau \ll 1$, $\arg G(j\omega) = -\arctan(\tau\omega) \approx \lim_{\omega \to 0^+}(-\arctan(\tau\omega)) = 0$

- **Step 2.** For $\omega\tau \gg 1$, $\arg G(j\omega) = -\arctan(\tau\omega) \approx \lim_{\omega \to +\infty}(-\arctan(\tau\omega)) = -\frac{\pi}{2}$

- **Step 3.** At the corner frequency, $\arg G(j\omega_c) = -\arctan(\omega_c\tau) = -\arctan 1 = -\frac{\pi}{4}$. See also, Exercise 11.7 below.

The Bode plot is shown in Figure 11.5.

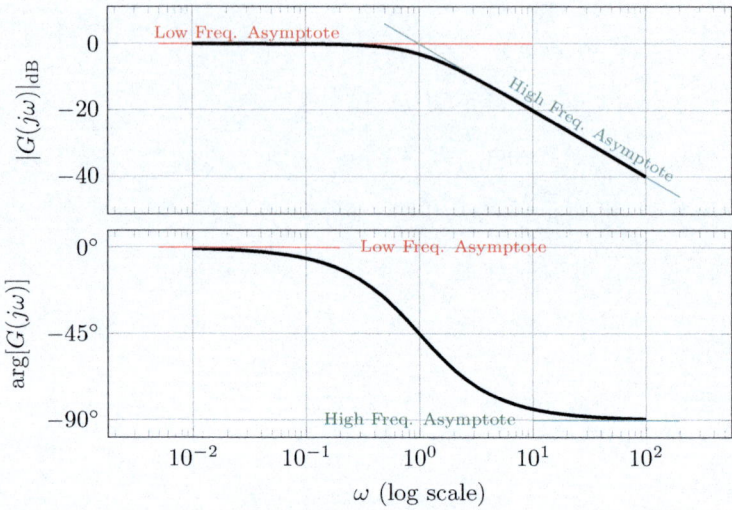

Figure 11.5: *Bode plot of first order system with $K = 1$ and $\tau = 1$. The HF asymptote of the magnitude has a slope of -20 dB in the magnitude plot: as we showed above, in Equation (11.85) at high frequencies the magnitude (in dB) drops at a constant rate of -20 dB per logarithmic unit of frequency: if the frequency is multiplied by 10, the magnitude drops by 20 dB.*

Exercise 11.7 (**Slope of phase lag in Bode plot**) What is the slope of $\arg[G(j\omega)]$ in the Bode phase plot at $\omega = \omega_c$ for the first-order system of Example 11.7?

✂ ....

Hint. We need to compute $\left. \dfrac{d \arg[G(j\omega)]}{d \log_{10} \omega} \right|_{\omega=\omega_c}$. This is equal to

$$\left. \dfrac{d \arg[G(j\omega)]}{d \log_{10} \omega} \right|_{\omega=\omega_c} = \left. \dfrac{\dfrac{d \arg[G(j\omega)]}{d\omega}}{\dfrac{d \log_{10} \omega}{d\omega}} \right|_{\omega=\omega_c} . \tag{11.86}$$

[This is because of the chain rule, which states that $\frac{dy}{dx} = \frac{dy}{dz}\frac{dz}{dx}$, so long as z is differentiable with respect to x and y and y is differentiable with respect to x.] Recall also that (see Equation (A.23))

$$(\log_{10} \omega)' = \dfrac{1}{\ln 10 \cdot \omega}. \tag{11.87}$$

Chapter 11. Frequency response and design of stable closed loops

Exercise 11.8 (Effect of K and τ on Bode plot) Sketch the Bode plots for three first-order systems with $K_1 = 1$, $\tau_1 = 1$, $K_2 = 1$, $\tau_2 = 10^{-3}$ and $K_3 = 100$, $\tau_3 = 1$. Observe how the static gain and the time constant affect the shape and position of the Bode plot.

Let us now construct the Bode plot of a simple integrator. We will follow the exact same procedure as above.

Example 11.8 (Simple integrator) Consider a simple integrator with transfer function
$$G(s) = \frac{1}{s}. \tag{11.88}$$

Then,
$$|G(j\omega)| = \left|\frac{1}{j\omega}\right| = \frac{1}{|j\omega|} = \frac{1}{\omega}, \tag{11.89}$$

and
$$\arg[G(j\omega)] = \arg\left[\frac{1}{j\omega}\right] = -\frac{\pi}{2}. \tag{11.90}$$

The common logarithm of $|G(j\omega)|$ is
$$\log_{10}|G(j\omega)| = -\log_{10}\omega. \tag{11.91}$$

It is straightforward to draw the Bode plot for this system.

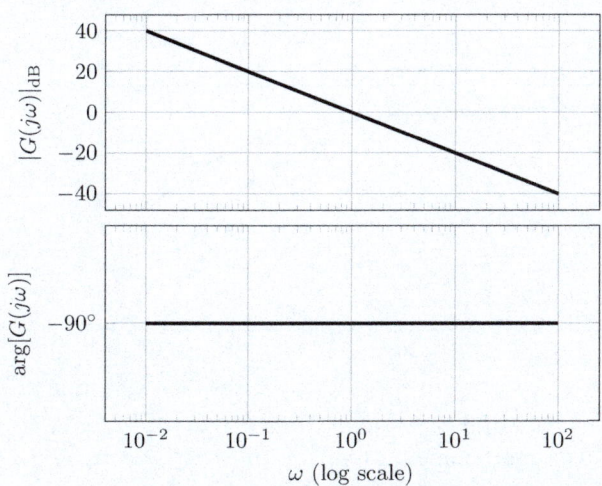

Figure 11.6: Bode plot of a simple integrator. Observe that the magnitude plot has a slope of -20 (the magnitude drops by $20\,\mathrm{dB}$ for every logarithmic unit of frequency).

We observe that the phase lag remains constant at $-90°$ and the magnitude, $|G(j\omega)|$, converges to 0 as $\omega \to \infty$ and goes to ∞ as $\omega \to 0^+$.

Exercise 11.9 (Double integrator) Sketch the Bode plot for a double integrator, that is, a system with transfer function $G(s) = \frac{1}{s^2}$.

Next, we will construct the Bode plot of a pure delay system, that is, a system with transfer function $G(s) = e^{-t_d s}$ for some $t_d > 0$. Recall that this is called a delay system because when excitated with an input $u(t)$ (that is, $U(s)$ in the s-domain), the output is $X(s) = \mathscr{L}^{-1}\{U(s)e^{-t_d s}\} = u(t-t_0)H_{t_0}(t)$, which is a "delayed" version of $u(t)$ (see Theorem 4.15).

Example 11.9 (Pure delay) Consider a pure delay system with transfer function
$$G(s) = e^{-t_d s}, \tag{11.92}$$
for some $t_d > 0$. Then, it is easy to verify that $|G(j\omega)| = 1$ (thus, $\log|G(j\omega)| = 0$, so $|G(j\omega)|_{\mathrm{dB}} = 0$ for all $\omega > 0$) and $\arg[G(j\omega)] = -t_d \omega$.

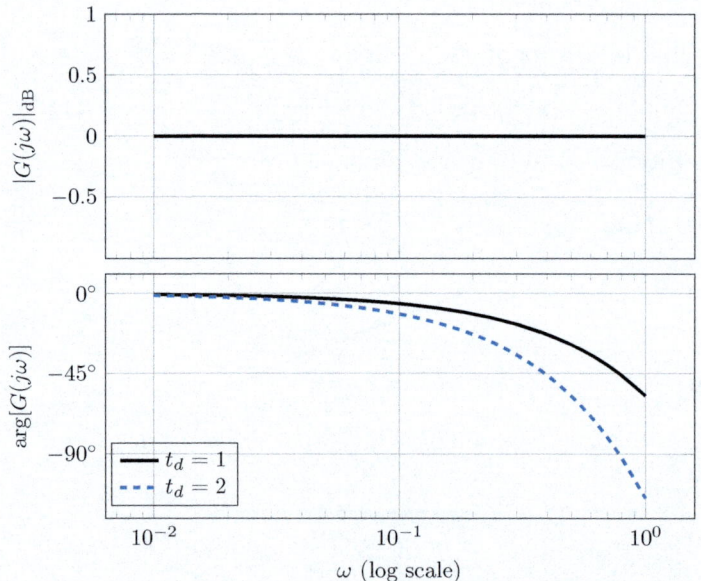

Figure 11.7: Bode plot of delay system with $G(s) = e^{-t_d s}$ for different values of t_d. The magnitude, $|G(j\omega)|$ is constant and $\arg G(j\omega)$ goes to $-\infty$ as $\omega \to \infty$ and converges to 0 as $\omega \to 0^+$.

Example 11.10 (Second-order systems) Consider a second-order system with transfer function
$$G(s) = \frac{K}{\tau^2 s^2 + 2\tau\zeta s + 1}. \tag{11.93}$$

Then,

$$|G(j\omega)| = \frac{K}{|\tau^2(j\omega)^2 + 2\tau\zeta(j\omega) + 1|}$$
$$= \frac{K}{\sqrt{(1-\tau^2\omega^2)^2 + (2\tau\zeta\omega)^2}}, \quad (11.94)$$

and

$$\arg[G(j\omega)] = -\arg[\tau^2(j\omega)^2 + 2\tau\zeta(j\omega) + 1]$$
$$= \operatorname{atan2}\left(-2\zeta\tau\omega, 1 - \tau^2\omega^2\right). \quad (11.95)$$

The common logarithm of $|G(j\omega)|$ is given by

$$\log_{10}|G(j\omega)| = \log_{10} K - \tfrac{1}{2}\log_{10}\left[(1-\tau^2\omega^2)^2 + (2\tau\zeta\omega)^2\right]. \quad (11.96)$$

The **low frequency asymptote** of $\log_{10}|G(j\omega)|$ is determined at frequencies $\omega \ll 1/\tau$; equivalently $\tau\omega \ll 1$. Then, $1 - \tau^2\omega^2 \approx 1$ and $2\zeta\tau\omega \approx 0$, therefore,

$$\log_{10}|G(j\omega)| \approx \log_{10} K. \quad (11.97)$$

The asymptote of the phase lag for $\omega \ll 1/\tau$ is $\arg[G(j\omega)] \approx \operatorname{atan2}(0,1) = 0$.

The **high frequency asymptote** of $\log_{10}|G(j\omega)|$ is determined at frequencies $\omega \gg 1/\tau$; equivalently $\tau\omega \gg 1$. Then, $1 - \tau^2\omega^2 \approx -\tau^2\omega^2$, so $(1-\tau^2\omega^2)^2 \approx \tau^4\omega^4$. Additionally, $\tau^4\omega^4 \gg (2\zeta\tau\omega)^2$, therefore,

$$\log_{10}|G(j\omega)| \approx \log_{10} K - \tfrac{1}{2}\log_{10}(\tau^4\omega^4) = \log_{10} K - 2\log_{10}(\tau\omega). \quad (11.98)$$

Observe that this result means that at high frequencies the magnitude drops at a constant rate of $-2 \cdot 20\,\mathrm{dB} = -40\,\mathrm{dB}$ per logarithmic unit of frequency as we can also observe in Figure 11.8. The high frequency asymptote of the phase lag, $\arg[G(j\omega)]$, is

$$\arg[G(j\omega)] = \operatorname{atan2}(-2\zeta\tau\omega, 1 - \tau^2\omega^2)$$
$$\approx \operatorname{atan2}(-2\zeta\tau\omega, -\tau^2\omega^2)$$
$$= \operatorname{atan2}(-2\zeta, -\tau\omega) = -\pi. \quad (11.99)$$

Here we used the property $\operatorname{atan2}(cy, cx) = \operatorname{atan2}(y, x)$ for $c > 0$ (see Equation (A.41)). The Bode plot of a second-order system with $K = 1$ and $\tau = 1$ is shown in Figure 11.8.

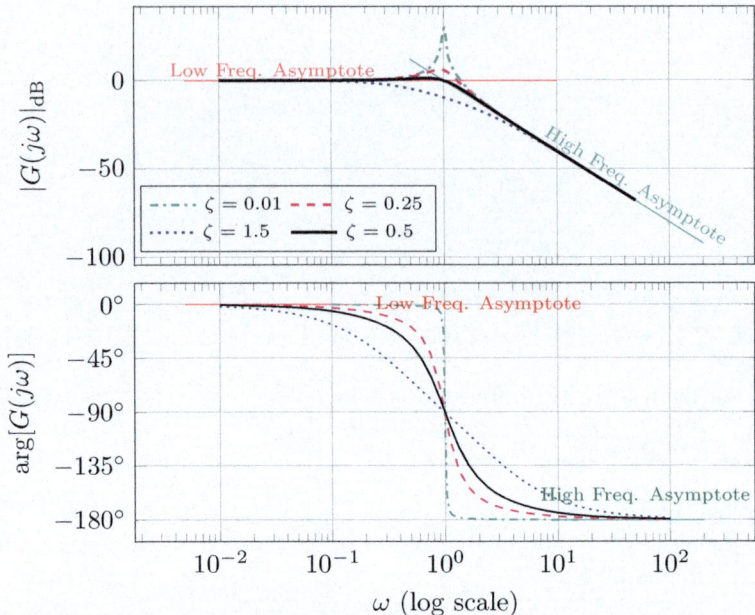

Figure 11.8: *Amplitude ratio and phase lag of second-order system for different values of ζ for $K = 1$ and $\tau = 1$. The HF asymptote of the magnitude has a slope of $-40\,\mathrm{dB}$ in the magnitude plot.*

As one can observe in Figure 11.8, for some ζ (adequately small) there is a frequency at which $|G(j\omega)|$ exhibits a local maximum for low values of ζ. This is known as **resonance frequency**, and we will denote it by ω_{res}. By Fermat's Theorem, at the local maximum ω_{res} we must have

$$\left. \frac{\mathrm{d}}{\mathrm{d}\omega} \log_{10} |G(j\omega)| \right|_{\omega_{\mathrm{res}}} = 0. \qquad (11.100)$$

Let us compute the derivative $\frac{\mathrm{d}}{\mathrm{d}\omega} \log_{10} |G(j\omega)|$:

$$\frac{\mathrm{d}}{\mathrm{d}\omega} \log_{10} |G(j\omega)| = \frac{\mathrm{d}}{\mathrm{d}\omega} \left[\log_{10} K - \tfrac{1}{2} \log_{10} \left[(1 - \tau^2 \omega^2)^2 + (2\tau\zeta\omega)^2 \right] \right]$$

$$= \underbrace{- \frac{1}{2 \ln 10} \frac{1}{(1 - \tau^2 \omega^2)^2 + (2\tau\zeta\omega)^2}}_{\neq 0}$$

$$\cdot \underbrace{\frac{\mathrm{d}}{\mathrm{d}\omega} \left[(1 - \tau^2 \omega^2)^2 + (2\tau\zeta\omega)^2 \right]}_{\clubsuit(\omega)}. \qquad (11.101)$$

But the first term (the highlighted part) cannot be zero, therefore

Chapter 11. Frequency response and design of stable closed loops

$\frac{d}{d\omega} \log_{10} |G(j\omega)||_{\omega_{\text{res}}} = 0$ if and only if ♣$(\omega_{\text{res}}) = 0$, that is

$$\frac{d}{d\omega}\left[(1-\tau^2\omega^2)^2 + (2\tau\zeta\omega)^2\right]\bigg|_{\omega_{\text{res}}} = 0.$$

$$\Leftrightarrow 4\omega_{\text{res}}\tau^2\left(\tau^2\omega_{\text{res}}^2 + 2\zeta^2 - 1\right) = 0, \qquad (11.102)$$

so the resonance frequency of a second order system is

$$\boxed{\omega_{\text{res}} = \frac{\sqrt{1-2\zeta^2}}{\tau},} \qquad (11.103)$$

which is defined for $0 < \zeta < \sqrt{2}/2$.

Remark: In Example 11.10 we computed the resonance frequency for second-order systems with $0 < \zeta < \sqrt{2}/2$ using Fermat's theorem. However, in principle, Fermat's theorem cannot be used alone to prove that ω_{res} is a local maximizer of $\log_{10}|G(j\omega)|$. We need to show that the second derivative $\frac{d^2}{d\omega^2}\log_{10}|G(j\omega)|$ at $\omega = \omega_{\text{res}}$ is negative. This is left to the reader as an exercise.

In Figure 11.8 we may observe that there is only one point where the derivative is zero. This supports, but does not prove, the claim that at ω_{res} is a point of local maximum of $\log_{10}|G(j\omega)|$ whenever $0 < \zeta < \sqrt{2}/2$. •

Example 11.11 (PI Controller) Here we will construct the Bode plot of a PI controller.
The transfer function of a PI controller has the form

$$G_c(s) = K_c\left(1 + \frac{1}{\tau_I s}\right) = K_c \frac{1 + \tau_I s}{\tau_I s}, \qquad (11.104)$$

therefore

$$|G_c(j\omega)| = K\frac{\sqrt{1+\tau_I^2\omega^2}}{\tau_I \omega}, \qquad (11.105\text{a})$$

$$\arg G_c(j\omega) = \arctan(\tau_I \omega) - \tfrac{\pi}{2}. \qquad (11.105\text{b})$$

The common logarithm of the modulus of $G_c(j\omega)$ is

$$\log_{10}|G_c(j\omega)| = \log_{10} K + \tfrac{1}{2}\log_{10}(1+\tau_I^2\omega^2) - \log_{10}(\tau_I\omega). \quad (11.106)$$

As before, we will estimate the high and low-frequency asymptotes of the transfer function G_c

- **Step 1.** We will construct the low-frequency asymptote (for $\omega\tau_I \ll 1$) where $\tau_I^2\omega^2 + 1 \approx 1$, therefore,

$$\log_{10}|G_c(j\omega)| \approx \log_{10} K - \log_{10}(\tau_I\omega).$$

This means that at low frequencies, the amplitude plot has slope -1; additionally, $\lim_{\omega \to 0^+} |G_c(j\omega)| = \infty$. For the argument of $G_c(j\omega)$ we see that $\lim_{\omega \to 0^+} \arg G_c(j\omega) = -\frac{\pi}{2}$

- **Step 2.** We will next construct the high-frequency asymptote (for $\omega \tau_I \gg 1$) where $\tau_I^2 \omega^2 + 1 \approx \tau_I^2 \omega^2$, so

$$\log_{10} |G_c(j\omega)| \approx \log_{10} K + \log_{10}(\tau_I \omega) - \log_{10}(\tau_I \omega) = \log_{10} K,$$

therefore, at high frequencies, the amplitude plot tracks a horizontal asymptote. The argument of $G_c(j\omega)$ clearly goes to 0, that is, $\lim_{\omega \to \infty} \arg G_c(j\omega) = 0$.

- **Step 3.** The corner frequency, where the two asymptotes meet, is $\omega = \tau_I$.

The Bode plot of a PI controller with $K = 1$ and different values of τ_I is given below.

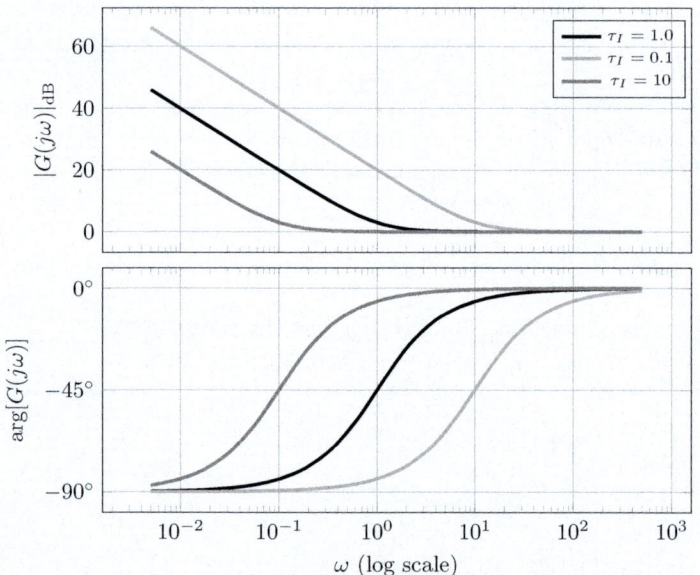

Figure 11.9: *Bode plot of PI controller.*

Exercise 11.10 (Low/High frequency asymptotes of PID[a]) Show that the low and high frequency asymptotes of a PID controller are given by the equations in Table 11.1. Then construct the Bode plot of a PID controller assuming $\tau_D < 1/\tau_I$.

---

[a] See also Exercise 11.5: the steady state frequency response of the PID controller is not of the form given in Theorem 11.2.

Chapter 11. Frequency response and design of stable closed loops 453

Table 11.1: Logarithm of magnitude, $\log|G(j\omega)|$, and phase angle, $\arg[G(j\omega)]$, for common transfer functions. The last two columns are the low-frequency and high-frequency asymptotes of $\log|G(j\omega)|$.

System	Transfer function	Magnitude, $\log|G(j\omega)|$	Phase, $\arg[G(j\omega)]$	LF $\log|G(j\omega)|$	HF $\log|G(j\omega)|$
1st order	$\dfrac{1}{\tau s + 1}$	$-\tfrac{1}{2}\log(1+\tau^2\omega^2)$	$-\arctan(\tau\omega)$	$0,\ \omega\tau \ll 1$	$-\log(\tau\omega),\ \omega\tau \gg 1$
2nd order	$\dfrac{1}{\tau^2 s^2 + 2\zeta\tau s + 1}$	$-\tfrac{1}{2}\log\left[(1-\tau^2\omega^2)^2 + (2\zeta\tau\omega)^2\right]$	$\operatorname{atan2}\left(-2\zeta\tau\omega,\ 1-\tau^2\omega^2\right)$	$0,\ \omega\tau \ll 1$	$-2\log(\tau\omega),\ \omega\tau \gg 1$
Integrator	$\dfrac{1}{s}$	$-\log\omega$	$-\dfrac{\pi}{2}$	—	—
Delay	$e^{-t_d s}$	0	$-t_d\omega$	—	—
PID	$1 + \dfrac{1}{\tau_I s} + \tau_D s$	$\tfrac{1}{2}\log\left[1+\left(\tau_D\omega-\dfrac{1}{\tau_I\omega}\right)^2\right]$	$\arctan\left(\tau_D\omega-\dfrac{1}{\tau_I\omega}\right)$	$-\log(\tau_I\omega),\ \tau_D\omega \ll {}^1\!/\!\tau_I\omega$	$\log(\tau_D\omega),\ \tau_I\omega \gg 1$
Oscillator	$\dfrac{1}{\omega^2 + b^2}$	$-\log|b^2-\omega^2|\ (\omega\neq|b|)$	$\begin{cases}0, & \text{if } \omega < b \\ -\pi, & \text{if } \omega > b\end{cases}$	$0,\ \omega \ll b$	$-2\log\omega,\ \omega \gg b$

11.2.2 Construction of Bode plots

In Section 11.2.1 we constructed the Bode plots of some elementary transfer functions. It turns out that the Bode plots of more complex transfer functions of the general form

$$G(s) = G_1(s)G_2(s) \cdots G_N(s)e^{-t_d s}, \qquad (11.107)$$

can be constructed using the elementary Bode plots of the previous section. The key property we use here is that $|z_1 \cdot z_2| = |z_1| \cdot |z_2|$, therefore $\log|z_1 \cdot z_2| = \log|z_1| + \log|z_2|$, and $\arg[z_1 \cdot z_2] = \arg z_1 + \arg z_2$. As a result

$$\log|G(j\omega)| = \log|G_1(j\omega)| + \ldots + \log|G_N(j\omega)|, \qquad (11.108a)$$
$$\arg[G(j\omega)] = \arg[G_1(j\omega)] + \ldots + \arg|G_N(j\omega)| - t_d\omega. \qquad (11.108b)$$

The pair of Equations (11.108) is often referred to as the **superposition principle** of Bode plots. We can choose G_i to be instances of the elementary transfer functions summarised in Table 11.1. In order to sketch the Bode plot of G we can then draw the Bode plots of all G_i, $i = 1, \ldots, N$, in the same axes and use Equations (11.108) to obtain the overall Bode plot. This is illustrated in the following example.

Example 11.12 (Product of first-order functions) Consider the transfer function
$$G(s) = \frac{2}{(s+1)(3s+1)}. \qquad (11.109)$$

We can write this as a product of two first-order functions, $G(s) = G_1(s)G_2(s)$, with

$$G_1(s) = \frac{2}{s+1}, \qquad (11.110a)$$
$$G_2(s) = \frac{1}{3s+1}. \qquad (11.110b)$$

These are two first-order systems with $K_1 = 2$, $\tau_1 = 1$ and $K_2 = 1$, $\tau_2 = 3$. The corner frequencies are $\omega_{c,1} = 1$ and $\omega_{c,2} = 1/3$. We can now plot the Bode plots of G_1 and G_2 (see Figure 11.10).

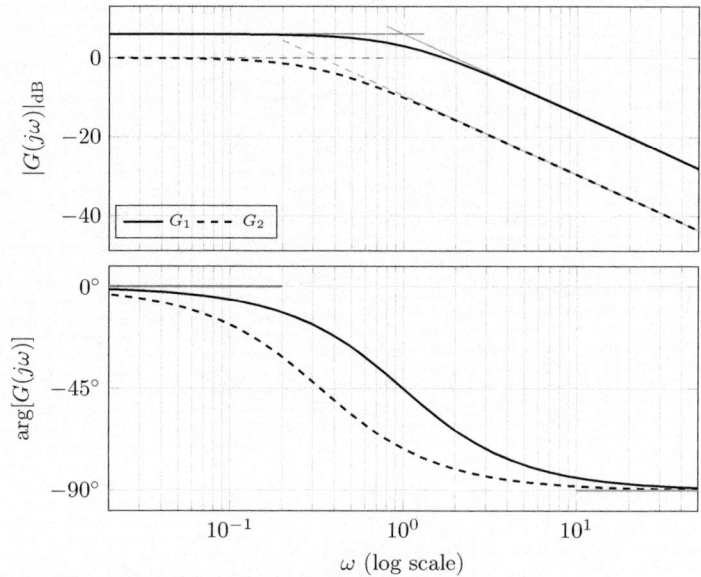

Figure 11.10: *Bode plot of G_1 and G_2.*

For G_1 we have that for $\omega \ll 1$, $\log_{10}|G_1(j\omega)| \approx \log_{10} K_1 = 0$ (low-frequency asymptote) and for $\omega \gg 1$, it is $\log_{10}|G(j\omega)| \approx \log_{10} K_1 - \log_{10}(\tau_1\omega) = -\log_{10}\omega$ (high-frequency asymptote). Similarly, for G_2 we have that for $\omega \ll 1/3$, $\log_{10}|G_2(j\omega)| \approx \log_{10} K_2 = \log_{10} 2$ and for $\omega \gg 1/3$, it is $\log_{10}|G(j\omega)| \approx \log_{10} K_2 - \log_{10}(\tau_2\omega) = -\log_{10} 2 - \log_{10}(3\omega)$.

By virtue of Equations (11.108), the low and high frequency asymptotes of $|G(j\omega)|$ will be

$$\text{LF Asymptote: } \log_{10}|G(j\omega)| \approx \log_{10} 2, \tag{11.111}$$

$$\text{HF Asymptote: } \log_{10}|G(j\omega)| \approx \log_{10} 2 - \log_{10}(3\omega) - \log_{10}\omega$$
$$= \log_{10} 2/3 - 2\log_{10}\omega. \tag{11.112}$$

Observe that at low frequencies, the magnitude tends to a constant (see Figure 11.11), while at high frequencies, the magnitude drops at a *constant rate* of $-2 \cdot 20\,\text{dB} = -40\,\text{dB}$ per logarithmic unit of frequency.

The phase lag of G will be

$$\arg[G(j\omega)] = -\arctan\omega - \arctan(3\omega), \tag{11.113}$$

so $\lim_{\omega \to 0^+} \arg[G(j\omega)] = 0$ and $\lim_{\omega \to \infty} \arg[G(j\omega)] = -\pi$. The Bode plot of G is shown below (see Figure 11.11).

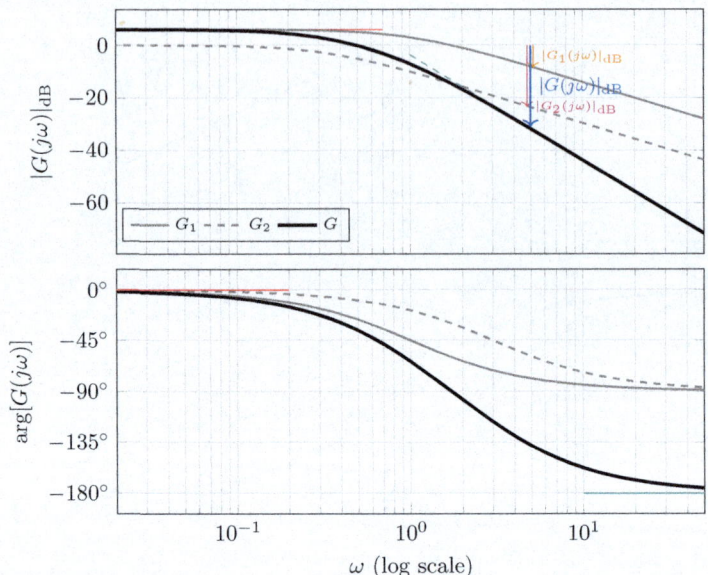

Figure 11.11: Bode plot of $G = G_1 G_2$. As shown in Equation (11.108), the Bode plot of G can be constructed by adding up the contributions of G_1 and G_2. Note that in the magnitude plot, the high-frequency asymptotes of G_1 and G_2 have a slope of -20, therefore, G has a slope of -40. This means that if the frequency, ω, is multiplied by 10, the magnitude, $|G(j\omega)|_{\mathrm{dB}}$, will drop by $40\,\mathrm{dB}$.

Exercise 11.11 (Construction of higher-order Bode plot) How would you construct the Bode plot of a system with transfer function

$$G(s) = \frac{K_1}{\tau_1 s + 1} \cdot \frac{K_2}{\tau_2 s + 1} \cdot \ldots \cdot \frac{K_N}{\tau_N s + 1}, \tag{11.114}$$

with $\tau_i, K_i > 0$ for all $i = 1, \ldots, N$?

By following this procedure we can sketch the Bode plot of any transfer function of the general form of Equation (11.107) (product of a rational with an exponential function). It is worth noting at this point that there is no need to approximate the exponential using a Padé — or other — approximation. Let us give another example.

Example 11.13 (Pole at zero, stable pole and delay) Consider the transfer function

$$G(s) = \frac{2e^{-0.01s}}{s(2s+1)}. \tag{11.115}$$

Chapter 11. Frequency response and design of stable closed loops

Again, we can write G as a product of elementary transfer functions:

$$G(s) = (2e^{-0.01s}) \cdot \frac{1}{s} \cdot \frac{1}{2s+1}, \qquad (11.116)$$

that is, $G_1(s) = 2e^{-0.01s}$, $G_2(s) = 1/s$ and $G_3(s) = 1/2s+1$. These are a pure delay, a simple integrator and a first-order system respectively. By Equation (11.107), the magnitude and phase lag of G are

$$\log_{10}|G(j\omega)| = \log_{10}|G_1(j\omega)| + \log_{10}|G_2(j\omega)| + \log_{10}|G_3(j\omega)|$$
$$= \underbrace{\log_{10} 2}_{\log_{10}|G_1(j\omega)|} + \underbrace{(-\log_{10}\omega)}_{\log_{10}|G_2(j\omega)|}$$
$$+ \underbrace{(-\tfrac{1}{2}\log_{10}(1+4\omega^2))}_{\log_{10}|G_3(j\omega)|}, \qquad (11.117a)$$

and

$$\arg[G(j\omega)] = \arg[G_1(j\omega)] + \arg[G_2(j\omega)] + \arg[G_3(j\omega)]$$
$$= \underbrace{-\omega}_{\arg[G_1(j\omega)]} + \underbrace{(-\pi/2)}_{\arg[G_2(j\omega)]} + \underbrace{(-\arctan(2\omega))}_{\arg[G_3(j\omega)]}. \qquad (11.117b)$$

At low frequencies, the magnitude is approximated by

$$\log_{10}|G(j\omega)| \approx \log_{10} 2 - \log_{10}\omega, \qquad (11.118)$$

while at high frequencies we have

$$\log_{10}|G(j\omega)| \approx \log_{10} 2 - \log_{10}\omega - \log_{10}(2\omega) = -2\log_{10}\omega. \qquad (11.119)$$

This means that at low frequencies, the slope of the magnitude plot is $-20\,\text{dB}$ per logarithmic unit of frequency and at high frequencies, the slope is $2 \cdot (-20) = -40\,\text{dB}$ per logarithmic unit of frequency. This can be observed in Figure 11.12. Additionally, the phase lag at low frequencies is $\lim_{\omega \to 0^+} \arg[G(j\omega)] = -\pi/2$, while at high frequencies it diverges to $-\infty$.

The Bode plot of G is shown below (see Figure 11.12).

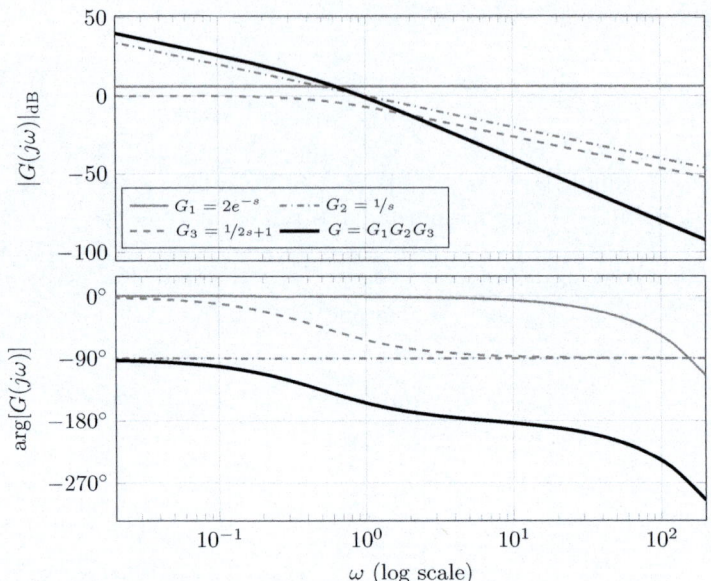

Figure 11.12: Bode plot of $G = G_1G_2G_3$.

Exercise 11.12 (Asymptotes of simple transfer functions) Determine the low and high frequency asymptotes and sketch the Bode plots of the following transfer functions

$$G(s) = \frac{5}{(5s+3)^3}, \tag{11.120}$$

and

$$F(s) = \frac{2}{s(s^2+s+1)(s+5)}e^{-s/2}. \tag{11.121}$$

What can you tell about the frequency response of these two systems by looking at their Bode plots?

We have so far treated cases where the numerator is rather simple. In many cases, however, the numerator of a transfer function may involve a polynomial of non-zero order, such as

$$G(s) = \frac{3s+1}{(s+1)(s+5)}e^{-0.5s}. \tag{11.122}$$

The procedure we follow in such cases is identical to the one we discussed above. We first need to write G as a product of simple functions, that is,

$$G(s) = (3s+1) \cdot \frac{1}{s+1} \cdot \frac{1}{s+5} \cdot e^{-0.5s}. \tag{11.123}$$

Chapter 11. Frequency response and design of stable closed loops

This way we define four simple transfer functions: $G_1(s) = 3s+1$, $G_2(s) = \frac{1}{s+1}$, $G_3(s) = \frac{1}{s+5}$ and $G_4(s) = e^{-0.5s}$. Here, G_2 and G_3 are first-order transfer functions and G_4 is a delay. Function G_1 is can be seen as the **reciprocal** of a first-order system, that is

$$G_1(s) = \frac{1}{\frac{1}{3s+1}}. \tag{11.124}$$

Let us define $\tilde{G}_1 = \frac{1}{3s+1}$ so that $G_1(s) = 1/\tilde{G}_1(s)$. This implies that

$$\log|G_1(j\omega)| = -\log|\tilde{G}_1(j\omega)| = \tfrac{1}{2}\log(1+3^2\omega^2), \tag{11.125a}$$

and

$$\arg[G_1(j\omega)] = -\arg[\tilde{G}_1(j\omega)] = \arctan(3\omega). \tag{11.125b}$$

Let us give an example with nontrivial numerators.

Example 11.14 (Nontrivial numerators) Consider the transfer function

$$G(s) = \frac{s(3s+1)}{(2s+1)(0.01s^2+0.002s+1)}. \tag{11.126}$$

This can be written as the product

$$G(s) = s \cdot (3s+1) \cdot \frac{1}{2s+1} \cdot \frac{1}{0.01s^2+0.002s+1}. \tag{11.127}$$

We define $G_1(s) = s$, $G_s(2) = 3s+1$, $G_3(s) = \frac{1}{2s+1}$ and

$$G_4(s) = \frac{1}{0.01s^2+0.002s+1}.$$

Note that G_1 is the reciprocal of an integrator (a **differentiator**), so its magnitude and phase lag are (see Table 11.1)

$$\log_{10}|G_1(j\omega)| = \log_{10}\omega, \tag{11.128a}$$
$$\arg[G_1(j\omega)] = \pi/2. \tag{11.128b}$$

Similarly, G_2 is the reciprocal of a first order system with time constant $\tau_2 = 7$, therefore

$$\log_{10}|G_2(j\omega)| = \tfrac{1}{2}\log_{10}(1+49\omega^2), \tag{11.129a}$$
$$\arg[G_2(j\omega)] = \arctan(7\omega). \tag{11.129b}$$

The low and high frequency asymptotes of $\log_{10}|G_2(j\omega)|$ are $\log_{10}|G_2(j\omega)| \approx 0$ and $\log_{10}|G_2(j\omega)| \approx \log_{10}(7\omega)$ respectively. The Bode plots of G_3 and G_4 can be constructed as discussed in the previous examples.

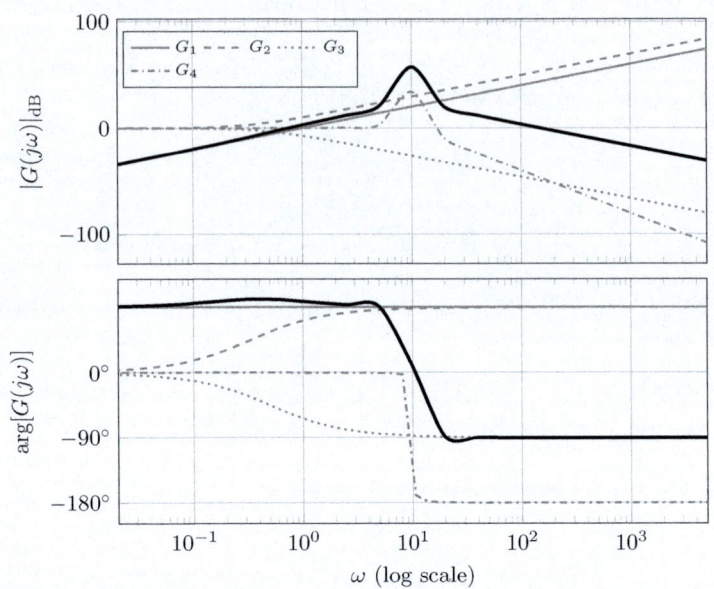

Figure 11.13: *Bode plot of G and its constituents.*

Exercise 11.13 (Sketch Bode plots) Sketch the Bode plots of the following transfer functions

$$G(s) = \frac{(s+1)(s+4)}{s(s^2+s+1)}, \tag{11.130}$$

and

$$T(s) = \frac{2}{(s+1)^2+1}e^{-0.03s}. \tag{11.131}$$

Exercise 11.14 (LF/HF asymptotes) A system is described by a rational transfer function

$$G(s) = \frac{P(s)}{Q(s)}, \tag{11.132}$$

where $\deg P < \deg Q$. Determine the slopes of the low and high frequency asymptotes of the magnitude plot.

✂ ··

Hint. Start by factorising P and Q: polynomial P has $\deg P$ (complex) roots, in particular n real roots, $a_1, \ldots, a_n$ (n is an integer between 0 and $\deg P$) and $m = \deg P - n$ complex roots, $b_v \mp c_v j$, $v = 1, \ldots, \frac{m}{2}$, which come in pairs of complex conjugate numbers). Polynomial P can be written as $P(s) = \prod_{i=1}^{n}(s+a_i)\prod_{v=1}^{\frac{m}{2}}(s+b_v)^2+c_v^2$. Then, do the same for polynomial Q and determine the low and high frequency asymptotes of $\log_{10}G(j\omega)$.

11.3 Bode stability criterion

11.3.1 Motivation for Bode's stability criterion

You might wonder why we need yet another stability criterion since Routh's tabulation can be used to tell whether a given rational transfer function is BIBO-stable.

In practice, typical control loops involve time delays. In Section 9.4 we saw that delay elements of the form $e^{-t_d s}$ can be approximated by a rational transfer function using the Padé approximation. It is reasonable to ask whether, since this approximation inevitably introduces some error, what impact this may have on the stability analysis of the system. We may use Routh's tabulation to tell whether the *approximate* system is BIBO-stable. Does that imply that the actual system is (or is not) BIBO-stable? The answer is negative, and we can easily find a counterexample.

Example 11.15 (Failure of Padé approximation) Consider the closed-loop system shown in the following figure:

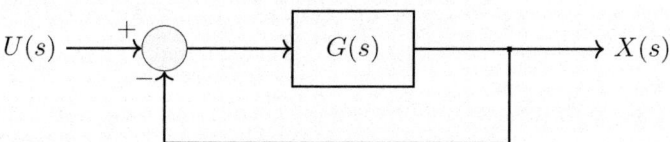

with
$$G(s) = \frac{2}{(s+0.5)^2 + 1} e^{-0.6s}. \qquad (11.133)$$

The closed-loop transfer function $G_c(s) = X(s)/U(s)$ is

$$G_{cl}(s) = \frac{G(s)}{1 + G(s)}. \qquad (11.134)$$

Since G involves an exponential function, Routh's criterion cannot be used to tell whether its is BIBO-stable. In order to use Routh's tabulation we need to use the Padé approximation of $e^{-0.6s}$ which is

$$e^{-0.6s} \approx \frac{1 - 0.3s}{1 + 0.3s}, \qquad (11.135)$$

so $G(s)$ can be approximated by

$$G_{\text{approx}}(s) = \frac{2}{(s+0.5)^2 + 1} \cdot \frac{1 - 0.3s}{1 + 0.3s}. \qquad (11.136)$$

Then, the corresponding closed-loop transfer function is approximated by

$$G_{\text{cl,approx}}(s) = \frac{G_{\text{approx}}(s)}{1 + G_{\text{approx}}(s)} = -\frac{80 - 24s}{12s^3 + 52s^2 + 31s + 130}. \qquad (11.137)$$

The reader can verify that $G_{\text{cl,approx}}(s)$ is a BIBO-stable transfer function using Routh's tabulation. However, the *actual* system with transfer function G is **not BIBO-stable**; indeed, the impulse response of G is not absolutely integrable as shown in Figure 11.14 (see Theorem 9.5).

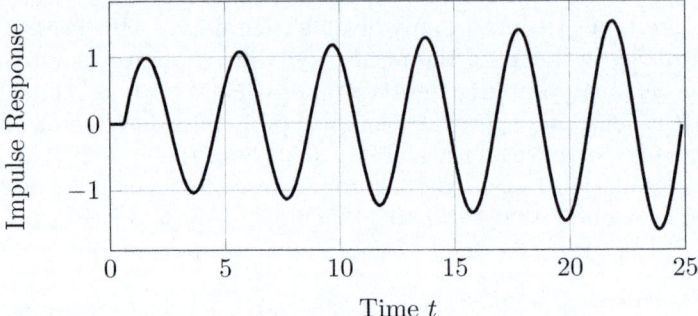

Figure 11.14: *Impulse response of G. Note that the impulse response does not converge to zero, therefore, it cannot be absolutely integrable.*

The key message from this counterexample is that Padé approximations can lead to erroneous results regarding the stability of the closed-loop system. Bode's stability criterion can be applied directly to systems with time delays without the need of a Padé, or other approximation.

11.3.2 Bode stability criterion

In order to state Bode's stability criterion, we first need to introduce two new definitions: the *open-loop transfer function* and its *phase crossover frequency*.

Definition 11.4 (Open-loop transfer function) Consider a closed-loop system which involves a controller with transfer function G_c, an actuator with transfer function G_a, a controlled system with transfer function G_s, and a sensor with transfer function G_m, where the loop is disconnected as shown in the following figure.

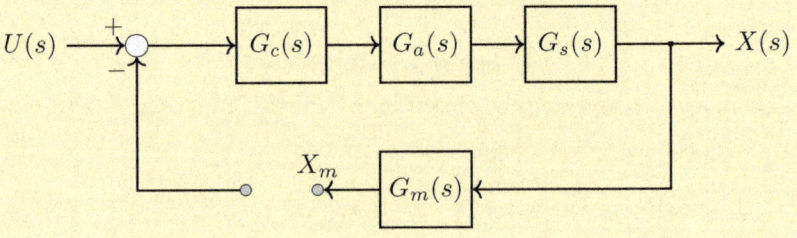

Chapter 11. Frequency response and design of stable closed loops

> The transfer function $G_{\text{ol}}(s) = X_m(s)/U(s)$ is called the **open-loop transfer function** of the system.

> **Definition 11.5 (Phase crossover frequency)** A frequency at which the phase lag of $G_{\text{ol}}(j\omega)$ is $-\pi$, if one exists, is called the **phase crossover frequency** of the open-loop system and is denoted by ω_{co}. It is
> $$\arg[G_{\text{ol}}(j\omega_{\text{co}})] = -\pi. \tag{11.138}$$
> Furthermore, we say that $\omega_{\text{co},k}$ is the k-th order phase crossover frequency, for $k = 1, 2, \ldots$, of the open-loop system if
> $$\arg[G_{\text{ol}}(j\omega_{\text{co},k})] = -(2k-1)\pi. \tag{11.139}$$

Note that a system may have none or multiple crossover frequencies. Bode's stability criterion applies to stable open-loop systems which have a phase crossover frequency.

> **Theorem 11.6 (Bode stability criterion [C14])** Let G_{ol} and G_{cl} denote the open and closed-loop transfer functions. Suppose that
>
> 1. G_{ol} has no poles located on or to the right of the imaginary axis, with the possible exception of a *single* pole at the origin,
>
> 2. The open-loop system has k-th order phase crossover frequencies, $\omega_{\text{co},k}$, for some $k \in \mathbb{N}$,
>
> 3. $|G_{\text{ol}}(j\omega_{\text{co},k})| < 1$ for all k-th order phase crossover frequencies $\omega_{\text{co},k}$.
>
> Then, the closed-loop system, with transfer function G_{cl}, is BIBO-stable.

Proof. This is a special case of the Nyquist stability criterion, which we state in the next chapter. □

The following corollary of the above stability criterion is rather straightforward to verify and very useful in some applications.

> **Corollary 11.7 (Bode stability criterion, special case)** Let G_{ol} and G_{cl} denote the open and closed-loop transfer functions. Suppose that
>
> 1. G_{ol} has no poles located on or to the right of the imaginary axis, with the possible exception of a *single* pole at the origin,
>
> 2. The open-loop system has exactly one phase crossover frequency, ω_{co},

> 3. $|G_{ol}(j\omega_{co})| < 1$,
>
> 4. $|G_{ol}(j\omega)|$ is a strictly decreasing function of ω for $\omega \geq \omega_{co}$.
>
> Then, the closed-loop system, with transfer function G_{cl}, is BIBO-stable.

Proof. By condition 2, the open-loop system has a zero-order phase crossover frequency, that is, ω_{co}. If it has no higher order phase crossover frequencies, the assertion follows trivially from Theorem 11.6. If there exist k-order phase crossover frequencies $\omega_{co,k}$ for some integers k, then $\omega_{co,k} > \omega_{co}$. By condition 4, $|G_{ol}(j\omega_{co,k})| < |G_{ol}(j\omega_{co})| < 1$ and the assertion follows from Theorem 11.6. □

Remark: *Bode's stability criterion (Theorem 11.6 and Corollary 11.7) is sufficient **but not necessary** for BIBO stability. There exist stable systems that do not satisfy the conditions of the Bode stability criterion. In other words, failure to meet the above stability conditions does not mean that the system is unstable. In such cases, we need to use the Nyquist stability criterion which we shall introduce in the next chapter.* •

Example 11.16 (Stable pole with multiplicity n) The open-loop transfer function of a system is

$$G_{ol}(s) = \frac{1}{(s+a)^n}, \qquad (11.140)$$

where $a > 1$ and $n \geq 3$. We will apply Bode's stability criterion to tell whether the *closed-loop* system is BIBO-stable. The first condition of Corollary 11.7 is satisfied: G_{ol} has the pole $-a < 0$ with multiplicity n, which is on the left-hand side of the imaginary axis. Let us now determine the phase crossover frequencies of G_{ol}. Following Definition 11.5, the phase crossover frequency, if any, must satisfy:

$$\arg[G_{ol}(j\omega_{co})] = -\pi \Leftrightarrow \arg \frac{1}{(j\omega_{co} + a)^n} = -\pi$$

$$\Leftrightarrow -\arg(j\omega_{co} + a)^n = -\pi$$

$$\Leftrightarrow n \arg(j\omega_{co} + a) = \pi$$

$$\Leftrightarrow \arctan \frac{\omega_{co}}{a} = \frac{\pi}{n}. \qquad (11.141)$$

We know that arctan is a continuous function with range $\left(-\frac{\pi}{2}, \frac{\pi}{2}\right)$. We can argue from the intermediate value theorem that since $n \geq 3$, it is $-\frac{\pi}{2} < \frac{\pi}{n} < \frac{\pi}{2}$, therefore, Equation has a solution and this is

$$\frac{\omega_{\text{co}}}{a} = \tan\frac{\pi}{n} \Leftrightarrow \omega_{\text{co}} = a\tan\frac{\pi}{n}. \tag{11.142}$$

Indeed, G_{ol} has *exactly one* phase crossover frequency, so the second condition of Corollary 11.7 is satisfied. The magnitude $|G_{\text{ol}}(j\omega)|$ is given by

$$|G_{\text{ol}}(j\omega)| = \left|\frac{1}{(j\omega + a)^n}\right| = \frac{1}{|j\omega + a|^n} = \frac{1}{(\omega^2 + a^2)^{n/2}}. \tag{11.143}$$

Since $a > 1$, it is clear that $|G_{\text{ol}}(j\omega)| < 1$ for all ω, therefore $|G_{\text{ol}}(j\omega_{\text{co}})| < 1$; we have shown that the third condition of Corollary 11.7 is satisfied. Observe that $|G_{\text{ol}}(j\omega)|$ is strictly decreasing for all $\omega > 0$ [check this as an exercise], therefore the fourth condition is also satisfied and the closed-loop system is BIBO-stable by virtue of Corollary 11.7.

Note that we would not be able to tell whether the closed-loop system were BIBO-stable in Example 11.16 for $n = 1$ and $n = 2$ as there would be no phase crossover frequencies. We shall address such cases in Proposition 11.8.

Example 11.17 (First-order system with delay) A closed-loop system is controlled with a PI controller with integral time $\tau_I = 1$ and proportional gain $K_P = 2$. The transfer function of the system is

$$G_s(s) = \frac{e^{-0.2s}}{3s + 1}. \tag{11.144}$$

Assuming that $G_a(s) = 1$ and $G_m(s) = 1$ we are interested in determining the phase crossover frequency of the open-loop system that will allow us to use Bode's criterion.

The transfer function of the PI controller is

$$G_c(s) = 2(1 + 1/s). \tag{11.145}$$

The open-loop transfer function of the system is

$$G_{\text{ol}}(s) = G_c(s)G_s(s) = 2(1 + 1/s)\frac{e^{-0.2s}}{3s + 1}. \tag{11.146}$$

We shall first give the Bode plot of the open-loop transfer function (see Figure 11.15).

In order to compute the phase crossover frequency of the system, we have

$$\arg[G_{\text{ol}}(j\omega)] = \arg(1 + 1/j\omega) + \arg e^{-0.2j\omega} - \arg(3j\omega + 1)$$

$$= \arg(1 + 1/j\omega) + \arg e^{-0.2j\omega} - \arg(3j\omega + 1)$$
$$= \arctan(-1/\omega) - 0.2\omega - \arctan(3\omega) \qquad (11.147)$$

The phase crossover frequency ω_{co} solves the equation $\arg[G_{ol}(j\omega_{co})] = -\pi$, equivalently $\arg[G_{ol}(j\omega_{co})] + \pi = 0$.

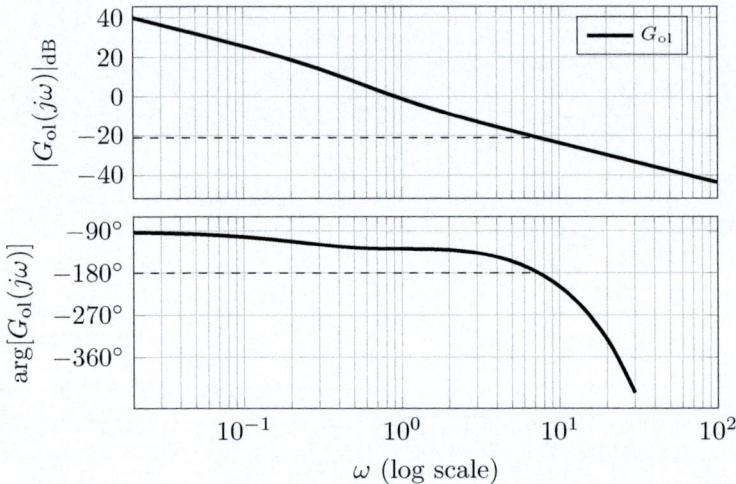

Figure 11.15: Bode plot of G_{ol}. There is a single phase crossover frequency at about 8 rad/s.

It is not possible (at least not obvious) to solve this equation analytically. We can determine however a solution numerically. In Python, we can use `fsolve` from `scipy.optimize` as in the following example

```
import numpy as np
from scipy.optimize import fsolve

def fun(x):
    f = np.arctan(-1/x) - 0.2*x \
        - np.arctan(3*x) + np.pi
    return f

wco = fsolve(fun, 8)
```

Similarly, in MATLAB we can use `fsolve` as follows

```
wco = fsolve(@(x) atan(-1/x) ...
             - 0.2*x - atan(3*x) + pi, 8)
```

Note that the second argument of `fsolve`, both in Python and in MATLAB, is the initial guess for the phase crossover frequency

Chapter 11. Frequency response and design of stable closed loops 467

which we can obtain from the Bode plot of G_{ol}.

The phase crossover frequency is found to be $\omega_{co} \approx 7.4079$. The amplitude gain at this frequency can be found to be

$$|G_{ol}(j\omega_{co})| = 2\,|1 + 1/j\omega_{co}|\,\frac{\left|e^{-0.2\omega_{co}}\right|}{|3\omega_{co} + 1|}$$

$$= 2\frac{\sqrt{1 + \frac{1}{\omega_{co}^2}}}{\sqrt{1 + 9\omega_{co}^2}} \approx 0.09072 < 1.$$

All conditions of Corollary 11.7 are satisfied:

- The open-loop transfer function, $G_{ol}(s) = 2\frac{s+1}{s}\frac{e^{-0.2s}}{3s+1}$, has one pole at the origin and one stable pole at $-1/3$

- The open-loop transfer function has exactly one phase crossover frequency (see Figure 11.15)

- $|G_{ol}(j\omega_{co})| < 1$

- $|G_{ol}(j\omega)|$ is strictly decreasing for $\omega \geq \omega_{co}$

therefore the **closed-loop system** is **BIBO-stable**.

Exercise 11.15 (**Phase crossover frequencies**) Determine the phase crossover frequencies (if any) up to order $k = 3$ of the following open-loop transfer functions:

1. $G_{ol}(s) = \frac{1}{s+1}$

2. $G_{ol}(s) = \frac{1}{s^2+0.5s+1}$

3. $G_{ol}(s) = \frac{2020}{(2s+1)^3}$

4. $G_{ol}(s) = \frac{2020}{(10s+1)^7}$

5. $G_{ol}(s) = \frac{e^{-0.1s}}{s+1}$

6. $G_{ol}(s) = \frac{1}{s(s+1)^2}$

※ ....

Answer. 1. None. 2. none. 3. one phase crossover frequency: $\omega_{co} = \frac{2}{1}\tan\frac{\pi}{3}$, 4. two phase crossover frequencies: $\omega_{co,1} = \frac{10}{\pi}\tan\frac{\pi}{7}$ and $\omega_{co,2} = \frac{10}{3\pi}\tan\frac{\pi}{7}$, 5. infinitely many phase crossover frequencies; the first three are approximately $\omega_{co,1} \approx 16.3199$, $\omega_{co,2} \approx 78.6999$ and $\omega_{co,3} \approx 141.4423$ (evaluated numerically). 6. one phase crossover frequency: $\omega_{co} = 1$.

In Example 11.17 we used Corollary 11.7 because the open-loop transfer function had a single phase crossover frequency and no higher order phase crossover frequencies. In the following example, taken from [C14], we present an example where we need to use Theorem 11.6 instead of

Corollary 11.7.

Example 11.18 (Hahn et al., 2001 [C14]) Consider a system with open-loop transfer function

$$G_{\text{ol}}(s) = 0.2 \frac{(1-s)(s+1)}{s(0.1s+1)(0.05s+1)} e^{-0.4s}. \quad (11.148)$$

All poles of G_{ol} are stable, except for a single pole at the origin, so we may proceed with our analysis using the Bode stability criterion. The Bode plot is given in Figure 11.16.

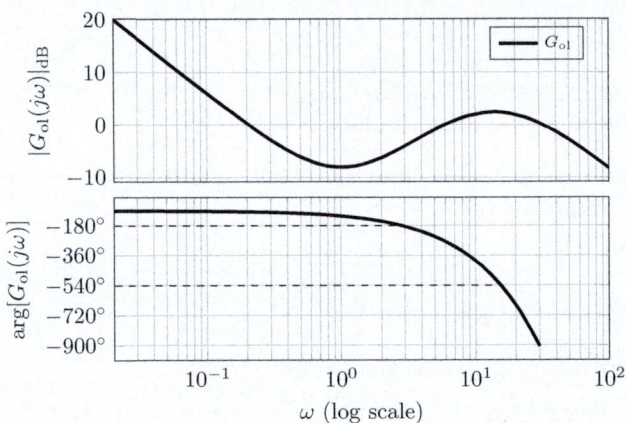

Figure 11.16: *Bode plot of G_{ol}. The first phase crossover frequency is at about $\omega_{\text{co}} \approx 3 \text{rad/s}$, and the second crossover frequency is about $\omega_{\text{co}} \approx 15 \text{rad/s}$.*

We may observe that there is a phase crossover frequency at about $\omega_{\text{co}} \approx 3 \text{rad/sec}$ where the corresponding amplitude gain seems to be less than one. However, $|G_{\text{ol}}(j\omega)|$ is not strictly decreasing after ω_{co}, therefore, we cannot use Corollary 11.7. We need to use Theorem 11.6: let us first determine the phase crossover frequency, ω_{co}. We have that

$$\arg[G_{\text{ol}}(j\omega)] = -\frac{\pi}{2} - \arctan(0.1\omega)$$
$$- \arctan(0.05\omega) - 0.4\omega. \quad (11.149)$$

We may now use `fsolve` to find an approximate solution of $\arg[G_{\text{ol}}(j\omega)] = -\pi$ (equivalently, $\arg[G_{\text{ol}}(j\omega)] + \pi = 0$)

```
import numpy as np
from scipy.optimize import fsolve

def fun(x):
    f = np.pi/2 - np.arctan(0.1*x) \
```

Chapter 11. Frequency response and design of stable closed loops

```
            - np.arctan(0.05*x) - 0.4*x
    return f

wco = fsolve(fun, 3)
```

The phase crossover frequency is $\omega_{co} \approx 2.8714$. At this frequency, the amplitude gain is

$$|G_{ol}(j\omega_{co})| = 0.2 \frac{1+\omega^2}{\omega\sqrt{1+0.1^2\omega^2}\sqrt{1+0.05^2\omega^2}} \approx 0.613 < 1. \tag{11.150}$$

We already pointed out to the fact that $|G_{ol}(j\omega)|$ is not strictly decreasing for $\omega \geq \omega_{co}$ as shown in Figure 11.16. The second phase crossover frequency is around $15^{\text{rad}}/\text{s}$. Using fsolve we can find that $\omega_{co,2} \approx 15.4931$, where the amplitude ratio is $|G_{ol}(j\omega_{co,2})| = 1.3339 > 1$, therefore, we cannot conclude that the system is BIBO-stable.

In the next chapter, we will be able to tell that this system is **unstable** using the Nyquist stability criterion (see Exercise 12.21).

The Bode stability criterion can be used to determine whether a given closed-loop system is stable. It can also be used to determine the values of the parameters of a controller that guarantee stability. In the following example, we shall determine the values of the gain K_c of a P-controller that makes the closed-loop system stable.

Example 11.19 (Stabilising P-controller design) A controlled system has the transfer function

$$G_s(s) = \frac{1}{s(s+1)} e^{-0.1s}, \tag{11.151}$$

and is controlled by a P-controller with transfer function $G_c(s) = K_c$ for some $K_c > 0$. Suppose that $G_m(s) = G_a(s) = 1$. The open-loop transfer function is

$$G_{ol}(s) = K_c \frac{1}{s(s+1)} e^{-0.1s}. \tag{11.152}$$

We compute

$$\arg[G_{ol}(j\omega)] = -\frac{\pi}{2} - \arctan\omega - 0.1\omega, \tag{11.153}$$

and

$$|G_{ol}(j\omega)| = K_c \frac{1}{\omega\sqrt{1+\omega^2}}. \tag{11.154}$$

The phase crossover frequency is the (unique) solution of $\arg[G_{\text{ol}}(j\omega_{\text{co}})] = -\pi$, that is

$$\tfrac{\pi}{2} - \arctan\omega - 0.1\omega = 0. \qquad (11.155)$$

(See also Exercise 11.23) Note that the phase crossover frequency is **independent of** K_c. We can solve (11.155) using `fsolve` as above; it is $\omega_{\text{co}} \approx 3.1105\,\text{rad/s}$.

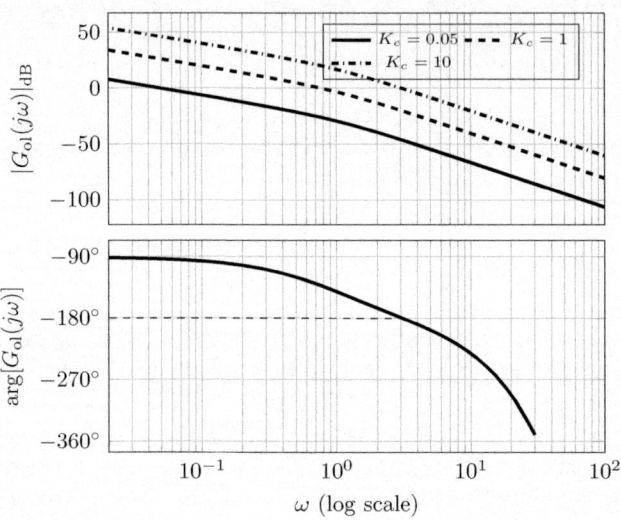

Figure 11.17: Bode plot of G_{ol}. The argument of $G_{\text{ol}}(j\omega)$ is independent of K_c.

As shown in Figure 11.17, $|G_{\text{ol}}(j\omega)|$ is strictly decreasing for $\omega \geq \omega_{\text{co}}$. We can therefore apply Corollary 11.7. The amplitude ratio at ω_{co} is

$$|G_{\text{ol}}(j\omega_{\text{co}})| = K_c \frac{1}{\omega_{\text{co}}\sqrt{1+\omega_{\text{co}}^2}}. \qquad (11.156)$$

Condition 3 of Corollary 11.7 is $|G_{\text{ol}}(j\omega_{\text{co}})| < 1$; equivalently

$$K_c \frac{1}{\omega_{\text{co}}\sqrt{1+\omega_{\text{co}}^2}} < 1 \Leftrightarrow K_c < \omega_{\text{co}}\sqrt{1+\omega_{\text{co}}^2} \approx 6.306. \qquad (11.157)$$

This means that if K_c is in the interval $(0, K_{c,\max})$ with $K_{c,\max} = 6.306$, the closed-loop system is BIBO-stable.

Remark: We have shown that for $0 < K_c < 6.306$, the closed-loop system is stable in the BIBO sense. This does not imply that for $K_c \geq 6.306$ the system is unstable because the Bode stability criterion provides only necessary and not sufficient conditions for stability. For this particular system, however, it is the case that $K_c \geq 6.306$ entails instability, but we

Chapter 11. Frequency response and design of stable closed loops

will not be able to prove this before we state the Nyquist stability criterion.•

Certain systems, such as first and second-order systems, do not have a phase crossover frequency. We may then use the following version of the Bode stability criterion, which, however, gives a rather conservative stability condition.

> **Proposition 11.8 (No phase crossover frequency)** Let G_{ol} and G_{cl} denote the open and closed-loop transfer functions. Suppose that G_{ol} is a meromorphic function[a] and
>
> 1. G_{ol} has no poles located on or to the right of the imaginary axis, with the possible exception of a single pole at the origin,
>
> 2. $|G_{ol}(j\omega)|$ is bounded for all $\omega > 0$ and there are no phase crossover frequencies (of any order) with $|G_{ol}(j\omega_{co,k})| > 1$.
>
> Then, the closed-loop system, with transfer function G_{cl}, is BIBO-stable.
>
> [a] Rational functions as well as rational-times-exponential functions are meromorphic. All transfer functions of practical interest, at least in the context of this book, are meromorphic.

Proof. The proof is a direct consequence of the Nyquist stability criterion (The two conditions of this proposition imply that $G_{ol}(j\omega)$ satisfies the conditions of Nyquist's criterion). □

Remark: *The attentive reader must have noticed that both Bode's stability criterion (Theorem 11.6) and its variants (Corollary 11.7, Proposition 11.8) accrue from the — significantly more general — Nyquist stability criterion, which gives necessary and sufficient stability conditions. We may in fact state more general versions of Proposition 11.8 for cases where the open-loop transfer function has no phase crossover frequencies. But we will refrain from doing so. We will revisit the problem of BIBO stability in the next chapter using a single and very powerful stability criterion.* •

Example 11.20 (Always stable system) Consider a first-order system with transfer function

$$G_s(s) = \frac{1}{\tau s + 1}, \qquad (11.158)$$

which is controlled by a P controller with transfer function $G_c(s) = K_c$. Suppose that $G_a(s) = 1$ and $G_m(s) = 1$. The open-loop transfer

function is
$$G_{\text{ol}}(s) = \frac{K_c}{\tau s + 1}, \tag{11.159}$$
and
$$|G_{\text{ol}}(j\omega)| = \frac{|K_c|}{\sqrt{1 + \tau^2 \omega^2}} \tag{11.160}$$

We consider two cases: (i) Suppose $K_c > 0$. Then
$$\arg G_{\text{ol}}(j\omega) = \arg K_c - \arctan(\tau\omega) = -\arctan(\tau\omega), \tag{11.161}$$

therefore, $-90° < \arg G_{\text{ol}}(j\omega) < 0$ for all $\omega > 0$. If $K_c < 1$, all conditions of Proposition 11.8 are satisfied and the closed-loop system is stable for all $0 < K_c < 1$.

(ii) Suppose $K_c < 0$. Then, $\arg K_c = \pi$, so
$$\arg G_{\text{ol}}(j\omega) = \arg K_c - \arctan(\tau\omega) = \pi - \arctan(\tau\omega), \tag{11.162}$$

therefore $90° < \arg G_{\text{ol}}(j\omega) < 180°$. If $K_c > -1$, all conditions of Proposition 11.8 are satisfied and the closed-loop system is stable for all $-1 < K_c < 0$.

Overall, we conclude that the closed-loop system is stable for all $-1 < K_c < 1$, including the value $K_c = 0$, for which we can see that the closed-loop system is trivially stable.

11.3.3 Stability margins

In Example 11.19 we computed the maximum value of the controller gain, $K_{c,\max} = 6.306$, that leads to a stable closed-loop. Any gain $0 < K_c < K_{c,\max}$ will make the closed-loop system stable. However, it is easily understood that the value $K_c = 6.3059$ will lead to a system too close to the maximum safe value for the controller gain — too risky a choice[4]. We have implicitly defined a metric that quantifies how stable or unstable a closed-loop system is: how much we need to change the open-loop gain before the system fails to meet the requirements of the Bode stability criterion. This quantity is referred to the **gain margin** of the system and is defined as follows

> **Definition 11.9 (Gain margin)** Suppose that the open-loop transfer function of a feedback control system satisfies conditions 1, 2 and 4 of Corollary 11.7. The **gain margin** of the closed-loop system is defined as $\mathrm{GM} = -|G_{\mathrm{ol}}(j\omega_{\mathrm{co}})|_{\mathrm{dB}}$.

Clearly, condition 3 of Corollary 11.7 is satisfied if and only if $\mathrm{GM} > 0$. The gain margin is a measure of robustness of the closed-loop system: the larger the gain margin, the more we can increase the gain of the open-loop function without destabilising it.

Note that the gain margin is measured in dB. In practice, reasonable values for the gain margin are between 2 and 5 dB (this is a form of controller tuning). If a system has a gain margin of 3 dB this means that we can introduce an additional gain into the closed loop up to a maximum of $10^{3/20} \approx 1.4125$ before the stabilising conditions of the Bode stability criterion cease to hold. Likewise, a gain margin of 5 dB means that we can introduce an additional element K into the closed-loop with $0 < K < 10^{5/20} \approx 1.7783$ before violating the stability conditions of Bode's criterion.

> **Exercise 11.16 (Gain margin)** Suppose that the open-loop transfer function of a feedback control system satisfies conditions 1, 2 and 4 of Corollary 11.7 and its gain margin is 1 dB. What is the percentage by which we can increase the gain of its open-loop transfer function until the gain margin becomes 0 dB?

> **Exercise 11.17 (GM-based tuning)** For the closed-loop system of Example 11.19, determine the value of K_c to have a gain margin of 3 dB.

---

[4]Suppose that we have misestimated the gain of the actuator and in reality it is $G_a(s) = 1.001$. If we choose $K_c = 6.3059$, the system will be unstable. Such a choice is not robust to the slightest misestimations of system parameters.

※

Answer. It is $K_c \approx 7.2$.

Another measure of robustness of a closed-loop system is the *phase margin*. The phase margin of a stable closed-loop system is the phase lag that the system can take before the stability conditions of the Bode criterion fail. Before we give its formal definition, let us give a motivating example.

Example 11.21 (Motivating the phase margin) Consider the system with open-loop transfer function $G_{ol}(s) = 2(1 + 1/s)\frac{e^{-0.2s}}{3s+1}$ that we studied in Example 11.17 and we proved that the corresponding closed-loop system is stable in the BIBO sense.

We are interested in determining the maximum additional phase lag that the system can take before it fails to satisfy the stabilising conditions of the Bode stability criterion. To this end, suppose that the open-loop transfer function is

$$\tilde{G}_{ol}(s) = 2(1 + 1/s)\frac{e^{-0.2s}}{3s+1}e^{-\theta s}, \qquad (11.163)$$

with $\theta > 0$.

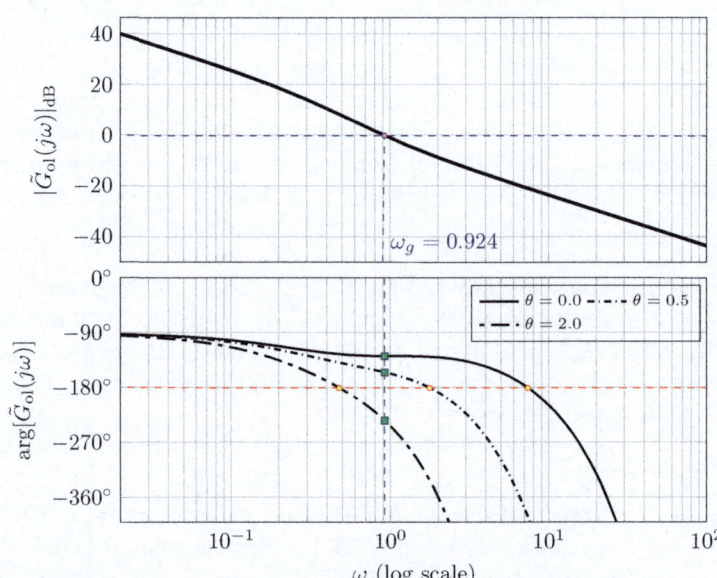

Figure 11.18: *Bode plot of G_{ol}. Note that $|\tilde{G}_{ol}(j\omega)|$ is independent of θ. In the phase plot, angles $\phi_{(1)}$ are denoted by squares and the phase crossover frequencies are denoted by circles.*

Chapter 11. Frequency response and design of stable closed loops

The introduction of this additional delay element does not affect the modulus of the transfer function which is $|\tilde{G}_{ol}(j\omega)| = |G_{ol}(j\omega)|$. Unfortunately, the phase crossover frequency of $\tilde{G}_{ol}$ cannot be expressed as a function of θ. However, given that conditions 1, 2 and 4 of Corollary 11.7 are satisfied for all θ, we can rewrite condition 3 without using the phase crossover frequency.

We first determine the (unique) frequency at which the amplitude ratio is equal to 1; we denote this frequency by ω_g. This is often referred to as the **gain crossover frequency**. It is

$$|G(j\omega_g)| = 1 \Leftrightarrow 2\frac{\sqrt{1+\frac{1}{\omega_g^2}}}{\sqrt{1+9\omega_g^2}} = 1. \qquad (11.164)$$

We can determine numerically (e.g., using fsolve) that $\omega_g \approx 0.924$. Then, as one may observe in Figure 11.18, condition 3 of Corollary 11.7 is equivalent to

$$\arg[G(j\omega_g)] > -\pi. \qquad (11.165)$$

We denote this special angle by $\phi_{(1)} := \arg[G(j\omega_g)]$. For example, we can see in Figure 11.18 that the closed-loop system is stable for $\theta = 0$ and $\theta = 0.5$ since $\phi_{(1)} > -\pi$, but for $\theta = 2.0$ we have $\phi_{(1)} < -\pi$, so stability is not guaranteed.

The distance of $\phi_{(1)}$ from $-\pi$ quantifies the phase lag that the system can take before it leaves the safety region of $\phi_{(1)} > -\pi$ (equivalently, $|G_{ol}(j\omega_{co})| < 1$).

The above example leads naturally to the definition of the *phase margin* of a system: the maximum phase a (stable) system can take before the stabilising conditions of Bode's criterion cease to hold.

> **Definition 11.10 (Phase margin)** Suppose that the open-loop transfer function of a feedback control system satisfies condition 1 of Corollary 11.7, G_{ol} has at most one phase crossover frequency, and $|G_{ol}(j\omega)|$ is strictly decreasing for all $\omega > 0$. Suppose that there is a frequency ω_g such that $|G_{ol}(j\omega_g)| = 1$ and let $\phi_{(1)} = \arg G_{ol}(j\omega_g)$. The **phase margin** of the system is defined as PM $= \phi_{(1)} + \pi$ (in radians).

The phase margin of a stable closed-loop controlled system is related to the maximum time delay that can be introduced into the loop before the stabilising conditions of Bode's criterion fail to hold — this is known as the *delay margin*. This is defined via the following proposition.

Proposition 11.11 (Delay margin) If G_{ol} satisfies the conditions of Definition 11.10, the closed-loop system is BIBO-stable with a phase margin PM, it will remain stable if we introduce a delay $t_d < \text{PM}/\omega_g$.

Proof. Suppose that G_{ol} satisfies the conditions of Bode's stability criterion. If we introduce a delay, $e^{-t_d s}$, into the open loop transfer function, the new open loop function becomes $G'_{ol}(s) = G_{ol}(s)e^{-t_d s}$. The crossover frequency corresponding to G'_{ol} is the same as the crossover frequency of G_{ol}. The maximum delay is attained when $G'_{ol}(j\omega_g) = -1$, that is

$$G_{ol}(j\omega_g)e^{-jt_{d,\max}\omega_g} = -1, \quad (11.166)$$

and solving for $t_{d,\max}$ completes the proof. $\square$

Definition 11.12 (Delay margin) If the phase margin and the crossover frequency are defined, the quantity[a]

$$\text{DM} = \frac{\text{PM}}{\omega_g}, \quad (11.167)$$

is called the **delay margin** of the system.

---
[a] Here PM should be measured in rad (not degrees) and ω_g should be in rad/s.

Remark: *It follows from the definition of the delay margin that if G_{ol} has a delay margin DM, then*

$$G'_{ol}(s) = G_{ol}(s)e^{-\text{DM}s}, \quad (11.168)$$

has a zero phase margin. •

Example 11.22 (Phase and delay margins) In Figure 11.18 we see that for $\theta = 0$, $\phi_{(1)} = -2.2343\,\text{rad} = -128.0134°$, so PM $= \phi_{(1)} + \pi = 0.9073\,\text{rad} = 51.9866°$. The gain crossover frequency is $\omega_g = 0.924\,\text{rad/s}$, so the delay margin is DM $= \frac{0.9073}{0.924} = 0.9820\,\text{s}$. Indeed, if we introduce a delay $\theta = 0.9820\,\text{s}$, then the phase margin becomes 0.

Example 11.23 (First-order open-loop transfer function: phase and delay margins) The open-loop transfer function of a closed-loop system is given by

$$G_{ol}(s) = \frac{K}{s+1}, \quad (11.169)$$

where $K > 0$. We will determine the phase and delay margins of the closed-loop system — if any.

The open-loop transfer function is BIBO-stable, does not have a phase crossover frequency and $|G_{ol}(j\omega)|$ is strictly decreasing for all $\omega > 0$ (see Example 11.7). We need to determine a frequency ω_g such that $|G_{ol}(j\omega_g)| = 1$:

$$|G_{ol}(j\omega_g)| = 1 \Leftrightarrow \frac{K}{\sqrt{\omega_g^2 + 1}} = 1$$

$$\Leftrightarrow \sqrt{\omega_g^2 + 1} = K$$

$$\Leftrightarrow \omega_g^2 = K^2 - 1, \qquad (11.170)$$

which has a (real) solution only if and only if $K > 1$, in which case

$$\omega_g = \sqrt{K^2 - 1}. \qquad (11.171)$$

Next we will determine $\phi_{(1)} = \arg G_{ol}(j\omega_g)$

$$\phi_{(1)} = \arg G_{ol}(j\omega_g) = -\arctan(\omega_g) = -\arctan\sqrt{K^2 - 1}, \qquad (11.172)$$

therefore, the phase margin of the closed-loop system is

$$\text{PM} = \pi - \arctan\sqrt{K^2 - 1}. \qquad (11.173)$$

For example, for $K = 10$, $\omega_g \approx 9.94987$, PM $\approx 95.7°$ and DM $\approx 0.168\,\text{s}$ and for $K = 2$, the phase margin is PM $= 120°$.

If $K \leq 1$, there is no frequency $\omega_g > 0$ satisfying (11.170). However, the reader can verify that the system will remain stable regardless of *any* time delay that might be introduced into the closed loop. In that case we say that the phase margin is $+\infty$.

Example 11.24 (Tuning based on phase margin) Consider the following open-loop transfer function

$$G_{ol}(s) = \frac{K}{(2s + 1)^4}. \qquad (11.174)$$

Our aim is to determine a value of $K > 0$ — if one exists — such that the phase margin is $30°$ (or $\pi/6$ rad).

The reader can verify that the open-loop transfer function is BIBO-stable, it has a single crossover frequency and $|G_{ol}(j\omega)|$ is strictly decreasing for all $\omega > 0$. Since we know that the phase margin is $30°$ we can determine ω_g as follows

$$\text{PM} = \frac{\pi}{6} \Leftrightarrow \phi_{(1)} + \pi = \frac{\pi}{6}$$

$$\Leftrightarrow \phi_{(1)} = -\frac{5\pi}{6}$$

$$\Leftrightarrow \arg G_{\text{ol}}(j\omega_g) = -\frac{5\pi}{6}$$

$$\Leftrightarrow \arg \frac{K}{(2\omega_g j + 1)^4} = -\frac{5\pi}{6}$$

$$\Leftrightarrow -4\arctan(2\omega_g) = -\frac{5\pi}{6}$$

$$\Leftrightarrow \arctan(2\omega_g) = \frac{5\pi}{24}$$

$$\Leftrightarrow \omega_g = \frac{1}{2}\tan\frac{5\pi}{24} \approx 0.3837 \,\text{rad}/\text{s}. \tag{11.175}$$

By definition of ω_g,

$$|G_{\text{ol}}(j\omega_g)| = 1 \Leftrightarrow \left|\frac{K}{(2\omega_g j + 1)^4}\right| = 1$$

$$\Leftrightarrow \frac{K}{(4\omega_g^2 + 1)^2} = 1$$

$$\Leftrightarrow K = (4\omega_g^2 + 1)^2 \approx 2.525. \tag{11.176}$$

Having determined the value of K that leads to the desired phase margin, we can determine the corresponding gain margin (see Definition 11.9) as well. In order to determine the gain margin, we need to determine the (here, unique) crossover frequency of the system, that is

$$\arg G_{\text{ol}}(j\omega_{\text{co}}) = -\pi$$
$$\Rightarrow -4\arctan(2\omega_{\text{co}}) = -\pi \Rightarrow \omega_{\text{co}} = \tfrac{1}{2}, \tag{11.177}$$

so the gain margin of the closed-loop system (in dB) is

$$\text{GM} = -20\log_{10}|G_{\text{ol}}(j\omega_{\text{co}})|$$
$$= -20\log_{10}\frac{2.525}{(2^2\omega_{\text{co}}^2 + 1)^2} \approx 4\,\text{dB}. \tag{11.178}$$

Lastly, the delay margin is

$$\text{DM} = \frac{\text{PM}}{\omega_g} = \frac{\frac{\pi}{6}}{0.3837} \approx 1.365\,\text{s}. \tag{11.179}$$

Chapter 11. Frequency response and design of stable closed loops

Exercise 11.18 (PM-based tuning) For the closed-loop system of Example 11.19, determine the value of K_c to have a phase margin of 45°. In practice, reasonable phase margin values are between 30° and 60°.

Exercise 11.19 (No phase crossover frequency) It is not necessary for G_{ol} to have a phase crossover frequency to define a phase margin. Find the phase margin of a closed-loop system with open-loop transfer function

$$G_{ol}(s) = \frac{20}{s+1}. \qquad (11.180)$$

Then, find the value of K so that the phase margin of the closed-loop system with open-loop transfer function

$$G_{ol}(s) = \frac{20K}{s+1}, \qquad (11.181)$$

is 120°.

※ ......

Answer. For $K = 0.1$ the phase margin is 120°.

Exercise 11.20 (GM from PM) The open-loop transfer function of a feedback control system is given by

$$G_{ol}(s) = \frac{K}{(3s+1)^4}, \qquad (11.182)$$

where K is a positive constant. The phase margin is 30°. Determine the gain margin (in dB).

※ ......

Answer. The crossover frequency is $\omega_{co} = \frac{1}{3}$ rad/s, $\omega_g = \frac{1}{3}\tan\frac{3\pi}{24} = 2.5246$. The gain is $K = 3.9973$ dB. GM is margin and the gain $\frac{5\pi}{24}$.

Remark: It may be tempting to think that a positive gain or phase (or delay) margin means that the closed-loop system is stable. However, it is important to check whether the conditions of Bode's stability criterion (Theorems 11.6 or 11.8) hold before we compute any margins, otherwise we may arrive at erroneous conclusions. This is particularly important when we use software such as Python or MATLAB to determine the gain, phase and delay margins of a system.

In summary, if a system is stable, its gain and phase margin — if defined — are positive. However, there are unstable margins for which

Python, MATLAB and other software will return falsely reassuring positive gain and phase margins. Some interesting counterexamples can be found in [C12, Section 6.3]. •

11.3.4 Necessary and sufficient stability conditions

The Bode stability criterion provides sufficient, but not necessary conditions for stability. It is natural to ask under what additional conditions on G_{ol}, the stabilising condition $|G_{\text{ol}}(j\omega_{\text{co}})| < 1$ becomes necessary for stability. Although the most definitive answer will be given in the next chapter by the Nyquist stability criterion, we can here give the following result.

> **Theorem 11.13** (Necessary and sufficient conditions for stability) Let G_{ol} and G_{cl} denote the open and closed-loop transfer functions. Suppose that
>
> 1. G_{ol} is *strictly proper*, i.e., it is a rational-times-exponential function and the degree of its numerator is less than the degree of its denominator
>
> 2. G_{ol} has no poles located on or to the right of the imaginary axis, with the possible exception of a single pole at the origin,
>
> 3. G_{ol} has a single phase crossover frequency, ω_{co}, i.e., equation $\arg[G_{\text{ol}}(j\omega)] = -\pi$ has a single solution
>
> 4. G_{ol} has a signle gain crossover frequency, ω_g, i.e., equation $|G_{\text{ol}}(j\omega)| = 1$ has a single solution
>
> Then, the closed-loop system, with transfer function G_{cl}, is BIBO-stable if and only if $|G_{\text{ol}}(j\omega_{\text{co}})| < 1$.

Proof. This is a special case of the Nyquist stability criterion, which we state in the next chapter. □

11.4 Effect of PID parameters on Bode plot

In this section we will give a few examples in order to understand the effect that the choice of parameter values of a PID controller have on the Bode plot of the open-loop transfer function. Our aim is to obtain a better understanding on how the choice of K_c, τ_I and τ_D affect the shape of the Bode curves. We will not be able to give general design guidelines that cover all possible transfer functions, but we will know what to expect when we change these values.

Suppose the open-loop transfer function is given by $G_{\text{ol}}(s) = G(s)G_c(s)$, where $G_c(s) = K_c(1 + \tau_D s + 1/\tau_I s)$, that is

$$G_{\text{ol}}(s) = K_c \left(1 + \tau_D s + \frac{1}{\tau_I s}\right) G(s). \tag{11.183}$$

11.4.1 Effect of proportional gain

If we keep τ_I and τ_D constant and increase K_c, assuming $K_c > 0$, the value of $|G_{\text{ol}}(j\omega)|$ will increase, while $\arg G_{\text{ol}}(j\omega)$ will remain unaffected. Changing K_c allows us to move the curve in the magnitude plot up and down, without affecting the argument plot. Indeed, for $K_c > 0$, $\arg[K_c G_{\text{ol}}(j\omega)] = \arg[G_{\text{ol}}(j\omega)]$. However, for $K_c < 0$, it is $\arg[K_c G_{\text{ol}}(j\omega)] = \arg K_c + \arg[G_{\text{ol}}(j\omega)] = \pi + \arg[G_{\text{ol}}(j\omega)]$, that is, multiplying by a negative number shifts the argument of G_{ol} by π (see Equation (A.4)).

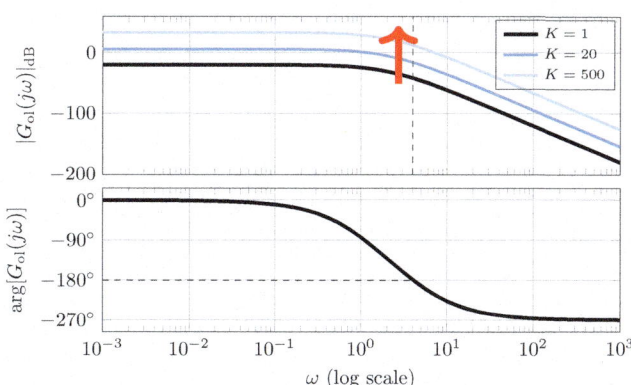

Figure 11.19: *Effect of $K_c > 0$ for a system with open-loop transfer function $G_{\text{ol}}(s) = \frac{K_c}{(s+1)(s+2)(s+5)}$. The systems with $K = 1$ and $K = 20$ are BIBO-stable, while the system with $K = 500$ is not. The critical value of the controller gain is $K_{c,\max} = 125$.*

We already know that there exist systems that are BIBO-stable for all $K_c > 0$, such as when G_{ol} is a first-order transfer function, and systems that are unsalvagable; they are unstable for all $K_c > 0$, for example $G_{\text{ol}}(s) =$

$K_c/((s-1)^2+1)$. An example of the effect of the proportional gain on the Bode plot's shape is given in Figure 11.19.

11.4.2 Effect of derivative time

Consider an open-loop transfer function that involves a PD controller with gain $K_c = 1$ and derivative time τ_D, $G_{\text{ol}}(s) = K_c\left(1 + \tau_D s\right)G(s)$. Suppose further that the proportional gain, K_c, is equal to 1; we want to study the effect that τ_D has on the shape of the Bode plot of G_{ol}.

At low frequencies, $|1+\tau_D j\omega|$ is approximately equal to 1, so neither the modulus nor the argument of G_{ol} will be affected; we will have $|G_{\text{ol}}(j\omega)| \approx |G(j\omega)|$ and $\arg G_{\text{ol}}(j\omega) \approx \arg[K_c G(j\omega)] = \arg G(j\omega)$.

At high frequencies, $1 + \tau_D j\omega$ will be approximately equal to $j\tau_D\omega$. As a result $|G_{\text{ol}}(j\omega)| \approx |G(j\omega)|\tau_D\omega$ and, by taking the logarithm on both sides

$$\log|G_{\text{ol}}(j\omega)| = \log|G(j\omega)| + \log(\tau_D\omega), \qquad (11.184)$$

therefore, at high frequencies we expect a linear increase in $\log|G_{\text{ol}}(j\omega)|$. For the argument of $G_{\text{ol}}(j\omega)$ we have

$$\arg G_{\text{ol}}(j\omega) = \arg G(j\omega) + \arctan(\tau_D\omega),$$

so, since $\lim_{\omega \to \infty} \arg \omega = \pi/2$, at high frequencies the argument of $G_{\text{ol}}(j\omega)$ is shifted by $+\pi/2$. Note that at low frequencies, the derivative element, $\tau_D s$, does not have an effect.

As an example, consider a system with transfer function

$$G(s) = \frac{1}{(s+1)^2}. \qquad (11.185)$$

The Bode plot of $G_{\text{ol}}(s) = K_c\left(1 + \tau_D\right)\frac{1}{s+1}$ is shown in Figure 11.20.

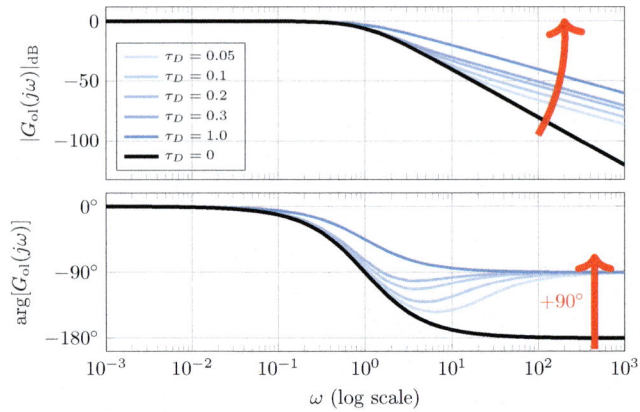

Figure 11.20: *Effect of τ_D on the shape of the Bode plot of G_{ol}. The closed-loop system remains stable for all $\tau_D > 0$ (see Prop. 11.8).*

Chapter 11. Frequency response and design of stable closed loops

The D element can have both a stabilising as well as a destabilising effect in the sense that it has the potential to increase the phase margin, but it may decrease the gain margin. For example, consider the following system

$$G(s) = \frac{e^{-0.1s}}{s+1}. \tag{11.186}$$

The system with open-loop transfer function $G_{\text{ol}}(s) = G(s)$ is stable with GM = 24.3 dB. However, by using a PD controller with $K_c = 1$ and $\tau_D = 0.5$ the gain margin drops to 6 dB. The system becomes unstable for $\tau_D \geq 1$.

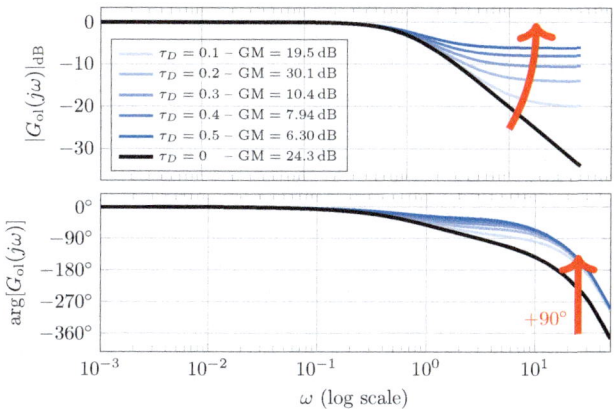

Figure 11.21: *Effect of τ_D on the shape of the Bode plot of $G_{\text{ol}}(s) = (1 + \tau_D s)G(s)$, where G is as in (11.186). The phase margin is always $-180°$.*

The D element can have a stabilising effect by increasing the phase margin, at the potential cost of reducing the gain margin. For example, consider the following system

$$G(s) = \frac{s+10}{s(10s+1)}, \tag{11.187}$$

which is controlled with a PD controlled with proportional gain $K_c = 1$ and derivative time τ_D. The open-loop transfer function is $G_{\text{ol}}(s) = (1 + \tau_D s)G(s)$ and its Bode plot for different values of τ_D is given below.

In Figure 11.22 we see that by increasing τ_D, we move the phase curve of the Bode plot upwards and away from $-180°$. By doing so we increase the phase margin. In this case since the open-loop transfer function does not have a phase crossover frequency and $\arg G(j\omega) > -\pi$ for all $\omega > 0$, the gain margin is infinite.

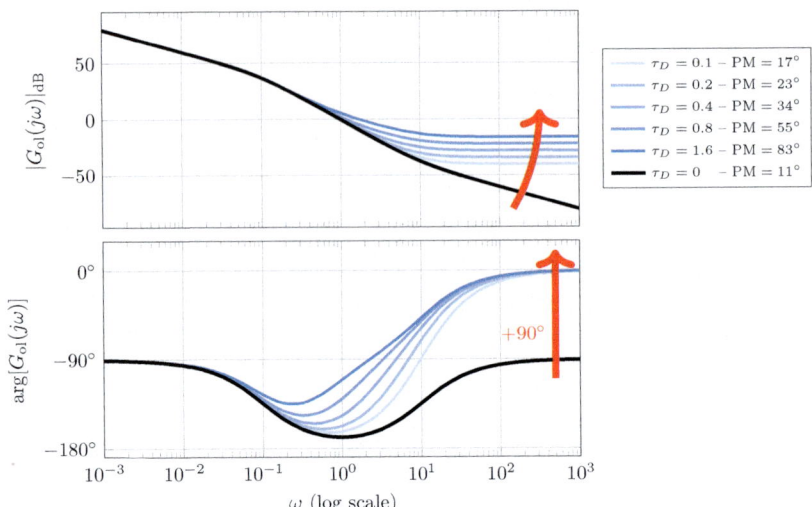

Figure 11.22: *Effect of parameter τ_D on the shape of the Bode plot of $G_{ol}(s) = (1 + \tau_D s)G(s)$, where G is given in Equation (11.187).*

11.4.3 Effect of integral time

In order to demonstrate the effect of the integral time, we will consider a PI controller with $K_c = 1$ and integral time τ_I. Suppose that the open-loop transfer function has the following form

$$G_{ol}(s) = \left(1 + \frac{1}{\tau_I s}\right) G(s). \qquad (11.188)$$

For large ω, the term $1 + \frac{1}{\tau_I j\omega}$ becomes approximately equal to 1, therefore, $|G_{ol}(j\omega)| \approx |G(j\omega)|$ and $\arg G_{ol}(j\omega) \approx \arg G(j\omega)$. This means that the presence of the PI controller does not affect the high-frequency behaviour of the open-loop transfer function.

For small ω, the term $1/j\omega = -j1/\omega$ becomes dominant and $1 - j1/\omega$ becomes approximately equal to $-j1/\omega$. Then,

$$|G_{ol}(j\omega)| = \left|1 + \frac{1}{j\omega}\right| \cdot |G(j\omega)| \approx \left|-j\frac{1}{\omega}\right| \cdot |G(j\omega)| = \frac{1}{\omega}|G(j\omega)|, \qquad (11.189)$$

therefore

$$\log |G_{ol}(j\omega)| = |G(j\omega)| - \log \omega. \qquad (11.190)$$

For the phase lag we have

$$\arg G_{ol}(j\omega) = \arg \left(1 + \frac{1}{j\omega}\right) + \arg G(j\omega) \approx -\frac{\pi}{2} + \arg G(j\omega). \qquad (11.191)$$

From (11.190) and (11.191) we have that at low frequencies, the PI controller increases the magnitude of $G_{ol}(j\omega)$ and introduces a phase shift of $-\pi/2$ (i.e., $-90°$).

Let us give an example. Consider the transfer function

$$G(s) = \frac{1}{(5s+1)^2(s+1)}. \qquad (11.192)$$

The Bode plot of G_{ol} for different values of τ_I is given in Figure 11.23. We may observe that at low frequencies the PI controller introduces a lag of $90°$ and leads to an increase in the magnitude of $G_{\text{ol}}(j\omega)$.

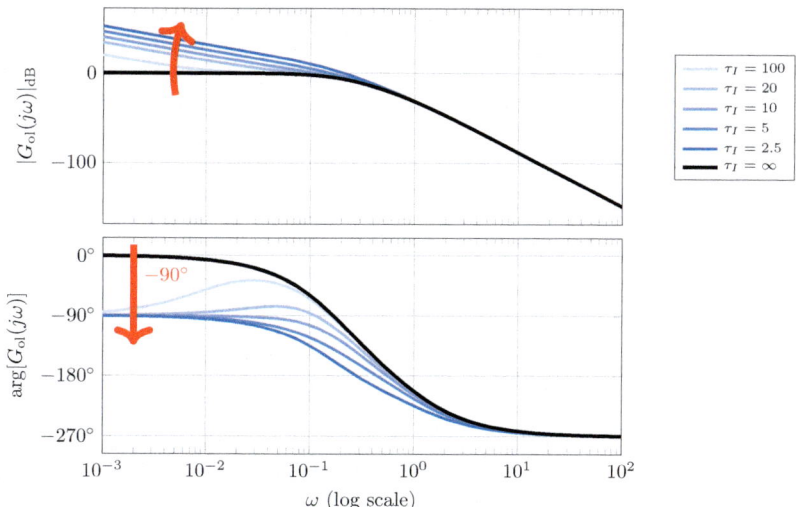

Figure 11.23: *Effect of parameter τ_I on the shape of the Bode plot of $G_{\text{ol}}(s) = (1 + 1/\tau_I s)G(s)$. The thick red arrows point to the direction of decrease of τ_I.*

11.5 Offset detection in Bode plots

We have so far used Bode plots to tell, under certain conditions, whether the closed-loop transfer function is BIBO-stable using the (simpler) open-loop transfer function. In this section we ask whether it is possible to determine the offset upon a unit step change of the set point using the Bode plot of G_{ol}.

Throughout the rest of this section we assume that

1. G_{ol} and G_{cl} are BIBO-stable

2. $G_m(0^+) = 1$, which means that the DC gain of the sensor is 1, in other words the sensor is *unbiased*,

3. G_m has no zeroes.

The second and third conditions are trivially satisfied if $G_m(s) = 1$ as well as in the common case where the sensor is modelled by a first-order transfer function.

Since G_{cl} is assumed to be BIBO-stable, we can apply the final value theorem to determine the final value of the step response of the closed-loop system. Let us denote this by $x^{\text{step}}(\infty)$ and let $X^{\text{step}}(s) = \frac{G_{cl}(s)}{s}$. This is

$$x^{\text{step}}(\infty) = \lim_{s \to 0^+} sX^{\text{step}}(s) = \lim_{s \to 0^+} G_{cl}(s). \tag{11.193}$$

In light of the third assumption, the closed-loop transfer function is

$$G_{cl}(s) = \frac{G_c(s)G_a(s)G_s(s)}{1 + G_c(s)G_a(s)G_s(s)G_m(s)} = \frac{1}{G_m(s)} \frac{G_{ol}(s)}{1 + G_{ol}(s)}, \tag{11.194}$$

(see Definitions 10.2 and 11.4).

The offset upon a unit step change of the set point is (see Definition 10.1)

$$\text{offset} = 1 - x^{\text{step}}(\infty)$$

$$= 1 - \lim_{s \to 0^+} G_{cl}(s) = 1 - \lim_{s \to 0^+} \frac{1}{G_m(s)} \frac{G_{ol}(s)}{1 + G_{ol}(s)}. \tag{11.195}$$

If $\lim_{s \to 0^+} G_{ol}(s)$ exists and is finite, then the offset becomes

$$\text{offset} = 1 - \frac{G_{ol}(0^+)}{1 + G_{ol}(0^+)}. \tag{11.196}$$

It is evident that the offset cannot be equal to zero if $G_{ol}(0^+)$ is finite. In fact, the offset is zero if and only if[5]

$$\lim_{s \to 0^+} |G_{ol}(s)| = \infty. \tag{11.197}$$

We now need to link $\lim_{s \to 0^+} |G_{ol}(s)|$ — which is taken over $s \in \mathbb{R}$ — to $\lim_{\omega \to 0^+} |G_{ol}(j\omega)|$. For general complex functions G_{ol}, these two limits need not be equal to one another[6]. However, in all cases of practical interest these will be indeed equal. For example, the reader can verify if G_{ol} is a rational function or the product of a rational with an exponential function, the two limits are equal. More generally, if G_{ol} can be written in the following form[7]

$$G_{ol}(s) = \frac{B(s)}{A(s)}, \tag{11.198}$$

where $A, B : D \to \mathbb{C}$, 0 is contained in D, and B is bounded in a neighbourhood of 0 relative to D. Condition (11.197) is equivalent to

$$\lim_{s \to 0^+} |G_{ol}(s)| = \infty \Leftrightarrow \lim_{s \to 0^+} \frac{|B(s)|}{|A(s)|} = \infty. \tag{11.199}$$

---

[5]This is because $\lim_{x \to \infty} \frac{x}{1+x} = 1$.

[6]As a counterexample, take $F(s) = \exp \frac{1}{s}$; then $\lim_{s \to 0^+} |F(s)| = \infty$, but $\lim_{\omega \to 0^+} |F(j\omega)| = 1$. This function exhibits what is called an *essential singularity* at $s = 0$. The point $s = 0$ is *not* a pole of the function and the function is not meromorphic on $\mathbb{C}$.

[7]This is a very general formulation. For example, every meromorphic function defined on an open set can be written as the quotient of two holomorphic functions (see [M2, Corollary 5.20]).

Given the local boundedness of B, this holds only if $\lim_{s \to 0^+} |A(s)| = 0$. If A is continuous[8] then $\lim_{s \to 0^+} |A(s)| = 0$ is equivalent to $\lim_{\omega \to 0^+} |A(j\omega)| = 0$. Again, in light of the local boundedness of B, this is equivalent to

$$\lim_{\omega \to 0^+} |G_{\mathrm{ol}}(j\omega)| = \infty. \qquad (11.200)$$

We have proven the following result:

> **Theorem 11.14 (Offset from Bode plot)** Suppose that following conditions are satisfied
>
> 1. G_{cl} is BIBO-stable,
>
> 2. $G_m(0^+) = 1$ and G_m has no zeroes,
>
> 3. The open-loop transfer function can be written in the form
>
> $$G_{\mathrm{ol}}(s) = \frac{B(s)}{A(s)}, \qquad (11.201)$$
>
> where A, B are defined in a neighbourhood of zero, B is locally bounded in a neighbourhood of zero and A is continuous at zero,
>
> then the offset of the closed-loop system upon a step change of the set point is zero if and only if
>
> $$\lim_{\omega \to 0^+} |G_{\mathrm{ol}}(j\omega)| = \infty. \qquad (11.202)$$

Proof. See above. □

The first condition of Theorem 11.14 can be checked using Bode's stability criterion. For that G_{ol} is required to satisfy additional conditions (see Theorem 11.6 and Corollary 11.7). The second condition is stated for simplicity; the reader can easily generalise this result when $G_m(0^+)$ is equal to a different finite value. The third condition is satisfied by all practically relevant transfer functions (to date).

To put it very simply: in the context of Theorem 11.14, the offset of a stable closed-loop system can only be eliminated if the open-loop transfer function has a **pole at zero**. This is why we often need to use a PI controller to eliminate the offset: it is because it introduces a pole at zero into the open-loop transfer function. This is also why in absence of a pole at zero, P and PD controllers cannot eliminate the offset.

---

[8] or at least if $\lim_{s \to 0} |A(s)|$ taken over $s \in \mathbb{C}$ exists

Exercise 11.21 (**Find the offset**) We are given the Bode plot of a stable open-loop transfer function and observe that as $\omega \to 0^+$ the magnitude, $|G_{\text{ol}}(j\omega)|$, goes to 40 dB. From the given Bode plot we can tell that the closed-loop system is BIBO-stable. If $G_m(s) = 1.1$, what is the offset?

✂ ....

Hint: Follow the procedure outline in the beginning of Section 11.5. You should find that the offset is approximately equal to 0.00991.

11.6 Bode plots using software

Bode plots can be easily constructed using Python or MATLAB.

11.6.1 Python

In Python, we can use `scipy` to plot the Bode plot of a given transfer function. Let us give an example for the function

$$G(s) = \frac{1 + 0.5s + 5}{s^3 + 0.2s^2 + 10s}. \tag{11.203}$$

```
from scipy import signal
import matplotlib.pyplot as plt

numerator = [1, 0.5, 8]
denominator = [1, 0.2, 10, 0]
transfer_function = signal.TransferFunction(
    numerator, denominator)
w, mag, phase = signal.bode(transfer_function)

plt.figure()
plt.subplot(2, 1, 1)
plt.semilogx(w, mag)       # Bode magnitude plot
plt.subplot(2, 1, 2)
plt.semilogx(w, phase)     # Bode phase plot
plt.show()
```

The same result can be achieved using the `control` module (install with `pip install control`)

```
import control
import matplotlib.pyplot as plt

numerator = [1 0.5 8]
denominator = [1 0.2 10 0]
```

```
sys = control.tf(numerator, denominator)
control.bode_plot(sys, dB=True)
plt.show()
```

Moreover, we can determine the gain margin, phase margin, and phase and gain crossover frequencies using `control.margin`

```
gm, pm, w_co, w_g = control.margin(sys)
```

where `gm` is the gain margin, `pm` is the phase margin, `w_co` is the phase crossover frequency (ω_{co}) and `w_g` is the gain crossover frequency ω_g. According to the documentation, "if more than one crossover frequency is detected, returns the lowest corresponding margin." Nonetheless, as discussed above, the conditions of Bode's stability theorem should be always checked.

A limitation of both `scipy` and `control` is that they do not support transfer functions with delays. However, using `control` we can determine a Padé approximation of $e^{-t_d s}$ using `control.pade`. Let us give an example for the function

$$G(s) = \frac{1 + 0.5s + 5}{s^3 + 0.2s^2 + 10s} e^{-0.1s}. \tag{11.204}$$

```
import control

t_delay = 0.1
pade_ord = 2
num = [1, 0.5, 8]
den = [1, 0.2, 10, 0]
tf_del = control.tf(*control.pade(t_delay, pade_ord))
tf = control.tf(num, den) * tf_del
```

11.6.2 MATLAB

MATLAB supports transfer functions with delays and allows us to plot the Bode plot of a transfer function and determine the stability margins. Here is an example of plotting a Bode plot using MATLAB's Control Systems Toolbox.

```
numerator = [1 0.5 8];
denominator = [1 0.2 10 0];
t_delay = 0.1;
transfer_function = tf(numerator, denominator, ...
    'InputDelay',t_delay);
bode(transfer_function);
```

MATLAB's `margin` can be used both to determine the stability margins of a system and produce a Bode plot similar to that of Figure 11.18 showing the crossover frequencies and the stability margins. The syntax is as simple as

```
margin(transfer_function);     % Bode plot with margins
[gm, pm, w_co, w_g] = margin(transfer_function);
```

In case of multiple crossover frequencies, one can use `allmargin` instead.

11.7 One-minute round-up

The asymptotic behaviour of the frequency response of stable linear systems (and some particular unstable systems) can be described by function $G(j\omega)$. In particular, $|G(j\omega)|$ and $\arg G(j\omega)$ give the asymptotic amplitude ratio and phase lag of the frequency response. This information can be presented in a Bode plot. The same plot can be used to study the stability properties of closed-loop systems using the corresponding open-loop transfer function.

Bode plots can be constructed and analysed easily. They can be used to analyse systems with time delays directly obviating the need for Padé — or other — approximations, which can lead to erroneous results. Moreover, Bode plots allow us to determine the stability margins of the closed-loop system, which quantify the system's *robustness*.

On the other hand, the Bode stability criterion comes with several limitations. It can only be applied to systems which have a unique crossover frequency (i.e., equation $\arg[G_{ol}(j\omega)] = -\pi$ has a single solution) and the criterion offers only sufficient, but not necessary conditions for stability unless we impose additional conditions on G_{ol}. Additionally, the Bode stability criterion, in all of its variants, requires that the open-loop transfer function has no poles on or to the right of the imaginary axis, with the possible exception of a single pole at the origin. This limits significantly the applicability of the criterion. Overall, the reader should be particularly wary when using the Bode stability criterion and make sure that all conditions are satisfied (see Exercise 11.29).

11.8 Exercises

Exercise 11.22 (True/False) True of false? Warning: tricky questions ahead!

	True	False
The number $|G(j\omega)|$ gives the amplitude of oscillation at large times when the input is $u(t) = \sin(\omega t)$	☐	☐
The gain margin is defined as $GM = -|G_{ol}(j\omega_{co})|_{dB}$	☐	☐
Suppose an open-loop transfer function has a gain margin $GM = 40\,dB$. This means that if we multiply the open-loop transfer function by $0 < K < 100$, the closed-loop system will remain BIBO-stable	☐	☐
If G is the transfer function of a BIBO-unstable system, $|G(j\omega)|$ is equal to infinity	☐	☐
For two nonzero complex numbers z and z', it is $\arg(z \cdot z') = \arg z + \arg z'$	☐	☐
For two nonzero complex numbers z and z', it is $\log|z \cdot z'| = \log|z| + \log|z'|$	☐	☐
If $G(s) = \frac{1}{s+1}e^{-0.01s}$, then $\lim_{\omega \to \infty} \arg[G(j\omega)] = -\infty$.	☐	☐
We can apply the Bode stability criterion to $G_{ol}(s) = \frac{1}{s-1}$	☐	☐
We can apply the Bode stability criterion to $G_{ol}(s) = \frac{1}{s+1}$	☐	☐
We can apply the Bode stability criterion to $G_{ol}(s) = \frac{2}{s^2+0.3s+1}$	☐	☐

✻ ......

Answer. F (Not if G is unstable), F (certain conditions must be satisfied), T, F, T, T, F, T, F, T, T.

Exercise 11.23 (Uniqueness of ω_{co} and strict monotonicity of $|G_{ol}(j\omega)|$) In Example 11.19 we found that

$$\arg[G_{ol}(j\omega)] = -\tfrac{\pi}{2} - \arctan\omega - 0.1\omega,$$

and
$$|G_{\text{ol}}(j\omega)| = K_c \frac{1}{\omega\sqrt{1+\omega^2}}.$$

Prove mathematically (without using the Bode plot) that

1. Equation $\arg[G_{\text{ol}}(j\omega)] = -\pi$ has a *unique* solution
2. $|G_{\text{ol}}(j\omega)|$ is strictly decreasing for all $\omega > 0$

✂ ··

Hint. 1. First we need to determine a frequency ω_{co} so that $\arg[G_{\text{ol}}(j\omega)] = -\pi$. In order to show that this is unique, it suffices to show that $\arg[G_{\text{ol}}(j\omega)]$ is strictly monotone; to do so it suffices to show that $\frac{d}{d\omega}\arg[G_{\text{ol}}(j\omega)]$ does not change sign over $\omega \in (0, \infty)$; 2. Similarly.

Exercise 11.24 (**Pair of complex conjugate poles**) Determine the magnitude and phase lag of $G(j\omega)$ for

$$G(s) = \frac{1}{(s+a)^2 + 1}, \qquad (11.205)$$

where $a \in \mathbb{R}$. What are the low and high frequency asymptotes of $\log_{10}|G(j\omega)|$? Sketch the Bode plot of G and verify its correctness using Python or MATLAB.

Exercise 11.25 (**Stability margins of system with delay**) The open-loop transfer function of a closed-loop system is

$$G_{\text{ol}}(s) = \frac{2e^{-0.05s}}{0.1s + 1}. \qquad (11.206)$$

Compute the gain, phase and delay margins.

✻ ··

Answer. The gain margin is 5.59 dB (the phase crossover frequency is 36.7 rad/s) and the phase margin is 70.383°. They delay margin is approximately 0.071 s.

Exercise 11.26 (**Newton-Raphson method**) A first-order system with time constant $\tau = 2$ and gain $K = 1$ is controlled by a PI controller with $K_p = 3$ and integral time constant $\tau_I = 0.5$. The transfer functions of the sensor and the actuator are given by $G_m(s) = e^{-0.1s}$ and $G_a(s) = \frac{1}{0.05s+1}$.

1. Construct the Bode plot of the open-loop transfer function in

Python or MATLAB

2. Write a script in Python or MATLAB to find an approximation of the phase crossover frequency of the system using the Newton-Raphson method; verify the correctness of your result using `fsolve`

3. Use the Bode stability criterion to tell whether the closed-loop system is stable

Recall that the Newton-Raphson method for finding a root of a function $f : \mathbb{R} \to \mathbb{R}$ starts from an initial guess $x^0 \in \mathbb{R}$ and iterates as follows

$$x^{\nu+1} = x^{\nu} - \frac{f(x^{\nu})}{f'(x^{\nu})}. \tag{11.207}$$

The method converges under certain conditions at a root r if f is twice continuously differentiable, $f'(r) \neq 0$ and x^0 is sufficiently close to r (a good initial guess can be obtained from the Bode plot).

Exercise 11.27 (TFs vs magnitude plots) Pair the following transfer functions with the corresponding plots

$$G_1(s) = \frac{2}{10s + 1}$$

$$G_2(s) = \frac{1}{s^2 + 1/2\,s + 1}$$

$$G_3(s) = \frac{5s+1}{(20s+1)(s^2+\frac{1}{2}s+1)}$$

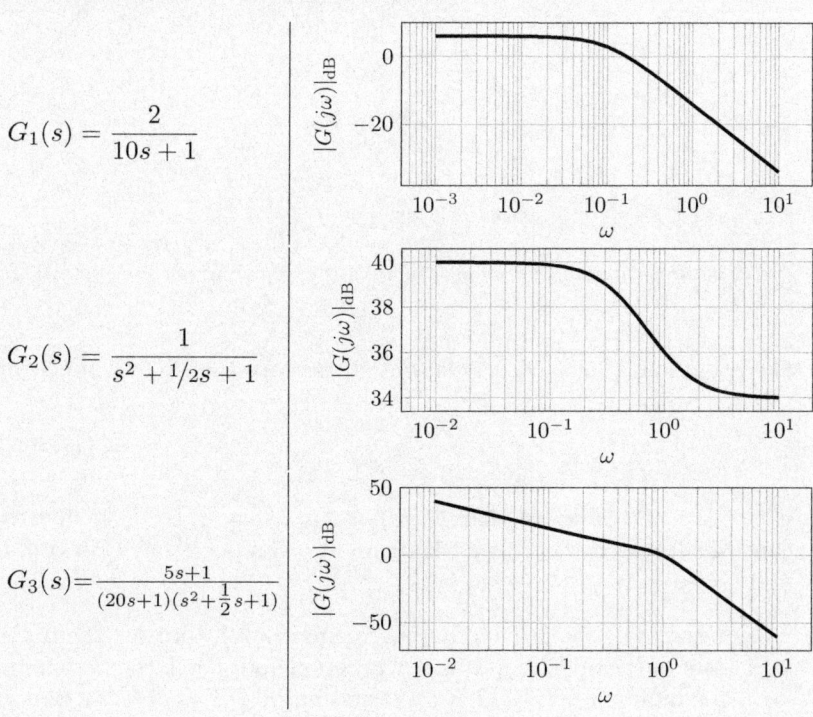

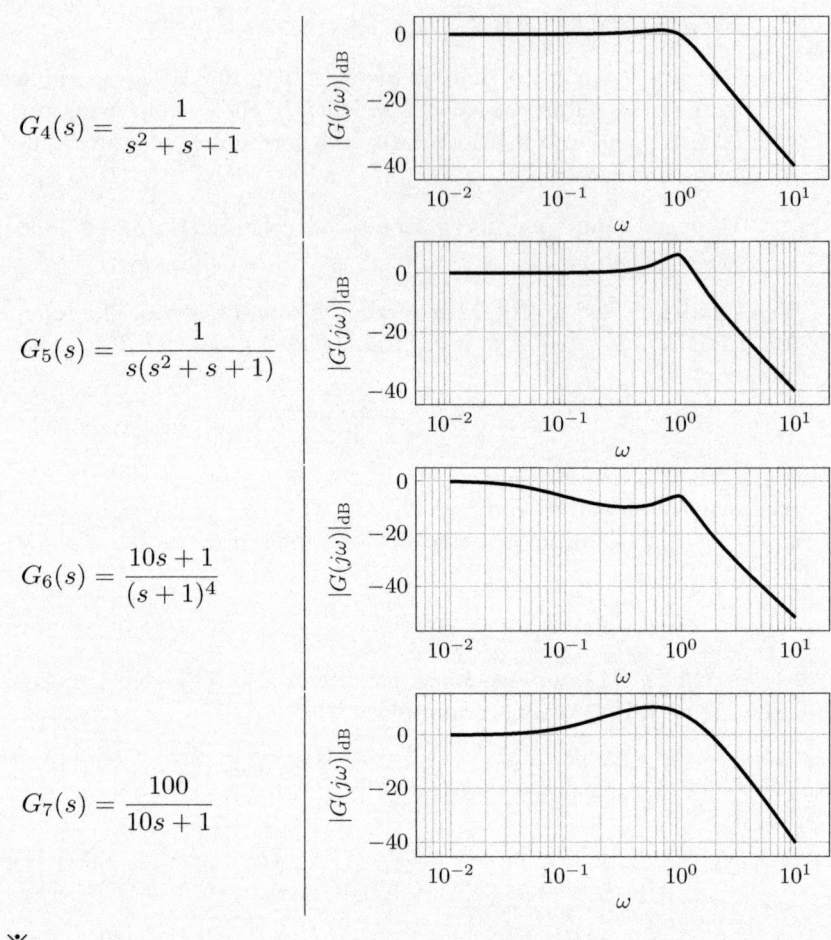

Answer. The magnitude plots in the right column, in the order of appearance, correspond to G_1, G_7, G_5, G_4, G_2, G_3 and G_6.

※ ....

Exercise 11.28 (Stability condition) A system has the transfer function

$$G_s(s) = \frac{3e^{-t_d s}}{s + 1/2}, \tag{11.208}$$

with $t_d > 0$. Suppose that $G_a(s) = G_m(s) = 1$. The system is controlled by a P controller with gain K_c. Suppose that $G_m(s) = 1$ and $G_a(s) = 1$.

1. Determine the low and high frequency asymptotes for the Bode plot of the open loop transfer function and sketch the Bode plot for different values of $K_c > 0$ and t_d,

Chapter 11. Frequency response and design of stable closed loops

2. Determine the conditions on K_c and t_d so that the stabilising conditions of Bode's criterion are satisfied.

3. For $t_d = 0.01$, what is the range of values of K_c so that the conditions of Bode's stability criterion are satisfied?

4. For $K_c = 100$, what is the maximum delay, t_d, so that the conditions of Bode's stability criterion are satisfied?

✂ ....

Hint. 1. Easy. 2. If we attempt to determine the phase crossover frequency we will end up with a complicated nonlinear equation that involves the inverse tangent function; instead we can check whether the phase margin is positive. Firstly, we need to make sure that the conditions of Definition 11.10 are satisfied so that the phase margin is well defined. Start by determining ω_g.

※ ....

Answer. 2. the condition is $\pi - t_d\sqrt{9K_c^2 - 1}/4 - \arctan(2\sqrt{9K_c^2 - 1}/4) > 0$

Exercise 11.29 (Bode criterion can lead to erroneous results if misused) Consider the following *unstable* open-loop transfer function

$$G_s(s) = \frac{0.5}{s-1}. \qquad (11.209)$$

Show that using the Bode stability criterion will lead to erroneous results.

※ ....

Answer. We cannot apply the Bode stability criterion. But if we did, and we constructed the Bode plot, we would wrongly conclude that the closed-loop system is stable. However, the denominator of the closed-loop transfer function

$$\chi(s) = 1 + G_{ol}(s) = \frac{s - 0.5}{s - 1}, \qquad (11.210)$$

which has a positive root at $s = 0.5$, therefore the closed-loop system is not stable.

Exercise 11.30 (Given PM, find GM) The open-loop transfer function of a system is

$$G_{ol}(s) = \frac{K}{s(s+1)^2}, \qquad (11.211)$$

and its phase margin is 30°. Determine its gain margin.

✂ ....

Hint. There is a catch in this exercise: it is not assumed that $K > 0$, therefore you need to examine both cases ($K > 0$ and $K < 0$). Recall that if $K < 0$, $\arg K = \frac{\pi}{2}$. However, you should observe that for $K > 0$ it is not possible to have a phase margin of $30°$.

※ ....

Answer. $K = \frac{4\sqrt{3}}{6}$.

Exercise 11.31 (**Similar Bode plots No. 1**) Give an example of two different strictly proper rational transfer functions, G_1, and G_2, which have identical magnitude plots, but different phase lag plots.

※ ....

Answer. Take $G_1(s) = \frac{1}{s+1}$ and $G_2(s) = \frac{1}{s-1}$.

Exercise 11.32 (**Similar Bode plots No. 2**) Give an example of two different rational transfer functions, G_1, and G_2, which have identical phase lag plots, but different magnitude plots. The obvious choice is that G_2 is a multiple of G_1, e.g., $G_2(s) = 5G_1(s)$. Can you think of a different example?

※ ....

Answer. Take $G_1(s) = \frac{s-1}{2s-1}$ and $G_2(s) = \frac{2s-1}{s+1}$.

Exercise 11.33 (**P controller tuning**) Let $N \geq 3$ and consider a dynamical system with input $U(s)$ and output $X(s)$, where the input-output dynamic relationship is described by the block diagram shown below.

$U(s) \longrightarrow \boxed{G_1(s)} \longrightarrow \boxed{G_2(s)} \longrightarrow \cdots \longrightarrow \boxed{G_N(s)} \longrightarrow X(s)$

We know that $N \geq 3$ and

$$G_i(s) = \frac{1}{3s+1}, \qquad (11.212)$$

for all $i = 1, \ldots, N$. Determine the gain of a P controller (as a function of N) so that the gain margin is $2\,\text{dB}$.

※ ....

Answer. It is $\log_{10} K_c = \frac{N}{2} \log_{10}(9\omega_\infty^2 + 1) - \frac{1}{10}$, where $\omega_\infty = \frac{1}{3} \tan \frac{\pi}{N}$.

Chapter 11. Frequency response and design of stable closed loops

Exercise 11.34 (Sketch the Bode plots) Sketch the Bode plots of the following functions.

$$G_1(s) = \frac{1}{5s+1}. \qquad (11.213)$$

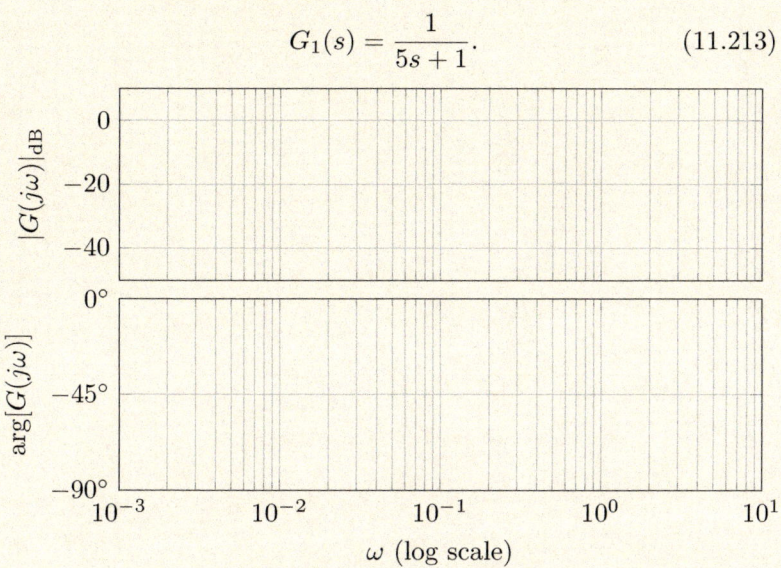

$$G_2(s) = \frac{3s}{10s+1}. \qquad (11.214)$$

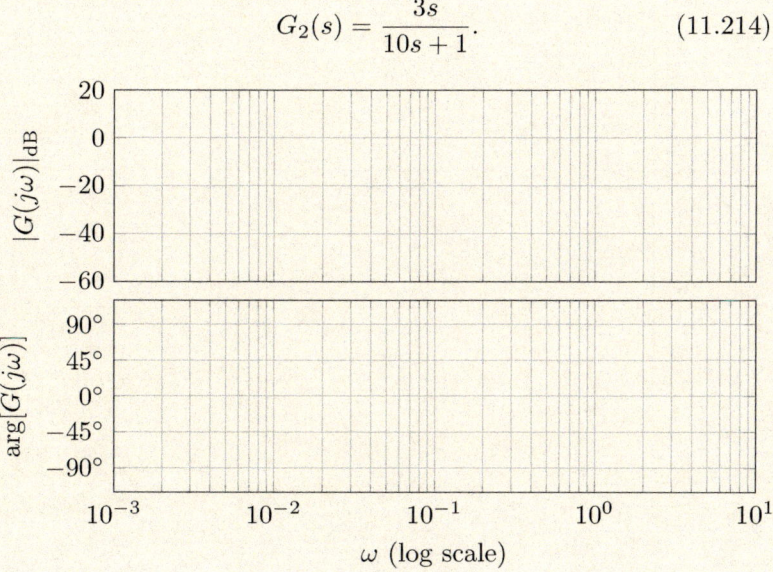

$$G_3(s) = \frac{2+s}{(5s+1)(s+4)}. \tag{11.215}$$

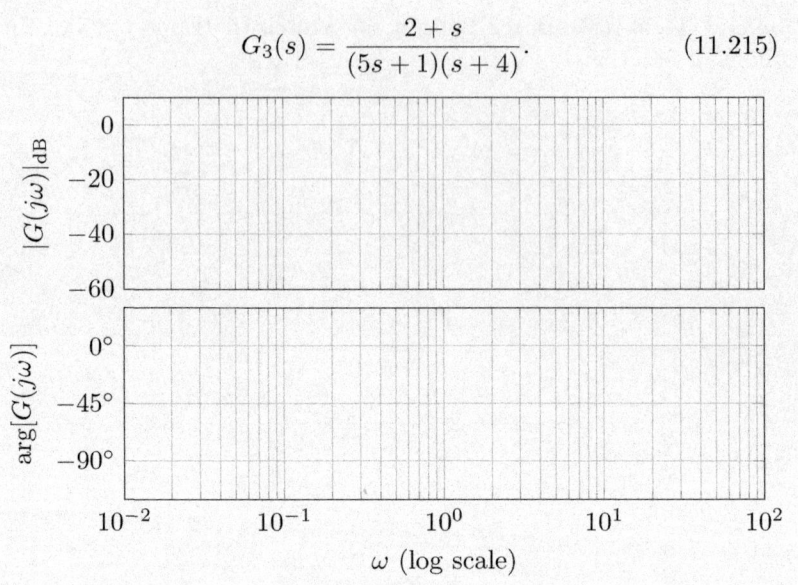

$$G_4(s) = \frac{3s}{(10s+1)^2}. \tag{11.216}$$

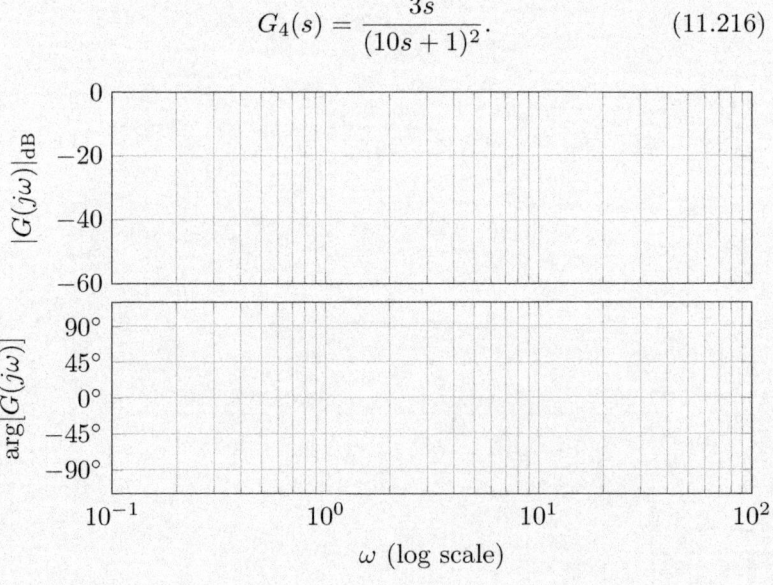

$$G_5(s) = e^{-s}, \text{ and } \tilde{G}_5(s) = \frac{1 - 1/2s}{1 + 1/2s} \tag{11.217}$$

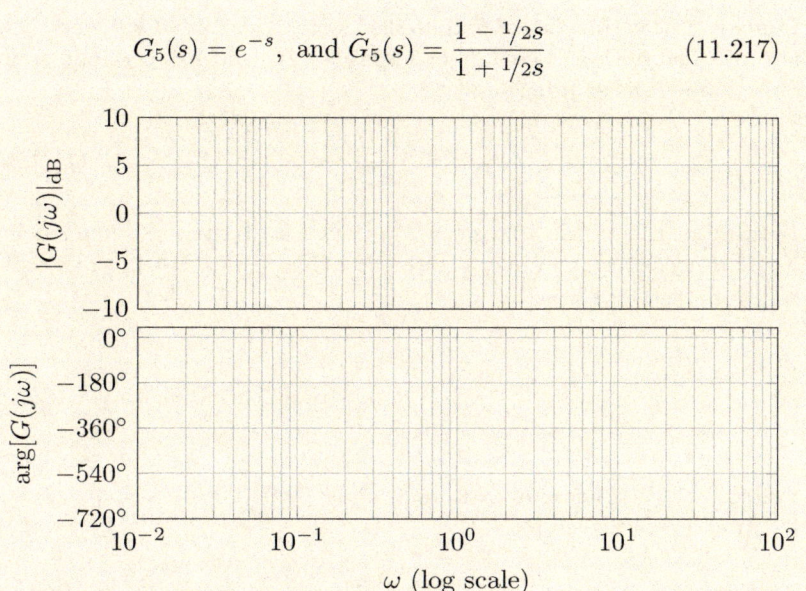

$$G_6(s) = \frac{s^2 + 1}{s(s^2 + 2s + 1)}. \tag{11.218}$$

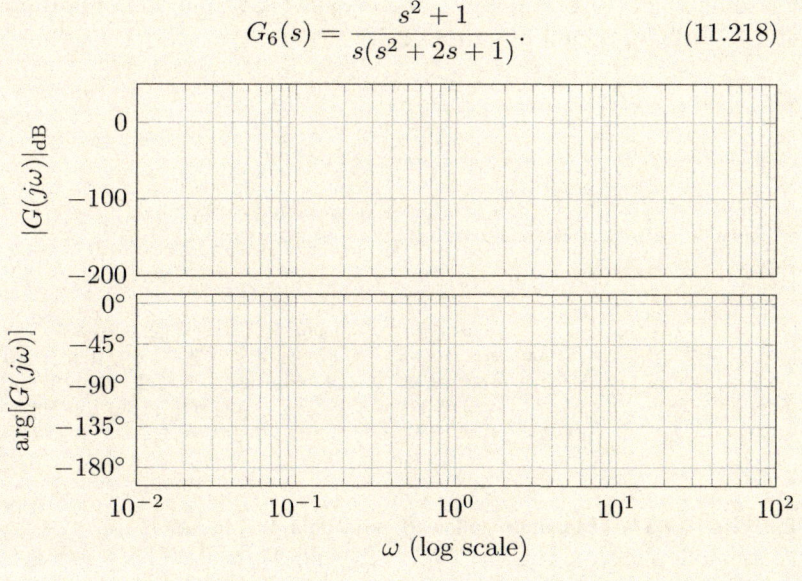

Exercise 11.35 (Margins of P-controlled system) A closed-loop system is controlled with a P controller with gain K_c and has the following open-loop transfer function:

$$G_{\text{ol}}(s) = \frac{K_c}{8s(s+1)^2}. \tag{11.219}$$

Determine the phase margin of the system if the gain margin is 6 dB. If necessary, you may use the fact that $x^3 + x = 1$ has one real solution: $x \approx 0.683$.

✳ ....

Answer. The phase margin is $21.4°$.

Exercise 11.36 (Choose K to achieve certain GM) The open-loop transfer function of a system is

$$G_{\text{ol}}(s) = \frac{K}{(3s+1)^n}, \tag{11.220}$$

for some $n \in \mathbb{N}$. Determine the value of K (as a function of parameter n) to have a gain margin of 2 dB.

✳ ....

Answer. If $n = 1$ or $n = 2$, there is no phase crossover frequency, and we need to resort to Routh's method — we will leave these two cases to the reader. Warning: The exercise does not state that K is necessarily positive. Case I: We are looking for $K > 0$. Then, if $n > 2$, the phase crossover frequency is $\omega_{co} = \frac{1}{3}\tan\frac{\pi}{n}$ and the gain that leads to a gain margin of 2 dB is

$$K = 10^{\frac{2}{10}\log_{10}(9\omega_{co}^2+1)^{-\frac{n}{2}} - \frac{1}{10}}. \tag{11.221}$$

If $K < 0$, then there is a phase crossover frequency only if $n > 3$, which is $\omega_{co} = \frac{1}{3}\tan\frac{3\pi}{n}$ and the value of K is given by

$$K = -10^{\frac{2}{10}\log_{10}(9\omega_{co}^2+1)^{-\frac{n}{2}} - \frac{1}{10}}. \tag{11.222}$$

Exercise 11.37 (Maximum allowed time delay) A closed-loop system is controlled with a P controller with gain $K_c > 0$. The controlled system has the transfer function

$$G_s(s) = \frac{1}{(2.87s+1)^3}. \tag{11.223}$$

Suppose that $G_a(s) = G_m(s) \stackrel{.}{=} 1$. Determine the value of K_c so that the gain margin is 3 dB. Suppose now that the actuator exhibits delay dynamics with $G_a(s) = e^{-t_d s}$. Determine the maximum value of t_d so that the closed-loop system is BIBO-stable.

✂ ......

Hint: (i) Function G satisfies the conditions of Theorem 11.13. (ii) For the second part of this exercise proceed as in Example 11.21.

Exercise 11.38 (Identify transfer function from Bode plot) Identify the rational transfer function, G, whose Bode plot is shown below.

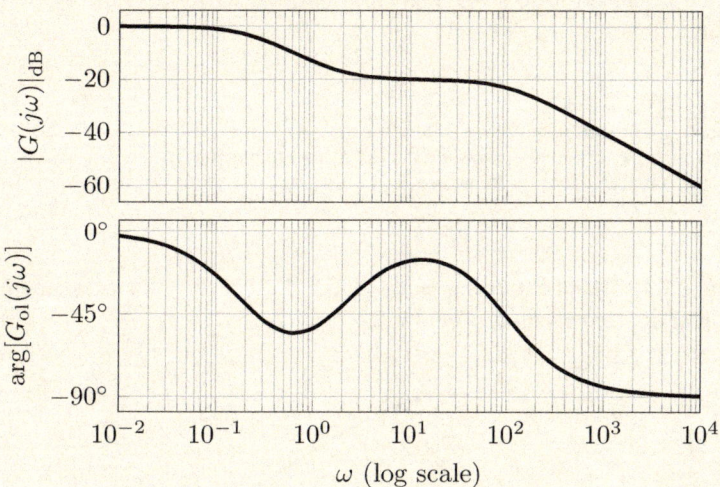

✳ ......

Answer: It is $G(s) = \frac{0.5s - 1}{(5s+1)(0.01s+1)}$

✂ ......

Hint: Start by observing the slope of the magnitude plot of the transfer function at different frequencies: at high frequencies it is -20, at around $\omega = 10^1$ rad/s it is zero, at around $\omega = 0.5$ rad/s it is -20 and at low frequencies it is 0. Make an assumption about the poles and zeros that may have lead to this pattern and verify your hypothesis using the phase plot.

Exercise 11.39 (Shortcomings of Bode stability criterion) Sketch the Bode plot of the following transfer function

$$G_{ol}(s) = 4e^{-1.5s} \frac{1.2s^2 + s + 0.2}{s(0.2s+1)} \qquad (11.224)$$

Show that there is a phase crossover frequency, ω_{co}, and

$|G_{\text{ol}}(j\omega_{\text{co}})| < 1$, but the stabilising conditions of Theorem 11.6 are not satisfied. Can you tell whether this system is stable using either Theorem 11.6 or Theorem 11.13?

Exercise 11.40 (Bode stability criterion not applicable) In some cases, the Bode stability criterion cannot be applied. Construct the Bode plot of

$$G_{\text{ol}}(s) = \frac{10}{s^3 + s^2 + s + 1}, \tag{11.225}$$

and confirm that the Bode criterion is not applicable. Can you tell whether the closed-loop system is BIBO-stable without using Bode's criterion? Assume $G_m(s) = 1$.

✂ ....

Hint. In order to construct the Bode plot of this open-loop transfer function we need to use the definition of atan2 (see Appendix A). Start by factorising the denominator of G_{ol} (you may use Horner's method, although it is not necessary).

Exercise 11.41 (Understanding Bode plots) The Bode plot of a linear dynamical system is shown below:

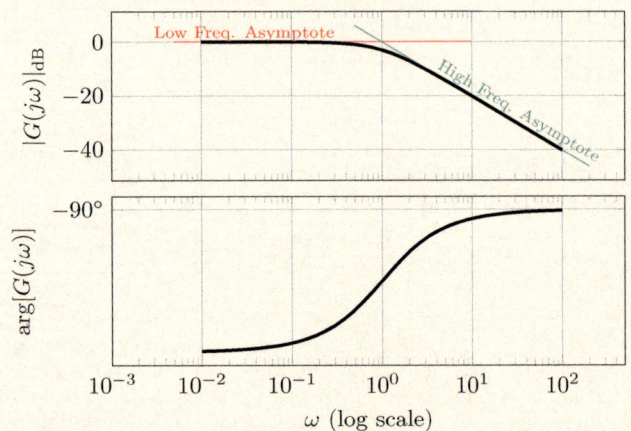

What is the steady state frequency response of the system at $\omega = 100 \,\text{rad/s}$?

✳ ....

Answer. One might be tempted to look at the Bode plot and see that $|G(j\omega)|_{\text{dB}}$ is $-40\,\text{dB}$ and that $\arg G(j\omega) \approx -90°$ and then use Theorem 11.1. However, this Bode plot corresponds to the *unstable* system with transfer function $G(s) = \frac{1-s}{1}$, so the steady state frequency response is not defined.

Chapter 11. Frequency response and design of stable closed loops

Exercise 11.42 (Bode plot of Padé approximation) Let us try to understand what happened in Example 11.15 where we showed that the Padé approximation can lead to erroneous results. Write a Python or MATLAB script to construct the Bode plots of $G(s)$ and $G_{\text{approx}}(s)$ and compare them.

Hint: Note that the phase plot of the Padé approximant is shifted by $+360°$. You should see that the magnitude plots of G and G_{approx} coincide, while the phase plots are in good agreement for $\omega < 1$, but start to diverge at higher frequencies. Note also that the phase crossover frequency of G is at 1.55 rad/s and that of G_{approx} is 1.59 rad/s.

Exercise 11.43 (Effect of PID parameters) In Section 11.4 we discussed the effect of the PID parameters on the shape of the Bode plot. Consider the open-loop transfer function

$$G_{\text{ol}}(s) = \underbrace{K_c\left(1 + \tau_D s + \frac{1}{\tau_I s}\right)}_{\text{PID controller}} \frac{s+1}{(s^2+s+1)^2}. \tag{11.226}$$

1. Let $K_c = 1$ and $\tau_I = 1$. Write a script to plot the Bode plot of G_{ol} for different values of τ_D in the interval $[0, 2]$ and determine for what values of τ_D the closed-loop system is BIBO-stable.

2. Let $K_c = 1$ and $\tau_D = 1$. Again, by plotting the Bode plot for values of τ_I between 0.1 and 2, tell for what values of τ_I the closed-loop system is BIBO-stable.

3. Let $\tau_I = 1$ and $\tau_D = 1$. Plot the Bode plot of G_{ol} with $K_c = 1$ and determine the gain margin.

Answer: 1. The system becomes stable for $\tau_D > 0.534$. 2. $\tau_I > 0.609$. 3. GM $= \infty$.

12
Nyquist plot and stability criterion

The Nyquist plot is a particular plot of the values of $G(s)$ on the complex plan as s moves along a contour, known as the Nyquist contour. The shape of the Nyquist plot of the open-loop transfer function can provide information regarding the stability properties of the corresponding closed-loop system.

We start by plotting $G(j\omega)$ for $\omega > 0$ on the complex plane for several simple transfer functions such as integrators, first and second-order systems, pure time delays and PID controllers. Unlike Bode plots, Nyquist plots of higher order systems are not easy to construct.

Next, we state Cauchy's principle of argument — the main theoretical tool from complex analysis that allows us to use the Nyquist plot of the open-loop transfer function to tell whether the corresponding closed-loop function is BIBO stable. The Nyquist plot is very easy to use, and its statement is a lot simpler compared to the Bode criterion.

12.1 Plotting $G(j\omega)$ on the complex plane

Given a complex function $G : \mathbb{C} \to \mathbb{C}$, defined over a domain $U \subseteq \mathbb{C}$, and a curve $\Gamma \subseteq U$, suppose that the complex variable s moves on Γ. Then, its image, $G(s)$, will create a certain path on the complex plane which is called a **plot** of G as s moves on Γ. In fact, assuming that s moves on Γ with a certain direction, the plot of G is equipped with a direction. We will denote this plot by $G(\Gamma)$.

In this chapter we are particularly interested in the plot of G as s moves along a path known as the **Nyquist contour**, which we will introduce later. The resulting plot is known as the **Nyquist plot** of G and is the definitive tool for analysing the stability of closed-loop systems using their open-loop transfer function without the limitations of the Bode stability criterion. Besides, the Bode stability criterion itself is a corollary of the Nyquist criterion.

As a warm-up, before we state the Nyquist stability criterion and discuss the properties of the namesake plot, we shall construct plots of $G(s)$ as s moves on the imaginary axis from 0^+ to $+j\infty$, that is, as it moves on the halfline $\Gamma = \{j\omega : \omega \in (0, \infty)\}$. We will denote this plot as $G(j(0,\infty))$ for brevity. The resulting plot is *part of* the Nyquist plot of G. In this section we will restrict our discussion to systems without imaginary poles. Throughout, we will draw the attention of the reader to the behaviour of $G(s)$ as s approaches the origin and as it escapes to infinity.

12.1.1 Integrators

A simple integrator has the transfer function

$$G(s) = \frac{1}{s}, \qquad (12.1)$$

and it is easy to verify that $|G(j\omega)| = 1/\omega$ and $\arg G(j\omega) = -\pi/2$. In fact, $G(j\omega) = \frac{1}{j\omega} = -\frac{1}{\omega}j$, that is, $G(j\omega)$ is an imaginary number. For the high and low-frequency asymptotes, we have

1. As $\omega \to 0^+$, $G(j\omega) \to -j\infty$ (i.e., $G(j\omega)$ escapes to infinity along the negative halfline of the imaginary axis as shown in Figure 12.1)

2. As $\omega \to \infty$, $G(j\omega) \to 0$

3. At intermediate frequencies $(0 < \omega < \infty)$ it is $\arg G(j\omega) = -\pi/2$.

The plot of $G(s) = 1/s$ as s moves on $\Gamma = \{js : s \in (0, \infty)\}$ is given in Figure 12.1.

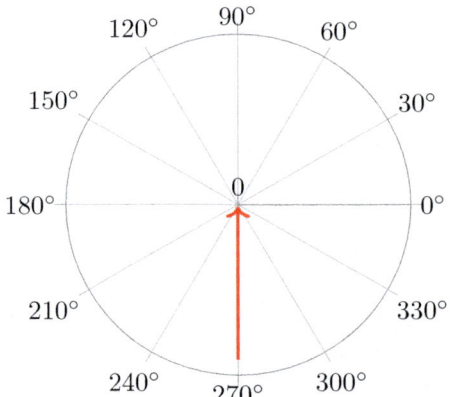

Figure 12.1: Plot of simple integrator, $G(s) = 1/s$, as s moves on $\Gamma = j \cdot (0, \infty)$.

12.1.2 First-order systems

A first-order system has the transfer function

$$G(s) = \frac{K}{\tau s + 1}. \qquad (12.2)$$

The modulus and argument of $G(j\omega)$ are

$$|G(j\omega)| = \frac{K}{\sqrt{1 + \tau^2 \omega^2}}, \qquad (12.3a)$$

$$\arg G(j\omega) = -\arctan(\tau\omega). \qquad (12.3b)$$

Observe that

1. As $\omega \to 0^+$, $|G(j\omega)| \to K$ and $\arg G(j\omega) \to 0$, that is $G(j\omega) \to K$,

2. As $\omega \to \infty$, $|G(j\omega)| \to 0$ and $\arg G(j\omega) \to -\pi/2$

3. For frequencies $\omega \in (0, \infty)$, it is

$$0 < |G(j\omega)| = \frac{K}{\sqrt{1 + \tau^2 \omega^2}} < K,$$

therefore, the plot of $G(j(0, \infty))$ is contained within a circle of radius K,

4. For all $\omega \in (0, \infty)$, it is $-\pi/2 < \arg G(j\omega) < 0$, therefore, the plot is contained in the fourth quadrant of the complex plane (with $\mathbf{re} G(j\omega) < 0$ and $\mathbf{im} G(j\omega) < 0$)

The reader can compare the plot given in Figure 12.2 with the corresponding Bode plots from Chapter 11. Exercise 12.12 will offer some additional insights into the plots of $G(j(0, \infty))$ for first-order systems.

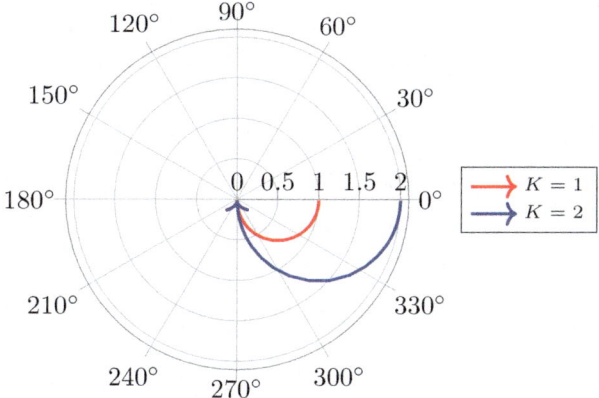

Figure 12.2: Plots of $G(j(0, \infty))$ for first-order systems with different values of the static gain, K. The value of τ does not affect the shape of the

plot. As $\omega \to 0^+$, the plot of $G(j(0,\infty))$ approaches K and as $\omega \to \infty$, it approaches the origin.

As one may observe in Figure 12.2, the $G(j(0,\infty))$-plot of a first-order system is a semicircle and in fact its shape does not depend on the value of τ. Let us prove this.

> **Proposition 12.1** ($G(j(0,\infty))$-plot of first-order system) The $G(j(0,\infty))$-plot of a first-order system with static gain K is a semicircle which lies in the fourth quadrant, is centred at $K/2$ and has radius $K/2$.

Proof. Let (x,y) be a point of the $G(j(0,\infty))$-plot of a first-order system with static gain K, that is, $x = \mathrm{re}G(j\omega)$ and $y = \mathrm{im}G(j\omega)$ for some $\omega > 0$. We have that $G(j\omega) = \frac{K}{\tau j\omega+1} = \frac{K(1-j\tau\omega)}{\tau^2\omega^2+1}$, therefore, $x = \frac{K}{\tau^2\omega^2+1}$ and $y = -\frac{K\tau\omega}{\tau^2\omega^2+1}$. The reader can then verify that

$$\left(x - \frac{K}{2}\right)^2 + y^2 = \left(\frac{K}{2}\right)^2,$$

which completes the proof. $\square$

> **Exercise 12.1** (Simlpe exercise) Prove that the $G(j(0,\infty))$-plot of the following transfer function
>
> $$G(s) = \frac{1}{(\tau_1 s + 1)(\tau_2 s + 1) \cdot \ldots \cdot (\tau_n s + 1)}, \quad (12.4)$$
>
> with $n \in \mathbb{N}$ and $\tau_1, \ldots, \tau_n > 0$, lies within the unit circle.
>
> ✂ ....
>
> *Hint.* We need to show that $|G(j\omega)| < 1$ for $\omega \in (0, \infty)$.

12.1.3 Second-order systems

A second-order system has the transfer function

$$G(s) = \frac{K}{\tau^2 s^2 + 2\zeta\tau s + 1}, \quad (12.5)$$

with $K, \tau, \zeta > 0$. We have already seen (see Example 11.10) that

$$|G(j\omega)| = \frac{K}{\sqrt{(1 - \tau^2\omega^2)^2 + (2\tau\zeta\omega)^2}} \quad (12.6\mathrm{a})$$

$$\arg G(j\omega) = \text{atan2}\left(-2\zeta\tau\omega, 1 - \tau^2\omega^2\right). \tag{12.6b}$$

From the second equation we conclude that $-\pi < \arg G(j\omega) < 0$, therefore, the $G(j(0,\infty))$-plot of the system is contained in the third and fourth quadrants of the complex plane (the region $\{z \in \mathbb{C} : \mathbf{re}(z) < 0\}$). As an exercise, the reader can compute the real part of $G(j\omega)$ and verify that it is negative for all $\omega > 0$.

We know that the amplitude ratio, $|G(j\omega)|$ does not exceed K if $\zeta \geq \sqrt{2}/2$ (see Example 11.10). If $0 < \zeta < \sqrt{2}/2$, the maximum amplitude ratio is attained at the resonance frequency given by

$$\omega_{\text{res}} = \frac{\sqrt{1 - 2\zeta^2}}{\tau}, \tag{12.7}$$

therefore, it is

$$|G(j\omega_{\text{res}})| = \frac{K}{\sqrt{(1 - \tau^2\omega_{\text{res}}^2)^2 + (2\tau\zeta\omega_{\text{res}})^2}} = \frac{K}{2\zeta\sqrt{1 - \zeta^2}},$$

as a result, as $\zeta \to 0^+$, $|G(j\omega_{\text{res}})| \to \infty$.

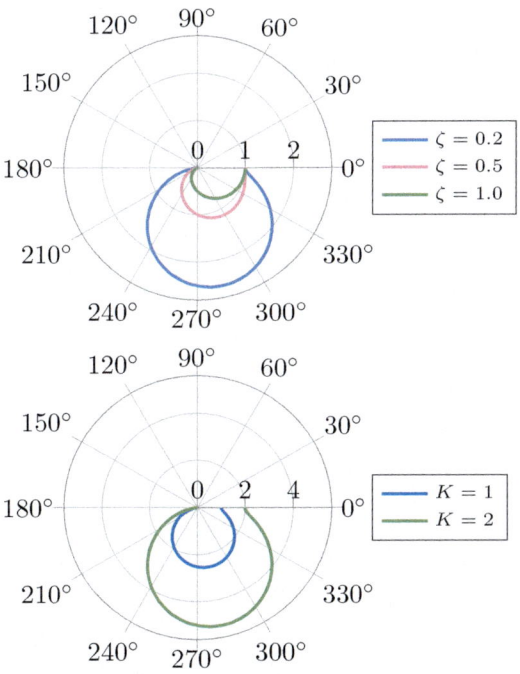

Figure 12.3: Plot of $G(j(0,\infty))$ of second-order system: (Up) Different values of ζ for $K = \tau = 1$, (Down) Different values of K for $\tau = 1$ and $\zeta = 0.2$. As $\omega \to 0^+$, the $G(j(0,\infty))$-plot approaches K and as $\omega \to \infty$, it approaches the origin.

As $\zeta \to (\sqrt{2}/2)^-$, $|G(j\omega_{\text{res}})| \to 1$. Note that the maximum amplitude ratio does not depend on τ. Overall, we have that the $G(j(0,\infty))$-plot of a second-order system is contained within a circle of radius K, if $\zeta \geq \sqrt{2}/2$ and a within a circle of radius $\frac{K}{2\zeta\sqrt{1-\zeta^2}}$ if $0 < \zeta < \sqrt{2}/2$.

The plot of $G(j(0,\infty))$ for different values of K and ζ is shown in Figure 12.3.

Exercise 12.2 (Intersection with unit circle) For a second-order system with the transfer function given in Equation (12.5), determine the frequency, or frequencies — if any exist — at which its $G(j(0,\infty))$-plot intersects the unit circle. What is the significance of such frequencies?

✂ ....

Hint. The plot of $G(j(0,\infty))$ intersects with the unit circle at a point $\omega \in (0,\infty)$ where $|G(j\omega)| = 1$.

12.1.4 Delay systems

A delay system is a dynamical system with transfer function $G(s) = e^{-t_d s}$ for some constant $t_d > 0$. We have already seen that $|G(j\omega)| = 1$ and $\arg G(j\omega) = t_d \omega$. This means that the $G(j(0,\infty))$-plot of a delay system is a circle centred at the origin of the complex plane with radius equal to 1 as shown in Figure 12.4. As $s = j\omega$ moves from $j0^+$ to $j\infty$, $G(j\omega)$ moves on the unit circle in the clockwise direction.

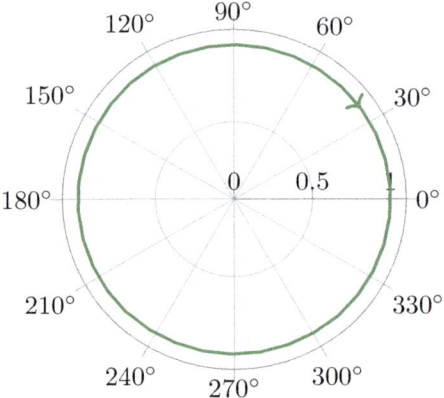

Figure 12.4: $G(j(0,\infty))$-plot of a delay system.

12.1.5 PID controllers

Let us examine the cases of P, PI and PID controllers separately. The case of a P controller with gain K_c is pretty trivial; clearly $G(j\omega) = K_c$, that is, $G(j\omega)$ is a real number, which does not depend on ω. For a PI controller with gain $K_c > 0$ and time constant $\tau_I > 0$ we have

$$G(j\omega) = K_c\left(1 + \frac{1}{j\omega\tau_I}\right) = K_c\left(1 - j\frac{1}{\omega\tau_I}\right), \quad (12.8)$$

therefore,
$$|G(j\omega)| = K_c\sqrt{1 + \frac{1}{\tau_I^2\omega^2}}, \qquad (12.9a)$$
$$\arg G(j\omega) = \arctan(-\frac{1}{\tau_I\omega}). \qquad (12.9b)$$

At the same time, we may easily observe that $\mathbf{re}G(j\omega) = K_c$ and $\mathbf{im}G(j\omega) = -\frac{K_c}{\tau_I\omega}$, therefore, the real part of $G(j\omega)$ is constant and does not depend on ω, whereas for the imaginary part of $G(j\omega)$ observe that (i) $G(j\omega) < 0$ for all ω, therefore, the $G(j(0,\infty))$-plot lies in the fourth quadrant of the complex plane, (ii) As $\omega \to 0^+$, $\mathbf{im}G(j\omega) \to -\infty$ and (iii) As $\omega \to \infty$, $\mathbf{im}G(j\omega) \to 0$. As we see in Figure 12.5b, the $G(j(0,\infty))$-plot of a PI controller is a half-line.

The transfer function of a PID controller is given by
$$G(s) = K_c\left(1 + \tau_D s + \frac{1}{\tau_I s}\right), \qquad (12.10)$$
with $K_c, \tau_I, \tau_D > 0$. We then have
$$\begin{aligned} G(j\omega) &= K_c\left(1 + \tau_D j\omega + \frac{1}{\tau_I j\omega}\right) \\ &= K_c\left(1 + j\left(\tau_D\omega - \frac{1}{\tau_I\omega}\right)\right) \end{aligned} \qquad (12.11)$$

Again, we see that $\mathbf{re}G(j\omega) = K_c$ and $\mathbf{im}G(j\omega) = K_c(\tau_D\omega - \frac{1}{\tau_I\omega})$, so the real part of $G(j\omega)$ is constant and does not depend on ω. For the imaginary part we observe that (i) As $\omega \to 0^+$, $\mathbf{im}G(j\omega) \to -\infty$, (ii) As $\omega \to \infty$, $\mathbf{im}G(j\omega) \to \infty$, (iii) If $\omega = 1/\sqrt{\tau_D\tau_I}$, then $\mathbf{im}G(j\omega) = 0$. We conclude that the $G(j(0,\infty))$-plot of a PID controller is a vertical line as shown in Figure 12.5.

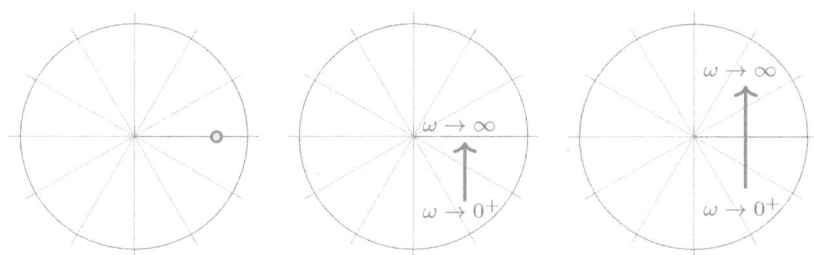

Figure 12.5: *Plots of $G(j(0,\infty))$ for (Left) a P controller, which is just a single point at K_c; (Middle) PI (Right) and PID controllers with the same static gain.*

12.1.6 Higher-order systems

The $G(j(0,\infty))$-plot of higher-order systems and systems with transfer functions of the rational-times-exponential form cannot be constructed from the $G(j(0,\infty))$-plots of their constituents. There are, however, some general rules that apply to all $G(j(0,\infty))$-plots.

> **Proposition 12.2 (Properties of $G(j(0,\infty))$-plots of rational functions)**
> Consider the transfer function
> $$G(s) = K\frac{(s-z_1)(s-z_2)\cdot\ldots\cdot(s-z_m)}{(s-p_1)(s-p_2)\cdot\ldots\cdot(s-p_n)}, \quad (12.12)$$
> with $m, n \in \mathbb{N}$, $K > 0$, $z_1, \ldots, z_m \in \mathbb{C}$ and $p_1, \ldots, p_n \in \mathbb{C}$. Then,
>
> 1. as $\omega \to \infty$, $\arg G(j\omega) \to (m-n)\frac{\pi}{2}$,
>
> 2. if G is strictly proper (that is, $n > m$), then its $G(j(0,\infty))$-plot approaches the origin as $\omega \to +\infty$,
>
> 3. if G is such that $n = m$, then its $G(j(0,\infty))$-plot approaches K as $\omega \to +\infty$.

Proof. For the first assertion, using the fact that $\arg(z_1 z_2) = \arg z_1 + \arg z_2$, for all $z_1, z_2 \in \mathbb{C}$ and $\arg(z_1/z_2) = \arg(z_1) - \arg(z_2)$, whenever $z_1 \neq 0$, we have that

$$\arg G(j\omega) = \sum_{i=1}^{m} \arg(j\omega - z_i) - \sum_{i=1}^{n} \arg(j\omega - p_i).$$

Suppose that $z_i = a_i + jb_i$, with $a_i, b_i \in \mathbb{R}$ for $i = 1, \ldots, m$ and $p_j = a'_j + jb'_j$, with $a'_j, b'_j \in \mathbb{R}$ for $j = 1, \ldots, n$. Then,

$$\arg G(j\omega) = \sum_{i=1}^{m} \operatorname{atan2}(\omega - b_i, -a_i) - \sum_{j=1}^{n} \operatorname{atan2}(\omega - b'_j, -a'_j).$$

Taking the limit as $\omega \to \infty$, we have $\lim_{\omega \to \infty} \arg G(j\omega) = (m-n)\frac{\pi}{2}$, which proves the first assertion.

For the second assertion, we need to prove that $\lim_{\omega \to \infty} |G(j\omega)| = 0$, when G is strictly proper ($m < n$). We have

$$|G(j\omega)| = |K|\frac{|j\omega - z_1|}{|j\omega - p_1|}\cdot\ldots\cdot\frac{|j\omega - z_m|}{|j\omega - p_m|}\cdot\ldots\cdot\frac{1}{|j\omega - p_{m+1}|}\cdot\ldots\cdot\frac{1}{|j\omega - p_n|} \quad (12.13)$$

The limits $\lim_{\omega \to \infty} \frac{|j\omega - z_i|}{|j\omega - p_i|}$ exist for all $i = 1, \ldots, m$ and are given by

$$\lim_{\omega \to \infty} \frac{|j\omega - z_i|}{|j\omega - p_i|} = \lim_{\omega \to \infty} \frac{|j(\omega - b_i) - a_i|}{|j(\omega - b'_i) - a'_i|} = \lim_{\omega \to \infty} \frac{\sqrt{(\omega - b_i)^2 + a_i^2}}{\sqrt{(\omega - b'_i)^2 + (a'_i)^2}}$$

$$= \lim_{\omega \to \infty} \sqrt{\frac{(\omega - b_i)^2 + a_i^2}{(\omega - b'_i)^2 + (a'_i)^2}},$$

for all $i = 1, \ldots, m$. Additionally, the following limit exists

$$\lim_{\omega \to \infty} \frac{(\omega - b_i)^2 + a_i^2}{(\omega - b_i')^2 + (a_i')^2} = \lim_{\omega \to \infty} \frac{2(\omega - b_i)}{2(\omega - b_i')} = \lim_{\omega \to \infty} \frac{1 - \frac{b_i}{\omega}}{1 - \frac{b_i'}{\omega}} = 1,$$

for all $i = 1, \ldots, m$, where in the first step we used L'Hôpital's rule. As a result, $\lim_{\omega \to \infty} \frac{|j\omega - z_i|}{|j\omega - p_i|} = 1$ (using the fact that the square root is a continuous function). The limits $\lim_{\omega \to \infty} \frac{1}{|j\omega - p_j|} = 0$ for all $j = m + 1, \ldots, n$, therefore, $\lim_{\omega \to \infty} |G(j\omega)| = 1 \cdot 0 = 0$. The third assertion can be proven easily and is left to the reader as an exercise. □

A direct consequence of Proposition 12.2 is that if G is strictly proper, then as $\omega \to \infty$ its $G(j(0, \infty))$-plot approaches the origin with angle $(m - n)\frac{\pi}{2}$.

According to Proposition 12.2, for first-order systems ($m = 0$, $n = 1$), we have that $G(j\omega)$ approaches the origin with angle $-\pi/2$. For second-order systems ($m = 0$, $n = 2$), $G(j\omega)$ approaches the origin with angle $-\pi$.

Remark: *The reader can easily show that Condition 2 of Proposition 12.2 holds true for strictly proper transfer functions of the form*

$$G(s) = K \frac{(s - z_1)(s - z_2) \cdot \ldots \cdot (s - z_m)}{(s - p_1)(s - p_2) \cdot \ldots \cdot (s - p_n)} e^{-t_d s}, \tag{12.14}$$

for any $t_d > 0$. •

12.2 Nyquist stability criterion

The Nyquist stability criterion relies on Cauchy's principle of argument which we present in Section 12.2.1. We will give several examples to fully appreciate how powerful a tool Cauchy's principle of argument is.

12.2.1 Cauchy's principle of argument

Cauchy's principle of argument describes the plot of $F(s)$ as s moves on Γ. It associates the number of times that the plots of F encircles the origin in the clockwise dimension (known as the **winding number** or **index number**), with the number of poles and zeroes of F enclosed by Γ. It will later become obvious that Cauchy's principle of argument can be used to determine the number of unstable poles of the closed-loop function of a controlled system using an appropriate contour Γ. The principle runs as follows.

> **Theorem 12.3 (Principle of argument)** Let $F : D \to \mathbb{C}$ be a *meromorphic function* (see Definition A.5) on an open domain $D \subseteq \mathbb{C}$ and let Γ be a *contour* (see Definition A.6) in D. Suppose that no poles or zeroes of F are on Γ. Then, as s moves on Γ in the clockwise direction, $F(s)$ encircles the origin in the clockwise direction N times, where
> $$N = Z - P, \qquad (12.15)$$
> where Z and P are the number of zeroes and poles of F inside the contour.

Proof. The proof can be found in [M14, Chapter IV]. A great introduction to the topic is available in [M9, Chapter 10]. □

Remark: *Following Theorem 12.3, if $N < 0$ the plot of $F(s)$ as s moves on Γ in the **counter-clockwise** direction. In general, if $N > 0$, $F(\Gamma)$ encircles 0 in the same direction as Γ, whereas if $N < 0$, the direction is reversed.* •

Example 12.1 (Cauchy's principle of argument) Consider the complex function
$$F(s) = \frac{s+2}{(s+1)(2s+1)}, \qquad (12.16)$$
which has one zero at $z_1 = -2$ and two poles at $z_1 = -1$ and $z_2 = -1/2$. Consider the contour $\Gamma = \{s = 1.5e^{j\theta}, \theta \in [0, 2\pi]\}$, which is a circle of radius 1.5 centred at the origin of the complex plane. This contour has a clockwise direction as shown in Figure 12.6.

Chapter 12. Nyquist plot and stability criterion

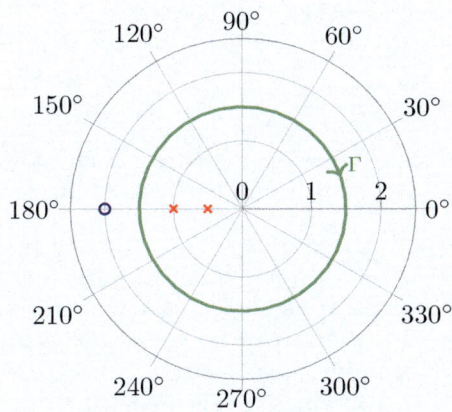

Figure 12.6: *Zeroes and poles of F and contour Γ which encircles no zeroes ($Z = 0$) and two poles ($P = 2$) of F.*

Since Γ encircles no zeroes ($Z = 0$) and two poles ($P = 2$) of F, by Cauchy's principle of argument, $F(s)$ will encircle the origin $N = Z - P = -2$ times, that is, two times in the counter-clockwise direction as shown in Figure 12.7.

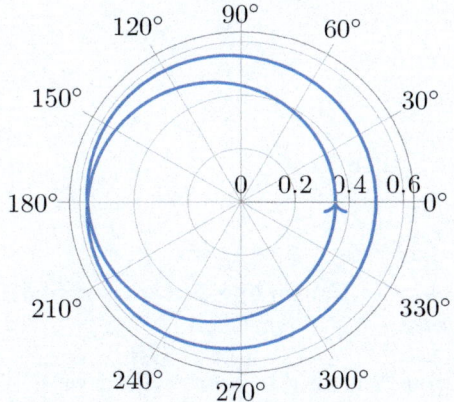

Figure 12.7: *Plot of $F(s)$ for $s \in \Gamma$ which encircles the origin twice in the counter-clockwise direction.*

Exercise 12.3 (Modification of Example 12.1) Write a Python or MATLAB script to plot $F(\Gamma)$ when F is as in Example 12.1 and $\Gamma = \{s = 2.5e^{j\theta}, \theta \in [0, 2\pi]\}$. How many times does the plot of $F(\Gamma)$ encircle 0 and in what direction (clockwise or counter-clockwise)?

Example 12.2 (Multiple revolutions) Consider the function

$$F(s) = \frac{(s-1)^2 + 1}{((s-0.5)^2 + 2)(s-1)^2(s+3)}, \qquad (12.17)$$

which has a pair of complex conjugate zeroes $z_{1,2} = 1 \pm j$ and the poles $p_{1,2} = 0.5 \pm j\sqrt{2}$, $p_{3,4} = 1$, $p_5 = -3$. Consider the contour

$$\Gamma = \{s = 2.5e^{j\theta}, \theta \in [0, 2\pi]\}, \qquad (12.18)$$

which is a circle of radius 2.5 centred at the origin of the complex plane. As illustrated in Figure 12.8, Γ encircles two zeroes ($Z = 2$) and four poles ($P = 4$) of F (counting multiplicities). The pole $p_5 = -3$ is not encircled by Γ. According to Theorem 12.3, the plot $\{F(s) : s \in \Gamma\}$ encircles the origin $N = Z - P = -2$ times in the clockwise direction, that is, 2 times in the counter-clockwise direction.

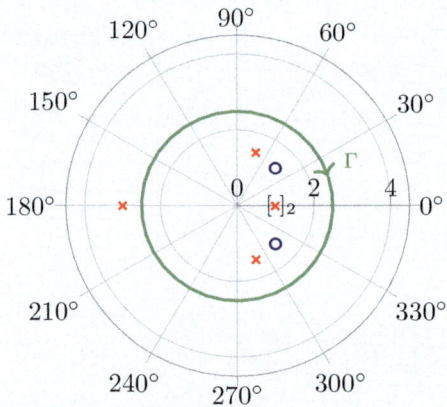

Figure 12.8: *Zeroes and poles of F and contour Γ which encircles two zeroes ($Z = 2$) and four poles ($P = 4$) of F.*

The plot of $F(s)$ as s moves on Γ is shown below where we can verify that $N = -2$.

It should be noted that function F in Equation (12.17) is very complex. Its plot in Figure 12.9 was computed using MATLAB. Constructing this plot analytically would be rather tedious. Nevertheless, using Cauchy's principle of argument we were able to tell, without drawing the plot of F, that it encircles the origin twice in the counter-clockwise direction.

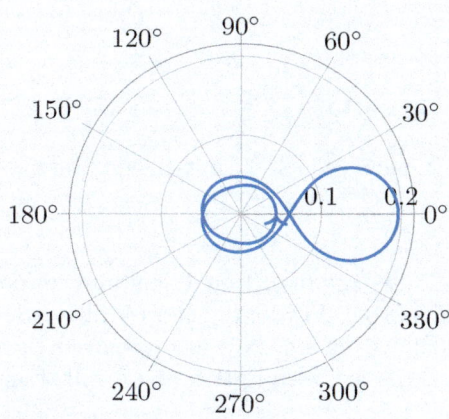

Figure 12.9: *Plot of $F(s)$ for $s \in \Gamma$ which encircles the origin twice in the counter-clockwise direction.*

The principle of argument gives the number of encirclements of $F(s)$ around the origin as s moves on Γ as well as the direction of these encirclements. This result can easily be extended to determine the number and the direction of encirclement around any point on the complex plane other than the origin. It suffices to note that for a given complex number $z_0 \in \mathbb{C}$, the plot of $z_0 + F(s)$ is precisely the plot of $F(s)$ translated by z_0. Therefore, the plot of $F(s)$ as s moves on a contour Γ encircles a point z_0 N times in the clockwise direction if the plot of $F(s) - z_0$ as s moves on Γ encircles the origin N times in the clockwise direction. Let us go through an example.

Example 12.3 (Revolution around a point $z_0 \neq 0$) In Example 12.1 we proved that the plot of F, as s moves on Γ, encircles the origin twice in the counter-clockwise direction. In Figure 12.7 we may observe that the plot of F encircles the point $z_0 = 0.4$ once in the counter-clockwise direction. It is possible to prove that without the need to sketch the plot.

As discussed above, we know that the plot of $F(s)$ encircles z_0 N times if and only if the plot of $F(s) - z_0$ encircles the origin N times. We shall therefore apply Cauchy's principle of argument on $F(s) - z_0$ as s moves on Γ.

Functions F and $F - z_0$ share the same poles. The zeroes of $F - z_0$, for $z_0 = 0.4$, are the solutions of the following equation

$$F(s) - 0.4 = 0$$
$$\Leftrightarrow \frac{s+2}{(s+1)(2s+1)} - 0.4 = 0$$

$$\Leftrightarrow s + 2 - 0.4(s+1)(2s+1) = 0$$
$$\Leftrightarrow s + 2 - 0.4(2s^2 + 3s + 1) = 0$$
$$\Leftrightarrow -0.8s^2 - 0.2s + 1.6 = 0,$$

which has two roots, $z_1 = -1.5447$ (which is not inside Γ) and $z_2 = 1.2947$ (which is inside Γ).

In the statement of Theorem 12.3 we required that Γ contains no poles of F. This requirement is essential so that $F(s)$ is defined for all $s \in \Gamma$ (If s_0 is a pole of F, then F is not defined at s_0). We also required that Γ contains no zeroes of F. This is a necessary assumption as we will show in the following example.

Example 12.4 (Contour runs through a zero of F) Consider the following transfer function
$$F(s) = \frac{s-1}{s+2}, \qquad (12.19)$$
which has one zero at $z = 1$ and one pole at $p = -2$. Consider the contour
$$\Gamma_R = \{Re^{js} : s \in [0, 2\pi]\}, \qquad (12.20)$$
for some $R > 0$, which runs through the zero of F only if $R = 1$.

By Theorem 12.3, since $\Gamma_{0.7}$ encloses no zeroes ($Z = 0$) and no poles ($P = 0$) of F, we conclude that the plot of $F(s)$, as s moves on $\Gamma_{0.7}$ will not encircle the origin. Similarly, $\Gamma_{1.5}$ encloses one zero ($Z = 1$) and no poles of F, therefore, as s moves on the contour $\Gamma_{1.5}$, $F(s)$ will encircle the origin once in the clockwise direction.

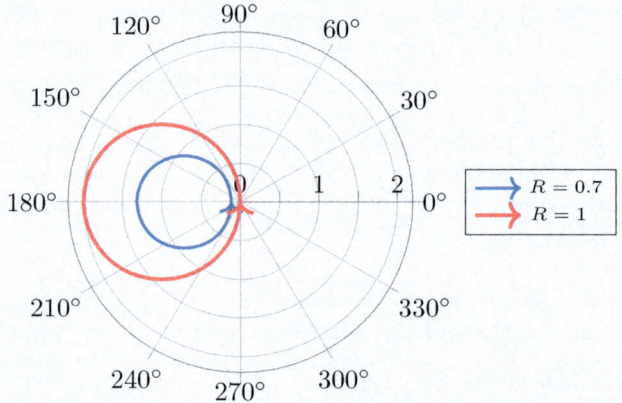

Figure 12.10: *Corner case: the contour contains a zero of F. If $R = 1$ the plot of F runs through the origin.*

However, Theorem 12.3 **cannot be applied** when $R = 1$ because Γ runs through a zero of F. As illustrated in Figure 12.10, the plot of $F(s)$, as s moves on Γ_1, runs through the origin. As a convention, we will not count this as an encirclement of the origin.

We, therefore, arrive at the following "converse" of Cauchy's Principle of Argument according to which we can know the difference between the number of zeroes and poles that are enclosed by Γ by observing the number of encirclements of the plot of $F(s)$ around zero.

Let $F : \mathbb{C} \to \mathbb{C}$ be a meromorphic function on an open domain D and let Γ be a closed bounded contour in D which does not contain any poles of F. Then

1. If the plot of $F(s)$, as s moves on Γ, runs through the origin, then Γ contains a zero of function F and this does not count as an encirclement of the origin. In that case Cauchy's principle of argument (Theorem 12.3) cannot be applied.

2. If the plot of $F(s)$, as s moves on Γ, encircles the origin N times, then Γ encloses Z zeroes of F and P poles of F which satisfy $N = Z - P$; N is positive if the encirclement has the same direction as Γ and negative if it has the opposite direction.

12.2.2 Nyquist stability criterion

Consider a closed-loop system involving a controlled system, an actuator, a controller and a sensor with transfer functions G_s, G_a, G_c and G_m respectively. Then, the open-loop transfer function is

$$G_{\text{ol}}(s) = G_a(s)G_c(s)G_s(s)G_m(s), \tag{12.21}$$

and the closed-loop transfer function is

$$G_{\text{cl}}(s) = \frac{G_a(s)G_c(s)G_s(s)}{1 + G_{\text{ol}}(s)}. \tag{12.22}$$

The *characteristic function* of the system is defined as the complex function

$$\chi(s) := 1 + G_{\text{ol}}(s). \tag{12.23}$$

Recall that the closed-loop system is BIBO stable if and only if the poles of G_{cl} are all in the open left side of the complex plane, i.e., have negative real parts (see Theorem 9.9). We, therefore, need to know whether χ has any roots (zeroes) with nonnegative real part[1], so it is expedient to construct a contour that **encloses the whole closed right half-plane**. Suppose for a moment that F has no poles on the imaginary axis and construct a

---
[1] It should be clear that the poles of G_{cl} are the zeroes of χ.

contour Γ_R as in Figure 12.11 (we will lift this assumption later).

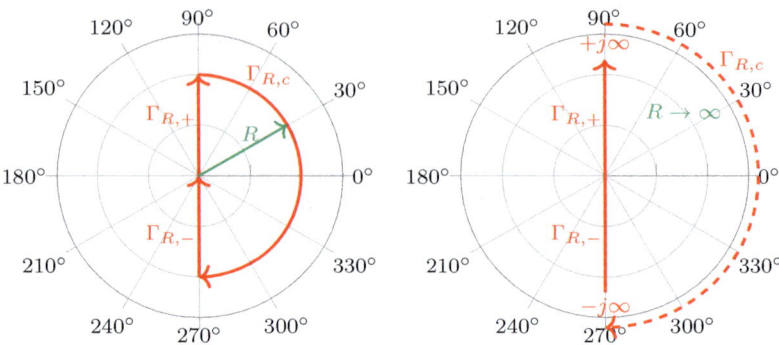

Figure 12.11: *(Left) Construction of Nyquist contour: $\Gamma_R = \Gamma_{R,+} + \Gamma_{R,c} + \Gamma_{R,-}$ (Right) As $R \to \infty$, the contour Γ_R covers the whole right half-plane.*

For $R > 0$, we define the contour $\Gamma_R = \Gamma_{R,+} + \Gamma_{R,c} + \Gamma_{R,-}$ as in Figure 12.11, where

$$\Gamma_{R,+} = \{js : s \in [0, R]\}, \tag{12.24a}$$

$$\Gamma_{R,c} = \{Re^{-js} : s \in [-\pi/2, \pi/2]\}, \tag{12.24b}$$

$$\Gamma_{R,-} = \{js : s \in [-R, 0]\}. \tag{12.24c}$$

Then, as $R \to \infty$, the contour Γ_R covers the whole closed right half-plane. Our objective is to determine **whether χ has any zeroes in Γ_R for any $R > 0$**, that is, whether χ has any poles in the closed right half-plane.

By Cauchy's principle of argument, as s moves on Γ_R, the plot of $\chi(s)$ revolves around the origin $N = Z - P$ times, where Z and P are the numbers of zeroes and poles of χ inside Γ_R. Equivalently, G_{ol} encircles the complex number $-1 + j0$ exactly $N = Z - P$ times. For G_{cl} to be a BIBO-stable transfer function, χ must have no zeroes in the closed right half-plane, i.e., it must be $Z = 0$ as $R \to \infty$. What is the same, a necessary and sufficient condition for stability is

$$N = -P, \tag{12.25}$$

as $R \to \infty$, meaning that G_{cl} is BIBO-stable if and only if the plot of $G_{\text{ol}}(s)$, as s moves on Γ_R and $R \to \infty$ encircles -1 in the counter-clockwise direction exactly P times — as many as the number of unstable poles of G_{ol}. This leads to the formulation of the Nyquist stability criterion.

> **Theorem 12.4 (Nyquist stability criterion)** Suppose that G_{ol} is a meromorphic function without any poles on the imaginary axis. Let P be the number of unstable poles of G_{ol}. Then G_{cl} is BIBO-stable if and only if the plot of $G_{\text{ol}}(s)$, as s moves on Γ_R, encircles -1 in

Chapter 12. Nyquist plot and stability criterion

the counter-clockwise direction P times as $R \to \infty$. In the special case that the Nyquist plot of G_{ol} runs through -1, the closed-loop system is not BIBO stable.

Proof. The proof follows by applying Theorems 12.3 to χ and the fact that χ encircles the origin N times if and only if G_{ol} encircles -1 the same number of times. □

Earlier we assumed that G_{ol} has no poles on the imaginary axis. This limitation can be easily lifted by slightly modifying the Nyquist contour to exclude these poles. Around each pole of G_{ol} on the imaginary axis we draw a semicircle or radius r with counter-clockwise direction which lies in the right half-plane. This way we construct a modified contour $\Gamma_{R,r}$ and take $r \to 0^+$ and $R \to \infty$.

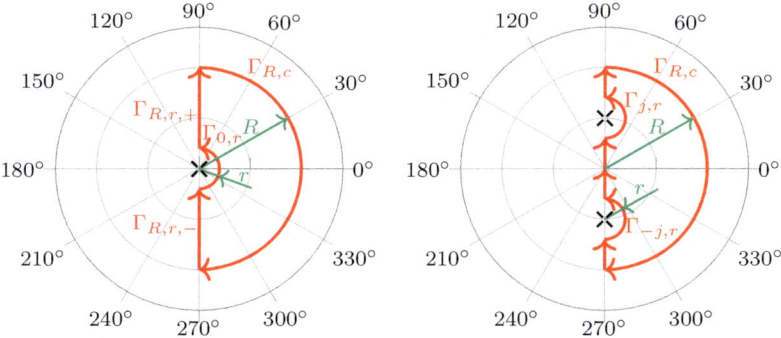

Figure 12.12: *Construction of Nyquist contour by excluding the poles of G_{ol} on the imaginary axis. (Left) A pole at the origin is excluded by introducing a semicircle $\Gamma_{0,r}$ of a radius $r > 0$. (Right) Two complex conjugate poles at $\pm j$ are excluded by taking two semicircles $\Gamma_{\pm j, r}$.*

The Nyquist stability theorem can be then restated as follows:

Theorem 12.5 (Nyquist stability criterion with poles on the imaginary axis) Suppose that G_{ol} is a meromorphic function. Let P be the number of poles of G_{ol} in the open right half-plane. Then G_{cl} is BIBO stable if and only if the plot of $G_{ol}(s)$, as s moves on $\Gamma_{R,r}$, encircles -1 in the counter-clockwise direction P times as $R \to \infty$ and $r \to 0^+$. In the special case that the Nyquist plot of G_{ol} runs through -1, the closed-loop system is not BIBO stable.

It remains to see how one can construct the Nyquist plot $G(\Gamma_{R,r})$ as $R \to \infty$ and $r \to 0^+$. We will then give examples where we will use the Nyquist stability criterion to determine whether a closed-loop system is BIBO-stable using its open-loop transfer function.

12.2.3 Nyquist plots

No poles on the imaginary axis

Let us first focus on transfer function without poles on the imaginary axis. The Nyquist contour is given in Figure 12.11. The contour consists of three parts: $\Gamma_{R,+}$, $\Gamma_{R,-}$ and $\Gamma_{R,c}$. In Section 12.1 we constructed plots of the form $G(\Gamma_{R,+})$ for $R \to \infty$. It turns out that the plot as s moves on $\Gamma_{R,-}$ can be readily constructed from the plot as s moves on $\Gamma_{R,+}$. We will further show that the plot of G over $\Gamma_{R,c}$, as $R \to \infty$, converges to a single point if G is *proper*, that is, it has no more zeroes than it has poles. This is a very weak assumption as all physically realisable transfer functions are proper.

Recall that since G_{ol} is either a rational or a rational-times-exponential function, it is (see Equation (A.26) in the Appendix)

$$G_{\text{ol}}(-js) = \overline{G_{\text{ol}}(js)}, \quad (12.26)$$

for all $s \in \mathbb{R}$. This means that the plot of $G_{\text{ol}}(j(-\infty, 0])$ is the reflection of the plot of $G_{\text{ol}}(j[0, \infty))$ on the real axis. A couple of examples are shown in Figure 12.13.

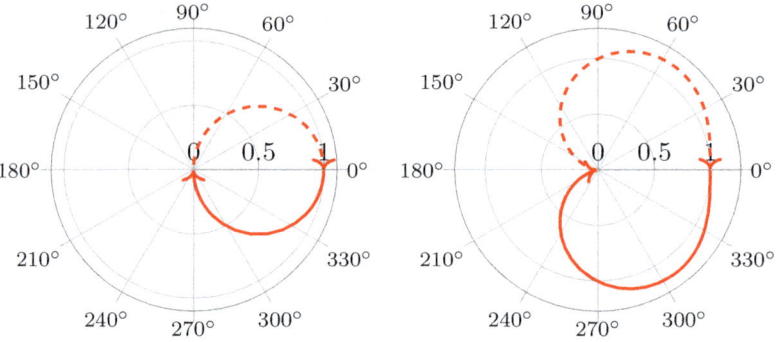

Figure 12.13: *(Left) Plot of $G(j[0,+\infty))$ (solid line) and $G(j(-\infty, 0])$ (dashed line) for a first-order system with $K = 1$, (Right) Plots for a second-order system with $K = 1$ and $\zeta = 0.5$. The dashed lines are symmetric to the solid lines with respect to the real axis.*

It remains to see how what $G(\Gamma_{R,c})$ maps to as $R \to \infty$. The reader can prove the following proposition, along the lines of Proposition 12.2 as an exercise.

> **Proposition 12.6 (Plot of $G(\Gamma_{R,c})$ as $R \to \infty$)** Consider a transfer function of the form
>
> $$G(s) = K\frac{(s-z_1)(s-z_2) \cdot \ldots \cdot (s-z_m)}{(s-p_1)(s-p_2) \cdot \ldots \cdot (s-p_n)} e^{-t_d s}, \quad (12.27)$$
>
> with $t_d > 0$ and suppose that it is *proper* ($n \geq 0$). Then, for any

$\theta \in \mathbb{R}$, $\lim_{R\to\infty} G(Re^{j\theta})$ exists, is independent of θ and

$$\lim_{R\to\infty} G(Re^{j\theta}) = \begin{cases} 0, & \text{if } n > m \\ K, & \text{if } n = m \end{cases} \quad (12.28)$$

Proof. Exercise. Hint: See the proof of Proposition 12.2. □

Example 12.5 (Nyquist criterion: first-order transfer function) Consider a controlled first-order system with gain $K_s = 1$ and time constant $\tau_s = 1$, which is controlled with a P controller. Suppose that $G_a(s) = 1$ and $G_m(s) = 1$. The open-loop transfer function is

$$G_{\text{ol}}(s) = \frac{K}{s+1}. \quad (12.29)$$

The Nyquist plot of G_{ol} is shown below.

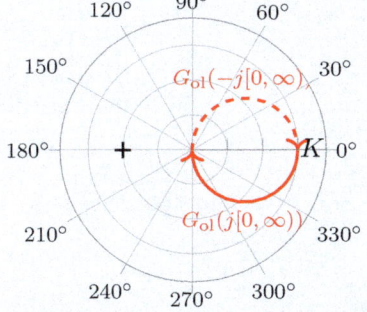

Figure 12.14: *The Nyquist plot of G_{ol} never encircles -1, that is, $N = 0$. Since G_{ol} does not have any unstable poles ($P = 0$), the number of unstable poles of G_{cl} is zero.*

We see that the Nyquist plot of G_{ol} does not encircle -1 ($N = 0$). Since G_{ol} does not have any unstable poles (in the closed right half-plane), the closed-loop system is BIBO-stable for all $K > 0$ (Theorem 12.4).

Example 12.6 (Unstable open-loop system) Consider a system with the following open-loop transfer function

$$G_{\text{ol}}(s) = \frac{K}{s-1}, \quad (12.30)$$

for some $K > 0$. The open-loop transfer function is unstable — it has one unstable pole at $1 + j0$ ($P = 1$). According to the Nyquist stability criterion (Theorem 12.4), the corresponding closed-loop system is stable if and only if the Nyquist plot of G_{ol} encircles the point $-1 + j0$ at least once in the counter-clockwise direction. This is the case only if $K > 1$ as shown in Figure 12.15.

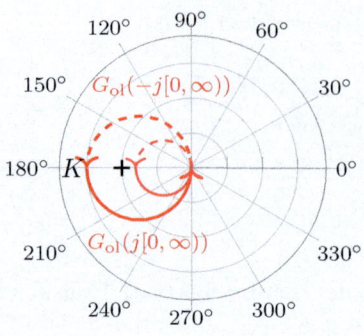

Figure 12.15: Nyquist plot of $G_{\text{ol}}(s) = \frac{K}{s-1}$: The large circle corresponds to $K = 1.5$ and the small circle to $K = 0.8$; the black cross mark is the point $-1 + 0j$.§ The Nyquist plot encircles -1 in the counter-clockwise direction only for $K > 1$. If (and only if) $K > 1$, the closed-loop system is BIBO-stable.

Example 12.7 (First-order transfer function with delay) Consider a controlled first-order system with gain $K_s = 1$ and time constant $\tau_s = 1$, which is controlled with a P controller. Suppose that $G_a(s) = e^{-0.5s}$ and $G_m(s) = 1$. The open-loop transfer function is

$$G_{\text{ol}}(s) = \frac{Ke^{-0.5s}}{s+1}. \tag{12.31}$$

We need to determine the maximum gain K for which the closed-loop system is BIBO-stable.

Before we construct the Nyquist plot of G_{ol} we can note that

1. As $\omega \to 0^+$, $G_{\text{ol}}(j\omega) \to K$

2. Since G_{ol} is strictly proper ($n = 1$, $m = 0$), $\lim_{R\to\infty} G_{\text{ol}}(Re^{j\theta}) = 0$ for all θ, therefore $G(\Gamma_{R,c})$ converges to 0 as $R \to \infty$ (see Proposition 12.6)

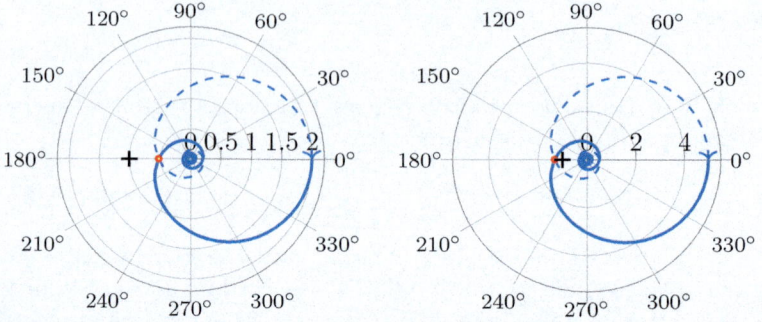

Figure 12.16: Nyquist plots of G_{ol}: (Left) for $K = 2$, where we observe that the Nyquist plot does not encircle -1 and (Right) for $K = 5$ where the Nyquist plot encircles -1 once in the clockwise direction. The red circles correspond to the phase crossover frequency.

As one may observe in Figure 12.6, for $K = 2$ (left) the Nyquist plot does not encircle -1, therefore, by the Nyquist Stability Criterion (Theorem 12.4) the closed-loop system is BIBO-stable. However, for $K = 5$ (right), the Nyquist plot encircles -1 once in the clockwise direction ($N = 1$), therefore, the closed-loop system is not BIBO-stable.

In order to determine the maximum gain to have a BIBO-stable closed-loop system, we need to require that the Nyquist plot intersects the real axis on the right of -1. In other words, we require that at the frequency where $\arg G_{\text{ol}}(j\omega) = -\pi$, the magnitude of $G_{\text{ol}}(j\omega)$ is less than 1. This frequency is precisely the *phase crossover frequency* of G_{ol} (see Definition 11.5) which is computed as in Example 11.17 using a numerical solver

$$\arg G_{\text{ol}}(j\omega_{\text{co}}) = -\pi \Leftrightarrow -0.5\omega_{\text{co}} - \arctan \omega_{\text{co}} = -\pi. \qquad (12.32)$$

The phase crossover frequency is $\omega_{\text{co}} \approx 3.8069$. By the Nyquist stability criterion we require that $|G(j\omega_{\text{co}})| < 1$, or equivalently

$$\left| \frac{Ke^{-jt_d\omega_{\text{co}}}}{j\omega_{\text{co}} + 1} \right| < 1 \Leftrightarrow K\frac{1}{\sqrt{1 + \omega_{\text{co}}^2}} < 1 \Leftrightarrow K < 3.8069. \qquad (12.33)$$

For this system, we could have well used the Bode stability criterion. In fact, we would have followed the same procedure to determine the phase crossover frequency and the maximum gain.

Remark: *Transfer functions involving time delays give rise to nice whirlpool-like Nyquist plots. But how are these related to the Nyquist plots of their non-delayed counterparts? For example, what is the relationship between the Nyquist plot of $G(s) = {1}/{s+1}$ and $\widetilde{G}(s) = {e^{-t_d s}}/{s+1}$?*

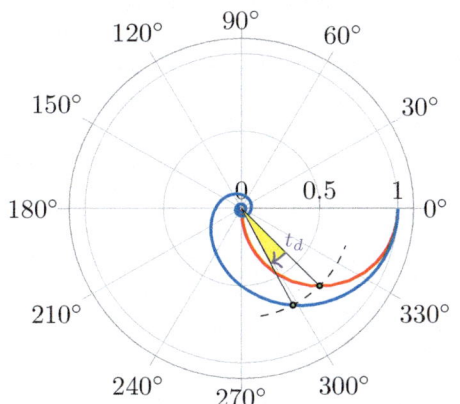

Figure 12.17: *Clockwise rotation of a Nyquist plot by the introduction of a delay, t_d. (Red line) Nyquist plot of $G(s) = {1}/{s+1}$ for $s = +j\omega$, $\omega \in [0, \infty)$, (Blue line) Nyquist plot of $G(s) = {e^{-0.3s}}/{s+1}$ for $s = +j\omega$, $\omega \in [0, \infty)$.*

Recall that for any nonzero complex number $z \in \mathbb{C}$, $|ze^{-jt_d\omega}| = |z|$ and $\arg(ze^{-jt_d\omega}) = \arg z - t_d\omega$. This means that the introduction of a delay **rotates** the original Nyquist plot clockwise, while it **preserves** its magnitude. This is illustrated graphically in the following figure. •

Exercise 12.4 (Nyquist plot) Construct the Nyquist plot of a system with transfer function

$$G(s) = \frac{e^{-\frac{\pi}{6}s}}{(s+1)(s+2)},$$

using the above remark (Firstly, construct the Nyquist plot of $\frac{1}{(s+1)(s+2)}$).

Exercise 12.5 (From Bode to Nyquist) Construct the Bode plot of G_{ol} in Example 12.7 and compare with the Nyquist plot in Figure 12.16.

In practice, Nyquist plots are constructed using software[2]. Unlike Bode plots, the Nyquist plot of a transfer function cannot be constructed from the Nyquist plots of its factors. However, we can extract information about Nyquist plots using the corresponding Bode plots. Let us give an example.

Example 12.8 (From Bode to Nyquist) Consider the following open-loop transfer function

$$G_{\text{ol}}(s) = \frac{Ks}{((s+1)^2 + 2)(s+2)}, \tag{12.34}$$

where $K > 0$. The degree of the numerator is $m = 1$ and the degree of the denominator is $n = 3$, that is, G_{ol} is strictly proper, therefore by Proposition 12.2 the Nyquist plot approaches the origin with angle $-\pi$ as s escapes to infinity on the imaginary axis. It is also easy to see that as $s \to 0^+$, $G_{\text{ol}}(s)$ approaches the origin because G_{ol} has a zero at the origin ($G_{\text{ol}}(0) = 0$). Other than that, it is not easy to extract any additional information about the shape of the Nyquist plot of G_{ol}. We may, however, use the Bode plot of G_{ol} which we can construct easily as we discussed in Chapter 11.

The Bode plot of G_{ol} is given in Figure 12.18. We observe that

1. The argument of $G_{\text{ol}}(j\omega)/K$ moves from $90°$ to $-180°$, therefore, the Nyquist plot of G_{ol}/K moves from the first to the fourth and then to the third quadrant in a clockwise direction,

2. The maximum magnitude of $G_{\text{ol}}(j\omega)/K$ is attained at about $1.53 \,\text{rad/s}$ and is equal to $-14.2\,\text{dB}$. This means that

---

[2]In Python, we can use the control.nyquist_plot function of the control toolbox. In MATLAB, we may use nyquist. A very useful MATLAB function is nyqlog, which can be downloaded from the MathWorks File Exchange; see [S4]

the Nyquist plot of G_{ol} is contained in a circle of radius $10^{-14.2/20} \approx 0.195$,

3. The argument of $G_{ol}(j\omega)/K$ becomes $0°$ at a certain frequency; at that point the magnitude of $G_{ol}(j\omega)/K$ is $-14.8\,\text{dB} \approx 0.18$. That is, the Nyquist plot of G_{ol}/K runs through $0.18 + 0j$,

4. The argument of $G_{ol}(j\omega)/K$ becomes $-90°$ at a certain frequency; at that point the magnitude of $G_{ol}(j\omega)/K$ is $-18.4\,\text{dB} \approx 0.12$. This means that the Nyquist plot of G_{ol}/K crosses $-0.12j$,

5. As $\omega \to 0^+$, the argument of $G_{ol}(j\omega)/K$ approaches $90°$,

6. Since G_{ol} is strictly proper, the mapping of $\Gamma_{R,c}$ via G_{ol} maps to the origin (Proposition 12.6).

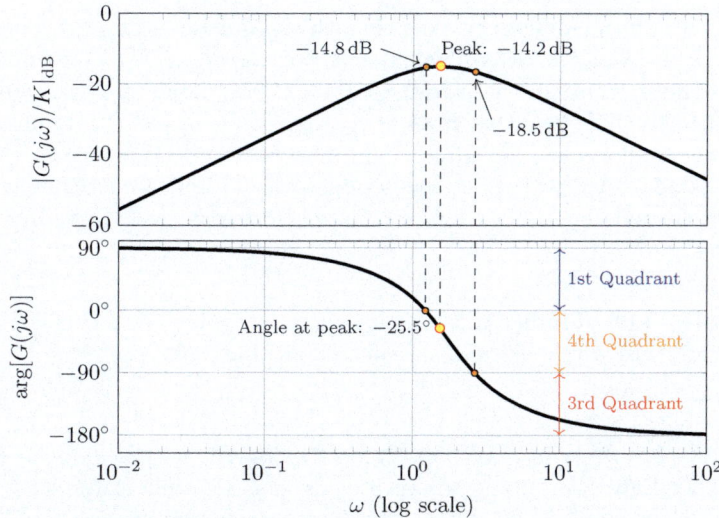

Figure 12.18: Bode plot of G_{ol}/K.

We may now construct the Nyquist plot of G_{ol}/K (see Figure 12.19) taking into account the above observations.

The Nyquist plot of G_{ol} does not intersect the real axis on the left of the origin, therefore the plot does not encircle -1 for all $K > 0$. By the Nyquist stability theorem, the closed-loop system is BIBO-stable for all $K > 0$.

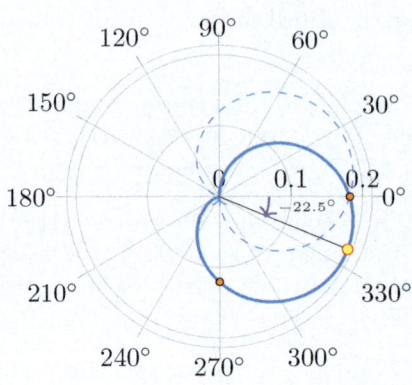

Figure 12.19: Nyquist plot of G_{ol}/K. (Solid blue line) Part of the Nyquist plot as s moves on the positive part of the imaginary axis ($j\omega$, $\omega > 0$), (Dashed blue line) part of the Nyquist plot for $s = -j\omega$, $\omega > 0$ (reflection of the solid line with respect to the real axis).

Poles on the imaginary axis

If G_{ol} has poles on the imaginary axis, then the Nyquist contour is modified to exclude them using the trick illustrated in Figure 12.12. It again suffices to focus on those points of the Nyquist contour. The large circle of radius R will map to a single point as $R \to \infty$ (Proposition 12.6). It remains to see what the small circles, of radius r will map to as $r \to 0^+$.

As discussed previously, if G_{ol} has a pole at the origin, we exclude it in the Nyquist contour by introducing a counter-clockwise circle or radius r,

$$\Gamma_{0,r} = \{re^{j\theta} : \theta \in [-\pi/2, \pi/2]\},$$

and we take $r \to 0^+$. Similarly, for each imaginary pole $j\omega^*$, we introduce a counter-clockwise circle or radius r centred as that pole, that is

$$\Gamma_{j\omega^*,r} = \{j\omega^* + re^{j\theta} : \theta \in [-\pi/2, \pi/2]\},$$

and again we take $r \to 0^+$. We can focus the imaginary poles above the real axis as we have been doing so far. Let us give a simple example.

Example 12.9 (Simple integrator) We will construct the Nyquist plot of a simple integrator with transfer function $G(s) = 1/s$.

Step 1. We will focus on $\Gamma_{R,r,+}$ which is the path on the imaginary axis, $s = j\omega$ for $\omega \in [r, R]$. This maps to

$$G(s) = G(j\omega) = \frac{1}{j\omega} = -\frac{j}{\omega}. \tag{12.35}$$

As s moves on the imaginary axis from rj to Rj, $G(s)$ moves on the imaginary axis from $-\frac{j}{r}$ to $-\frac{j}{R}$. As $R \to \infty$ and $r \to 0^+$, this converges to the halfline $j(-\infty, 0]$.

Step 2. Since G is strictly proper ($m = 0$, $n = 1$),

$\lim_{R\to\infty} G(Re^{j\theta}) = 0$ for any $\theta \in \mathbb{R}$. This means that $\Gamma_{R,c}$ maps to the origin as $R \to \infty$.

Step 3. The blue semicircle in Figure 12.20 is the set of points $s = re^{j\theta}$ for $\theta \in [-\pi/2, \pi/2]$. These points map to

$$G_{\text{ol}}(s) = \frac{1}{re^{j\theta}} = \frac{1}{r}e^{-j\theta}, \qquad (12.36)$$

that is, $G_{\text{ol}}(s)$ is a clockwise semicircle centred at the origin with radius $1/r$. As $r \to 0^+$, its radius goes to infinity and the semicircle encompasses the whole right half-plane in the clockwise direction as shown in Figure 12.20.

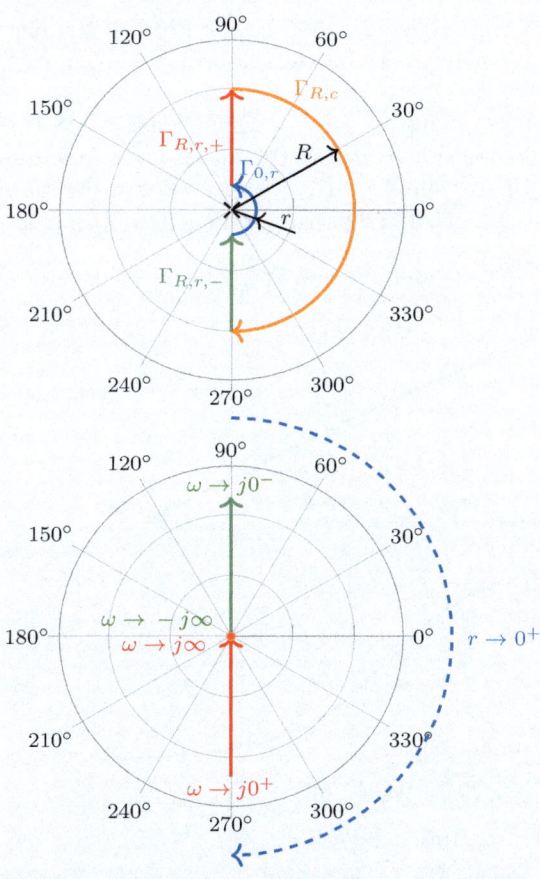

Figure 12.20: *Nyquist plot of simple integrator: (Up) Construction of Nyquist contour excluding the pole of G_{ol} at the origin, (Down) Nyquist plot of G_{ol}. The colours of the segments of the Nyquist contour match the colours of their mappings (as $R \to \infty$ and $r \to 0^+$).*

By the Nyquist stability criterion (Theorem 12.5), if a closed-loop system has the open-loop transfer function $G_{\text{ol}}(s) = 1/s$, then it is BIBO-stable.

Let us give yet another example of a transfer function with a pole at the origin with multiplicity 5. We will see that the Nyquist plot encompasses the whole complex plane two and a half times.

Example 12.10 (Pole at origin with multiplicity 5) Consider a system with open-loop transfer function

$$G_{\text{ol}}(s) = \frac{1}{s^5(s+1)}. \tag{12.37}$$

Since G_{ol} has a pole at the origin, we need to modify the Nyquist contour to exclude it. We will proceed as in the previous example.

Step 1. Again, we first focus on $\Gamma_{R,r,+}$ which consists of the complex numbers $s = j\omega$ for $\omega \in [r, R]$. Since this transfer function does not fall into any of the simple cases we presented in Section 12.1, it is useful to first construct the corresponding Bode plot.

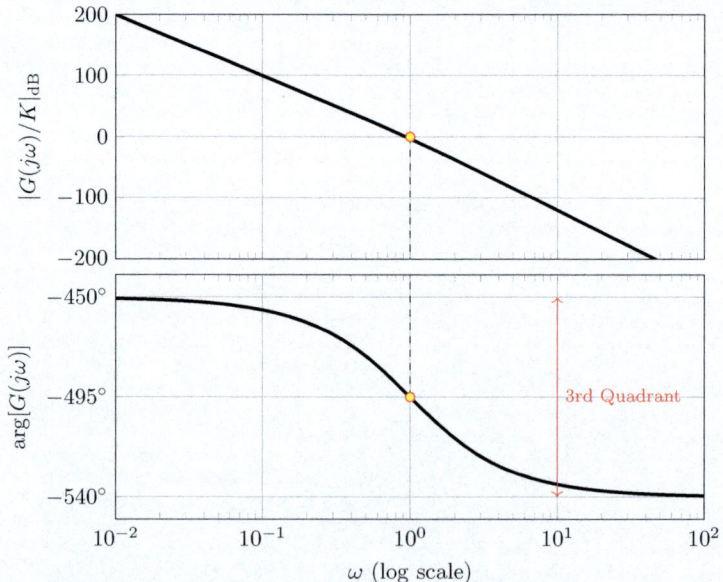

Figure 12.21: *Bode plot of G_{ol} in Equation (12.37). The Nyquist plot of G_{ol} lies in the third quadrant and intersects the unit circle at a point with argument $-495°$.*

Step 2. Since G_{ol} is strictly proper ($m = 0$, $n = 6$), $\lim_{R \to \infty} G(Re^{j\theta}) = 0$ for any $\theta \in \mathbb{R}$. This means that $\Gamma_{R,c}$

maps to the origin as $R \to \infty$.

Step 3. The curve $\Gamma_{0,r}$ consists of points $s = re^{j\theta}$ for $\theta \in [-\pi/2, \pi/2]$. These points map to

$$G_{\text{ol}}(s) = G_{\text{ol}}(re^{j\theta}) = \frac{1}{(re^{j\theta})^5(re^{j\theta}+1)}. \quad (12.38)$$

But since r is very small (we will take the limit $r \to 0^+$), $1 + re^{j\theta}$ is approximately equal to 1, therefore, as $r \to 0^+$, $G_{\text{ol}}(re^{j\theta})$ can be asymptotically approximated by

$$G_{\text{ol}}(re^{j\theta}) \approx \frac{1}{r^5 e^{j5\theta}} = \frac{1}{r^5} e^{-j5\theta}. \quad (12.39)$$

Therefore, as θ travels from $-\pi/2$ to $\pi/2$, $G_{\text{ol}}(re^{j\theta})$ revolves around the origin, at a distance $1/r^5$, from an initial angle $-5\pi/2$ to $5\pi/2$. This corresponds to 2.5 revolutions. As $r \to 0^+$, $G_{\text{ol}}(re^{j\theta})$ revolves around the whole complex plane 2.5 times.

The Nyquist plot of G_{ol} is shown in Figure 12.22.

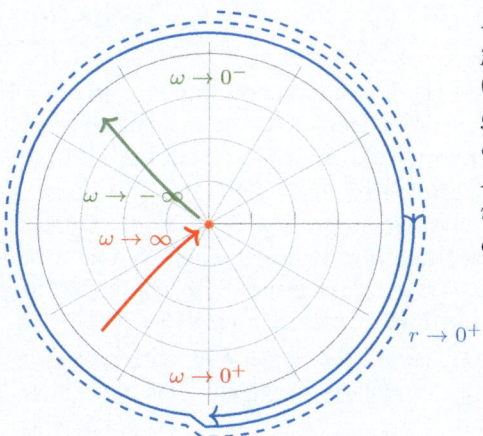

Figure 12.22: Nyquist plot of G_{ol} in Equation (12.37). The green line goes to $+j\infty$ and red line comes from $-j\infty$. The Nyquist plot can be seen to revolve two and a half times around -1.

The reader should think that as $\omega \to 0^+$ the Nyquist plot of G_{ol} goes to $-j\infty$ and connects with the blue line in Figure 12.22. Similarly, as $\omega \to 0^-$, the green line in Figure 12.22 extends to $+j\infty$ and connects with the dashed blue line. This way, the Nyquist plot can be thought of as a closed line that encircles -1 *two and a half times*. Therefore, by the Nyquist stability criterion (Theorem 12.5), the closed-loop system is not BIBO-stable.

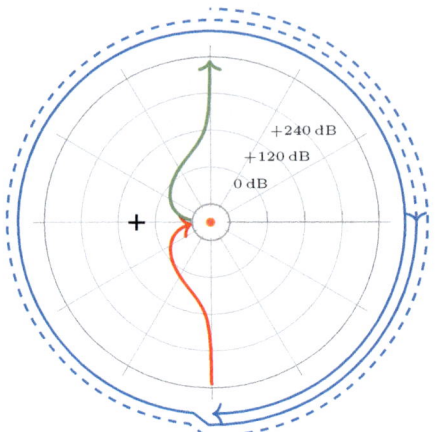

Figure 12.23: Nyquist plot in log-polar coooordinates for the transfer function G_{ol} of Equation (12.37).

In Example 12.10 we constructed the Bode and Nyquist plots of a function with a multiple pole at the origin. In the Bode plot (Figure 12.21) we can see that as $\omega \to 0^+$, the argument of $G(j\omega)$ goes to $-540°$. However, this is not well visible in the Nyquist plot in Figure 12.22. One may be falsely led to believe that the argument of $G(j\omega)$ converges to an angle between $210°$ and $240°$. The remedy is to draw the Nyquist plot in **log-polar coordinates** (see Figure 12.23); this allows us to properly zoom in close to zero and infinity. Such a log-polar Nyquist plot is shown on the left. Here the radius is plotted in dB. In this figure it is evident that $G_{\text{ol}}(j\omega)$ goes to $\mp j\infty$ as ω goes to $\pm\infty$.

Note that in the above figure there is a circle around the origin, which corresponds to a gain of $-120\,\text{dB}$. We consider this to be a value very close to zero (10^{-6}). In a log-polar plot — and in a Cartesian logarithmic plot — we cannot plot quantities of arbitrarily small gain.

The log-polar form of the Nyquist plot should be used whenever we want to inspect how the transfer function behaves when its magnitude is either very close to zero or very large. It is generally a good idea to plot the Nyquist plot in log-polar coordinates (see Exercise 12.15).

We will now give a couple of examples of systems with poles on the imaginary axis. The Bode and Nyquist plots of such systems have two distinctive characteristics

1. The argument of $G(j\omega)$ is discontinuous,

2. The modulus of $G(j\omega)$ diverges to infinity at a finite frequency (the resonance frequency of the system).

Indeed, if in the denominator of an open-loop transfer function there is a term of the form $H(s) = s^2 + b^2$, for some $b > 0$, then $|H(j\omega)|$ is

$$|H(j\omega)| = |(j\omega)^2 + b^2| = |b^2 - \omega^2|, \tag{12.40}$$

which goes to $+\infty$ as $\omega \to b$. The argument of $H(j\omega)$ is

$$\arg H(j\omega) = \arg(b^2 - \omega^2) = \begin{cases} 0, & \text{if } 0 \leq \omega < b \\ -\pi, & \text{if } 0 \leq \omega > b \end{cases}, \quad (12.41)$$

which is discontinuous. Let us give a simple example.

Example 12.11 (Purely imaginary poles) Consider the following open-loop transfer function

$$G_{\text{ol}}(s) = \frac{1}{(s^2 + 1)(2s + 1)}. \quad (12.42)$$

Function G_{ol} has a (stable) pole at $-1/2$ and a pair of purely imaginary poles at $\pm j$, therefore, we expect that $|G_{\text{ol}}(j\omega)|$ diverges to infinity as $\omega \to 1$ and $\arg G_{\text{ol}}(j\omega)$ is discontinuous at $\omega = 1$. In order to facilitate the construction of the Bode plot we can first construct the Bode plot.

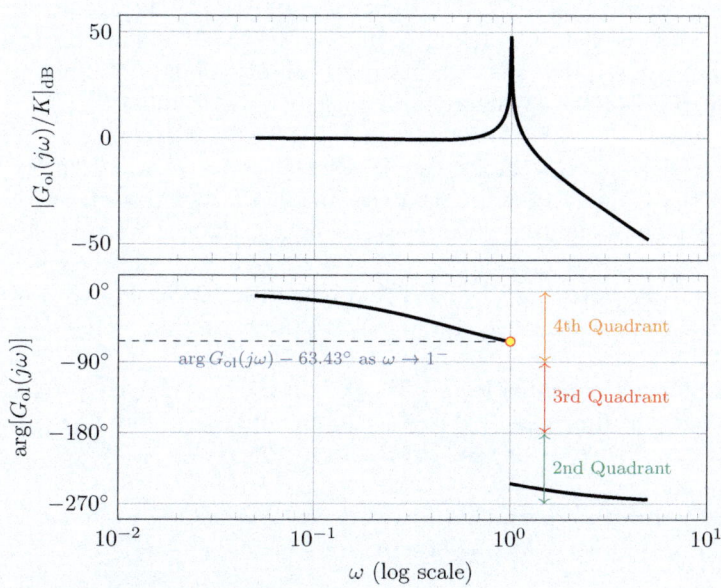

Figure 12.24: *Bode plot of G_{ol} in Equation (12.42).*

Firstly, we observe that $G_{\text{ol}}(j\omega)$ lies entirely in the fourth and second quadrants. In particular, at low frequencies it is in the fourth quadrant and at high frequencies it lies in the second one.

Low Frequencies. At low frequencies $G_{\text{ol}}(j\omega)$ approaches the real number 1 and its argument approaches $0°$. Moreover, for frequencies on the left of the resonance frequency and not too close to it, $|G_{\text{ol}}(j\omega)|$ is approximately equal to 1 while its argument decreases. This means that for these frequencies $G_{\text{ol}}(j\omega)$ moves approximately on the unit circle in the clockwise direction.

High frequencies. At high frequencies, $G_{\text{ol}}(j\omega)$ approaches the origin and its argument approaches $-270°$ (this is indeed confirmed by Proposition 12.2).

Resonance frequency. As ω approaches the resonance frequency ω_{res} from the left, the argument of $G_{\text{ol}}(j\omega)$ approaches $-63.43°$, while its argument escapes to infinity. Indeed, for $0 < \omega < 1$,

$$\arg G_{\text{ol}}(j\omega) = -\arg(1-\omega^2) - \arg(1+2j\omega) = -\arctan(2\omega), \quad (12.43)$$

so $\lim_{\omega \to 1^-} \arg G_{\text{ol}}(j\omega) = \lim_{\omega \to 1^-} -\arctan(2\omega) = -\arctan 2 = -63.43$.

This means that the Nyquist plot of G_{ol} escapes to infinity along a halfline at an angle of $-63.43°$. Similarly, as $\omega \to 1^+$, the argument of $G_{\text{ol}}(j\omega)$ approaches $-63.43° - 180° = -243.43°$. This means that the Nyquist plot of G_{ol} comes from infinity along a halfline at an angle of $-243.43°$. We may now construct the Nyquist plot.

Step 1: ($s = j\omega$, not close to j). Take $\omega > 0$. We will start by plotting the Nyquist plot for s on the imaginary axis, $s = j\omega$ with $\omega > 0$, but not excluding an interval around j, that is, we will plot $G_{\text{ol}}(j\omega)$ for $\omega \in [0, 1-r] \cup [1+r, R]$ and take $R \to \infty$ and $r \to 0^+$. This corresponds to the thick red line in Figure 12.25. For $\omega < 0$, the Nyquist plot is the reflection of the red line over the real axis.

Step 2: (large semicircle, $s \in \Gamma_{R,c}$). Since G_{ol} is strictly proper ($n = 3$, $m = 0$), $\lim_{R \to \infty} G_{\text{ol}}(Re^{j\theta}) = 0$ for all θ, therefore $G(\Gamma_{R,c})$ converges to 0 as $R \to \infty$ (see Proposition 12.6).

Step 3: (small semicircles around $\pm j$). Consider a semicircle around j of radius r which is described by

$$s = j + re^{j\theta}, \quad (12.44)$$

for $\theta \in [-\pi/2, \pi/2]$. The transfer function becomes

$$G_{\text{ol}}(s) = \frac{1}{[(j+re^{j\theta})^2 + 1](2(j+re^{j\theta}) + 1)}$$

$$= \frac{1}{(\cancel{-1} + r^2 e^{2j\theta} + 2jre^{j\theta} + \cancel{1})(2(j+re^{j\theta})+1)}$$

$$= \frac{1}{r\left(re^{2j\theta} + 2e^{j\left(\theta + \frac{\pi}{2}\right)}\right)(2j + 2re^{j\theta} + 1)}. \quad (12.45)$$

The argument of $G_{\text{ol}}(s)$ becomes

$$\arg G_{\text{ol}}(s) = -\arg\left(re^{2j\theta} + 2e^{j(\theta+\frac{\pi}{2})}\right) - \arg\left(2j + 2re^{j\theta} + 1\right), \tag{12.46}$$

and as r goes to 0 this can be asymptotically approximated by

$$\begin{aligned}\arg G_{\text{ol}}(s) &= -\arg\left(e^{j(\theta+\frac{\pi}{2})}\right) - \arg\left(2j + 1\right) \\ &= -\theta - \tfrac{\pi}{2} - \arctan 2 = -153.43° - \theta,\end{aligned} \tag{12.47}$$

where θ moves from $-90°$ to $90°$. This way, $\arg G_{\text{ol}}(s)$ travels from $-63.43°$ to $-243.43°$ in the clockwise direction.

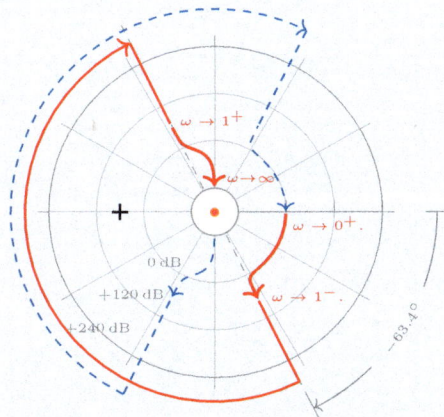

Figure 12.25: *Nyquist plot of G_{ol} in Equation (12.42) in log-polar coordinates.*

By the Nyquist stability criterion, Theorem 12.5, we conclude that this system is unstable.

Table 12.1: Nyquist plots of first and second-order transfer functions with P, PD and PI controllers

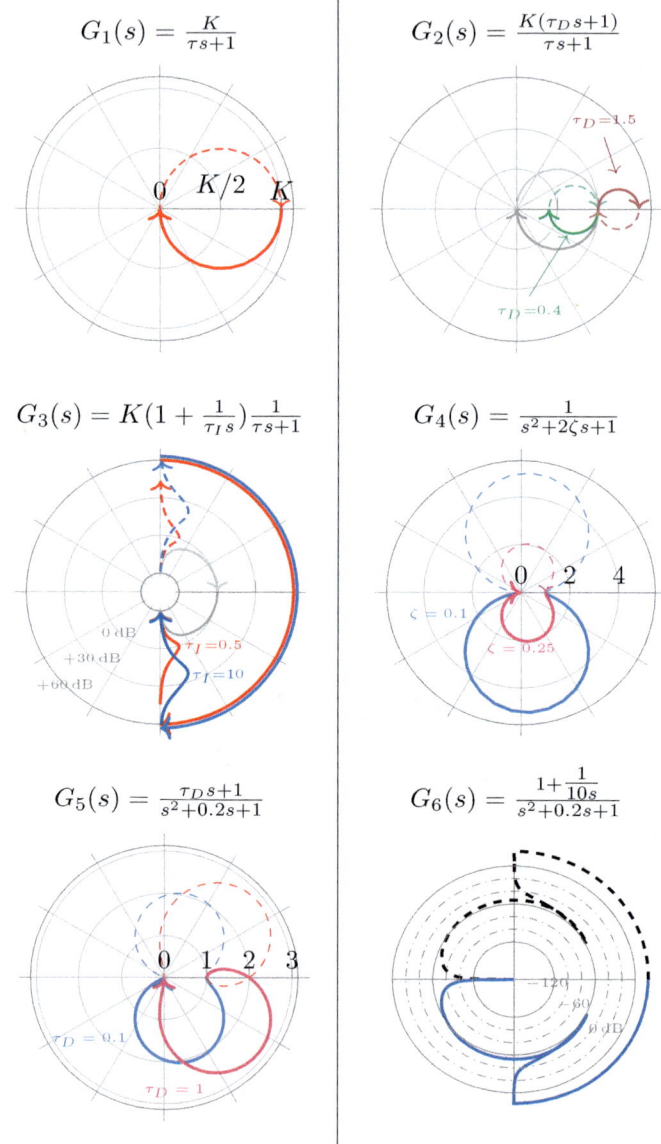

12.3 Stability margins

12.3.1 Gain, phase, and delay margins

When we first gave the definitions of the gain margin (Definition 11.9) and phase margin (Definition 11.10) in the previous chapter, we imposed certain requirements such as the existence of a unique phase crossover frequency. These are all because of the limitations of the Bode plot as a stability analysis tool.

> **Definition 12.7 (Gain margin)** Let G_{ol} be an open-loop transfer function so that the closed-loop system is BIBO stable. The maximum gain, $K > 1$, so that the system with open-loop transfer function KG_{ol} is BIBO stable, measured in dB, is called the gain margin of the closed-loop system. If the system remains stable for all $K > 1$, the gain margin is infinite.

If G_{ol} is such that the closed-loop system is unstable, then the gain margin is the maximum $K \in (0, 1)$, measured in dB, so that the system with open-loop transfer function KG_{ol} becomes BIBO stable.

> **Definition 12.8 (Phase margin)** Let G_{ol} be an open-loop transfer function so that the closed-loop system is BIBO stable. The phase margin is the maximum delay that can be introduced into the loop so that the closed loop remains BIBO-stable.

The gain and phase margins can be determined on a Nyquist plot. We need to determine two characteristic points: the point where the Nyquist plot intersects the negative real halfline and the point where it intersects the unit circle. The *first* frequency at which the Nyquist plot crosses the negative real halfline is the *phase crossover frequency*, that is,

$$\arg G_{ol}(j\omega_{co}) = -\pi. \tag{12.48}$$

and the point(s) at which the Nyquist plot intersects the unit circle corresponds to ω_g (see Definition 11.10).

> **Example 12.12 (Phase and gain margin of third-order system)** Consider the open-loop transfer function
>
> $$G_{ol}(s) = \frac{0.5}{(s^2 + 0.25s + 1)(0.5s + 1)}. \tag{12.49}$$
>
> The amplitude ratio, $|G_{ol}(j\omega)|_{dB}$, is shown in Figure 12.26. Note that $|G_{ol}(j\omega)|_{dB}$ is not monotonically decreasing for all ω, therefore, the definition of the phase margin we gave in the context of the Bode stability criterion does not apply here (see Definition 11.10).

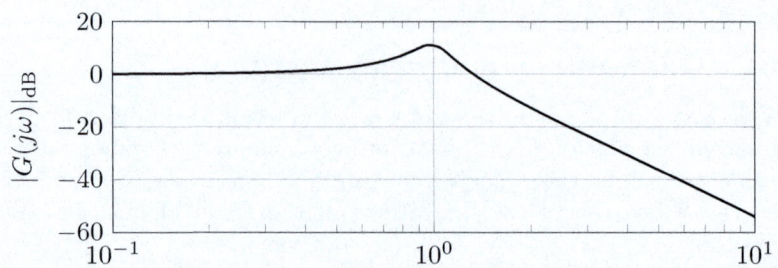

Figure 12.26: *Amplitude ratio of G_{ol}*

However, the new definitions of the stability margins we gave above (Definitions 12.7 and 12.8 can be applied. The stability margins of the system are show in Figure 12.27.

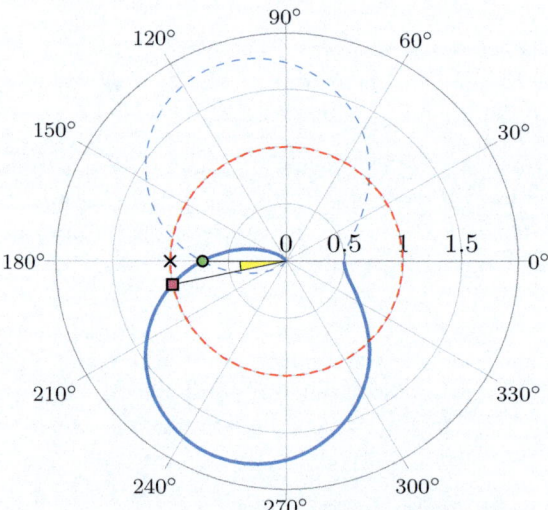

Figure 12.27: *Nyquist plot of G_{ol} and the phase and gain margins of the closed-loop system. (Dashed red circle - - -) circle of radius 1, ●: Point corresponding to ω_{co} where $\arg G_{ol}(j\omega_{co}) = -\pi$, ■: Point corresponding to ω_g where $|G_{ol}(j\omega_g)| = 1$, and ×: Point $(-1, 0)$.*

By the Nyquist stability criterion, the closed-loop system is BIBO stable. The coordinates of ● are $(-0.7269, 0)$, therefore, the system will remain stable if we multiply the open-loop transfer function by $K \in (0, 1/0.7269)$. This means that the *gain margin* of the closed-loop system is $1/0.7269 = 1.3756$ or $2.77\,\text{dB}$.

The angle between point ■ and the real axis is $11.7°$. This is the *gain margin* of the closed-loop system. Indeed, the closed-loop system

Chapter 12. Nyquist plot and stability criterion

will remain BIBO-stable if we rotate clockwise the Nyquist plot of Figure 12.27 by no more than 11.7°.

Remark: *Be careful: the gain margin is defined as the maximum $K > 1$ (larger than one!) so that KG_{ol} remains BIBO stable (see Definition 12.7). However, it is possible that a gain of 0.999 will destabilise the system. We would advise the reader to always check the Nyquist and Bode plots to make better sense of these stability margins. We strongly encourage the reader to try the following exercise.* •

Exercise 12.6 (Gain margin peculiarity) The open-loop transfer function of a system is

$$G_{ol}(s) = \frac{1.1(1+s)}{1-s^2}. \qquad (12.50)$$

The Nyquist plot of G_{ol} is shown below.

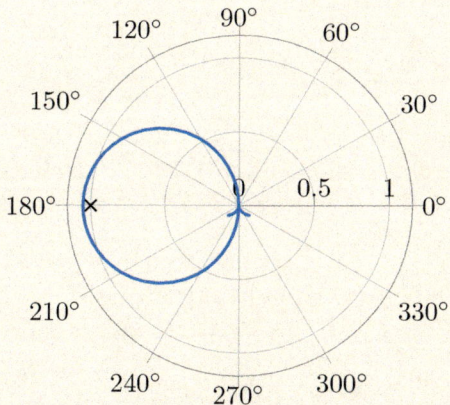

Figure 12.28: *Nyquist plot of the open-loop transfer function in Equation (12.50). The point $-1 + 0j$ is indicated with a ×.*

We can see that the Nyquist contour of G_{ol} encircles -1 in the counter-clockwise direction, so the closed-loop system is stable (why?).
What is the gain margin of this system? What happens if we introduce a gain $K = 0.9$?

❋ ....

Answer. Firstly, note that G_{ol} has one unstable pole. Following the definition, $GM = \infty$. If we introduce the gain $K = 0.9$, the closed-loop system becomes unstable.

12.3.2 Sensitivity

Let us start with a motivating example.

> **Example 12.13** (High GM and PM, but the system is not so robust)
> Consider a closed-loop system with open-loop transfer function
>
> $$G_{\text{ol}}(s) = \frac{0.52((s+0.045)^2 + 0.74^2)}{s(s+1)((s+0.04)^2 + 0.7^2)}. \tag{12.51}$$
>
> The Nyquist plot of G_{ol} is shown in Figure 12.29.

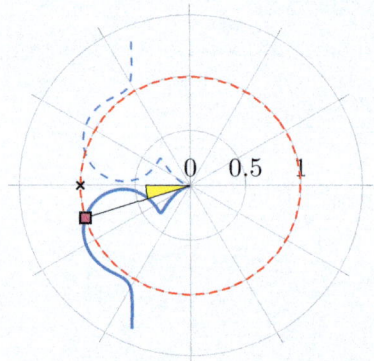

Figure 12.29: *Nyquist plot of G_{ol} in Equation* (12.51). *(Dashed red circle* --- *) circle of radius 1,* ■: *Point corresponding to ω_g where* $|G_{\text{ol}}(j\omega_g)| = 1$, *and* ×: *Point* $(-1, 0)$.

Let us have a look at the Bode plot of G_{ol} in Figure 12.30.

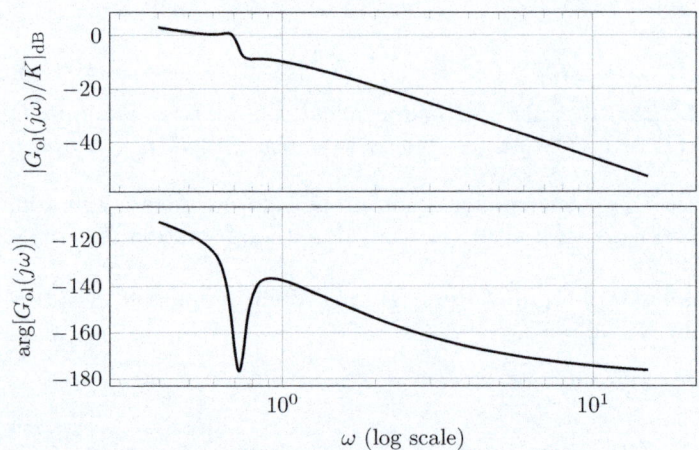

Figure 12.30: *Bode plot of G_{ol} in Equation* (12.51).

From both plots we can see that G_{ol} has an *infinite* gain margin and

Chapter 12. Nyquist plot and stability criterion

a phase margin of $17.4° \approx 0.3\,\text{rad}$. However, we see that the Nyquist plot of G_{ol} is very close to -1. For example, the reader can verify that if we multiply G_{ol} by $20e^{-0.1s}$, that is if we introduce both an additional gain and a phase lag into the loop, the system becomes unstable.

Example 12.13 demonstrates that we should be careful when interpreting what the gain and phase margins mean. Moreover, it became evident that proximity of the Nyquist plot of G_{ol} to -1 can serve as a third "stability margin." The distance between a point of the Nyquist plot, $G_{\text{ol}}(j\omega)$, and $-1 + 0j$ is $|G_{\text{ol}}(j\omega) + 1|$, therefore, the minimum distance is $\inf_{\omega \in \mathbb{R}} |G_{\text{ol}}(j\omega) + 1|$. Because of the symmetry of the Nyquist plot along the real axis, this is equal to $\inf_{\omega \geq 0} |G_{\text{ol}}(j\omega) + 1|$. The smaller this number is, the less robust the closed-loop system is. This leads naturally to the definition of the *sensitivity peak*.

Definition 12.9 (Sensitivity peak) The scalar[a]

$$M_S = \sup_{\omega \geq 0} \frac{1}{|G_{\text{ol}}(j\omega) + 1|} = \sup_{\omega \geq 0} \left|\frac{1}{G_{\text{ol}}(j\omega) + 1}\right|, \qquad (12.52)$$

is called the **sensitivity peak** of the controlled system.

---
[a]We denote the *supremum* by sup. See Section A.9.

Alongside we introduce a special transfer function.

Definition 12.10 (Sensitivity transfer function) The transfer function

$$S(s) = \frac{1}{1 + G_{\text{ol}}(s)}, \qquad (12.53)$$

is defined as the **sensitivity transfer function**

Recommended values for the sensitivity peak are between 1.3 and $2\,\text{dB}$.

Example 12.14 (Sensitivity peak of G_{ol} of previous example) The Bode plot of the sensitivity function of G_{ol} in Equation (12.51) is shown in Figure 12.31.

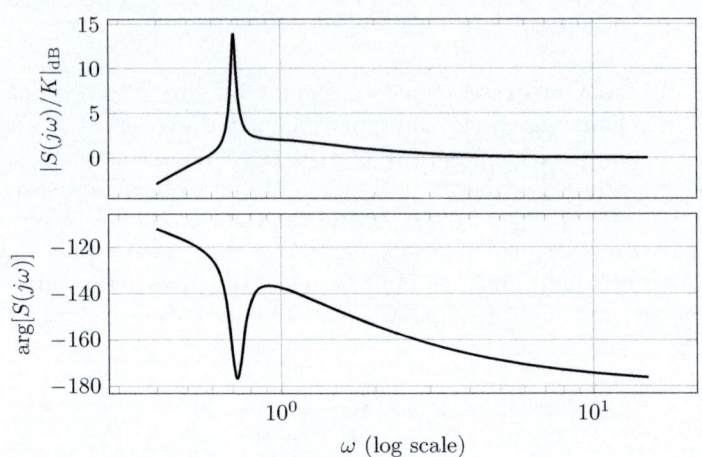

Figure 12.31: *Bode plot of the sensitivity function. The magnitude has a peak of* $13.9\,\text{dB}$ *at* $0.705\,\text{rad/s}$.

The sensitivity peak (in dB) is $13.9\,\text{dB}$, which is an excessively high value.

Exercise 12.7 (Step response of sensitive system) Let G_{ol} be the transfer function in Equation (12.51). Use Python or MATLAB to determine the step response of the closed-loop transfer function

$$G_{\text{cl}}(s) = \frac{G_{\text{ol}}(s)}{1 + G_{\text{ol}}(s)}. \tag{12.54}$$

What do you observe?

✳ ....

Answer. The step response converges to 1, so there is no offset (indeed, note that G_{ol} has a pole at the origin), but the response is very oscillatory.

There is a second interpretation of the sensitivity function, S. Let us have a look at the unit-feedback system of Figure 12.32. This system has two inputs: the set-point, Y^{sp}, and a disturbance, D, which is acting on the output.

Chapter 12. Nyquist plot and stability criterion

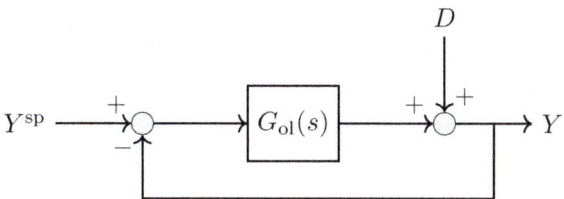

Figure 12.32: Feedback system with unit feedback.

The transfer function from D to Y (assuming $Y^{\text{sp}} = 0$) is exactly the sensitivity function of the closed-loop system, S. A high sensitivity peak means that disturbance signals of certain frequencies will be amplified significantly. For instance, as we discussed in Example 12.14, a disturbance of frequency $0.705\,\text{rad/s}$ (approx. $8.9\,\text{Hz}$) will be amplified by $13.9\,\text{dB}$.

12.4 One-minute round-up

The Bode stability criterion — useful and elegant as it may be — comes with several limitations and shortcoming and so do the definitions of the two stability margins. The **Nyquist stability criterion** is a powerful stability criterion that, unlike the Bode criterion, provides necessary and sufficient stability conditions. In this chapter

1. We learned how to construct the Nyquist plots of simple transfer functions and discussed some of their properties (see for example, Proposition 12.2)

2. We stated Cauchy's principle of argument: a result of paramount importance in complex analysis, and we discussed how we can use it to derive the Nyquist stability criterion

3. We applied the Nyquist stability criterion in a great variety of dynamical systems, including ones with time delays (see for instance, Example 12.7) and systems with imaginary poles

4. We redefined the gain and phase margins in the context of this new stability criterion, and we saw that it is not always a good idea to trust them

5. Lastly, we discussed the concept of sensitivity and defined the sensitivity transfer function and the sensitivity peak.

The Nyquist stability criterion removes the monotonicity and open-loop stability requirements of Bode's stability criterion and offers a necessary and sufficient stability condition.

12.5 Exercises

Exercise 12.8 True of false?

	True	False
The poles of the open-loop transfer function are the zeroes of the characteristic function (assume that $G_m = 1$)	☐	☐
If G_{ol} has a pole at zero, then its Nyquist plot is unbounded (cannot be enclosed in a circle)	☐	☐
If G_{ol} is strictly proper, then $\lim_{R \to \infty} G_{ol}(Re^{3j}) = 0$	☐	☐
If $G_{ol}(s) = e^{-t_d s} P(s)/Q(s)$ with $\deg P = m$, $\deg Q = n$ and $n > m$, then $\lim_{\omega \to \infty} \arg G_{ol}(j\omega) = (m-n)\pi/2$	☐	☐
The Nyquist plot of $G_{ol}(s) = \frac{5s}{10s+1}$ approaches 5 as s escapes to $+j\infty$ moving on the imaginary axis	☐	☐
The Nyquist plot of $G_{ol}(s) = -\frac{1}{s+1}$ runs through -1, so we consider that it does not encircle -1	☐	☐
The Nyquist plot of a first-order system with static gain $K = 1$ encircles $1/2$ in the clock-wise direction	☐	☐
The Nyquist plot of a first-order system is a semi-circle	☐	☐
The Nyquist plot of a first-order system with static gain K is a circle centred at $K/2$ and has radius $K/2$	☐	☐
The Nyquist plot of $G_{ol}(s) = \frac{e^{-t_d s}}{s+1}$ encircles -1 in the clockwise direction for adequately large t_d	☐	☐
If the Nyquist plot of G_{ol} runs through -1, G_{cl} is unstable	☐	☐
If G_{ol} has no unstable poles and its Nyquist plot encircles -1 in the clockwise direction, then G_{cl} is unstable	☐	☐

※

Answer: F, T, T, F, T, T, F, F, T, T, T, T (see Proposition 12.6), T, T, F, T.

Exercise 12.9 (Same Nyquist plot) Find two rational transfer functions without delay and with a first-order denominator that have the same Nyquist plot.

✻ ....

Answer. Take for example $G_1(s) = 1/s+1$ and $G_2(s) = s/s+1$. Note that the Bode plots of these functions are different.

Exercise 12.10 (Nyquist plot) Construct the Nyquist plot of $G(s) = 1/s^3$ (exclude the pole at the origin). Show that the plot encompasses the whole complex plane 1.5 times. Then show that the Nyquist plot of $G(s) = 1/s^n$, for some $n \in \mathbb{N}$, encompasses the whole complex plane $n/2$ times.

Exercise 12.11 (Complete the proof) Complete the proof of Proposition 12.2.

Exercise 12.12 (Tangential lines of Nyquist plot of FOS) For a first-order system with static gain K and time constant τ verify that the following equation is correct

$$G(j\omega) = \frac{K}{1+\tau^2\omega^2} + j\frac{-K\tau\omega}{1+\tau^2\omega^2}. \tag{12.55}$$

Determine the tangential lines for every $\omega > 0$. An example is shown in the following figure for a first-order system with $K = 1$ and $\tau = 1$; the tangential line corresponds to the frequency $\omega = 0.5$.

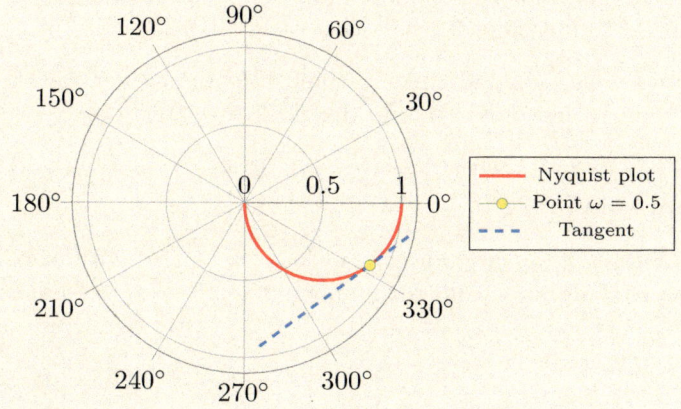

What happens as $\omega \to 0^+$ and as $\omega \to \infty$?

✂ ··

Hint. Recall that when a curve on the x-y plane is given in a parametric form $x = f(\tau)$ and $y = g(\tau)$, where f and g are smooth functions, the tangent line that corresponds to the point $\tau = \tau_0$ is given by $y - y(\tau_0) = \frac{dy}{dx}\big|_{\tau_0} \cdot (x - x(\tau_0))$, where $\frac{dy}{dx}\big|_{\tau_0} = \frac{g'(\tau_0)}{f'(\tau_0)}$, provided that $f'(\tau_0) \neq 0$. If $f'(\tau_0) = 0$, then, at τ_0 there is a vertical asymptote.

✂ ··

Hint. There are two ways to determine the tangential lines of this Nyquist plot. The **geometric** approach relies on the fact that this particular Nyquist plot is a semicircle of radius $K/2$ centred at $K/2$ (on the real axis). The tangents can be determined via simple geometric arguments. On the other hand, the **analytic** approach is based on the fact that the Nyquist plot is a parametric curve (reG(jω), imG(jω)), for $\omega > 0$, on the complex plane.

Exercise 12.13 (Nyquist plots) For a second-order system with static gain K, damping factor ζ and time constant τ verify that the following equation is correct

$$G(j\omega) = \frac{K(1 - \tau^2\omega^2)}{(1 - \tau^2\omega^2)^2 + (2\zeta\tau\omega)^2} + j\frac{-2K\zeta\tau\omega}{(1 - \tau^2\omega^2)^2 + (2\zeta\tau\omega)^2}. \quad (12.56)$$

Determine the tangential lines of the Nyquist plot of this system for every $\omega > 0$. What happens as $\omega \to 0^+$ and as $\omega \to \infty$?

Suppose that $K = 1$ and $\zeta = 0.1$ and $\tau = 1$. Write a Python or MATLAB script to plot the Nyquist plot

Exercise 12.14 (Cauchy's principle of argument) Use Cauchy's principle of argument to determine the number of encirclements of the origin by $F(s)$ for $s \in \Gamma$ in each of the following cases

1. Function F is as in Example 12.2 and Γ is a circle centred at the origin with radius $r = 4$, in the clockwise direction

2. $F = \frac{s^2-1}{(s^2+4)(s+1/2)^2}$ with $\Gamma = \Gamma_R$ with $R = 1.5$, where Γ_R is defined as in (12.24)

3. $F = \frac{1}{(s-0.5)(s+0.8)(s+1.4)}$ and Γ is a clockwise contour of square shape with edges the complex numbers $1 + j$, $1 - j$, $-1 - j$ and $-1 + j$.

Chapter 12. Nyquist plot and stability criterion

Exercise 12.15 (**Log-polar plots**) Plot the Nyquist plot of a system with open-loop transfer function

$$G(s) = \frac{K}{(s+0.1)^2(s+20)}, \qquad (12.57)$$

for $K = 1$ (You can use Python or MATLAB to do so). Can you see that the Nyquist plot intersects the real axis? Plot the Nyquist plot in log-polar coordinates. What do you observe?

Using the Nyquist stability criterion, can you tell whether the closed-loop system is BIBO stable for $K = 1$? What is the maximum value of K so that the system is stable? What is the phase margin of the system?

Exercise 12.16 (**Nyquist plots and number of unstable poles**) Use the Nyquist stability criterion to show that the closed-loop system with open-loop transfer function

$$G_{ol}(s) = \frac{10(s+4)}{s^2(s+3)}, \qquad (12.58)$$

is unstable. How many poles does G_{cl} have in the right half plane?

Exercise 12.17 (**Nyquist plots in log-polar coordinates**) Construct the Nyquist plots of the following open-loop transfer functions and tell how many unstable poles the corresponding closed-loop functions have.

1. $G(s) = \dfrac{e^{-0.1s}}{7s+1}$

2. $G(s) = \dfrac{e^{-0.05s}}{(s+1)(s+3)}$

3. $G(s) = \dfrac{1}{s^2+4s+6}$

4. $G(s) = \dfrac{e^{-0.01s}}{s^2+0.1s+1}$

5. $G(s) = \dfrac{s-3}{(s+3)^2}$

6. $G(s) = \dfrac{1}{(s+1)(3s+1)}$

7. $G(s) = \dfrac{4s+1}{s^2(s+1)(2s+1)}$

In some cases, sketching the Bode plot first will be helpful.

Exercise 12.18 (Nyquist plots) Construct the Nyquist plots of the following open-loop transfer functions in log-polar coordinates

1. $G(s) = \dfrac{1}{s(s+1)}$

2. $G(s) = \dfrac{s\left(1+\frac{1}{10s}\right)}{(s+1)(s+2)}$

3. $G(s) = \dfrac{1}{s^2+a^2}$, $a \in \mathbb{R}$

4. $G(s) = \dfrac{1}{s(s^2+2)}$

5. $G(s) = \dfrac{1}{s^2(s^2+1)}$

6. $G(s) = \dfrac{1}{s(s^2+1)^2}$

7. $G(s) = \dfrac{s-1}{s((s-1)^2+1)}$

Exercise 12.19 (PD controller design) A controlled system has the transfer function

$$G_s(s) = \dfrac{1}{(s-0.1)^2+10}. \tag{12.59}$$

Suppose that it is controlled with a controller with transfer function G_c and the actuator and sensor in the closed loop have transfer functions

$$G_a(s) = \dfrac{1}{0.4s+1}, \tag{12.60}$$

and $G_m(s) = 1$ respectively.

1. Plot the Nyquist plot of the open-loop transfer function of the system for $G_c(s) = 1$. Is the closed-loop system BIBO-stable? Show that the system cannot be stabilised with a P-controller, $G_c(s) = K_c$, for any $K_c > 0$.

2. Suppose that the system is controlled by a PD controller with gain $K_c = 1$ and differential time τ_D. Plot the Nyquist plot of the open-loop transfer function for different values of τ_D. Find a value of τ_D that renders the closed-loop system BIBO-stable.

Exercise 12.20 (Match the TFs with the Nyquist plots) Match the transfer functions

$$G_1(s) = \dfrac{1}{s+1}, G_2(s) = \dfrac{1}{(s+1)^2}, G_3(s) = \dfrac{1}{(s+1)^3},$$

Chapter 12. Nyquist plot and stability criterion

$$G_4(s) = \frac{1}{s(s+1)}, G_5(s) = \frac{\exp(-0.5s)}{s+1}, G_6(s) = \frac{1+s}{s^2+0.5s+1},$$

to the following Nyquist plots:

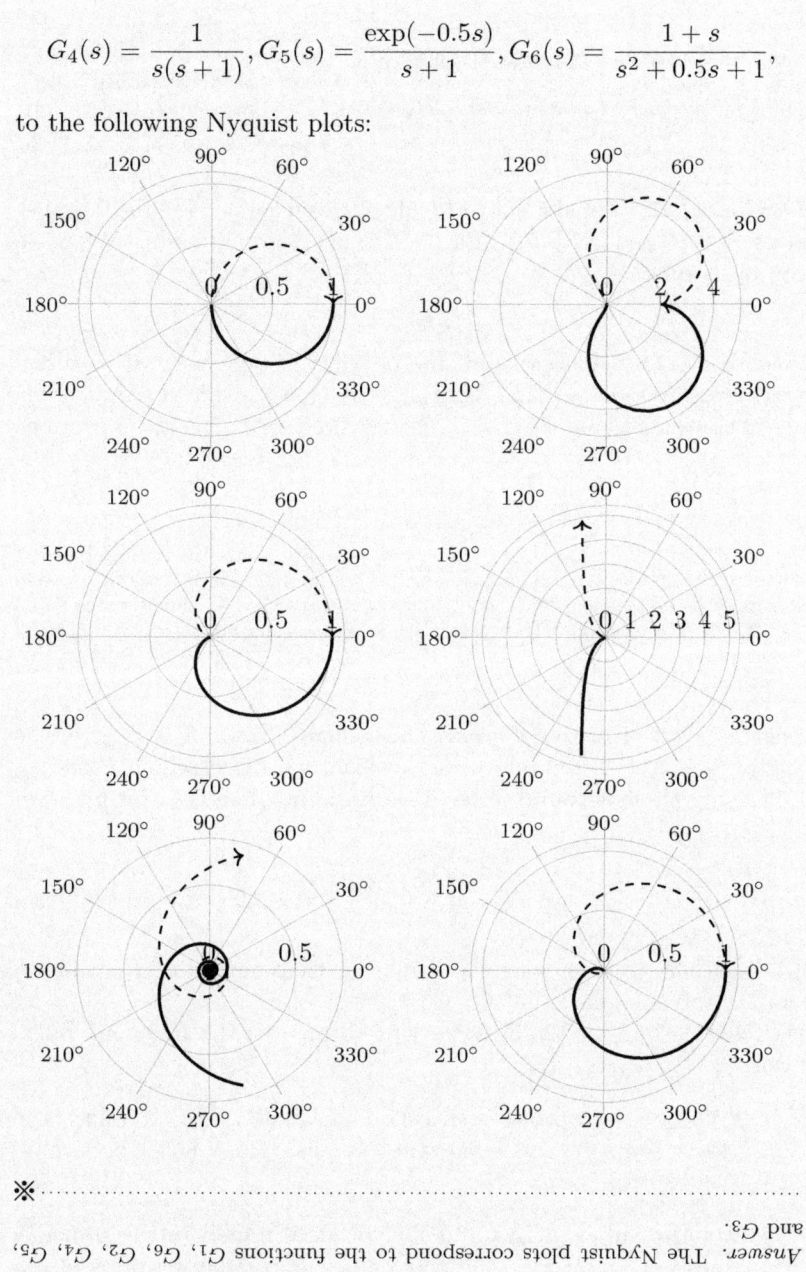

Answer. The Nyquist plots correspond to the functions G_1, G_6, G_2, G_4, G_5, and G_3.

Exercise 12.21 (Example 11.18 revisited) In Example 11.18 we were not able to use Bode's stability criterion to tell whether the closed-loop system is stable. Can you use the Nyquist criterion?

※ ....

Answer. We can use Python or MATLAB to plot the Nyquist plot of the open-loop transfer function. We can see that the Nyquist plot encircles the point -1 four times in the clockwise direction. As a result the closed-loop system is unstable.

Exercise 12.22 (Nyquist plots of Padé approximants) Plot the Nyquist plots of e^{-s} and its [1/1] and [2/2] Padé approximants. What do you observe?

Exercise 12.23 (Application of the principle of argument) How many roots does $F(s) = \exp((1 + 0.1j)s) - 3s^3$ have inside the unit circle of the complex plane. You may use Python or MATLAB to produce a plot.

※ ....

Answer. Take the contour $\Gamma = \{s = e^{jt}; t \in [0, 2\pi]\}$, which is a positively oriented circle. Then, $F(\Gamma)$ encircles the origin 3 times in the clockwise direction, it does not intersect the origin, and F is holomorphic (and has no poles), therefore, F has 3 roots inside the unit circle.

Exercise 12.24 (Control of inverted pendulum) We want to control the angle, θ, inverted pendulum of the form we discussed in Example 2.11 using the horizontal force F as an input. The transfer function of the (linearised) system is

$$G(s) = -\frac{6.25}{s^2 - 25.8}. \qquad (12.61)$$

To that end we use a PD controller with parameters K_c and τ_D respectively. Assume that $G_m(s) = 1$.
Write a Python or MATLAB script to answer the following questions:

1. Choose appropriate controller parameters to stabilise the closed-loop system using the Nyquist plot of the open-loop transfer function.

2. For the values of K_c and τ_D you determined in the previous question, determine the gain, phase and delay margins of the closed-loop system and

3. plot the response of the closed-loop system to a step change of the set point.

4. Lastly, determine the gain margin of the controlled system.

Chapter 12. Nyquist plot and stability criterion 551

✂ ....

Hint. Here K_c should be taken to be negative. Try, for example, $K_c = -5$ and $\tau_D = 0.1$. If you use the margin function of MATLAB you will get wrong results; if K_p and τ_D are such that the closed-loop system is stable, then the gain margin (see Definition 12.7) is infinite.

Exercise 12.25 (Fractional-order systems) We know that $s^n X(s)$, with $n \in \mathbb{N}$, is a term associated with the Laplace transform of the n-th order derivative of $x(t)$ (see Theorem 4.21). Based on this we can associate terms of the form $s^a X(s)$, $a \geq 0$, ($a \in \mathbb{R}$) with derivatives of order a — these are known as *fractional-order derivatives*, and you can read more about them in [M20].

Write a program in Python or MATLAB to produce the Bode and Nyquist plots of the following transfer functions

1. $G_1(s) = \frac{1}{s^a+1}$, $a = 0.5, 0.7, 0.9, 1, 1.1$

2. $G_2(s) = \frac{1}{s^{2a}+s^a+1}$, $a = 0.5, 1, 1.5$

3. $G_3(s) = K_c(1 + \frac{1}{\tau_I s^\lambda} + \tau_D s^\mu)$, for various values of $\lambda, \mu > 0$ and fixed $K_c = 1$, $\tau_I = 1$, and $\tau_D = 1$. This is known as a PI$^\lambda$D$^\mu$ controller [C27].

Exercise 12.26 (Modified Nyquist criterion) In Section 9.3.5 we used Routh's stability criterion to tell whether all the poles of a given transfer function are in the region D_θ shown below.

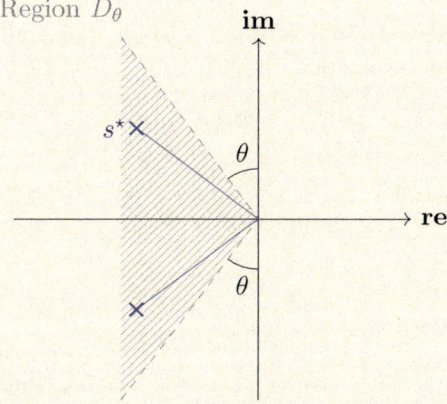

However, this approach can only be applied to cases where the transfer function is rational (it cannot be applied to cases where there are delays).

Propose a modification of the Nyquist stability criterion that can be used to tell whether all the poles of a given transfer function are in D_θ for a given θ. Then write a Python or MATLAB script to implement this.

To the best of the author's knowledge, this has not been done before, so the first person to implement this, can release a useful software.

Exercise 12.27 (Modified Nyquist criterion) In Section 9.3.4 we used Routh's stability criterion to tell whether all the poles of a given transfer function have a sufficiently negative real part ($\mathbf{re}(s_i) < -c$). Modify the Nyquist criterion (by choosing a different contour) to tell whether all poles of the closed-loop system have a real part less than $-c$. Take care of possible poles of G_{ol} with a real part equal to $-c$.

Exercise 12.28 (Control of n-th order integrator) What is the largest integer n such that the system with transfer function

$$G(s) = \frac{1}{s^n}, \qquad (12.62)$$

can be controlled with a PID controller (with some parameters K_c, τ_I and τ_D)?

※ ⋯⋯

Answer. It is $n = 2$.

Exercise 12.29 (Beyond PID control) In Exercise 12.28 we saw that it is not possible to control a third-order integrator, $G(s) = 1/s^3$, with PID controller. Can you propose an alternative controller that can stabilise this system?

Then, tune the proposed controller in such a way so that the step response of the closed-loop system exhibits an overshoot no larger than 25%.

✂ ⋯⋯

Hint. Try $G_c(s) = K_c \left(1 + \tau_D s + \tau_a s^2\right)$ (we do not need an integrator in the controller).

Mathematical background

This appendix gives an overview of some key definitions and results that are used throughout this book. Special emphasis is given on complex numbers and functions and polynomials, which play an important role in classical control theory.

Throughout this book, we denote by $\mathbb{N}$, $\mathbb{Z}$, $\mathbb{R}$, and $\mathbb{C}$ the sets of natural, integer, real and complex numbers respectively. The set of real vectors of dimension n is denoted by $\mathbb{R}^n$. The symbol $:=$ means "is defined as." The symbol $\approx$ means "is approximately equal to," and the symbol $\ll$ means "is significantly less than."

A.1 Complex numbers

A.1.1 Definitions

A **complex number** is a number of the form $z = a + jb$, where $a, b \in \mathbb{R}$ and j solves $x^2 = -1$ and is called an **imaginary unit**. Number a is called the **real part** of z and is denoted by $\mathbf{re}(z)$. Number b is called the **imaginary part** of z and is denoted by $\mathbf{im}(z)$. We say that two complex numbers, $z = a + jb$ and $w = c + jd$ are **equal** if their real and imaginary parts are equal, that is, if $a = b$ and $c = d$. The **conjugate** of a complex number $z = a + bj$ is the complex number $\bar{z} = a - bj$.

A.1.2 Basic operations with complex numbers

Complex numbers can be added, subtracted, multiplied and divided similarly to real numbers. In particular, let $z = a + jb$ and $w = c + jd$ be two complex numbers. Then,

1. **Addition.** $z + w = (a + jb) + (c + jd) = (a + c) + j(b + d)$

2. **Subtraction.** $z - w = (a + jb) - (c + jd) = (a - c) + j(b - d)$

3. **Multiplication.** $zw = (a+jb)(c+jd) = (ac-bd) + j(bc+ad)$. Note that for any complex number $z = a+bj \in \mathbb{C}$, the product $z\bar{z} = a^2 + b^2$ is always a nonnegative real number.

4. **Division.** Suppose $w \neq 0$. In order to perform the operation z/w we multiply the numerator and the denominator of the fraction by $\bar{w}$. Then,

$$\frac{z}{w} = \frac{z\bar{w}}{w\bar{w}} = \frac{(a+jb)(c-jd)}{c^2+d^2} = \frac{1}{c^2+d^2}\left[(ac+bd)+j(bc-ad)\right].$$

A.1.3 Modulus, argument and exponential form

The **modulus** of a complex number $z = a + jb$ is the nonnegative real number

$$|z| = \sqrt{a^2 + b^2}. \tag{A.1}$$

This is the counterpart of the absolute value. Indeed, if $z \in \mathbb{R}$ (i.e., if $b = 0$), then $|z|$ is the absolute value of z. The modulus has the following properties

1. For $z \in \mathbb{C}$, $|z| \geq 0$ and $|z| = 0$ implies that $z = 0$,

2. For $z, w \in \mathbb{C}$, $|zw| = |z||w|$,

3. For $z, w \in \mathbb{C}$ and $w \neq 0$, $|z/w| = |z|/|w|$.

A property of central importance in complex analysis is Euler's formula:

$$e^{jx} = \cos x + j \sin x. \tag{A.2}$$

The **argument**[1] of a complex number $z = a + jb \neq 0$ is the angle

$$\arg z = \text{atan2}(b, a), \tag{A.3}$$

where atan2 is the two-argument inverse tangent function defined in Section A.4.2. If $z \in \mathbb{R}$, then

$$\arg z = \begin{cases} 0, & \text{if } z > 0, \\ \pi, & \text{if } z < 0 \end{cases} \tag{A.4}$$

The argument of 0 is not defined.

---

[1] To be more precise, this is the **principal value** of the argument of z; the argument of z is multivalued and is equal to $\arg z + 2k\pi$ for $k \in \mathbb{Z}$. Here we use the notation $\arg z$, but in some texts the notation $\text{Arg } z$ is used, while $\arg z$ denotes the (multivalued) argument of z. The principal value of the argument of z, given in Equation (A.3) is a value in $(-\pi, \pi]$.

Appendix A. Mathematical background

A complex number $z = a + jb$ has an equivalent representation which is very useful in practice. This is

$$z = |z|(\cos \phi + j \sin \phi), \qquad (A.5)$$

where $\phi = \arg z$ is the argument of z[2]. This can be written as

$$z = |z|e^{j\phi}, \qquad (A.6)$$

and is known as the **exponential form** or **polar form** of z.

The complex conjugate has the property

$$z\bar{z} = |z|^2, \qquad (A.7)$$

that is, $z\bar{z}$ is a real number. Additionally,

$$|\bar{z}| = |z|, \qquad (A.8a)$$
$$\arg \bar{z} = -\arg z, \qquad (A.8b)$$

and for $z, s \in \mathbb{C}$ and $k \in \mathbb{N}$,

$$\overline{z + s} = \bar{z} + \bar{s}, \qquad (A.9a)$$
$$\overline{z - s} = \bar{z} - \bar{s}, \qquad (A.9b)$$
$$\overline{zs} = \bar{z}\bar{s}, \qquad (A.9c)$$
$$\overline{z/s} = \frac{\bar{z}}{\bar{s}} \qquad (A.9d)$$
$$\overline{z^k} = \bar{z}^k. \qquad (A.9e)$$

Lastly, for every $z \in \mathbb{C}$,

$$z + \bar{z} = 2\mathbf{re}(z), \qquad (A.10a)$$
$$z - \bar{z} = 2j\mathbf{im}(z). \qquad (A.10b)$$

A.1.4 Operations via the exponential form

We may also perform operations between two complex numbers using their exponential form. In particular, for two complex numbers $u = |u|e^{j\alpha}$ and $v = |v|e^{j\beta}$,

- **Multiplication.** $uv = |u||v|e^{j(\alpha+\beta)}$, therefore,

$$|uv| = |u||v|, \qquad (A.11a)$$
$$\arg(uv) = \arg u + \arg v. \qquad (A.11b)$$

---

[2] To be more precise, $\arg z$ is an argument of z which may not necessarily be in $(-\pi, \pi]$.

- **Division.** Suppose $v \neq 0$. Then, $u/v = \frac{|u|}{|v|}e^{j(\alpha-\beta)}$, therefore,

$$\left|\frac{u}{v}\right| = \frac{|u|}{|v|}, \tag{A.12a}$$

$$\arg \frac{u}{v} = \arg u - \arg v. \tag{A.12b}$$

The exponential form of complex numbers allows us to compute powers of complex numbers. This is

$$z^\gamma = (|z|e^{j\phi})^\gamma = |z|^\gamma e^{j\gamma\phi}. \tag{A.13}$$

For example, for a complex number $z = a + jb$ with modulus $|z| = \sqrt{a^2 + b^2}$ and argument $\phi = \operatorname{atan2}(b, a)$, it is $z = |z|e^{jb}$, so

$$z^\gamma = (a^2 + b^2)^{\gamma/2} e^{j\gamma\phi}. \tag{A.14}$$

In particular

$$\sqrt{z} = \sqrt[4]{a^2 + b^2}\, e^{j\frac{\phi}{2}} = \sqrt[4]{a^2 + b^2}\left(\cos\frac{\phi}{2} + j\sin\frac{\phi}{2}\right). \tag{A.15}$$

For example,

$$\sqrt{j} = (e^{j\frac{\pi}{2}})^{1/2} = e^{j\frac{\pi}{4}} = \cos\frac{\pi}{4} + j\sin\frac{\pi}{4} = \frac{\sqrt{2}}{2}(1+j). \tag{A.16}$$

A.1.5 Complex functions

Consider a function of the form $f : \mathbb{C} \to \mathbb{C}$; such functions are often simply referred to as *complex functions*. A complex function f maps a complex number $z = a + jb$ to

$$f(z) = f(a + jb) = u(a, b) + jv(a, b), \tag{A.17}$$

where $u, v : \mathbb{R} \times \mathbb{R} \to \mathbb{R}$ are real functions. In other words, a complex function, can be identified by such a pair of real functions.

Continuity is defined in exactly the same way as for real functions (see Definition A.13)

> **Definition A.1 (Continuous complex function)** A complex function is *continuous* at z_0 if $\lim_{z \to z_0} f(z) = f(z_0)$.[a] Function f is continuous over a subset A of its domain if it is continuous at each $z \in A$.
>
> ---
> [a] Naturally, we say that a sequence of complex numbers $(z_n)_{n \in \mathbb{N}}$ converges to a limit $z^* \in \mathbb{C}$ if $|z_n - z^*| \to 0$ as $n \to \infty$.

Differentiability is also defined similarly

Appendix A. Mathematical background

Definition A.2 (Derivative of complex function) Function f is *differentiable* at z_0 if the limit

$$f'(z_0) = \lim_{z \to z_0} \frac{f(z) - f(z_0)}{z - z_0}, \qquad (A.18)$$

exists. The complex number $f'(z_0)$ is the derivative of f at z_0.

A concept of central importance in complex analysis is that of a *holomorphic* or *analytic* function.

Definition A.3 (Holomorphic function) Consider a function $f : D \to \mathbb{C}$ where $D \subseteq \mathbb{C}$ is a nonempty open set. We say that f is **holomorphic** (also known as **analytic**) on D if it is differentiable at every $z \in D$.

A complex number z such that $f(z) = 0$ is called a **zero** of f. A **pole** of a complex function f is a special form of (benign) singularity around which the function behaves regularly. Let us give the definition of a pole.

Definition A.4 (Pole) Consider a function $f : U \to \mathbb{C}$ defined on a nonempty open set U. We say that $z_0 \in U$ is a pole of f if there is an integer $n \in \mathbb{N}$ so that $(z - z_0)^n f(z)$ is holomorphic on a neighbourhood of z_0. The smallest such integer n is called the order of the pole.

For example, function $f(z) = 1/z$, defined for $z \in \mathbb{C} \setminus \{0\}$ has a pole at $z_0 = 0$ of order 1. Indeed, function $zf(z) = 1$ is holomorphic on $\mathbb{C}$. Similarly, function $f(z) = \frac{z}{(z-2)(z-3)^2}$, defined for $z \in \mathbb{C} \setminus \{2,3\}$, has a pole at $z_0 = 3$. Indeed, function $(z-3)^2 f(z) = \frac{z}{z-2}$ is holomorphic is a neighbourhood of z_0 and the order of this pole is 2. The reader can verify that this function has yet another pole at $z_1 = 2$ of order 1.

Functions with poles are not holomorphic in neighbourhoods of their poles. Functions with no other singularities than a finite set of poles are called *meromorphic*.

Definition A.5 (Meromorphic function) Consider a function $f : U \to \mathbb{C}$ defined on a nonempty open set U with finitely many poles. We say that f is **meromorphic** on U if it is holomorphic on U with the possible exception of finitely many points, which are the poles of f.

Rational and rational-times-exponential functions are typical examples of meromorphic functions we encounter in this book.

The integral of a function $f : [a,b] \to \mathbb{C}$, given by $f(z) = u(z) + jv(z)$ where $u, v : [a,b] \to \mathbb{R}$, is defined by

$$\int_a^b f(z) \mathrm{d}z = \int_a^b u(z) \mathrm{d}z + j \int_a^b v(z) \mathrm{d}z. \qquad (A.19)$$

Complex functions can be integrated along directed smooth curves. A directed smooth curve is defined by a function $\gamma : [a, b] \to \mathbb{C}$ with a continuous derivative with the property $\gamma(t) \neq \gamma(t')$ whenever $t \neq t'$ for all $t, t' \in (a, b)$. As t traverses the interval $[a, b]$ from a to b, $\gamma(t)$ moves on the curve $\Gamma = \{\gamma(t) : t \in [a, b]\} \subseteq \mathbb{C}$ and defines its *direction*.

Consider a complex function $f : U \to \mathbb{C}$, defined on a nonempty set $U \subseteq \mathbb{C}$ which contains a directed smooth curve Γ. The integral of f over Γ is defined as

$$\int_\Gamma f(z) dz = \int_a^b f(\gamma(\tau)) \gamma'(\tau) d\tau, \tag{A.20}$$

and the right hand side integral is of the form we gave in Equation (A.19).

A contour is a type of piecewise-smooth closed curve on the complex plane. Let us give the definition.

Definition A.6 (Contour) Consider a function $\gamma : [a, b] \to \mathbb{C}$ such that (i) γ is continuous on $[a, b]$, (ii) γ has a continuous derivative on (a, b), except for a finite number of points, (iii) γ is one-to-one on (a, b), (iv) $\gamma(a) = \gamma(b)$. Then, the set $\Gamma = \{\gamma(t) : t \in [a, b]\}$ is called a **contour**.

A contour can be constructed by joining multiple curves $\Gamma_i = \{\gamma_i(t) : t \in [a_i, b_i]\}$ provided that $\gamma_i(b_i) = \gamma_{i+1}(a_{i+1})$. In that case, we denote $\Gamma = \Gamma_1 + \Gamma_2 + \ldots + \Gamma_n$.

A.2 Real logarithm

The logarithm of base $b > 0$, denoted by $\log_b : (0, \infty) \to \mathbb{R}$, is a function defined via the following equivalence

$$\log_b(x) = y \Leftrightarrow b^y = x, \tag{A.21}$$

for $x > 0$. The number b is called the **base** of the logarithm. For example, $\log_{10} 1000 = 3$ (because $10^3 = 1000$), $\log_2 64 = 6$ (because $2^6 = 64$), $\log_3 243 = 5$ (because $3^5 = 243$) and so on. The logarithm of zero is not defined. The logarithm of a positive number is a real number.

We denote the *natural logarithm* by ln and the common logarithm by $\log_{10}$. Recall that the natural logarithm is the logarithm with base e.

We will use the notation log for a wildcard logarithm, which can be replaced by $\log_{10}$, ln or any other logarithm with positive base. We will heavily use the following properties of the logarithm:

$$\log(xy) = \log x + \log y, \tag{A.22a}$$
$$\log(x/y) = \log x - \log y, \tag{A.22b}$$
$$\log(x^y) = y \log x. \tag{A.22c}$$

The following formula allows us to change the base of logarithms

$$\log_b x = \frac{\log_c x}{\log_c b}, \tag{A.23}$$

for $b, c > 0$.

Remark: *The logarithm can be generalised to nonzero complex numbers. For $z \in \mathbb{C}$, $z \neq 0$, we say that $\ln z = w$ if $e^w = z$. If $z = \rho e^{j\theta}$ (for some $\rho, \theta > 0$; cf. Equation (A.6)), then $\ln z = \ln \rho + \ln e^{j\theta} = \ln \rho + j\theta$.* •

A.3 Polynomials

A.3.1 Basic concepts

A **monomial** is a function f of the form $f(t) = at^n$, where $n \in \mathbb{N}$. If $a \neq 0$, we call the integer n the **degree** of the monomial. A polynomial is a finite sum of monomials, that is, a function of the form

$$p(t) = a_0 + a_1 t + \ldots + a_n t^n, \tag{A.24}$$

where $a_0, \ldots, a_n$ are real or complex numbers known as the **coefficients** of p. The **degree** of a polynomial is the highest degree of its monomials. The degree of a polynomial p is denoted by $\deg p$. The polynomial $p(t) = 0$ does not have a degree. Constant polynomials, $p(t) = p_0$ with $p_0 \neq 0$, have degree zero. Polynomials of degree 1 have the form $p(t) = a_0 + a_1 t$.

We say that two non-zero polynomials

$$p(t) = a_0 + a_1 t + \ldots + a_n t^n, \tag{A.25a}$$
$$q(t) = b_0 + b_1 t + \ldots + b_n t^n, \tag{A.25b}$$

are **equal** if (i) they have the same degree, (ii) the corresponding coefficients are equal to each other, i.e., $a_0 = b_0$, $a_1 = b_1$, ..., $a_n = b_n$.

A complex number $t^* \in \mathbb{C}$ is called a **root** of a polynomial p if $p(t^*) = 0$. We say that a root has **multiplicity** κ if p can be written as

$$p(t) = (t - t^*)^\kappa q(t),$$

where q is a polynomial of degree $n - \kappa$ and $q(t^*) \neq 0$. A root is called **simple** if its multiplicity is equal to 1. The **fundamental theorem of algebra** states that

> **Theorem A.7 (Fundamental theorem of algebra)** *A polynomial of degree $n \geq 1$ with real (or complex) coefficients has exactly n (complex) roots (counting multiplicities).*

Another notable result is the complex conjugate root theorem:

Theorem A.8 (Complex conjugate root theorem) If $z = a + bj$ is a (complex) root of a polynomial p with real coefficients, then the complex conjugate of z, $\bar{z} = a - bj$ is also a root of p.

Exercise A.1 (Roots of polynomials) Find the roots of the following polynomials and their multiplicity: (i) $p_1(x) = (x-1)^3(x+1)^2(2x+5)$ (ii) $p_2(x) = x(3x+5)^2$ (iii) $p_3(x) = (x^2 + 2x + 1)(x^2 - 1)$.

Exercise A.2 (Equality of polynomials) Let $p(x) = \alpha x^2 + \alpha\beta x + 1$ and $h(x) = (x+1)^2$ be two polynomials. If p and h are equal, find α and β.

✂ ....

Hint: Expand h and use the definition of equality of polynomials.

A useful property is that if p is a polynomial with real coefficients (that is, $a_0, \ldots, a_n \in \mathbb{R}$) then
$$p(-jb) = \overline{p(jb)}, \qquad (A.26)$$
for all $b \in \mathbb{R}$. The notation $\overline{p(jb)}$ refers to the complex conjugate of $p(jb)$ (see Section A.1.3). More generally, we can use Equation (A.9e) to prove that
$$\overline{p(z)} = p(\bar{z}), \qquad (A.27)$$
for all $z \in \mathbb{C}$.

A.3.2 Horner's scheme

For every two polynomials, P and $S \neq 0$ (i.e., S is not the zero polynomial), with $\deg S < \deg P$, there are polynomials Q and R, with either $\deg R < \deg Q$, or $R = 0$, so that
$$P = QS + R. \qquad (A.28)$$
There is unique pair of Q and R that satisfies this property. The process of determining Q and R given P and S is called **Euclidean division of polynomials**. We call Q and R the **quotient** and **remainder** of the division, respectively. If $R = 0$, i.e., if $P = QS$, then we say that S **divides** P.

Horner's scheme is a Euclidean division method that allows us to divide and polynomial P (with $\deg P \geq 1$) with a monomial, $x - \rho$. We will explain how this algorithm works with an example:

Suppose we want to divide $P(x) = 3x^5 + 3x^4 - 2x^2 + 1$ by $S(x) = x - 2$. We first write the coefficients of P in a matrix followed by ρ as shown in the following matrix

Appendix A. Mathematical background

x^5	x^4	x^3	x^2	x^1	x^0	
3	3	0	-2	0	1	2

The first step is to copy the leading coefficient down to the third row as shown below:

3	3	0	-2	0	1	2
↓						
3						

Next, we multiply this number by the root of the monomial and write the result north east of it, as shown below

3	3	0	-2	0	1	2
	6=3·2					
3						

We now add the numbers in the second column and write the result in the last row

3	3	0	-2	0	1	2
	6					
3	9=3+6					

and we repeat the same procedure

3	3	0	-2	0	1	2
	6	18	36	68	136	
3	9	18	34	68	137	

The last number is the residual of the division of P with $x - 2$; the rest are the coefficients of the quotient, that is,

$$P(x) = (3x^4 + 9x^3 + 18x^2 + 34x + 68)(x - 2) + 137. \qquad (A.29)$$

Additionally, the remainder of the division of P by $x - 2$ is equal to $P(2)$. Horner's scheme involves fewer operations for the computation of the value of P at a point x_0 than substituting that value in P directly. Horner's scheme can be used to check whether a given number is a root of a polynomial.

Example A.1 (Horner's scheme) We want to find the roots of $P(x) = x^3 - 6x^2 + 11x - 6$. P is a **monic polynomial**, i.e., the coefficient of its higher-order term is 1. Therefore, if it has integer roots they must be any of the divisors of its constant term, that is $\pm 1, \pm 2, \pm 3$ or ± 6. Let us check whether $a = 1$ is a root of P using Horner's scheme

1	-6	11	-6	1
	1	-5	6	
1	-5	6	0	

Indeed, the remainder of the division of P by $x - 1$ is zero, therefore, $a = 1$ is a root of P and

$$P(x) = (x - 1)(x^2 - 5x + 6). \tag{A.30}$$

We can easily find the roots of the binomial $S(x) = x^2 - 5x + 6$ using its discriminant. However, let us do this using Horner's scheme. Again, if Q has integer roots, these are among the divisors of its constant term. Let us check whether $b = 2$ is a root of Q

1	-5	6	2
	2	-6	
1	-3	0	

Therefore, $Q(x) = (x-2)(x-3) =$, so $P(x) = (x-1)(x-2)(x-3)$.

Remark: We can use Horner's scheme to check whether a number is a root of a polynomial. Horner's scheme cannot be used to find the roots of any polynomial per se. If P is a monic polynomial, any integer roots it may have, are divisors of its constant term. If P is not monic, then any rational roots, $x^* = a/b$, it may have, are such that a is divisor of the constant term and b is a divisor of its leading coefficient. •

Exercise A.3 (Division with monomial) Divide

$$P(x) = x^5 + 5x^4 + 8x^3 + 7x^2 + 3x - 5, \tag{A.31}$$

by $S(x) = x + 1$.

※ ....

Answer. $P(x) = (x^4 + 4x^3 + 4x^2 + 3x)S(x) - 5$.

Exercise A.4 (Another division with monomial) Divide

$$P(x) = 2x^5 - 5x^3 + 7x^2 - 5, \tag{A.32}$$

Appendix A. Mathematical background

by $S(x) = 2x + 1$.

※

Answer. $P(x) = \left(x^4 - \frac{x}{x^3} - \frac{2}{9x^2} + \frac{4}{27x} - \frac{8}{27 \cdot 4x} + \frac{16}{37}\right) S(x) - \frac{16}{43}$.

Exercise A.5 (Polynomial roots) Find all the roots of

$$P(x) = 18x^4 + 33x^3 + 11x^2 - 8x - 4. \tag{A.33}$$

※

Answer. First observe that $x_1 = -1$ is a root of P, that is, $P(-1) = 0$. Using Horner's method we can find that $P(x) = (18x^3 + 15x^2 - 4x - 4)(x + 1)$. Define $Q(x) = 18x^3 + 15x^2 - 4x - 4$. Then, verify that $\frac{2}{3}$ is a root of Q. By Horner's method once again, $Q(x) = (18x^2 + 24x + \frac{6}{3})(x - \frac{2}{3})$. Lastly, $18x^2 + 24x + 8 = 2(9x^2 + 12x + 4) = 2((3x))^2 + 2 \cdot 2 \cdot 3x + 2^2) = 2(3x + 2)^2$. The four roots of P are:
$\{-1, \frac{2}{3}, -\frac{2}{3}, -\frac{2}{3}\}$.

A.4 Trigonometric identities

A.4.1 Sines, cosines and tangents

Some essential identities involving sums and differences of angles

$$\sin(a+b) = \sin a \cos b + \cos a \sin b, \tag{A.34a}$$
$$\sin(a-b) = \sin a \cos b - \cos a \sin b, \tag{A.34b}$$
$$\cos(a+b) = \cos a \cos b - \sin a \sin b, \tag{A.34c}$$
$$\cos(a-b) = \cos a \cos b + \sin a \sin b, \tag{A.34d}$$
$$\tan(a+b) = \frac{\tan a + \tan b}{1 - \tan a \tan b}, \tag{A.34e}$$
$$\tan(a-b) = \frac{\tan a - \tan b}{1 + \tan a \tan b}. \tag{A.34f}$$

Identities involving double angles:

$$\sin(2a) = 2 \sin a \cos a = \frac{2 \tan a}{1 + \tan^2 a}, \tag{A.35a}$$
$$\cos(2a) = \cos^2 a - \sin^2 a, \tag{A.35b}$$
$$= 2 \cos^2 a - 1, \tag{A.35c}$$
$$= 1 - 2 \sin^2 a, \tag{A.35d}$$
$$= \frac{1 - \tan^2 a}{1 + \tan^2 a}, \tag{A.35e}$$

$$\tan(2a) = \frac{2\tan a}{1 - \tan^2 a}. \tag{A.35f}$$

Identities involving squares:

$$\sin^2 a + \cos^2 a = 1, \tag{A.36a}$$
$$\sin^2 a = \frac{1 - \cos(2a)}{2}, \tag{A.36b}$$
$$\cos^2 a = \frac{1 + \cos(2a)}{2}. \tag{A.36c}$$

The following identity is very useful:

$$a \sin x + b \cos x = c \sin(x + \phi_0), \tag{A.37}$$

where

$$c = \sqrt{a^2 + b^2}, \tag{A.38a}$$
$$\phi_0 = \operatorname{atan2}(b, a), \tag{A.38b}$$

where atan2 is the two-argument inverse tangent function (see Section A.4.2).

The trigonometric functions $\sin x$ and $\cos x$ are **periodic** with a period $T = 2\pi$. This means that $\sin(x + 2\pi) = \sin x$ and $\cos(x + 2\pi) = \cos x$ for all $x \in \mathbb{R}$. The functions $F_1(x) = \sin(\omega x)$ and $F_2(x) = \cos(\omega x)$ are also periodic with period $T = 2\pi/\omega$.

The **frequency** of a periodic signal is defined as the inverse of its period, that if, $f = 1/T$.

A.4.2 Inverse trigonometric functions

The inverse trigonometric functions are inverses of trigonometric functions on a certain domain on which these functions can be inverted. For example, the inverse of the sine is defined for $|\theta| \leq \pi/2$ and satisfies the properties

$$\arcsin(\sin \theta) = \theta, \text{ for } |\theta| \leq \pi/2, \tag{A.39a}$$
$$\sin(\arcsin x) = x, \text{ for } |x| \leq 1. \tag{A.39b}$$

Similarly, the inverse cosine is defined for $0 \leq \theta \leq \pi$ and $|x| \leq 1$ and the inverse tangent is defined for angles $|\theta| < \pi/2$ and $x \in \mathbb{R}$.

Appendix A. Mathematical background

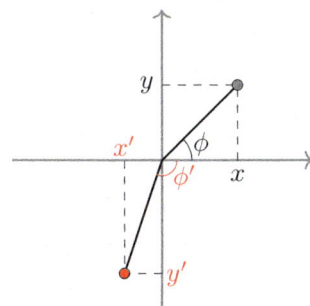

Of particular importance is the **two-argument inverse tangent function**, which is denoted as atan2(y, x). Unlike arctan, the range of atan2 is $(-\pi, \pi]$. Given a two-dimensional vector (x, y), as shown in the figure on the left, can be identified by its norm, $r = \sqrt{x^2 + y^2}$, and its angle, θ, which is given by

$$\text{atan2}(y, x) = \theta.$$

In the open right half-plane (for $x > 0$, or, $|\theta| < \pi/2$), we have

$$\text{atan2}(y, x) = \arctan\left(\frac{y}{x}\right).$$

The two-argument inverse tangent function is defined over all four quadrants as follows

$$\text{atan2}(y, x) = \begin{cases} \arctan y/x, & \text{if } x > 0 \\ \arctan y/x + \pi, & \text{if } x < 0 \text{ and } y \geq 0 \\ \arctan y/x - \pi, & \text{if } x < 0 \text{ and } y < 0 \\ \pi/2, & \text{if } x = 0 \text{ and } y > 0 \\ -\pi/2, & \text{if } x = 0 \text{ and } y < 0 \end{cases} \quad (A.40)$$

It is undefined for $x = y = 0$. It is useful to note that

$$\text{atan2}(cy, cx) = \text{atan2}(y, x), \quad (A.41)$$

whenever $c > 0$. Additionally, it follows from the definition that

$$\text{atan2}(-y, x) = -\text{atan2}(y, x), \quad (A.42)$$

for all $x, y \in \mathbb{R}$ and at least one of them is nonzero. Similarly, we can verify that

$$\text{atan2}(y, -x) + \text{atan2}(y, x) = \pm\pi, \quad (A.43)$$

where the right-hand side is $-\pi$ only when $y < 0$.

Lastly, atan2 has the following property

$$\text{atan2}(x, y) + \text{atan2}(y, x) = \pi/2, \quad (A.44a)$$

if x, y are not both negative and

$$\text{atan2}(x, y) + \text{atan2}(y, x) = \pi/2 - 2\pi, \quad (A.44b)$$

if $x, y < 0$. However, we should note that the angles $\pi/2$ and $\pi/2 - 2\pi$ are equivalent in the sense that they have the same sine and cosine values and correspond to the same points on the unit circle.

A.5 Limits

A.5.1 Limits of sequences and functions

A sequence of real or complex numbers is an ordered collection of infinitely many elements $a = (a_0, a_1, a_2, \ldots)$. Sequences are often denoted by $(a_n)_{n \in \mathbb{N}}$, or, simply, $(a_n)_n$. For example, consider the sequence of real numbers $a = (1, \frac{1}{2}, \frac{1}{3}, \frac{1}{4}, \ldots)$. It is clear that the elements of this sequence become closer and closer to zero; we can say that the sequence **converges** to zero, and we denote $a_n \to 0$ as $n \to \infty$.

A sequence may not converge to a limit. Take for example the sequence $a = (0, 1, 0, 1, \ldots)$. This sequence oscillates between 0 and 1 without converging. Take $b = (0, 1, 2, 3, \ldots)$. The elements in this sequence grow without converging to a particular limit. In both examples, we say that the sequences diverge. In the second case, we say that the sequence diverges to infinity.

Let us give the formal definition of the limit of a converging sequence.

> **Definition A.9 (Convergent real/complex sequence)** We say that a sequence $a_0, a_1, \ldots$ of real or complex numbers converges to a (real or complex, respectively) limit L if for every $\epsilon > 0$ there is an index n_0 so that $|a_n - L| < \epsilon$ for all $n \geq n_0$.
>
> Then, L is called the limit of the sequence. This is denoted by $L = \lim_n a_n$ or "$a_n \to L$ as $n \to \infty$" or, simply, "$a_n \to L$".

> **Definition A.10 (Infinite limit)** We say that a sequence $a_0, a_1, \ldots$ of real numbers diverges to infinity if for every $M > 0$, there is an index n_0 so that $|a_n| \geq M$ for all $n \geq n_0$. This is denoted by $\lim_n a_n = \infty$ or $a_n \to \infty$.
>
> We say that the sequence diverges to $-\infty$ if the sequence $-a_0, -a_1, \ldots$ diverges to infinity.

For sequences of functions convergence can be defined in several ways (known as *modes* of convergence). The two most popular notions of convergence are *pointwise* and *uniform* convergence.

> **Definition A.11 (Pointwise convergence)** Let $(f_n)_n$ be a sequence of functions $f_n : \mathbb{R} \to \mathbb{R}$, or, more generally, $f_n : \mathbb{C} \to \mathbb{C}$. We say that f_n converges pointwise to a function $f : \mathbb{R} \to \mathbb{R}$ (or $f : \mathbb{C} \to \mathbb{C}$ respectively) over a set $A \subseteq \mathbb{R}$ (or $A \subseteq \mathbb{C}$) if
>
> $$\lim_{n \to \infty} f_n(x) = f(x), \qquad (A.45)$$
>
> for all $x \in A$.

For example, consider the sequence of functions $f_n(x) = \frac{1}{n}x$ with $x \in \mathbb{C}$. Then, for every $x \in \mathbb{C}$, $f_n(x) = \frac{1}{n}x$ converges to 0 as $n \to \infty$, therefore, f_n converges to $f(x) = 0$, $x \in \mathbb{C}$, pointwise. However, according to the following definition, $(f_n)_n$ does not converge uniformly.

> **Definition A.12 (Uniform convergence)** Let $(f_n)_n$ be a sequence of functions $f_n : \mathbb{R} \to \mathbb{R}$, or, more generally, $f_n : \mathbb{C} \to \mathbb{C}$. We say that f_n converges uniformly to a function $f : \mathbb{R} \to \mathbb{R}$ (or $f : \mathbb{C} \to \mathbb{C}$ respectively) over a set $A \subseteq \mathbb{R}$ (or $A \subseteq \mathbb{C}$) if for every $\epsilon > 0$ there is an index $n_0 \in \mathbb{N}$, such that
>
> $$|f_n(x) - f(x)| < \epsilon, \tag{A.46}$$
>
> for all $n \geq n_0$ and $x \in A$. We say that f_n converge locally uniformly to f over $A \subseteq \mathbb{R}$, if they converge uniformly over all compact sets $K \subseteq A$.

Uniform convergence is stronger than pointwise. The prime example of a sequence of functions that converges pointwise, but not uniformly is the sequence of $f_n : [0,1] \to [0,1]$ defined by $f_n(t) = t^n$. A useful property of uniform convergence is that it transfers continuity: if $(f_n)_n$ is a sequence of continuous functions which converges uniformly to a function f, then f is continuous.

A.5.2 Limits of rational functions at infinity

Let p and q be two polynomials with degrees n and m respectively. If $n < m$, then

$$\lim_{x \to \pm\infty} \frac{p(x)}{q(x)} = 0. \tag{A.47}$$

To prove this, let us write $p(x) = a_0 + a_1 x + \ldots + a_n x^n$ and $q(x) = b_0 + b_1 x + \ldots + b_n x^n + b_{n+1} x^{n+1} + \ldots + b_m x^m$, so

$$\lim_{x \to \pm\infty} \frac{p(x)}{q(x)} = \lim_{x \to \pm\infty} \frac{a_0 + a_1 x + \ldots + a_n x^n}{b_0 + b_1 x + \ldots + b_n x^n + b_{n+1} x^{n+1} + \ldots + b_m x^m}$$

$$= \lim_{x \to \pm\infty} \frac{\frac{a_0}{x^n} + \frac{a_1}{x^{n-1}} + \ldots + \frac{a_{n-1}}{x} + a_n}{\frac{b_0}{x^n} + \frac{b_1}{x^{n-1}} + \ldots + \frac{b_{n-1}}{x} + b_n + b_{n+1} x + \ldots + b_m x^{m-n}} = 0.$$

Likewise, we can show that if $m > n$, the limit diverges to ∞ or $-\infty$ and if $m = n$, it converges to a_n/b_n.

A.6 Differentiation and integration

In this section we present some key results on the differentiation and integration of real-valued functions. Some of these can be extended to complex functions. For example, juxtapose Definitions A.1 and A.13. Definite integrals of complex functions can be defined via Equation (A.19) via definite integrals of *real* functions. We need to be more careful with differentiation when it comes to complex functions (see Definition A.2), which possess some interesting and exotic properties, but this is beyond the scope of this book.

A.6.1 Derivative and its properties

Let us recall the definition of a continuous function.

> **Definition A.13 (Continuous function)** A function $f : A \to \mathbb{R}$, where $A \subseteq \mathbb{R}$, is a called **continuous at a point** $x_0 \in A$ if the limit $\lim_{x \to x_0} f(x)$ exists and
> $$\lim_{x \to x_0} f(x) = f(x_0). \tag{A.48}$$
> It is called **continuous** on A if it is continuous at all points $x_0 \in A$.

The definition is analogous for functions $f : A \to \mathbb{R}^n$, where $A \subseteq \mathbb{R}^m$. Let us now introduce the following definition of a piecewise continuous function. Note that in different contexts, piecewise continuity may be defined differently. In this book, we use the following definition.

> **Definition A.14 (Piecewise continuous)** A function $f : A \to \mathbb{R}$ is called **piecewise continuous** on $A \subseteq \mathbb{R}$ if A can be partitioned into a finite number of subintervals, $A_1, A_2, \ldots, A_n$, and f is continuous on the interior of each subinterval and has finite limit points at the boundaries of each subinterval.

As an example, consider the function

$$g(x) = \begin{cases} 1, & \text{for } 0 \leq x \leq 1 \\ 2, & \text{for } 1 < 1 \leq 2 \end{cases} \tag{A.49}$$

which is defined on $A = [0, 2]$. We can partition A into $A_1 = [0, 1]$ and $A_2 = [1, 2]$. Then, g is continuous on $(0, 1)$ and on $(1, 2)$ and the limits $\lim_{x \to 0+} g(x)$, $\lim_{x \to 1-} g(x)$, $\lim_{x \to 1+} g(x)$ and $\lim_{x \to 2-} g(x)$ are finite. Therefore, g is piecewise continuous. Lastly, consider the following function

$$f(x) = \frac{1}{x}, \tag{A.50}$$

which is defined on $A = (0, \infty)$; this function is **not** piecewise continuous because $\lim_{x \to 0+} f(x) = \infty$.

Appendix A. Mathematical background

Definition A.15 (Derivative) Let $f : A \to \mathbb{R}$ be a function defined over the open set $A \subseteq \mathbb{R}$. Suppose that f is continuous over A. If the limit

$$f'(x_0) = \lim_{x \to x_0} \frac{f(x) - f(x_0)}{x - x_0}, \tag{A.51}$$

exists, then we say that f is **differentiable** at x_0, and we call $f'(x_0)$ its **derivative** at x_0. We say that f is **differentiable** on A if it is differentiable at all $x_0 \in A$.

There exist rules for the differentiation of common functions. A few are listed below

$$\alpha' = 0, \tag{A.52a}$$
$$x' = 1, \tag{A.52b}$$
$$(\alpha x)' = \alpha, \tag{A.52c}$$
$$(x^2)' = 2x, \tag{A.52d}$$
$$(\sqrt{x})' = \frac{1}{2\sqrt{x}}, x > 0, \tag{A.52e}$$
$$(e^x)' = e^x, \tag{A.52f}$$
$$(\ln x)' = \frac{1}{x}, x > 0, \tag{A.52g}$$
$$(\sin x)' = \cos x, \tag{A.52h}$$
$$(\cos x)' = -\sin x, \tag{A.52i}$$
$$(\arcsin x)' = \frac{1}{\sqrt{1-x^2}}, \tag{A.52j}$$
$$(\arccos x)' = -\frac{1}{\sqrt{1-x^2}}, \tag{A.52k}$$
$$(\arctan x)' = \frac{1}{1+x^2}. \tag{A.52l}$$

Moreover, the derivative of compositions of functions can be computed using the **chain rule**:

$$(f(g(x)))' = f'(g(x))g'(x). \tag{A.53}$$

For example, the function $F(x) = \sin(x^3 + 2x + 5)$ can be seen as the composition of $f(x) = \sin(x)$ with $g(x) = x^3 + 2x + 5$, so the derivative of F is

$$(\sin(x^3 + 2x + 5))' = \cos(x^3 + 2x + 5) \cdot (3x^2 + 2). \tag{A.54}$$

Lastly, for two differentiable functions $f, g : \mathbb{R} \to \mathbb{R}$, the **product rule** holds, that is, for $h(x) = f(x)g(x)$ we have

$$h(x)' = f'(x)g(x) + f(x)g'(x). \tag{A.55}$$

If, additionally, $g(x) \neq 0$, then the **quotient rule** holds, that is, for $h(x) = f(x)/g(x)$ we have

$$h'(x) = \frac{f'(x)g(x) - f(x)g'(x)}{g(x)^2} \tag{A.56}$$

The derivative of a function may not always be continuous. For example, if we define $f(x) = x^2 \sin(1/x)$ for $x \neq 0$ and $f(0) = 0$, then f is differentiable, but its derivative is discontinuous at $x = 0$. We give the following definition

> **Definition A.16 (Continuously differentiable)** Let $f : A \to \mathbb{R}$ be a function defined over the open set $A \subseteq \mathbb{R}$. We say that f is **continuously differentiable at** $x_0 \in A$ if it is differentiable at x_0 and its derivative, f', is continuous at x_0. We say that f is **continuously differentiable** on A if it is continuously differentiable at each $x_0 \in A$.

When we need to clarify that a derivative is taken with respect to a particular variable we write $\frac{\mathrm{d}f(x)}{\mathrm{d}x}$, that is, $f'(x) = \frac{\mathrm{d}f(x)}{\mathrm{d}x}$. This is useful in presence of additional parameters; for example if $f(x) = yx^2$, we prefer to write $\frac{\mathrm{d}f(x)}{\mathrm{d}x} = 2yx$ to avoid confusion with y. If f is a function of several variables, $f(x_1, x_2, \ldots, x_n)$ then we use the notation $\frac{\partial f}{\partial x_i}(x_1, x_2, \ldots, x_n)$ to denote the derivative with respect to a particular variable. For example, take

$$f(a, b, c) = ab^2 + ce^{a-2b}.$$

The notation f' would be confusing. We write

$$\frac{\partial f(a, b, c)}{\partial a} = b^2 + ce^{a-2b},$$

and

$$\frac{\partial f(a, b, c)}{\partial b} = 2ab - 2ce^{a-2b},$$

and

$$\frac{\partial f(a, b, c)}{\partial c} = e^{a-2b}.$$

A.6.2 Definite integrals

Conceptually, the integral of a function $f : [a, b] \to \mathbb{R}$ over the interval $[a, b]$ is the *signed* area that is enclosed between the curve of f and the x-axis as shown in Figure A.1.

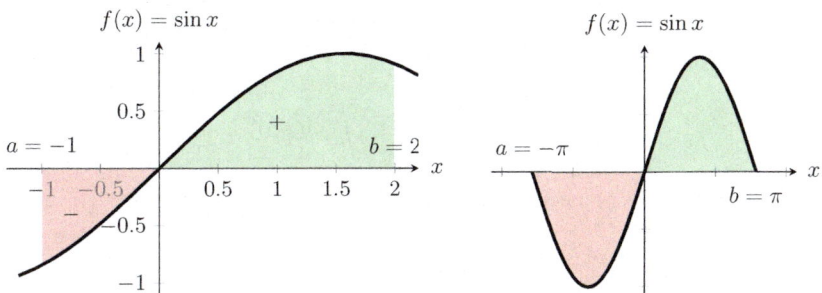

Figure A.1: *(Left) The concept of the integral of a function f over $[a, b]$: The red part has a negative sign because it lies below the x-axis. The green part, instead, has a positive sign. (Right) Demonstration of the fact that the integral of $f(x) = \sin x$ from $a = -\pi$ to $b = \pi$ is 0.*

The integral of f from point $x = a$ to $x = b$ is denoted as

$$\int_a^b f(x)\mathrm{d}x. \qquad (A.57)$$

The "$\mathrm{d}x$" means that the integration is carried out with respect to x. This is important for functions with multiple arguments. For example by writing $\int_a^b f(x, y)\mathrm{d}x$ it is clear that the integration is carried out with respect to the first argument of x.

A few properties of the integral of a function are listed below:

1. (Linearity) For all $c, d \in \mathbb{R}$, and integrable functions f, g on $[a, b]$

$$\int_a^b (cf(\xi) + dg(\xi))\mathrm{d}\xi = c\int_a^b f(\xi)\mathrm{d}\xi + d\int_a^b g(\xi)\mathrm{d}\xi.$$

2. (Splitting) For $a, b, c \in \mathbb{R}$, with $c \in [a, b]$ and an integrable function f on $[a, b]$,

$$\int_a^b f(\xi)\mathrm{d}\xi = \int_a^c f(\xi)\mathrm{d}\xi + \int_c^b f(\xi)\mathrm{d}\xi.$$

3. (Inversion of limits) For $a, b \in \mathbb{R}$ and an integrable function f on $[a, b]$,

$$\int_a^b f(\xi)\mathrm{d}\xi = -\int_b^a f(\xi)\mathrm{d}\xi.$$

4. (Zero) For all $a \in \mathbb{R}$ and a function f

$$\int_a^a f(\xi)\mathrm{d}\xi = 0.$$

5. (Sign preservation) If f is an integrable function on $[a, b]$ with $f(\xi) \geq 0$ for all $\xi \in [a, b]$, then

$$\int_a^b f(\xi)\mathrm{d}\xi \geq 0.$$

6. (Order preservation) If f and g are integrable functions on $[a,b]$ and $f(\xi) \leq g(\xi)$ for all $\xi \in [a,b]$, then

$$\int_a^b f(\xi) \mathrm{d}\xi \leq \int_a^b g(\xi) \mathrm{d}\xi,$$

which can be proven using the sign preservation property on $f - g$.

7. (Absolute value) If f is an integrable function on $[a,b]$ then

$$\left| \int_a^b f(\xi) \mathrm{d}\xi \right| \leq \int_a^b |f(\xi)| \mathrm{d}\xi,$$

which can be proven using the order preservation property and the fact that $-|f(\xi)| \leq f(\xi) \leq |f(\xi)|$.

Lastly, we give **Leibniz's rule** for differentiation under integrals which states that

> **Theorem A.17 (Leibniz's rule)** Suppose that $f(x,t)$ is a real-valued function, differentiable in x, and f and $\mathrm{d}f/\mathrm{d}x$ are continuous in both x and t for all $t \in [a,b]$ and $x \in [c,d]$. Then $\int_a^b f(x,\tau) \mathrm{d}\tau$ is differentiable with respect to x and
>
> $$\frac{\mathrm{d}}{\mathrm{d}x} \int_a^b f(x,\tau) \mathrm{d}\tau = \int_a^b \frac{\partial}{\partial x} f(x,\tau) \mathrm{d}\tau, \qquad (A.58)$$
>
> for all $x \in [c,d]$.

A.6.3 Improper integrals

An **improper integral** is simply the limit of a definite integral as its endpoints approach either a number, ∞ or $-\infty$. We use the following shorthand notation:

$$\int_a^\infty f(x) \mathrm{d}x := \lim_{b \to \infty} \int_a^b f(x) \mathrm{d}x, \qquad (A.59a)$$

$$\int_{-\infty}^b f(x) \mathrm{d}x := \lim_{a \to -\infty} \int_a^b f(x) \mathrm{d}x, \qquad (A.59b)$$

$$\int_{-\infty}^\infty f(x) \mathrm{d}x := \lim_{y \to \infty} \int_{-y}^y f(x) \mathrm{d}x, \qquad (A.59c)$$

$$\int_a^{b+} f(x) \mathrm{d}x := \lim_{y \to b^+} \int_a^y f(x) \mathrm{d}x, \qquad (A.59d)$$

$$\int_a^{b-} f(x) \mathrm{d}x := \lim_{y \to b^-} \int_a^y f(x) \mathrm{d}x, \qquad (A.59e)$$

Appendix A. Mathematical background

assuming the expressions on the right-hand side are well-defined and finite.

There exist two main types of improper integrals. Improper integrals of the **first kind**, which are integrals over an unbounded domain, such as the ones in Equations (A.59a), (A.59b) and (A.59c) where f is assumed to be well defined and finite for all x in within the integration limits. Examples of such improper integrals are

$$\int_0^\infty e^{-x} \mathrm{d}x := \lim_{b \to \infty} \int_0^b e^{-x} \mathrm{d}x$$
$$= \lim_{b \to \infty} \left(-e^{-x}\big|_0^b\right) = \lim_{b \to \infty} \left(-e^{-b} + e^0\right) = 1, \qquad (A.60)$$

where function $f(x) = e^{-x}$ is well defined and finite over all intervals $[0, b]$, for all $b \in \mathbb{R}$. A second example is

$$\int_1^\infty \frac{1}{x^2} \mathrm{d}x = \lim_{b \to \infty} \int_1^b \frac{1}{x^2} \mathrm{d}x = \lim_{b \to \infty} \left. -\frac{1}{x^2} \right|_1^b = 1, \qquad (A.61)$$

where function $g(x) = 1/x^2$ is well defined and finite for all intervals $[1, b]$ for all $b \in \mathbb{R}$.

The second kind of improper integrals has the unsurprising name **improper integral of the second kind**; these are integrals where the integrand, that is function $f(x)$, has a vertical asymptote at either of the endpoints of the integral. For example,

$$\int_{0^+}^1 \ln x \, \mathrm{d}x, \qquad (A.62)$$

is an improper integral because $\ln x$ has a vertical asymptote at zero. In other words $\ln x$ goes to minus infinity as $x \to 0^+$. Let us give the following definitions

> **Definition A.18 (Improper integrals of the first kind)** Suppose that $f : \mathbb{R} \to \mathbb{R}$ does not have any vertical asymptotes and for a given $a \in \mathbb{R}$ the integrals $\int_a^b f(x)\mathrm{d}x$ exist for all $a, b \in \mathbb{R}$. Then, for $a \in \mathbb{R}$, the integrals
>
> $$\int_a^\infty f(x)\mathrm{d}x := \lim_{b \to \infty} \int_a^b f(x)\mathrm{d}x, \qquad (A.63)$$
>
> and
>
> $$\int_{-\infty}^a f(x)\mathrm{d}x := \lim_{b \to \infty} \int_{-b}^a f(x)\mathrm{d}x, \qquad (A.64)$$
>
> are called **improper integrals of the first kind**.

> **Definition A.19 (Improper integrals of the second kind)** Suppose that $f : (a, b] \to \mathbb{R}$ has a vertical asymptote at $x = a$, that is $\lim_{x \to a^+} f(x) =$

∞ or $\lim_{x \to a^+} f(x) = -\infty$. Then, the integral

$$\int_{a^+}^b f(x)\mathrm{d}x := \lim_{y \to a^+} \int_y^b f(x)\mathrm{d}x, \qquad (A.65)$$

is called **improper integrals of the second kind**.

Here is an example of an improper integral of the second kind

Example A.2 (Improper integral of the second kind) We will compute

$$I = \int_{0^+}^1 \ln x \, \mathrm{d}x = \lim_{a \to 0^+} \int_a^1 \ln x \, \mathrm{d}x, \qquad (A.66)$$

we now use the substitution $x = e^u$, therefore, $\mathrm{d}x = e^u \mathrm{d}u$. Additionally, regarding the endpoints of the integral, we have that as $x = e^u \to 0^+$ we have that $u \to -\infty$ and $x = e^u = 1$ corresponds to $u = 0$, so the integral becomes

$$I = \lim_{y \to -\infty} \int_y^0 u e^u \, \mathrm{d}u = \lim_{y \to -\infty} e^u(u-1)\big|_y^0 = -1. \qquad (A.67)$$

A.6.4 Integrals of sequences (and series)

It is important to underline that the linearity and splitting properties of the integral stated above cannot be extended to infinitely many terms without additional conditions. For example, we should **not** assume that

$$\int_a^b \lim_{n \to \infty} f_n(\xi)\mathrm{d}\xi = \lim_{n \to \infty} \int_a^b f_n(\xi)\mathrm{d}\xi, \qquad (A.68)$$

or that

$$\int_a^b \sum_{n=0}^\infty f_n(\xi)\mathrm{d}\xi = \sum_{n=0}^\infty \int_a^b f_n(\xi)\mathrm{d}\xi. \qquad (A.69)$$

Conditions under which we can interchange the integral and the infinite sum are stated in the following theorem which we give without a proof.

Theorem A.20 (Dominated convergence theorem (DCT)[a]) Suppose $f_n : \mathbb{R} \to \mathbb{R}$ is a sequence of functions with $\int_a^b |f_n(\xi)|\mathrm{d}\xi < \infty$ and for each $x \in \mathbb{R}$, $\lim_{n \to \infty} f_n(x) = f(x)$. If there is a function $g : \mathbb{R} \to \mathbb{R}$, with $|f_n(x)| \leq g(x)$ for all $x \in \mathbb{R}$ and $\int_a^b |g(\xi)|\mathrm{d}\xi < \infty$ then

$$\lim_{n \to \infty} \int_a^b f_n(\xi)\mathrm{d}\xi = \int_a^b f(\xi)\mathrm{d}\xi. \qquad (A.70)$$

> [a] For a detailed discussion and proof, see [M8, Theorem 11.32].

We can apply this to series since $\sum_{n=0}^{\infty} f_n(x) = \lim_{N \to \infty} \sum_{n=0}^{N} f_n(x)$. We may then define the sequence of functions $F_N(x) = \sum_{n=0}^{N} f_n(x)$ and apply the dominated convergence theorem. Therefore, if

1. For $x \in \mathbb{R}$, $\lim_{N \to \infty} F_N(x)$ exists and equals $F(x) = \sum_{n=0}^{\infty} f_n(x)$,

2. For each $N \in \mathbb{N}$, $\int_a^b |F_N(\xi)| d\xi < \infty$, and

3. There is a function $g : \mathbb{R} \to \mathbb{R}$ with $\left|\sum_{n=0}^{N} f_n(x)\right| \le g(x)$,

then,

$$\int_a^b \sum_{n=0}^{\infty} f_n(\xi) d\xi = \sum_{n=0}^{\infty} \int_a^b f_n(\xi) d\xi. \tag{A.71}$$

The conditions of the dominated convergence theorem (Theorem A.20) are necessary. Let us give a counterexample where the conditions of DCT are not satisfied and (A.70) fails to hold.

Example A.3 (Counterexample to DCT) Consider the sequence of functions

$$f_n(x) = n^2 x^n (1-x), \tag{A.72}$$

defined over $[0, 1]$. We have that for all $x \in [0, 1]$ it is

$$\lim_{n \to \infty} f_n(x) = 0, \tag{A.73}$$

(pointwise), but

$$\int_0^1 f_n(\xi) d\xi = 1. \tag{A.74}$$

A.6.5 Antiderivative & the fundamental theorem of calculus

We give with the definition of an antiderivative:

> **Definition A.21 (Antiderivative)** An **antiderivative** of a function $f : \mathbb{R} \to \mathbb{R}$ is a differentiable function $F : \mathbb{R} \to \mathbb{R}$ whose derivative is equal to f, that is $F'(x) = f(x)$ for all $x \in \mathbb{R}$.

Every continuous function f has an antiderivative — in fact, infinitely many because if F is an antiderivative of f, then $F + c$ is also an antiderivative for all $c \in \mathbb{R}$. In fact, if F_1 and F_2 are both antiderivatives of f, then there is a constant $c \in \mathbb{R}$ such that $F_2 = F_1 + c$. The **fundamental theorem of calculus** states that

Theorem A.22 (Fundamental theorem of calculus) Let $f : [a, b] \to \mathbb{R}$ be a continuous function on $[a, b]$. We define a function $F : [a, b] \to \mathbb{R}$ as

$$F(x) = \int_a^x f(\xi)\mathrm{d}\xi, \tag{A.75}$$

then F is continuous on $[a, b]$, differentiable on (a, b) and

$$F'(x) = f(x). \tag{A.76}$$

Additionally, if $\bar{F}$ is an antiderivative of f on $[a, b]$ (that is, if $\bar{F}'(x) = f(x)$ for all $x \in (a, b)$), then

$$\int_a^b f(\xi)\mathrm{d}\xi = \bar{F}(b) - \bar{F}(a). \tag{A.77}$$

Proof. (1) We have that $F(x+h) = \int_a^{x+h} f(\xi)\mathrm{d}\xi$, therefore,

$$\frac{F(x+h) - F(x)}{h} = \frac{\int_a^{x+h} f(\xi)\mathrm{d}\xi - \int_a^x f(\xi)\mathrm{d}\xi}{h} = \frac{1}{h}\int_x^{x+h} f(\xi)\mathrm{d}\xi. \tag{A.78}$$

From the mean value theorem of integral calculus[3], we know that there exists a $c \in [x, x+h]$, which depends on x and h, so that $\int_x^{x+h} f(\xi)\mathrm{d}\xi = f(c)h$, therefore,

$$\frac{F(x+h) - F(x)}{h} = f(c), \tag{A.79}$$

and taking the limit as $h \to 0$ on both sides, we obtain

$$F'(x) = \lim_{h \to 0} \frac{F(x+h) - F(x)}{h} = \lim_{h \to 0} f(c). \tag{A.80}$$

Given that $x \leq c \leq x + h$, we have that $c \to x$ as $h \to 0$ (because of the sandwich theorem).

(2) Let $\bar{F}$ be an antiderivative of f. Given that $F(x) = \int_a^x f(\xi)\mathrm{d}\xi$ is also an antiderivative, we have that $\bar{F}' = F'$, so $F = \bar{F} + \sigma$, for some constant $\sigma \in \mathbb{R}$. We have

$$\bar{F}(a) + \sigma = F(a) = \int_a^a f(\xi)\mathrm{d}\xi = 0, \tag{A.81}$$

so $\sigma = -\bar{F}(a)$, that is $F(x) = \bar{F}(x) - \bar{F}(a)$, therefore,

$$\int_a^b f(\xi)\mathrm{d}\xi = F(b) = \bar{F}(b) - \bar{F}(a). \tag{A.82}$$

$\square$

---

[3]The mean value theorem of integral calculus states that if $f : [a, b] \to \mathbb{R}$ is a continuous function, then, there exists $c \in [a, b]$ such that $\int_a^b f(\xi)\mathrm{d}\xi = f(c)(b - a)$.

The fundamental theorem of calculus gives us two important facts: first, it states that functions defined by integrals are continuous and differentiable, and, second, it links the integral with the antiderivative. This is why an antiderivative of f is often denoted by $\int f(x)\mathrm{d}x$.

Since $\int_a^x f(\xi)\mathrm{d}\xi$ is continuous on $[a,b]$, then, by the definition of continuity (at point a), we have that $\lim_{x \to a} \int_a^x f(\xi)\mathrm{d}\xi = \int_a^a f(\xi)\mathrm{d}\xi = 0$.

The **substitution rule** of integration allows us to change the integration variable. In particular, if $u : [a,b] \to \mathbb{R}$ is a differentiable function on (a,b) then

$$\int_{u(a)}^{u(b)} f(y)\mathrm{d}y = \int_a^b f(u(x))u'(x)\mathrm{d}x. \qquad (A.83)$$

A.6.6 L'Hôpital's rule

L'Hôpital's rule allows us to determine certain indeterminate limits of the form $0/0$ and ∞/∞.

Theorem A.23 (L'Hôpital's rule) Let $f, g : (a,b) \to \mathbb{R}$ and let $c \in (a,b)$ where f and g are differentiable on (a,b) with the possible exception of c. Assume that $g'(x) \neq 0$ for all $x \in (a,b) \setminus \{c\}$. Suppose also that the following limit exists and is finite

$$L := \lim_{x \to c} \frac{f'(x)}{g'(x)}. \qquad (A.84)$$

Suppose also that one of the following conditions holds

1. $\lim_{x \to c} f(x) = \lim_{x \to c} g(x) = 0$, or

2. $\lim_{x \to c} |f(x)| = \lim_{x \to c} |g(x)| = \infty$.

Then,

$$\lim_{x \to c} \frac{f(x)}{g(x)} = L. \qquad (A.85)$$

Proof. The proof can be found in [M8, Theorem 5.13]. □

A.7 Asymptotic notation

Asymptotic notation is a mathematical notation used to describe the asymptotic behaviour of functions as their arguments tends towards a certain value or infinity. This notation is used in this book to describe the linearisation error and the rate at which it converges to zero. In particular, we use the Big-O and little-o notation which we define below.

A.7.1 Big-O notation

For two function $f, g : \mathbb{R} \to \mathbb{R}$, where g is a strictly positive function, we use the notation "$f(x) = \mathcal{O}(g(x))$ as $x \to \infty$" to express that the order of magnitude of f is no larger that the order of magnitude of g for large enough x (as $x \to \infty$). Let us give the formal definition:

> **Definition A.24** ($f(x) = \mathcal{O}(g(x))$ as $x \to \infty$) For two functions $f, g : \mathbb{R} \to \mathbb{R}$ we say that[a]
>
> $$f(x) = \mathcal{O}(g(x)), \text{ as } x \to \infty, \tag{A.86}$$
>
> if there is an x_0 and an $M > 0$ so that
>
> $$|f(x)| \leq M|g(x)|, \tag{A.87}$$
>
> for all $x \geq x_0$.
>
> ---
> [a]Some authors use the notation "$f(x) \in \mathcal{O}(g(x))$", instead of $f(x) = \mathcal{O}(g(x))$. This might be a more appropriate notation as it means that f belongs to the family of functions which are of the same order as g as $x \to \infty$. The reader should not assume that $\mathcal{O}(g(x))$ is a function. In the Big-O notation, "=" should not be interpreted as an equality of two functions.

The Big-O notation in Equation (A.86) means that $f(x)$ is of the same or lower magnitude as $g(x)$ for large enough x (as $x \to \infty$) or, equivalently, that the growth of f as $x \to \infty$ is controlled by g. One can easily verify that if $f(x) = \mathcal{O}(g(x))$ as $x \to \infty$ and $g(x) = \mathcal{O}(h(x))$ as $x \to \infty$, then $f(x) = \mathcal{O}(h(x))$. Let us give an example where f and g are polynomial functions.

Example A.4 (Order of magnitude of polynomials) We will show that for a polynomial function, $f(x) = x^4 + x^3 + 5x^2 + 2x + 3$, its order of magnitude for large x (as $x \to \infty$) is determined by its highest-order term. In this case, will show that $f(x) = \mathcal{O}(x^4)$. Indeed, take $x_0 = 1$. Then, for $x > x_0$,

$$\begin{aligned}|f(x)| &= |x^4 + x^3 + 5x^2 + 2x + 3| \\ &\leq |x^4| + |x^3| + 5x^2 + 2|x| + 3 \\ &\leq x^4 + x^4 + 5x^4 + 2x^4 + 3x^4 = 12x^4,\end{aligned}$$

therefore, $f(x) = \mathcal{O}(x^4)$ (according to Definition A.24, with $M = 12$).

We can generalise the result of Example A.4 to show that for any polynomial P of degree n, it is $P(x) = \mathcal{O}(x^n)$ as $x \to \infty$, that is, the order of magnitude of the value $P(x)$ of a polynomial is determined by its highest-

Appendix A. Mathematical background

order term, x^n, for large x (as $x \to \infty$). For example, as $x \to \infty$

$$x^2 - 3x + 1 = \mathcal{O}(x^2),$$
$$x^5 + x^2 - 11 = \mathcal{O}(x^5),$$
$$x^{10} + 555 = \mathcal{O}(x^{10}),$$
$$8x^3 - 3x^2 + 200x + 10^{100} = \mathcal{O}(x^3).$$

Definition A.24 can be naturally modified to describe the asymptotic behaviour of functions as $x \to a$, where $a \in \mathbb{R}$. We give the following definition.

> **Definition A.25** ($f(x) = \mathcal{O}(g(x))$ as $x \to a$) For two functions $f, g : \mathbb{R} \to \mathbb{R}$ and $g(x) > 0$ for adequately large x we say that
>
> $$f(x) = \mathcal{O}(g(x)), \text{ as } x \to a, \quad (A.88)$$
>
> if for every sequence $x_n \to a$ it is
>
> $$f(x_n) = \mathcal{O}(g(x_n)), \text{ as } n \to \infty. \quad (A.89)$$

Example A.5 (Function of order $\mathcal{O}(x)$) Consider the polynomial function

$$f(x) = x + x^2 + x^3. \quad (A.90)$$

Then we can show that $f(x) = \mathcal{O}(x)$ as $x \to 0$. Indeed, consider a sequence $x_n \to 0$ and, without loss of generality, suppose that $|x_n| < 1$ for all $n \in \mathbb{N}$. Then

$$\begin{aligned}|f(x_n)| &= |x_n + x_n^2 + x_n^3| \leq |x_n| + |x_n^2| + |x_n^3| \\ &\leq |x_n| + |x_n| + |x_n| = 3|x_n|,\end{aligned} \quad (A.91)$$

meaning that $f(x_n) = \mathcal{O}(x_n)$ as $n \to \infty$, therefore, $f(x) = \mathcal{O}(x)$ as $x \to 0$.

Example A.6 (Function of order $\mathcal{O}(x^3)$) Consider the function

$$f(x) = 1 + x^3 + \tfrac{1}{x}, \quad (A.92)$$

then we can show that $f(x) = \mathcal{O}(x^3)$ as $x \to \infty$. This means that for large x, the order of magnitude of $f(x)$ is mainly determined by the term x^3. All other terms become insignificant compared to x^3. Indeed, take $x_0 = 1$. Then, for $x > x_0$

$$|f(x)| = \left|1 + x^3 + \tfrac{1}{x}\right| \leq 1 + |x^3| + \left|\tfrac{1}{x}\right| \leq 3|x^3|. \quad (A.93)$$

Likewise, we can show that $f(x) = \mathcal{O}(1/x)$ as $x \to 0$. This means that for x close to 0, the order of magnitude of $f(x)$ is chiefly determined by $1/x$. Indeed, take a sequence $x_n \to 0$ and without loss of generality assume that $|x_n| < 1$ for all $n \in \mathbb{N}$. Then,

$$|f(x_n)| = \left|1 + x_n^3 + \frac{1}{x_n}\right| \le 1 + |x_n^3| + \left|\frac{1}{x_n}\right| \le 3\left|\frac{1}{x_n}\right|, \quad (A.94)$$

which means that $f(x_n) = \mathcal{O}(1/x_n)$ as $n \to \infty$. Equivalently, $f(x) = \mathcal{O}(1/x)$ as $x \to 0$.

A.7.2 Little-o notation

We sometimes need to describe the growth of a function as x approaches (plus or minus) infinity or a finite limit by comparing it with the growth of some other function. Loosely speaking, the notation $f(x) = o(g(x))$ as $x \to \infty$ means that for large enough x, g grows *much faster* compared to f. For example, after we state the formal definition we will see that $x^2 = o(x^3)$ as $x \to \infty$.

> **Definition A.26 (Little-o notation as $x \to \infty$)** Let $f, g : \mathbb{R} \to \mathbb{R}$ be such that the limit $\lim_{x \to \infty} f(x)/g(x)$ exists. We say that $f(x) = o(g(x))$ as $x \to \infty$ if
> $$\lim_{x \to \infty} \frac{f(x)}{g(x)} = 0. \quad (A.95)$$

As an example, we can see that the exponential function $f(x) = e^x$ grows much faster compared to any polynomial $P(x)$. Indeed, we have that $\lim_{x \to \infty} P(x)/e^x = 0$, therefore, $P(x) = o(e^x)$ as $x \to \infty$. Likewise, we may show that $x = o(x^2)$, $x^2 = o(x^3 + x)$, $e^{-x} = o(x)$ and $e^{-x} = o(1)$. The definition of little-o at a finite limit is given below.

> **Definition A.27 (Little-o notation as $x \to a$)** We say that $f(x) = o(g(x))$ as $x \to a$, where $a \in \mathbb{R}$, if for every sequence $x_n \to a$
> $$f(x_n) = o(g(x_n)), \text{ as } n \to \infty. \quad (A.96)$$

A.8 Hyperbolic functions

The following functions are commonly used in practice

$$\sinh x = \frac{e^x - e^{-x}}{2}, \quad (A.97a)$$

$$\cosh x = \frac{e^x + e^{-x}}{2}, \quad (A.97b)$$

$$\tanh x = \frac{\sinh x}{\cosh x} = \frac{e^x - e^{-x}}{e^x + e^{-x}}, \quad (A.97c)$$

and they are respectively called the hyperbolic sine, hyperbolic cosine and hyperbolic tangent. These functions have interesting properties when applied to imaginary numbers:

$$\sinh(jx) = j\sin(x), \tag{A.98a}$$
$$\sin(jx) = j\sinh(x), \tag{A.98b}$$
$$\cosh(jx) = \cos(x), \tag{A.98c}$$
$$\cos(jx) = \cosh(x). \tag{A.98d}$$

Additionally,

$$\sinh(-x) = -\sinh(x), \tag{A.99a}$$
$$\cosh(-x) = \cosh(x). \tag{A.99b}$$

The derivatives of hyperbolic functions are

$$\sinh(x)' = \cosh(x), \tag{A.100a}$$
$$\cosh(x)' = \sinh(x), \tag{A.100b}$$
$$\tanh(x)' = \frac{1}{\cosh^2(x)}. \tag{A.100c}$$

The inverse hyperbolic functions $\sinh^{-1}(x) := \ln(x + \sqrt{x^2 + 1})$, $x \in \mathbb{R}$, and $\tanh^{-1}(x) := \frac{1}{2}\ln(\frac{1+x}{1-x})$, $x \in (-1, 1)$ have the following derivatives

$$(\sinh^{-1}(x))' = \frac{1}{\sqrt{x^2 + 1}}, \text{ for } x \in \mathbb{R}, \tag{A.101a}$$
$$(\tanh^{-1}(x))' = \frac{1}{1 - x^2}, \text{ for } x \in (-1, 1). \tag{A.101b}$$

A.9 Infimum and supremum

Given a set $A \subseteq \mathbb{R}$, we say that $\underline{a}$ is a *lower bound* of A if $\underline{a} \leq a$ for all $a \in A$. For example, $\underline{a} = 1$ is a lower bound of $A = [1, 2]$; $\underline{a} = 0$ is another lower bound. A set may have no lower bounds; for example $A = (-\infty, 0]$. A lower bound a^* is the *infimum* of A if $a^* \geq \underline{a}$ for all lower bounds $\underline{a}$. The infimum of A is denoted by $\inf A$. For example, the infimum of $A = (0, 1)$ is $\inf A = 0$. Note that the infimum of A is not inside A in this example.

If the infimum of a set is inside the set, it is called a *minimum* and is denoted by $\min A$. We then say that the minimum is *attained*. For example $\min[0, 1] = 0$.

Likewise, for a set $A \subseteq \mathbb{R}$, we say that $\overline{a}$ is an *upper bound* of A if $\overline{a} \geq a$ for all $a \in A$. For example, $\overline{a} = 2$ is an upper bound of $A = [1, 2]$; $\overline{a} = 10$ is another upper bound. A set may have no upper bounds; for example $A = [5, \infty)$. An upper bound is the *supremum* of A if $a^* \leq \overline{a}$ for all upper bounds $\overline{a}$. The supremum of A is denoted by $\sup A$. For example, $\sup(0, 1) = 1$. If the supremum of A is an element of A then it is called the *maximum* of A. Note that the set $A = (0,1)$ has a supremum (sup $A = 1$), but it does not have a maximum.

Next, given a function $f : D \to \mathbb{R}$, $(D \subseteq \mathbb{R})$ and a set $A \subseteq D$ on which f is defined, we define the set $f(A) = \{f(x), x \in A\}$. For example, if $f(x) = x^2$ and $A = [-1,1]$, then $f(A) = [0,1]$. Similarly, for $f(x) = e^x$ and $A = \mathbb{R}$, $f(A) = (0, \infty)$. We can now define the infimum/supremum of a function f over a set A as

$$\inf_{x \in A} f(x) := \inf f(A), \tag{A.102}$$

and likewise we define $\sup_{x \in A} f(x)$. If the minimum exists, we define

$$\min_{x \in A} f(x) := \min f(A).$$

The reader can verify that $\inf_{x \in \mathbb{R}} e^x = 0$, but the minimum of e^x over $x \in \mathbb{R}$ is not defined.

A.10 One-minute round-up

In this chapter we learned the following

1. What is a polynomial, when two polynomials are equal, the concept of Euclidean division of polynomials and how to use Horner's scheme to divide a polynomial by a monomial

2. A few trigonometric identities; in particular, identities involving the sum of two angles and the double of an angle; Equation (A.37) is very useful, and we invoke it in Chapter 5

3. The two-argument inverse tangent function which is used in the definition of the argument of a complex number

4. The notions of a continuous, differentiable and continuous differentiable function

5. The concept of the integral and the antiderivative, the fundamental theorem of calculus, the substitution rule of integration and the trick of integration by parts

6. Complex numbers and some basic properties thereof.

Bibliography

This is a collection of bibliographic references and other resources, organised by topic.

B.1 Control

C1 K.J. Åström and P.R. Kumar. Control: a perspective. *Automatica*, 50:3–43, 2014

C2 K.J. Aström and R.M. Murray. *Feedback Systems: An Introduction for Scientists and Engineers*. Princeton University Press, 2008

C3 S. Engelberg. *A Mathematical Introduction to Control Theory*. Imperial College Press, 2005

C4 R.C. Dorf and R.H. Bishop. *Modern Control Systems*. Prentice-Hall, Inc., Upper Saddle River, NJ, USA, 9th edition, 2000

C5 K. Ogata. *Modern Control Engineering*. Prentice Hall PTR, Upper Saddle River, NJ, USA, 4th edition, 2001

C6 H. Ozbay. *Introduction to Feedback Control Theory*. CRC Press, Inc., Boca Raton, FL, USA, 1st edition, 1999

C7 B.C. Kuo and F. Golnaraghi. *Automatic Control Systems*. John Wiley & Sons, Inc., New York, NY, USA, 8th edition, 2002

C8 N.S. Nise. *Control Systems Engineering*. Wiley, 2011

C9 G. Stephanopoulos. *Chemical Process Control: An Introduction to Theory and Practice*. PTR Prentice Hall, 1984

C10 P.N. Paraskevopoulos. *Modern Control Engineering (Automation and Control Engineering)*. Marcel Dekker Inc., New York, 2002

C11 G.F. Franklin, J.D. Powell, and A. Emami-Naeini. *Feedback Control of Dynamical Systems*. Pearson, 7th edition, 2015

C12 Y. Bafava-Toosi. *Introduction to Linear Control Systems*. Academic Press, 2017

C13 A. Lewis. A mathematical approach to classical control: Single-input, single-output, time-invariant, continuous time, finite-dimensional, deterministic, linear systems, 2003. Lecture notes available online at https://bit.ly/3wfa7XB

C14 J. Hahn, T. Edison, and T.F. Edgar. A note on stability analysis using Bode plots. *Chemical Engineering Education*, pages 208–211, 2001: A paper that sheds some light on some common misconceptions about the application of the Bode stability criterion

C15 J. Kong, M. Pfeiffer, G. Schildbach, and F. Borrelli. Kinematic and dynamic vehicle models for autonomous driving control design. In *2015 IEEE Intelligent Vehicles Symposium (IV)*, pages 1094–1099, June 2015

C16 K.J. Aström. Control system design, lecture notes for ME 155A, 2002. https://bit.ly/3TOweuU

C17 P.H. Jonathan and E. Frazzoli. Feedback control systems, lecture notes for MIT undergraduate course 16.30/16.31, 2010. https://bit.ly/3wdQ6B0

C18 K. Iqbal. Introduction to control systems, 2020. Online book available at https://eng.libretexts.org/

C19 K.J. Åström and T. Hägglund. *PID Controllers: Theory, Design, and Tuning*. ISA, 1995

C20 K.J. Åström and T. Hägglund. *Advanced PID Control*. ISA, 2006

C21 K.H. Lundberg, H.R. Miller, and D.L. Trumper. Initial conditions, generalized functions, and the Laplace transform troubles at the origin. *IEEE Control Systems Magazine*, 27(1):22–35, 2007: In this paper, the authors advocate that the Laplace transform of a function f should rather be defined as

$$(\mathscr{L}f)(s) = \int_{0^-}^{\infty} f(\tau)e^{-s\tau}\,\mathrm{d}\tau, \tag{B.1}$$

so that $(\mathscr{L}f')(s) = sF(s) - f(0^-)$ and $f(0^+) = \lim_{s\to\infty} sF(s)$; these are convenient properties for solving initial value problems, especially when f is a generalised function such as the Dirac delta. Of course, f needs to be defined for $t \in \mathbb{R}$ (or at least on $(-c, \infty)$ for some $c > 0$) rather than for $t \geq 0$ for this approach to make sense.

However, in Section 4.1, the integral of the Laplace transform can be improper at 0 (for example $f(t) = \ln t$, $t > 0$ as in Exercise 4.43), and this cannot be remedied by changing the limit to 0^-. Instead, we use Definition 4.2, with 0^+ if necessary, and apply the Laplace transform to *functions* (the Dirac delta is not a function), that is, we always have[1]

$$\int_0^{0^+} f(\tau) e^{-s\tau} \mathrm{d}\tau = 0. \tag{B.3}$$

The case of the Dirac delta is treated separately in Section 4.4.4.

C22 D.E. Davison, J. Chen, ott R. Ploen, and D.S. Bernstein. What is your favorite book on classical control? responses to an informal survey. *IEEE Control Systems Magazine*, 27(3):89–99, 2007: In this paper, several experts from the academia and the industry talk about their favourite textbook on classical control theory

C23 M.-T. Ho, A. Datta, and S.P. Bhattacharyya. An elementary derivation of the Routh-Hurwitz criterion. *IEEE Transactions on Automatic Control*, 43(3):405–409, March 1998

C24 G. Meinsma. Elementary proof of the Routh-Hurwitz test. *Systems & Control Letters*, 25(4):237–242, 1995: This paper gives a very short and elegant proof of the Routh-Hurwitz stability criterion without using Sturm's theorem.

C25 E. Gluskin. Let us teach this generalization of the final-value theorem. *European Journal of Physics*, 24(6):591–597, Sep 2003

C26 P B Guest. *Laplace transform and an introduction to distributions.* Ellis Horwood Series in Mathematics & Its Applications. Ellis Horwood Ltd, Publisher, Harlow, England, October 1991: an advanced text that offers a mathematically rigorous discussion of the Laplace transform using distributions

C27 I. Podlubny. Fractional-order systems and $PI^\lambda D^\mu$ controllers. *IEEE Transactions on Automatic Control*, 44(1):208–214, 1999

C28 N. Minorsky. Directional stability of automatically steered bodies. *Journal of the American Society for Naval Engineers*, 34(2):280–309, March 1922

---

[1]According to [M12, Exercise 7.6] if a function $f : \mathbb{R} \to \mathbb{R}$ is integrable (it does not have to be continuous), then the function $F : \mathbb{R} \to \mathbb{R}$ defined by

$$F(x) = \int_0^x f(\tau) \mathrm{d}\tau, \tag{B.2}$$

is a continuous function. We know that $F(0) = 0$. By the continuity of F, $F(0^+) = F(0) = 0$, and Equation (B.3) follows suit.

B.2 Physics

Φ1 D. Morin. *Introduction to Classical Mechanics: With Problems and Solutions*. Cambridge University Press, first edition edition, 2010: One of the most comprehensive textbooks on classical mechanics with lots of solved problems

Φ2 D. Kleppner and R. Kolenkow. *Introduction to Mechanics*. Cambridge University Press, second edition, 2014: Rigorous and comprehensive presentation of Newtonian mechanics; ideal for high school and undergraduate students. It covers most standard topics in classical mechanics except Lagrangian mechanics

B.3 Mathematics

M1 L.V. Ahlfors. *Complex Analysis*. McGraw-Hill, Inc., New York, third edition edition, 1979: An excellent book to start a journey into complex analysis.

M2 J.B. Conway. *Functions of one complex variable*. Springer-Verlag, New York, first edition edition, 1973: Another exceptional book on complex analysis with focus on complex functions.

M3 J.C.P. Campuzano. *Complex analysis: A Visual and Interactive Introduction*. LibreTexts, 2022. Available online at https://complex-analysis.com/: This is an online book on complex analysis with interactive tools that will boost your understanding of complex functions.

M4 G. Strang. *Calculus Online Textbook*. MIT, Spring 2005. RES.18-001, MIT OpenCourseWare, License: CC BY-NC-SA, Available at https://bit.ly/3PzyXZl: A textbook that offers a comprehensive treatment of introductory calculus. It is freely available online, contains numerous illustrations and covers all an engineering student needs to know about derivatives and integrals.

M5 T. Tao. *Analysis II*. Springer Singapore, 2016: A more advanced and theoretical, yet accessible, textbook on analysis

M6 T. Needham. *Visual Complex Analysis*. Clarendon Press, 1999: A highly accessible text on complex analysis.

M7 L. Perko. *Differential Equations and Dynamical Systems*. Springer New York, 2001: If you would like to learn more about linearisation and the Hartman-Grobman theorem, Perko's book offers an in-depth discussion.

M8 W. Rudin. *Principles of Mathematical Analysis*. International Series in Pure and Applied Mathematics. McGraw-Hill Inc., third edition, 1976: A classic textbook on analysis.

M9 W. Rudin. *Real and Complex Analysis*. McGraw-Hill Inc., third edition, 1987: If you want to understand the Nyquist stability criterion, you first need to understand Cauchy's principle of argument, but first you need an introduction to complex analysis and in particular an understanding of the basic properties of holomorphic functions. Chapter 10 of this book (entitled "Elementary Properties of Holomorphic Functions") is a great place to start.

M10 W.R.L. Page. *Complex Variables and the Laplace Transform for Engineers*. McGraw-Hill, 1961: This book offers a rigorous and detailed presentation of the Laplace transform.

M11 E. Freitag and R. Busam. *Complex analysis*. Springer, 2005

M12 R.F. Bass. *Real Analysis for Graduate Students*. Createspace Ind. Pub., 2013

M13 W. Rudin. *Functional Analysis*. McGraw-Hill Inc., second edition, 1976

M14 S. Lang. *Complex Analysis*. Springer, New York, fourth edition, 2003

M15 C.H. Edwards. *Advanced calculus of several variables*. Avademic Press, New York and London, 1973

M16 A. Cohen. *Numerical methods for Laplace transform inversion*. Springer, 2007

M17 R.V. Churchill. *Operational Mathematics*. McGraw-Hill, 1958

M18 G. Simmons. *Differential Equations with Applications and Historical Notes*. CRC Press, 2016: One of the best introductory books on differential equations with a chapter on existence and uniqueness.

M19 R. Coleman. *Calculus on Normed Vector Spaces*. Springer, 2012: Among other interesting topics, this book offers a rigorous treatment of the concept of differentiability in general normed (Banach and Hilbert) spaces. Chapter 5 is on Taylor's theorem in general normed spaces.

M20 I. Podlubny. *Fractional Differential Equations: An Introduction to Fractional Derivatives, Fractional Differential Equations, to Methods of Their Solution and Some of Their Applications*. ISSN. Elsevier Science, 1998

M21 E. Norbert. Linear operators and the general solution of elementary linear ordinary differential equations. *CODEE Journal*, 11(9), 2012. Available at: http://scholarship.claremont.edu/codee/vol9/iss1/11

M22 H. Cartan, K. Maestro, and J. Moore. *Differential Calculus on Normed Spaces: A Course in Analysis*. CreateSpace Independent Publishing Platform, 2 edition, 2017: Henri Cartan was a founding member of a group of mathematicians who wrote under the collective *nom de plum* "Nicolas Bourbaki." This book was previously published by Kershaw Publishing Company under the title "Differential Calculus", but it was out of print for several years. Henri Cartan offers a rigorous treatment of Taylor's approximation theorem in general normed (Banach) spaces (see Theorem 3.6). Just be warned that the typesetting of the book is horrible (at least its January 2023 version).

M23 Vladimir A. Dobrushkin. *Applied Differential Equations with Boundary Value Problems*. CRC Press, 2017

B.4 Software

S1 M. Pilgrim and S. Willison. *Dive Into Python 3*, volume 2. Springer, 2009: Note that all examples in this book have been tested in Python 3.8 using `matplotlib` 3.5.1, `numpy` 1.22.2, `scipy` 1.8.0, `sympy` 1.9 and `control` 0.9.1 and executed from within a virtual environment using `virtualenv` version 20.4.2.

S2 G. Van Rossum and F.L. Drake. *Python 3 Reference Manual*. CreateSpace, ScottsValley, CA, 2009

S3 MATLAB. *version 9.9.0.1570001 (R2020b) Update 4*. The MathWorks Inc., Natick, Massachusetts, 2020

S4 Trond Andresen (2022). Nyquist plot with logarithmic amplitudes (https://bit.ly/nyqlog), MATLAB Central File Exchange. Retrieved April 24, 2022.

S5 H.-P. Halvorsen. Python for control engineering, 2020. Available online at https://www.halvorsen.blog/documents/programming/python/python.php: A free online book on Python for control applications. The book is accompanied by a series of lectures on YouTube.

S6 D. Xue and Y. Chen. *Modeling, Analysis and Design of Control Systems in MATLAB and Simulink*. World Scientific, 2014

Index

Absolutely integrable, 342
Affine system, 88
Amplitude ratio
    first-order systems, 302
    second-order systems, 330
Analytic function, 557
Angular
    acceleration, 45
    velocity, 45
Antiderivative, 575
Arc length theorem, 54
Argument, 554
    Principal value, 554
Asymptotic notation, 577
atan2, 565

Bernoulli's principle, 66
BIBO instability, 342
BIBO stability, 22
    definition, 342
    instability quick check, 351
    theorem (impulse response), 344
    theorem (poles), 350
    with delays, 353
Big O, 89, 578
Block diagrams, 251
    feedback, 252
    parallel, 252
    series, 252
Bode plot sketch, 454

Bode plots, 444
    delay, 448
    first order, 444
    second order, 448
    simple integrator, 447
Bode stability criterion, 461, 462
Bounded signal, 341

Capacitance, 59
Capacitor, 59
Cauchy's principle of argument, 514
Chain rule, 569
Characteristic equation, 348
Characteristic polynomial, 348
Charge, 59
Classical control theory, 22
Closed-loop transfer function, 391
Complex
    argument, 553
    conjugate, 553
    exponential form, 555
    function, 556
    modulus, 553
    number, 553
    polar form, 555
Complex integration, 557
Continuous, 568
    piecewise, 568
Contour, 558
Control actions, 18

Control Systems, 29
Convergence, 566
    of sequence, 566
        Pointwise, 182, 566
        Uniform, 158, 567
Convolution
    definition, 208
    Laplace transform, 211
    properties, 209
Critically damped, 315

Damping factor, 314
Decay ratio, 324
Delay, 203
Derivative operator, 260
Deviation variables, 96
Differentiable, 568
    continuously, 570
Dirac delta, 168
Disturbance, 19
Dominated convergence, 574
Drag, 38
Dynamical system, 29

Envelopes (upper/lower), 324
Equilibrium point, 93
Euclidean division, 560
Euler's formula (fluid dynamics), 68
Euler's gamma function, 139
Euler-Mascheroni constant, 184
Exponential order, 132

Feedback control, 19
Final value theorem, 164
First-order system, 295
Frame (rotation), 45
Free body diagram, 50
Frequency response, 242, 427
    BIBO-stable, 427
    imaginary poles, 440
    pole at zero, 435
Friction, 38
    kinetic, 38
    static, 38
fsolve, numerical solver, 466

Gain crossover frequency, 475

Gain margin, 473, 537
Gamma function, 139
Gang of Four, 289
Gradient, 99

Heaviside function, 136
High-pass filter, 27, 306
Holomorphic function, 557
Hooke's law, 39
Horner's scheme, 560
Hyperbolic functions, 580

Impedance, 272
Impulse response, 242
Index number, 514
Inductance, 59
Infimum, 581
Initial value problem, 30, 216, 250
Initial value theorem, 161
Input variables, 18, 29
Integral, 570
    improper, 572
    limit of, 574
Integration by parts, 138
Inverse Laplace Transform
    rational functions, 188
    rational times exponential, 203
    rational, equal degrees, 202
    solve ODEs and IVPs, 216
    uniqueness, 186
Inverse Laplace transform, 185
Inverse trigonometric, 564

Kernel (linear operator), 262
Kirchhoff's laws, 59

L'Hôpital's rule, 577
Laplace transform
    definition, 131
    derivative of, 157
    Dirac functional, 169
    exponential function, 137
    Heaviside function, 136
    integral of, 157, 159
    linearity, 141
    multiple branches, 146
    of $f(t)/t$, 159

Index 591

of t^ν, $\nu \in \mathbb{N}$, 138
of t^a, $a > -1$, 139
of $tf(t)$, 157
of derivative, 148
of high order derivative, 149
of integral, 150
periodic, 152
polynomial, 138
properties, 141
shifted/delayed functions, 142
sine and cosine, 139
time scaling, 146
Leibniz rule, 158, 572
Lerch's theorem, 186
Limit, 566
    infinite, 566
Linear differential operator, 261
Linear system, 87
little O, 89, 181
little o, 580
Load transfer function, 391
Log-polar, 532
Lorenz system, 81
Low-pass filter, 27, 305, 306

Mass balance, 64
Meromorphic function, 557
MIMO systems, 23
Modern control theory, 23
Monomial, 559

Newton's laws, 36
Newton-Raphson method, 492
Nonlinear system, 87
Nyquist contour, 505, 522
Nyquist plot, 522
Nyquist stability criterion, 520, 521

ode45, 74
Ohm's law, 59
Open-loop transfer function, 462
Overdamped, 315
Overshoot, 321

Padé approximation, 369
Peak time, 320
Pendulum

inverted, 49, 82, 108, 285, 550
simple, 47, 84, 127, 226
Periodic function, 152
Phase crossover frequency, 463
Phase lag
    first-order systems, 302
    second-order systems, 330
Phase margin, 475, 537
PID controller, 383
Pole, 165, 188, 557
Pole-zero map, 280, 282
Poles, 348
    imaginary, 440
    multiple, complex, 197
    multiple, real, 191
    simple, complex, 192
    simple, real, 188
    zero, 190, 435
Polynomial, 559
    coprime, 188
    degree, 559
    Euclidean division, 560
    monic, 561
    root, 559
    root multiplicity, 559
Product rule, 569
Proper TF, 349
Pulley, 78
Pulse function, 168

Quotient rule, 570

Rational function, 188
Rectangle pulse function, 168
Region D_θ, 367
Resistance, 59
Resonance, 221, 442
    frequency, second-order, 450
Rise time, 323
Routh's stability criterion, 355
Routh's tabulation, 354

Second-order system, 314
Sensitivity, 540
    peak, 541
    transfer function, 541
Sequence, 566

Set point, 18, 271, 383
Settling time, 325
Sine integral, 160
SISO systems, 22
Solution of ODE, 30
Spring-mass system, 39
Stability margins, 473, 537
State space, 30
State variables, 29
Static gain
    First-order system, 294
    Integrator, 294
    Second-order system, 314
Step response, 242
Stiffness, 39
Strictly proper, 162, 349
Substitution rule, 577
Supremum, 541, 581
Symbolic differentiation, 116

Tanks
    interconnected, 71
    non-cylindrical, 70
    tandem, independent, 67
Theorem
    best linear approximation
        one variable, 90
        several variables, 100
    Cauchy's principle of argument, 514
    Complex conj. root, 560
    dominated convergence, 574
    final value, 164
    frequency response, 427
    fundamental, of algebra, 559
    fundamental, of calculus, 575
    initial value, 161
    Lerch, 186
    Taylor, 89, 143, 180
Time constant
    First-order system, 294
    Second-order system, 314
Torque, 45
Torricelli's law, 66
Transfer function
    closed-loop, 391
    load, 391
    open-loop, 462
Tuning, 390
    manual, 409

Underdamped, 315
Unsteady flow, 68

Winding number, 514

Yo-Yo, 57, 84

Zero, 188, 557
Zero-pole-gain, 277
Ziegler-Nichols
    first method, 410
    ultimate sensitivity, 411

Printed in Great Britain
by Amazon